MOLECULAR BIOLOGY
INTELLIGENCE
UNIT

Cell-Cell Channels

Frantisek Baluska, Ph.D.

Institute of Cellular and Molecular Botany
Rheinische Friedrich-Wilhelms University of Bonn
Bonn, Germany

Dieter Volkmann, Ph.D.

Institute of Cellular and Molecular Botany
Rheinische Friedrich-Wilhelms University of Bonn
Bonn, Germany

Peter W. Barlow, Ph.D.

School of Biological Sciences
University of Bristol
Bristol, U.K.

LANDES BIOSCIENCE / EUREKAH.COM
GEORGETOWN, TEXAS
U.S.A.

SPRINGER SCIENCE+BUSINESS MEDIA
NEW YORK, NEW YORK
U.S.A.

CELL-CELL CHANNELS

Molecular Biology Intelligence Unit

Landes Bioscience / Eurekah.com
Springer Science+Business Media, LLC

ISBN: 0-387-36058-1   Printed on acid-free paper.

Springer Science+Business Media, LLC, 233 Spring Street, New York, New York 10013, U.S.A.
http://www.springer.com

Please address all inquiries to the Publishers:
Landes Bioscience / Eurekah.com, 810 South Church Street, Georgetown, Texas 78626, U.S.A.
Phone: 512/ 863 7762; FAX: 512/ 863 0081
http://www.eurekah.com
http://www.landesbioscience.com

Printed in the United States of America.

9 8 7 6 5 4 3 2 1

Library of Congress Cataloging-in-Publication Data

Cell-cell channels / [edited by] Frantisek Baluska, Dieter Volkmann, Peter W. Barlow.
    p. ; cm. -- (Molecular biology intelligence unit)
  Includes bibliographical references and index.
  ISBN 0-387-36058-1 (alk. paper)
  1. Cell junctions. 2. Cell interaction. 3. Cells--Permeability. 4. Cell membranes. I.
Balu°ka, F. II. Volkmann, Dieter. III. Barlow, Peter W. IV. Series: Molecular
biology intelligence unit (Unnumbered)
  [DNLM: 1. Cell Communication. 2. Intercellular Junctions. QU 350 C393 2006]
QH603.C4.C45 2006
571.6--dc22

                                    2006018962

About the Editors...

FRANTISEK BALUSKA is lecturer and researcher in the Institute of Cellular and Molecular Botany at the University of Bonn, Germany. His main interest is in cell biology of plants, especially as related to the cytoskeleton, endocytosis, vesicle trafficking, and polarity. He has investigated root apices for more that 20 years now and made original contributions to the understanding of the root apex organization by the discovery of a transition zone interpolated between the apical meristem and rapid cell elongation region. Frantisek Baluska is also interested in response of roots to environmental factors such as gravity as well as in emerging neurobiological aspects of plants. Finally, he is interested in conceptual analysis of the cell theory. He obtained his first degree from the Comenius University in Bratislava and his Ph.D. from the Slovak Academy of Sciences in Bratislava, Slovakia.

DIETER VOLKMANN is apl. Professor at the Institute of Cellular and Molecular Botany, Division of Plant Cell Biology, at the University of Bonn, Germany. His main research interest is in the field of plant responses to environmental factors, in particular responses related to gravity and microgravity at the cellular level. He made contributions to knowledge of the gravity controlled signal transduction chain as well as to the process of cell-cell-communication. He received a Ph.D. in botany at the University of Bonn and is member of the Faculty of Science at this university.

PETER BARLOW is Senior Research Fellow in the School of Biological Sciences at the University of Bristol, Bristol, U.K. His present research interests lie in the field of theoretical systems that contribute to biology, especially those which specify cellular patterns and differentiation, as well as in their possible cytological counterparts. He has also made contributions in the fields of plant cytology, human cytogenetics and mammalian embryology, and has also worked in a consultative capacity for the European Space Agency. Peter Barlow received a Bachelor of Science degree in botany from the University of St. Andrews, Scotland, from which he later received the degree of D.Sc. He also was awarded a D. Phil. degree from Oxford University, U.K.

CONTENTS

EDITORS

Frantisek Baluska
Institute of Cellular and Molecular Botany
Rheinische Friedrich-Wilhelms University of Bonn
Bonn, Germany
Email: baluska@uni-bonn.de
Chapters 1, 8

Dieter Volkmann
Institute of Cellular and Molecular Botany
Rheinische Friedrich-Wilhelms University of Bonn
Bonn, Germany
Email: unb110@uni-bonn.de
Chapters 1, 8

Peter W. Barlow
School of Biological Sciences
University of Bristol
Bristol, U.K.
Email: P.W.Barlow@bristol.ac.uk
Chapter 1

CONTRIBUTORS

Yoselin Benitez Alfonso
Cold Spring Harbor Laboratory
Cold Spring Harbor, New York, U.S.A.
Chapter 6

John R. Barnett
School of Plant Sciences
University of Reading
Whiteknights, Reading, U.K.
Email: j.r.barnett@reading.ac.uk
Chapter 9

Gianfranco Bazzoni
Department of Biochemistry
and Molecular Pharmacology
Istituto di Ricerche Farmacologiche
Mario Negri
Milano, Italy
Email: bazzoni@marionegri.it
Chapter 18

Anatoly Georgievich Bogdanov
A.N. Belozersky Institute
of Physico-Chemical Biology
and
Biological Department
Moscow State University
Moscow, Russia
Chapter 17

Laurence Cantrill
School of Biological Sciences
University of Sydney
Sydney, NSW, Australia
Chapter 6

Nigel Chaffey
Department of Biology
Bath Spa University
Newton St. Loe, U.K.
Chapter 8

Eric S. Cole
Biology Department
St. Olaf College
Northfield, Minnesota, U.S.A.
Email: colee@stolaf.edu
Chapter 3

Zhan Feng Cui
Department of Engineering Science
University of Oxford
Oxford, U.K.
Chapter 8

Georgy Natanovich Davidovich
A.N. Belozersky Institute
of Physico-Chemical Biology
and
Biological Department
Moscow State University
Moscow, Russia
Chapter 17

Jean-Jacques Fournié
Groupe d'Etude des Antigènes
Non-Conventionnels
Département Oncogénèse
and Signalisation dans les Cellules
Hématopoiétiques
INSERM Unité 563
Centre de Physiopathologie
de Toulouse Purpan
Toulouse, France
Email: fournie@toulouse.inserm.fr
Chapter 20

Svetlana Ivanovna Galkina
A.N. Belozersky Institute
of Physico-Chemical Biology
and
Biological Department
Moscow State University
Moscow, Russia
Email: galkina@genebee.msu.ru
Chapter 17

Eduardo Garcia
Department of Dermatology
and Venereology
University Hospital of Geneva
Geneva, Switzerland
Chapter 22

Hans-Hermann Gerdes
Institute for Biochemistry
and Molecular Biology
University of Bergen
Bergen, Norway
and
Interdisciplinary Center of Neuroscience
Institute of Neurobiology
University of Heidelberg
Heidelberg, Germany
Email:
Hans-hermann-gerdes@biomed.nib.no
Chapter 14

Julie Gertner
Groupe d'Etude des Antigènes
Non-Conventionnels
Département Oncogénèse
and Signalisation dans les Cellules
Hématopoiétiques
INSERM Unité 563
Centre de Physiopathologie
de Toulouse Purpan
Toulouse, France
Chapter 20

Elisabeth Grohmann
Department of Environmental
Microbiology
University of Technology Berlin
Berlin, Germany
Email: Elisabeth.grohmann@tu-berlin.de
Chapter 2

Manfred Heinlein
Institut Biologie Moléculaire des Plantes
CNRS UPR 2357
Strasbourg, France
Email:
manfred.heinlein@ibmp-ulp.u-strasbg.fr
Chapter 10

Andrej Hlavacka
Institute of Cellular
   and Molecular Botany
University of Bonn
Bonn, Germany
Chapter 8

Harold J. Hoops
Department of Biology
State University of New York at Geneseo
Geneseo, New York, U.S.A.
Chapter 4

Jean-René Huynh
Institut Jacques-Monod, CNRS
Universités Paris
Paris, France
Email: huynh@ijm.jussieu.fr
Chapter 16

David Jackson
Cold Spring Harbor Laboratory
Cold Spring Harbor, New York, U.S.A.
Email: jacksond@cshl.edu
Chapter 6

Jan Jasik
Max Planck Institute for Plant
   Breeding Research
Cologne, Germany
Chapter 8

Alain Joliot
Homeoprotein Cell Biology
CNRS UMR8542
Ecole Normale Supérieure
Paris, France
Email: joliot@biologie.ens.fr
Chapter 21

David L. Kirk
Department of Biology
Washington University
St. Louis, Missouri, U.S.A.
Email: kirk@biology.wustl.edu
Chapter 4

Alexis Maizel
Department of Molecular Biology
Max Planck Institute
   of Developmental Biology
Tübingen, Germany
Email: maizel@tuebingen.mpg.de
Chapter 12

Fabio Mammano
Venetian Institute
   of Molecular Medicine
University of Padova
Padua, Italy
Email: fabio.mammano@unipd.it
Chapter 13

Luca Manzi
Department of Biochemistry
   and Molecular Pharmacology
Istituto di Ricerche Farmacologiche
   Mario Negri
Milano, Italy
Chapter 18

Diedrik Menzel
Institute of Cellular
   and Molecular Botany
University of Bonn
Bonn, Germany
Chapter 12

William A. Mohler
Department of Genetics
   and Developmental Biology
University of Connecticut Health Center
Farmington, Connecticut, U.S.A.
Email: wmohler@neuron.uchc.edu
Chapter 23

Ichiro Nishii
Department of Biology
Washington University in St. Louis
St. Louis, Missouri, U.S.A.
and
Nishii Initiative Research Unit
RIKEN Institute
Saitama, Japan
Chapter 4

Vincent Piguet
Department of Dermatology
and Venereology
University Hospital of Geneva
Geneva, Switzerland
Email:
vincent.piguet@medecine.unige.ch
Chapter 22

Mary Poupot
Groupe d'Etude des Antigènes
Non-Conventionnels
Département Oncogénèse
and Signalisation dans les Cellules
Hématopoiétiques
INSERM Unité 563
Centre de Physiopathologie
de Toulouse Purpan
Toulouse, France
Chapter 20

Alain Prochiantz
Development and Neuropharmacology
Group
CNRS UMR8542
Ecole Normale Supérieure
Paris, France
Email: prochian@biologie.ens.fr
Chapter 21

Nick D. Read
Fungal Cell Biology Group
Institute of Cell Biology
University of Edinburgh
Edinburgh, U.K.
Email: Nick@fungalcell.org
Chapter 5

M. Gabriela Roca
Fungal Cell Biology Group
Institute of Cell Biology
University of Edinburgh
Edinburgh, U.K.
Chapter 5

Amin Rustom
Interdisciplinary Center of Neuroscience
Institute of Neurobiology
University of Heidelberg
Heidelberg, Germany
Chapter 14

Jozef Samaj
Institute of Cellular
and Molecular Botany
University of Bonn
Bonn, Germany
and
Institute of Plant Genetics
and Biotechnology
Slovak Academy of Sciences
Nitra, Slovak Republic
Email: jozef.samaj@uni-bonn.de
Chapter 8

Corinne Schmitt-Keichinger
Institut Biologie Moléculaire des Plantes
CNRS UPR 2357
Strasbourg, France
Email:
corinne.keichinger@ibmp-ulp.u-strasbg.fr
Chapter 11

Radu V. Stan
Dartmouth Medical School
Angiogenesis Research Center
Department of Pathology
Hanover, New Hampshire, U.S.A.
Email: Radu.V.Stan@Dartmouth.edu
Chapter 19

Galina Fedorovna Sud'ina
A.N. Belozersky Institute
of Physico-Chemical Biology
and
Biological Department
Moscow State University
Moscow, Russia
Chapter 17

Uday Tirlapur
Department of Engineering Science
University of Oxford
Oxford, U.K.
Chapter 8

Aart J.E. van Bel
Plant Cell Biology Group
Institute of General Botany
Justus-Liebig-University
Giessen, Germany
Email: aart.v.bel@bot1.bio.uni-giessen.de
Chapter 7

Jan W.M. van Lent
Laboratory of Virology
Department of Plant Sciences
Wageningen University
Wageningen, The Netherlands
Email: jan.vanlent@wur.nl
Chapter 11

Sami Ventelä
Departments of Otorhinolaryngology,
  Head and Neck Surgery,
  and Physiology and Pediatrics
University of Turku
Turku, Finland
Email: satuve@utu.fi
Chapter 15

PREFACE

The biological sciences are dominated by the idea that cells are the functionally autonomous, physically separated, discrete units of life. This concept was propounded in the 19th century by discoveries of the cellular structuring of both plants and animals. Moreover, the apparent autonomy of unicellular eukaryotes, as well as the cellular basis of the mammalian brain (an organ whose anatomy for a long while defied attempts to validate the idea of the cellular nature of its neurons), seemed to provide the final conclusive evidence for the completeness of 'cell theory', a theory which has persisted in an almost dogmatic form up to the present day. However, it is very obvious that there are numerous observations which indicate that it is not the cells which serve as the basic units of biological life but that this property falls to some other, subcellular assemblage. To deal with this intricate problem concerning the fundamental unit of living matter, we proposed the so-called *Cell Body* concept which, in fact, develops an exceedingly original idea proposed by Julius Sachs at the end of the 19th century.

In the case of eukaryotic cells, DNA-enriched nuclei are intimately associated with a microtubular cytoskeleton. In this configuration—as a *Cell Body*—these two items comprise the fundamental functional and structural unit of eukaryotic living matter. The *Cell Body* seems to be inherent to all cells in all organisms. In effect, the cell is an elaboration of the *Cell Body*: the cell's periphery is a result of a *Cell-Body*-organized secretion which can also produce an extracellular matrix within which cells are embedded to form tissues. From an evolutionary perspective, it is proposed that the nuclear component of the *Cell Body* is descended from the first endosymbionts which were the invaders of an ancient host structure, and which later became specialized for the storage and distribution of DNA molecules to daughter nuclei after a binary nuclear division. It follows that the eukaryotic nucleus corresponds to the entire cell of a prokaryote. Although prokaryotes have a different organization, these organisms can also be accomodated within our present version of *Cell Body* theory. This is because, in prokaryotic cells, just as in eukaryotes, both the organization and the partitioning of DNA molecules are dependent upon a physical linkage between DNA and cytoskeletal elements.

Contemporary prokaryotes and eukaryotes have an inherent property of being joined together via temporary or stable cell-cell channels. Sometimes these channels even lead to the complete fusion of cells. In bacteria, these channels serve for conjugative transfer of DNA while in ciliates they permit the conjugative transfer of whole nuclei. Similarly, whole nuclei can pass through cell-cell channels in fungal cells. In the case of plants, cells are linked via permanent or semi-permanent channels known as plasmodesmata. These sometimes enlarge into pores which can then transfer whole nuclei from cell to cell. Also, animal cells can be linked with membraneous cell-cell channels which can allow the intercellular passage of whole organelles. Intriguingly, plasmodesmata of plant cells are similar in many respects to nuclear pores, suggesting that nuclear pores are representatives of a prototypic cell-cell channels that already existed, in their primordial form, between the original guest and host cells which merged together in the early formation of the eukaryotic cell. All these considerations call for an update of the traditional 'cell theory'.

This book covers the topic of cell-cell channels at all levels of biological organization, starting with bacteria and unicellular ciliates, via algal, fungal and plant cells, up to and including the diverse cell types of animals. We hope that this book will help update the traditional cell theory and will also stimulate new discussions concerning the basic units of life.

Frantisek Baluska
Dieter Volkmann
Peter W. Barlow
Bonn – Bristol, 6 April 2006

CHAPTER 1

Cell-Cell Channels and Their Implications for Cell Theory

Frantisek Baluska,* Dieter Volkmann and Peter W. Barlow

Abstract

Cells show diverse appearances and sizes, ranging from some 30 nanometers up to several meters in length. Besides the classical prokaryotic and eukaryotic cells, there are also very bizarre cells such as the highly reduced symbiotic mitosomes which lack DNA. Other examples of extremely small cells in the nanometre range are the mycosomes and nanobacteria. On the other hand, there are huge eukaryotic cells, the size of which can reach up to several meters. Most of these are multinucleate (coenocytic) due to mitotic divisions not having been followed by cytokinesis. Moreover, cells at all levels of cellular complexity show an inherent tendency to form cell-cell channels. The most conspicuous example is the plant 'supercell' where all the cells of the plant body are permanently connected via plasmodesmata. In the last year, the first reports of similar cell-cell channels between animal cells have been published. Moreover, fungal cells fuse together into supracellular mycelia, even exchanging their motile nuclei. This phenomenon is also known for plant cells. Intriguingly, transcellularly moving fungal nuclei communicate with their mating partners via pheromone-like signaling mechanisms.

Already in 1892 Julius Sachs was aware of most of the problems associated with the Cell Theory, which in fact survive until the present. Sachs proposed the Energide concept, postulating that it is the nucleus and its protoplasm which represent the vital unit of living matter within a supracellular construction, while the cell periphery is only a secondary structure generated by the active Energide for its shelter and protection. Recently, we elaborated the Cell Body concept which explains how and why the nucleus and the microtubular cytoskeleton have become merged together to build a coherent and universal unit of eukaryotic life which is autonomous and can synthesize the rest of the cell. However, there are several problems with the term Cell Body as it is sometimes used in other meanings. Here we show that the Energide concept of Sachs can be united with the Cell Body concept. Moreover, we agree with Julius Sachs that the term Energide better invokes the unique properties of this universal unit of supracellular living matter endowed with the vital energy.

Introduction

Current biological thinking is dominated by cells. All living things are defined as either being composed of cells or are cells themselves—membraneous compartments filled with organic molecules in a way that allows them to be autonomous self-replicating units of life. The cell concept originates from Robert Hooke's seminal observations of plant tissue in 1665 when he saw small chambers (in fact, dead and empty cells of cork tissue) which he called 'cells'.[1] These were later, during the next 200 years (when living cells were being examined), shown to contain a nucleus and

*Corresponding Author: Frantisek Baluska—Institute of Cellular and Molecular Botany, Rheinische Friedrich-Wilhelms University of Bonn, Kirschallee 1, D 53115 Bonn, Germany. Email: baluska@uni-bonn.de

Cell-Cell Channels, edited by Frantisek Baluska, Dieter Volkmann and Peter W. Barlow.
©2006 Landes Bioscience and Springer Science+Business Media.

all the other now familiar organelles. In 1838, Matthias Schleiden and Theodor Schwann proposed that both plants and animals are built of cells; Rudolf Virchow announced his famous dictum *'omnis cellula e cellula' in 1858.*[2] Virchow also wrote that: *'every animal appears as the sum of vital units, each of which bears in itself the complete characteristics of life'.*[3] After the cellular nature of brain tissue was confirmed,[4] cells became synonymous with life. Cell Theory has been ever since the most influential dogma of contemporary biology.[2-7]

The Permanent Crisis Surrounding Cell Theory

There are fundamental problems with the Cell Theory. In fact, many of these problems started to be apparent soon after Schleiden and Schwann proposed the cellular basis of plants and animals.[8,9] First of all, multinuclear cells were found. Next, prokaryotic cells were discovered which lacked organelles, and their simple architecture was taken as evidence for their ancient or archaic nature. Nowadays, however, several genome-based data have emerged which suggest just the opposite—that the contemporary prokaryotic cells are not so ancient and primitive after all.[10,11] More importantly, the eukaryotic cell is defined as a structurally coherent and physically autonomous entity being completely separated from its surroundings by a plasma membrane. However, there are numerous examples, listed in the next sections of this chapter as well as in other chapters of this volume, of cells which are interlinked via cell-cell channels (Fig. 1), cells which contain many nuclei either due to nuclear divisions not followed by cell divisions, or which are generated by fusions between cells.

Permanent cell-cell channels are typical of plant tissues. Similar supracellular constructions which allow the exchange of organelles occur also in the animals kingdom (see other chapters in this volume).[12,13] Furthermore, classical cell organelles such as mitochondria and plastids have finally been demasked, after many years of discussion, as being cells too. They are highly reduced descendents of once free-living cells. Besides these smaller organelles, nuclei can also be considered to

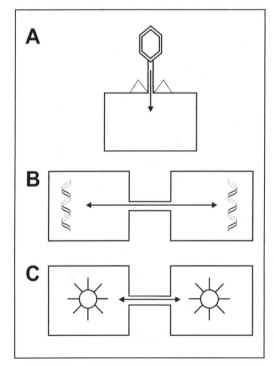

Figure 1. Cell-cell channels in viruses (A), bacteria, organelles, eukaryotic cells (B). In eukaryotic cells, whole Energides (C) can move from cell-to-cell via microtubule-based motility.

Figure 2. 61 years old Julius Sachs, taken one year after the publishing of his Energide concept and four years before his death.

Box 1. Energide (Cell Body) at a glance

An Energide consists of a nucleus, centrosome, microtubules, endoplasmic reticulum (ER), Golgi apparatus (GA), GA-based secretory vesicles, and ribosomes. It gains control over the original cell periphery-based structures, such as the actin cytoskeleton and endosomes. This provides the Energide with the tools necessary to regulate its own motility, the distribution of organelles and the reproduction of its enclosing cell. The Energide can form a new cell periphery from endosomes filled with cell wall molecules and enclosed by membranes derived from the plasma membrane. Thus, the Energide rapidly generates a new cell surface in the case of accidental damages to the cell periphery.

The Energide produces the rest of the cell by means of its synthetic activities. It resides in a space derived from the ancient host cell, now known as cytoplasmic space, which provides the Energide with a unique environmental niche. Some autonomy is still maintained by other guest cells, mitochondria and plastids. The cell periphery consists of the plasma membrane, cell wall, and actin cytoskeleton. All host DNA has been transferred into the nucleus. Nevertheless, the cell periphery still maintains some structural independence as it is essential for sheltering the Energide, providing it with information about the environment outside the cell, and generating the endosomes, lysosomes, as well as vacuoles in order to provide the Energide with nutrition and to store nutritive and energy-rich compounds. Also, this system is responsible for detoxification of toxic compounds entering cells. Early in cellular evolution, this endosomal system was adjusted for rapid repair of damaged cell periphery and for final events in cytokinesis. Later in the evolution of multicellular organisms, this endosomal and recycling machinery was used for synaptic cell-to-cell communication, the formation of immunity, cell/organism individuality (self, nonself recognition), the brain and consciousness. Importantly, the inherent feature of cells at all levels of cellular complexity is the formation of cell-cell channels. If these are sufficiently large, even the whole of an Energide can move from its original cell into an adjacent cell. Intriguingly, transcellularly moving fungal nuclei communicate with their mating partners via pheromone-like signaling mechanisms. This is the ultimate evidence that the Energide is an autonomous organism representing the universally valid unit of supracellular eukaryotic life.

---

represent vestiges of ancient endosymbionts.[12,14-20] In fact, the nucleus emerges as the first endosymbiont to have invaded the ancient host cell.[15] Recently, we have reviewed numerous data in favour of the Cell Body concept which considers not the cell but rather the nucleus with its associated microtubules, the so-called Cell Body, to represent the smallest unit of life in the construction of multicellular or, more exactly, supracellular, organisms.[12,21] However, by 1892, Julius Sachs (Fig. 2) had already discussed these problems from the perspective of giant cells, such as multinucleate coenocytes and syncytia, which did not obviously conform with the Cell Theory. In order to resolve this problem, he proposed the 'Energide' concept.[6,22,23]

Below, we discuss the Energide concept of Sachs in relation to our Cell Body concept and we conclude that both concepts can easily be combined, resulting in an ultimate concept that explains the smallest autonomous unit of supracellular eukaryotic life (summarized in Box 1 and Fig. 3). As the Cell Body is not as suitable term, from now on we will use 'Energide' instead of 'Cell Body' for this universal unit of eukaryotic life.

More Problems with Cell Theory: What Is a Cell?

Obviously, the definition of a cell is very vague. If the cell should be the smallest unit of life showing all the attributes of living matter, how can Cell Theory accommodate the fact that cellular organelles are, or have been, cells also? Even recently evolved eukaryotic cells have been internalized and enslaved by a recent endosymbiotic event and now act as chloroplasts.[24,25] There are other obvious examples of 'cells within cells' such as bacterial and yeast spores, as well as pollen tubes and embryo sacs in plants.[9,21] Moreover, as Sachs noted more than a hundred years ago, the structural organization of multinucleate cells of coenocytic algae, for instance, those of *Ventricaria ventricosa*,[26] cannot be explained by the Cell Theory.[9,21-23]

Cell-merging is one of the most important factors shaping the evolution of more complex eukaryotic cells.[15] Not only are these fusions examples of secondary endosymbiotic events, but the enslaved cells accomplish dramatic reductions which do not permit their recognition as cells any more.[9] Hydrogenosomes are endosymbiotic organelles related to mitochondria, and some of them, like those of *Trichomonas* and *Neocallimastix*, lack any DNA.[27] Moreover, another type of mitochondria-derived relict organelles of human pathogen, *Entamoeba histolytica*, the so-called mitosomes, have sizes below 500 nm and lack any DNA.[28] Nevertheless, these mini-cells can reproduce themselves as their numbers are maintained at about 250 for one host cell.[29] Both mitosomes and hydrogenosomes provide us with convincing evidences that DNA can be completely removed from one partner cell during an endosymbiotic partnership. Further examples of exotic cellular organisms are mycosomes (0.1–1.0 mm in diameter) isolated from plant plastids and which can give rise to yeast-like cells.[30] Even smaller are nanobacteria which are about 30-100 nm in diameter. These tiny spheres are predicted to cause several human diseases.[31] The bacterium *Spiroplasma* is one of the smallest of cells, comparable in size to the diameter of a bacterial flagellar bundle. Despite its extremely small size, it is equipped with a quite complex cytoskeleton.[32] On the other hand, there are some viruses which are larger than the smallest bacteria.[33] Even more, there are large eukaryotic cells which are visible to the naked eye. Avian egg cells are the largest cells by volume organized by a single nucleus. Plant pollen cells grow several centimeters in length to deliver sperm cells to ovules deeply buried within flower tissues, and multinucleate laticifers may extend the whole length of a *Euphorbia* plant. The marine algae, *Caulerpa* and *Acetabularia*, are single cells which can attain up to several meters in length (in the case of *Caulerpa*). They are filled with numerous nuclei, each organizing its own microtubular array and cytoplasmic domain. One of the most dramatic examples of huge cells is the mammalian placental syncytiotrophoblast whose surface area can reach up to ten square meters.[34]

In the classical cell concept, the limiting membrane plays the dominant role whereas the internal parts are not as important and vary from having no DNA, such as in the above-mentioned mitosomes and some hydrogenosomes, to having just a few DNA molecules encoding several proteins within highly reduced organelles, up to the extremely complex cellular interiors encompassing hundreds of nuclei and enormous numbers of other organelles (also cells) with a cytoplasm. In contrast to this diffuse and vague situation with cells, the structure and morphology of the Energide (Cell Body) units are essentially the same throughout the eukaryotic superkingdom (see below), implying that they are much more useful for defining the basic vital units of supracellular eukaryotic life. As we discuss below, cell-cell channels are inherent for all these cell types, starting with the prokaryotic cells and symbiotic cellular organelles, up to eukaryotic cells covering all kingdoms including fungi, plants and animals. If we really need a basic unit of supracellular eukaryotic life, then we must define a new one.

Julius Sachs: Energide as the Basic Unit of Eukaryotic Life Endowed with Vital Energy

Julius Sachs was well aware of all these problems of Cell Theory 110 years ago. He proposed the Energide concept in which it was not the cell, but rather the nucleus with a surrounding sphere of protoplasmic influence ('Wirkungssphäre') that defined the autonomous vital unit bearing all the attributes of life.[22,23] Energide does not encompass the plasma membrane ('Zellhaut') and cell wall which, in contemporary cells, are postulated to be of a secondary nature produced by the active Energide to protect itself from the hostile environment.[23] Importantly, while the Energide contains the 'vital energy' ('Lebenskraft') and divides by a template-based duplication followed by fission, the cell periphery cannot multiply in this way and grows rather via fusion of vesicles generated by the secretory activities of the Energide ('...von ihm selbst erzeugten Gehäuse'). For Julius Sachs, the cell is only a chamber ('Zellstoffkammer') which can harbour one, several, or many Energides. Recently, we discussed numerous modern data confirming most of the predictions made by Julius Sachs more than 100 years ago.[9,21] In order to convey his line of arguments to contemporary cell biologists as well as to a broader community of scientists, we translate from German into English the most important parts of his text.

Julius Sachs, Flora 1892; 75:57-67.

Pages 57, 59:

'Unter einer Energide denke ich mir einen einzelnen Zellkern mit dem von ihm beherrschten Protoplasma, so zwar, dass ein Kern und das ihn umgebende Protoplasma als ein Ganzes zu denken sind und dieses Ganze ist eine Organische Einheit, sowohl im morphologischen wie im physiologischem Sinne. Den Namen Energide wähle ich, um damit die Haupteigenschaft diese Gebildes zu bezeichnen: dass es nämlich innere Thatkraft, oder wenn man will: Lebenskraft besitzt.' ...die einzelne Energide für sich frei leben kann, ohne von einer Zellhaut oder Zelle umgeben zu sein...' 'Zum Begriff der Energide gehört also die Zellhaut nicht; die Sache liegt vielmehr so, dass jede einzelne Energide sich mit einer Zellhaut umgeben kann, oder aber mehrere Energiden zusammen bilden eine Zellhaut...'

'Energide is represented by a nucleus associated with its protoplasm in such a way that the nucleus and surrounding protoplasm form an organic unit, both from the morphological and physiological perspectives. The name Energide is chosen to stress the major property of this structure: endowment with the inner power for action or, if one wants: endowment with a vital power.' '...an individual Energide is capable of living freely, without being enclosed within the cell skin or cell chamber...' 'The term Energide does not encompass the cell skin; the case is more that each individual Energide is able to enclose itself by a cell skin, or that several Energides together can enclose themselves with one single cell skin ...'

Page 60:

"...dagegen beginnt die Wissenschaft von den lebendingen Dingen mit einem Wort, welches vor mehr als 200 Jahren infolge eines Irrthums entstanden und dann beibehalten worden ist: dem Wort Zelle.

Bekantlich ist das Wort Zelle als Terminus technicus der Botanik nur historisch zu verstehen, insofern Robert Hooke 1667 die innere Configuration des Korkes und der Holzkohle eine zellige, im Sinn einer Bienenwabe, nannte. Auch die Zootomie hat später dieses unglückliches Wort aufgegriffen und für die Elementartheile des thierischen Organismus verwendet, obgleich es dort noch weinger Sinn hatte, als bei denn Pflanzen. – In der 40er Jahren erkannten die Botaniker, dass das Wesentliche der Pflanzenzelle nich ihr Gehäuse, sondern ihr Inhalt, wie wir jetzt sagen, das Protoplasma mit dem Kern ist, und so unterschied man Zelle und Zellinhalt."

'...the science which deals with living matter commenced with a controversial word, originating as a mistake more than 200 years ago, and then maintained up to the present day: it is the word Cell.'

'It is well-known that the word Cell should be understood only from a historical perspective as Robert Hooke referred to the inner structures of cork as being cellular because they closely resembled hexagonal wax cells of honeycomb. Later, zoology also accepted this unfortunate term, despite this word being even more controversial when applied to animals. In the forties [i.e., 1840s], botanists realized that the true living matter is not the shell, but that it resides inside the Cell, and is called protoplasm and nucleus. One should therefore be careful in distinguishing the Cell's interior from its external boundary'.

Energide versus Cell Periphery

Julius Sachs stressed that although the Energide can undergo autonomous template-based growth followed by binary fission, its boundary structures ('Gehäuse' or 'Zellstoffkammer') can divide only after the accomplishment of Energide-based growth and Energide-instructed division. Importantly, Energide division does not need to be followed by cell periphery division, and this results in the formation of coenocytes. Even more importantly, the opposite situation, that cell divison could occur without a preceding mitosis, was never recorded. Moreover, 'cells' lacking their Energides, so-called cytoblasts, cannot divide. On the other hand, Energides can be released from ruptured coenocytic algae, whereupon each Energide regenerates both its plasma membrane and cell wall to form a new complete cell.[9] The Energide is the primary unit of supracellular life as it can escape from its cellular boundaries and, later, it can autonomously regenerate these boundary structures. Cell periphery structures can not undergo sustained growth and binary fission without being supported by an Energide.

Recently, we have discovered that cytokinesis in plant cells is driven by the fusion of endosomes which are enclosed by the plasma membrane-derived membrane and enriched with cell wall pectins.[35,36] These endosomes are recruited to the division site by the Energide (Cell Body) microtubules.[36] In animal cells, too, Energide microtubules specify cytokinesis both spatially and temporally.[37] Similarly, sporulating yeast cells generate their plasma membrane de novo from early endosomes enriched with PI(3)P.[38] Also, in this situation, the Energide and its microtubules play the essential role of enclosing the cell with a plasma membrane.[39] Last, but not least, de novo formation of the plasma membrane was demonstrated during the fragmentation of coenocytic algae, releasing 'naked' Energides which then enveloped themselves with boundary structures, thus regenerating new cells.[9] All these examples, from throughout the eukaryotic superkingdom, convincingly document the secondary nature of the boundary structures, including the plasma membrane and cell wall/extracellular matrix. This is evident also from the mode of their growth and division. In contrast to the Energide, cell periphery grows and divides by fusion of vesicles.

Adaptability versus Complexity in Cellular Evolution

The formation of eukaryotic cells represented a great leap forward in cellular complexity.[40] The Darwinian selection-based evolution towards increased fitness, which acted even on ancient proto-cells early in the development of life,[41,42] cannot explain the dramatic increase in cellular complexity.[40] In fact, both viruses and prokaryotic cells are far better adapted to harsh environments, and this is the reason why they have been so successful up to the present time. They have been selected by their environment to be efficient mutators and replicators. In several aspects, these simple cells are not primitive at all, and they clearly dominate the biosphere of the Earth. Darwinian evolution is strong in explaining how their adaptability comes about, driven by variation, competition, and selection. However, Darwinian *adaptive* evolution is weak in explaining satisfactorily why cells should proceed from the less complex towards the more complex forms of cellular organization. It may be a question of selection creating the most efficient combinations of molecular modules at the subcellular level,[43] whereas at the cellular level evolution is a question of finding the best combination of symbiotic associations.

In fact, we know several examples where complex cells undergo 'regressive' evolution towards a less complex type of cell if they are enslaved by other cells or exposed to extremely challenging environments. Recent advances in the study of symbiogenesis and its role in cellular evolution allowed a breakthrough in our understanding of this paradigm. It was primarily Lynn Margulis[15,45-48] who revived this old concept, after more than 70 years of oblivion.[44] It appears that the most powerful source of innovation with regard to cellular complexity is the merging of individual cells.[15] Of course, this cellular merging is a by-product of cellular competition. But this time it is not the *adaptive* competition in the sense of Darwinian *adaptive* evolution, but it rather is *predative* competition[9] in the sense of Margulisian *symbiotic* evolution which is based on serial endosymbiosis.[15,48] In other words, *adaptive* evolution towards increased fitness is driven by the conflict of organisms within a harsh selectionist environment, whereas *predative* evolution that drives towards increased cellular complexity is powered by nutritive conflicts between heterotrophic unicellular organisms entertaining a predator/prey based life style. Sometimes, a balance of 'forces' is reached between predator and prey cells allowing the development of stable symbiotic relationships between the host and its guest. This ultimately leads to the generation of more complex cells.[15,48]

Recent analysis of genomes and their genes have revealed that the last universal common ancestor of eukaryotic cells was formed by cell-cell fusion.[14,49] This universal ancestor enslaved endosymbiotic cells (organelles) via predatory phagocytosis.[9,15-20,45-48,50-52] Thus there are two processes which pave the way towards more complex eukaryotic cells and which disobey Darwinian evolution: lateral gene transfer and endosymbiosis.[14,49] Both these processes played a crucial role in evolution of complex eukaryotic cells endowed with nuclei and Energides.[9,15]

Recently, even viruses have come into consideration as having invented the first nucleus in ancient protocells. Viruses have the inherent ability to invade cells and there are several intriguing similarities between nuclei and viruses.[52] Nuclei and viruses lack the protein synthesis machinery and lipid-producing pathways; both transcribe DNA but do not translate RNA, and they contain linear chromosomes equipped with primitive telomeres rather than the circular chromosomes found in bacteria, mitochondria, and plastids.[52] Furthermore, viruses are well-known for their ability to induce cell-cell fusion,[53] a phenomenon which was essential for the early events required for the formation of ancient eukaryotic cells.[14]

Duality of Eukaryotic Cells

Eukaryotic cells are characterized by a distinct duality at the structural level (Energide and cell periphery), at the level of genome organization (eubacterial and archaebacterial features), cytoskeleton (actin and tubulin), membrane flow (exocytosis and endocytosis), and cell division (mitosis and cytokinesis).[9] These features strongly suggest that the most ancient eukaryotic cell was generated from at least two cellular organisms fusing together and generating the ancient nucleus by transformation of the host cell via recruitment of all the host DNA.[9] Recent analysis of diverse genomes has revealed that the eukaryotic genome was, in fact, generated from a fusion of at least two prokaryotic-like genomes, an event which massively speeded-up horizontal gene transfer.[14,49] Due to this, all host DNA was transferred into the guest cell which was then transformed into a eukaryotic nucleus.[9,15] Moreover, some mitochondria-related hydrogenosomes have also lost all of their DNA.[27] This new twist in the evolutionary origin of the contemporary eukaryotic cell is in accordance with the symbiotic origin of eukaryotic nuclei,[9,14-20] and it can also explain the absence of DNA from the plasma membrane. Very convincing examples of the complete loss of DNA from endosymbiotic cells are mitosomes.[28] In fact, any DNA introduced into a eukaryotic cell is transported into the nucleus and integrated into the genome, allowing for the easy transformability of eukaryotic cells.

As mentioned above, endosymbiosis and lateral gene transfer are not only the most important processes shaping evolution of complex eukaryotic cells, but also these are phenomena which disobey Darwinian evolution.[49] Thus, the mechanisms shaping the evolution of eukaryotic cells are extremely important for understanding biological evolution in a much broader perspective. Evolution of complex eukaryotic cells provides us with an important paradigm for the elusive nature of living matter and why it should evolve from low to high complexity. The next big issue to be solved by cell biology is the topic of this book—cells at all levels of complexity are characterized by cell-cell channels. Why and what for? This feature must be very ancient and of extreme importance because it is evident in all contemporary cellular organisms, at all levels of cellular complexity. It seems that both RNA and DNA molecules could be relevant in this respect. Importantly, viewing nuclear pores as cell-cell channels not only supports the endosymbiotic origin of nuclei but also explains the perplexing similarity between nuclear pores and plasmodesmata.[54]

Cell-Cell Channels: Supracellularity Is Found at All Levels of Cellular Organization

As is apparent throughout this book, cells which have been created by division also seem to have an inherent tendency to fuse into supracellular assemblies. The first breakthrough in this respect was the early discovery of plasmodesmata by Eduard Tangl in 1867.[55] It is perhaps typical of biology that this discovery was not fully accepted until some 100 years later when electron microscopy allowed the submicroscopic resolution of these cell-cell channels.[55] Much more recently, cell-cell channels of animal cells have been discovered.[13,56-58] Hopefully it will not take a further 100 years to have this discovery accepted (or definitely rejected) by mainstream biology.

Cell-Cell Channels in Viruses, Prokaryotes and in Eukaryotic Organelles: Something Special about Cellular End-Poles

Cell-cell channels can be found at all levels of cellular complexity, suggesting that this feature is inherent to cellular organization. Bacterial viruses inject phage DNA into host cells via a channel which is still not characterized in detail (Fig. 1A).[59] This ability of viruses suggests that they should be considered as highly reduced cells specialized for a unique form of cellular parasitism. Bacteria also develop a wide spectrum of transcellular injection machines for cell-cell delivery of their virulence factors into host cells (Fig. 1B).[60] Interestingly, the nongrowing cellular end-poles assemble these injection organelles which export DNA to other prokaryotic or eukaryotic cells.[61-63] In *Streptomyces* colonies, pore-like structures assemble at end-poles for the conjugative spread of plasmids throughout the colony.[64] These cells interacting via their end-poles are also equipped with pores secreting slime and motile hair-like filaments known as pili.[65-68] Could this polarisation of cell-cell interactions at the end-poles have implications for the siting of plant plasmodesmata?

Dynamic pili protruding from the bacterial end-poles can rapidly extend and then retract, exerting pulling and pushing forces.[66] These might contribute to the alignment of cells into filamentous rows[65,67] resembling the cell files of plant tissues.[69] In the cyanobacteria, *Phormidium* and *Anabaena*, junctional pore complexes are similarly formed at the end-poles of cells which are also arranged into cell files.[70] These end-poles are traversed by cell-cell channels which allow direct transport of metabolites as well as communication between heterocysts specialized for nitrogen fixation and other cells accomplishing photosynthesis.[6] Similarly, end-poles/cross-walls of plant cells are abundantly traversed by plasmodesmata.[69] Intriguingly, prokaryotic cell-cell channels can be targeted by movement proteins of plant viruses which then modify these prokaryotic cell-cell channels in a manner known from virus-affected plant plasmodesmata.[71] This finding strongly suggest that prokaryotic and eukaryotic cell-cell channels are evolutionarily related. Intriguingly in this respect, nongrowing end-poles traversed by cell-cell channels, play a central role in polarity of both bacteria and plant cells.[69] Furthermore, viruses are well-known to induce cell-cell channels and cell-cell fusions in eukaryotic cells.[53]

In conformity with free-living bacteria, bacteria-like descendent organelles of eukaryotic cells, such as plastids, mitochondria, and peroxisomes, are all known to fuse together, an event which necessitates the existence of fusion channels. Plastids generate stroma-filled long tubules, so-called stromules. These are highly dynamic structures interconnecting adjacent plastids and allowing an exchange of molecules as large 560 kDa.[72-78] Mitochondria are also well-known to undergo fusion and fission processes.[79-82] The fusion machinery for plastids are not known yet, but there are data accumulating on the fusion machineries of mitochondria[83] and peroxisomes.[84] It is obvious that these 'cell-cell' fusion mechanisms differ from those which drive membrane fusion in exocytic and endocytic pathways.[85]

The outer leaflet of the nuclear envelope extrudes endoplasmic reticulum (ER) whereas the inner leaflet is anchored via the nuclear lamina, to the chromatin complex. ER pervades the whole cytoplasmic space, budding as a Golgi Apparatus (GA) and secretory vesicles (Fig. 3), as well as interacting with other organelles.[86-88] In plants, ER elements extend from cell-to-cell across the plasmodesmata,[54-55] but this is likely to be a secondary event following from the primary contact and fusion of adjacent cell peripheries. Microinjection of 3 kDa dye into the lumen of ER enables visualization of dye movement into adjacent cells where it first appears within the nuclei.[89] Obviously, Energides from adjacent plant cells are now able to interact via the physical continuity of their ER membranes.

Nuclear Pores as Prototypic Cell-Cell Channels

If the endosymbiotic origin of nuclei can be definitely confirmed, then nuclear pores also can be viewed as specialized cell-cell channels resembling plant plasmodesmatal cell-cell channels.[54] These are optimized for effective cell-cell transport of proteins, as well as RNA and DNA molecules. Intriguingly, nuclear pores are highly selective with respect to the macromolecules which they transport. Particles as large as ribosomes can pass through nuclear pores whereas much smaller molecules may be actively excluded.[90,91] Similarly, plant plasmodesmata support transport of certain large proteins whereas the small auxin molecule, for example, is excluded from direct cell-cell transport.[92] Recently, both plasmodesmata[93-95] and nuclear pores[96,97] were revealed to be closely associated with molecules driving endocytosis. Viruses, RNA and DNA molecules all are internalized

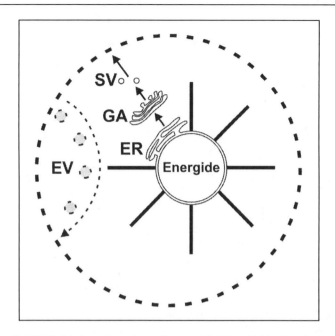

Figure 3. Energide (Cell Body), shown in continuous lines, settled within a highly reduced guest cell, shown in dotted lines, which serves predominantly as a shelter. ER: endoplasmic reticulum, GA: Golgi apparatus, SV: secretory vesicles, EV: endocytic vesicles.

into endosomes which then deliver them into either the ER or the nuclear pores.[98,99] Many cell periphery and endocytic proteins use this endosomal route for their transport into and out of the nucleus.[100-102] Interestingly in this respect, endosomes are the only type of organelle which invades and travels through cell-cell channels of animal cells.[13] Furthermore, the acrosomal compartment which drives the cell-cell fusion between sperm cells and oocytes, has endosomal features; and it is also enriched with synaptic molecules which recycle via secretory endosomes.[103-104]

Cell-Cell Channels in Filamentous Fungi, Plants and Animals

Cell-cell channels are found in all multicellular organisms, including filamentous fungi, plants and animals. As their presence, structure, and functions are extensively discussed in several chapters of this volume, we limit ourselves to just some basic information. Fungal hyphae are tip-growing tubules which inherently fuse together. As a result, complex mycelial networks are formed to allow the exchange of large masses of protoplasm, organelles and, importantly, the highly motile Energides.[105-109] All cells of plant bodies are well-known to be interconnected via plasmodesmata,[55,71,87] and if they are not formed as an accompaniment of cytokinesis, they can form later as so-called secondary plasmodesmata. But plant tissues also form much larger cell-cell channels, known as cytoplasmic channels or intercellular bridges.[110,111] These macro-channels allow transfer not only of large organelles but also of nuclei (Energides). This spectacular process was documented both in situ[110,112] and in vivo.[113] In addition, plasmodesmata can be transformed into the massive pores of sieve plates, thus allowing the mass flow of metabolites along phloem elements.[114]

There are numerous examples of cell-cell channels in animals. Well studied are fusomes and intercellular bridges of germinal cells of numerous taxa. They form as a result of incomplete cytokinesis.[115-118] Fusomes develop from spherical vesicular compartments into large cell-cell channels which not only nurse developing oocytes but also determine their polarity due to the anchoring of the Energide microtubules.[119] Similar to plasmodesmata of plant cells, fusomes provide direct

Energide-Energide connectivity via ER elements.[120] Recently, exciting observations have revealed thin (50-200 nm) nanotubules between cultured PC12 rat neural cells and kidney cells.[13] These actin-based cell-cell channels allow transcellular transport of myosin-associated endosomes. Similar nanotubes were found also at immunological synapses where exchange of GPI-anchored proteins and class I MHC protein was noted.[56,57] Even larger (100 nm-5 μm in width, and up to 100 μm in length) cell-cell tubules were reported between DU 145 human prostate cancer cells. These are positive for alpha-tubulin and transport very large membrane vesicles (up to 3 μm in diameter).[58]

Cell-Cell Channels and the Energide: Implications for Cell Theory

Microtubules radiating from the Energide/Cell Body represent an ideal instrument by which to delimit the cytoplasmic, cellular boundary. Moreover, this limit is somehow controlled by the DNA mass, or amount, stored within the nucleus.[9] The dynamic instability of microtubules allows the Energide to measure continuously the size and shape of its 'sphere of influence' (cell). Also, the microtubules collectively serve as a highly effective tool for Energide motility within the confines of its cell.[9,122] But it is mysterious how DNA, irrespective its coding information, can determine both the size of the nucleus (Energide) and the size of its cell. Concerning the DNA amount/size of the nucleus correlation, the skeletal DNA hypothesis has been proposed[121] while the Energide/Cell Body concept can explain the tight correlation between DNA amount and cell size.[9,123]

Energide/Cell Body Microtubules Act as a Tool to Prevent Their Fusion, to Obtain Information Which Is Acquired and Processed at the Plasma Membrane, to Explore the Cellular Space and to Invade Adjacent Cells, as Well as to Repair Old and Generate New Cell Periphery

What features are so special about the Energide (Cell Body) that confer upon it properties which have allowed the Energide to act as the vital unit of supracellular eukaryotic organisation involving constructions composed of millions of Energides? We argue that this feature is due to the unique association between the DNA-based nuclei and the tubulin-based microtubular cytoskeleton.[9] Julius Sachs obviously could not incorporate microtubules into his Energide concept. So, by merging the Energide concept with the Cell Body concept,[9,21] we should be able to unravel the full scope of the unique properties of the Sachsian Energide.

First of all, whereas cells show an extremely wide range of sizes and structural organizations, Energides are basically constant structures, always consisting of nucleus sheathed within perinuclear radiating microtubules. These are instrumental for numerous properties of a vital Energide. Energide microtubules are essential for preventing accidental nuclear fusions which would easily happen either during cytokinesis or after cell-cell fusions.[9] In fact, depolymerization of Energide microtubules during mitosis results in the immediate fusion of daughter Energides and the subsequent formation of a polyploid nucleus.[124] This inherent tendency of adjoining nuclei to fuse together, if they are not ensheathed with microtubules or cell periphery boundaries, is a very strong argument for cellular nature and endosymbiotic origin of nuclei. Energide-Energide fusions can be observed in some ephemeral plant tissues like endosperm.[21,123-125] This is because radiating microtubules from adjacent Energides, which interact via plus ends of their microtubules so that they usually keep their distance from each other, under some circumstances can fail to be effective organizers of their cytoplasmic 'spheres' of influence.[9,21]

Next, Energides of syncytial and coenocytic cells are regularly distributed in their cytoplasmic space. This is also due to the physical interactions between plus ends of microtubules from the adjacent Energides.[9,21] The claim upon protoplasmic space, the volume of which is directly related to the amount of DNA stored within a given nucleus, leads to the phenomenon known as the nucleo-cytoplasmic ratio.[9,123] However, also important is the way in which DNA is packed within the nuclei: highly condensed generative nuclei of pollen grains form small Energides whereas decondensed large vegetative nuclei form large Energides, despite the fact that the DNA amount in both these nuclei can be the same.[123] The mechanisms behind these phenomena include both nucleus-stored molecules having MTOC properties[123] and the Energide's ability to generate longer microtubules when more MTOC molecules are available.[9,21] Microtubules initiated at the nuclear surface, and having their dynamic plus ends interacting with the plasma membrane, are ideally suited for the transfer towards the Energide nucleus of information acquired and processed at the cell periphery and plasma membrane. Importantly, Energide-associated dynamic microtubules can act as vehicles allowing the movement of whole Energides/Cell Bodies through cellular space.[122]

Finally, these microtubules can gain control over the distribution of cell periphery-derived endosomes filled with internalized cell wall material and enclosed by a membrane, which are ideally suited for cell periphery repair or for rapid generation of a new cell periphery.[35,36,39] In plants, Energides are actively recruited to cell periphery sites which have been compromised by pathogen attack or which accomplish rapid and highly polarized cell growth via tip-growth.[126,127] Similar tipward localization of Energides is typical also of filamentous fungi.[109,128,129]

Energide/Cell Body Can Use Cell-Cell Channels to Travel across Cellular Boundaries

As briefly mentioned above, Energides (Cell Bodies) not only explore the confines of their own cells[122] but can also move to an adjacent cell if the connecting channels are sufficiently large. This phenomenon, which is clearly incompatible with the current version of Cell Theory, is the final proof for the autonomous primary nature of the Energide. In plants, there are numerous ultrastructural observations of this process, known also as cytomixis.[110,112] However, these findings have been criticized as representing nothing more than aberrant cytokinesis where daughter nuclei have been squeezed thtough a maturing cell plate. But in vivo studies provided the final proof for a genuine cytomixic cell-cell movement of nuclei (Energides).[105,106,109,113] Moreover, other well studied examples of nuclear transfer involve a structure called the conjusome, a cell-cell channel specialized for conjugation in *Tetrahymena*.[130,131]

A rather exotic cell-cell transfer of nuclei occurs between the parasitic alga, *Choreocolax*, and cells of its red-alga host, *Polysiphonia*.[132] This resembles the situation in fusing fungal hyphae in which genetically different Energides co-exist and are exchanged.[107,108,133,134] Intriguingly, the fungal Energides can recognize their respective mating Energide partners.[135] This mysterious process is accomplished via a pheromone/receptor system operating at the level of Energides (Cell Bodies) of the opposite mating types.[136,137] This pheromone-like communication between individual Energides is a further very strong argument for the cellular and endosymbiotic nature of nuclei. When a spindle pole body is inserted into the nuclear envelope, it acts as an MTOC and interacts with the DNA. This DNA-spindle pole body complex was proposed to be responsible for the recognition of self and nonself at the level of individual Energides.[138]

Energide microtubules are extremely sensitive to mechanical treatments. Thus, it is not surprising to find an extremely low (typically between 0-4%) success rate in experiments of animal cloning which have involved nuclear transfer.[139-141] We can safely envision that if more care were taken over the intactness of the whole Energide during this process, the success rate (as judged by the frequency of developing embryos) would increase considerably and would help optimize the cloning process. To achieve this, one needs to be aware that the whole intact Energide (Cell Body), including its ensheathing microtubules, must be transferred into the recipient cell, not just the nucleus alone.

Impacts of Cell-Cell Channels on a Hypothetical Scenario of Eukaryotic Cell Evolution

From the currently available data, two major phases in the evolution of complex eukaryotic cells can be envisioned. An initial 'communal' phase[142,143] dominated by repeated fusion events between proto-cells allowed the early cells to increase their amounts of DNA. As a vestige of this primordial phase, the cell periphery of contemporary cells still shows an inherent tendency to fuse together thereby forming cell-cell channels. Some ancient proto-cells were more effective in this fusion process than others, which resulted in their larger cell size. These large cells suffered from osmotic imbalance resulting in the rupture of the early plasma membrane.[144,145] The frequent bursting of these proto-cells forced them to develop a means of rapid repair of the cell periphery as well as an ability to assemble more robust peripheries by enveloping the plasma membrane with a cell wall. These events heralded the second 'predatory' phase which, in fact, continues up to the present time. A convincing example of this is, the mutual enslavement of eukaryotic algae thereby generating eukaryote-eukaryote chimaeras.[24,25]

The onset of the 'predatory' phase was preceded by the generation of the most ancient eukaryotic cell. Now it is clear that this last common ancestor of eukaryotic cells was a complete cells equipped with nucleus, cytoskeleton, mitochondria, endoplasmic reticulum, Golgi apparatus, and vesicle trafficking.[146,147] Merging of cells was accomplished presumably with the invasion of a large and soft-periphery host cell, still suffering from the frequent bursting events, by a more advanced smaller guest cell already equipped with a more rigid-periphery.[9] These ancient nucleated proto-cells then went on to 'invent' actin-based phagocytosis which allowed them not only to initiate an active predatory lifestyle but also to accomplish a very efficient cell periphery repair process using recycling secretory endosomes filled with extracellular matrix components and enclosed by a plasma membrane-derived membrane.[35,36] Similarly, a phagocytosis-like process is known to be accomplished in the bacterium *Bacillus subtilis* during the formation of spores.[148,149] The cellular prey internalized by these proto-cells active in phagocytosis, and perhaps not fully equipped with all the enzymes needed to digest prey, often survived and developed into organelles such as peroxisomes, mitochondria, and plastids which are now features of eukaryotic cells.[9] During evolution, lateral gene transfer resulted in loss of DNA from developing organelles (DNA-losing ancient guest cells) and their accumulation within nuclei (DNA-gaining ancient guest cells). In some situations, all DNA can be lost: for example, mitosomes are remnants of mitochondrion-related organelles that lack any detectable DNA.[28] Furthermore, some hydrogenosomes also lost all of their DNA. Similarly, all DNA from the host cell (the cell contributing the plasma membrane) was transferred into the guest cell (the cell which became the nucleus), and it might well be that peroxisomes and centrosomes also represent highly reduced guest cells, having lost all DNA via lateral gene transfer.[150] For instance, the cell wall of plant cells appears to be descended from endosymbiotic cyanobacteria which, during their evolutionary transformation into plastids, have donated their cell wall genes to the guest cell.[151]

Conclusions and Outlook

The most important message of this chapter (and the whole book) is that cell-cell channels are an inherent feature of cells at all levels of complexity, from relatively simple bacteria up to large and complex plant and animal cells. In fact, plants are supracellular organisms because most of their cells (Energides) are interconnected by plasmodesmatal channels.[55,87] This might turn out to be true for animals. Conclusive evidence would be if the sensitive and apparently short-lived nanotubes[13,56-58] were found in intact tissues. There are some intriguing generalities with respect to cell-cell channels. First, viruses and molecules of RNA and DNA seem to have the ability to induce cell-cell channels. Both viruses and bacteria—and here for plant cells, consider *Agrobacterium tumefaciens*[152]— can induce cell-cell channels, enabling them to invade host cells and then to pirate the nuclear pores and thereby gain access to the DNA replication machinery within the host nuclei. After their replication, they escape from the nucleus and return to the cytoplasm; they then either kill the cell to gain access to other cells (animals) thereby inducing new cell-cell channels, or they gate the existing channels to allow cell-cell spread of virus or bacterium throughout the whole organism (plants). Besides viruses and bacteria, small RNA molecules also can spread from cell-to-cell, and this ability is used for transcellular and global silencing of specific genes. This transfer of small RNAs across cell boundaries is an ancient feature, allowing application of the RNA interference (RNAi) technology for both animals and plants.[153-155] The original discovery of RNAi was in *C. elegans*[156] where the transport of RNAs across cellular borders is dependent on the transmembrane protein SID1.[157]

Darwinian *adaptive* evolution that tends towards increased fitness is still the major force with respect to the adaptation of cells and other more complex units of life, like multicellular organisms, to the ever-changing ambient environment. However, it operates in parallel with the Margulisian *symbiotic* evolution which generates an increased complexity of evolving systems such as eukaryotic cells and their vital units of supracellular life, which are here called Energides. This is true both for the evolution of complex eukaryotic cells, as well as for their assembly into multi-cellular or, better, supra-cellular, organisms. Cell-cell fusion emerges as important factor determining animal cell behavior.[158] Cell-cell channels are inherent not only to plants[87] but also to animals.[13,56-58] Both unicellular[15] as well as supra-cellular organisms[87] should be considered as complex ecosystems composed of numerous modified cellular micro-organisms.[175] Finally, in contrast to cells which are widely heterogeneous in their size (ranging from 30 nanometres up to several metres), architecture, and mechanism of their divisions; Energides are extremely conservative in all these respects. This makes the Energide an ideal candidate for the universal unit of supracellular eukaryotic life. Appreciation of Energides as vital units endowed with all attributes of life is critical for our understanding of still enigmatic processes and phenomena of contemporary cell biology,[158,159] such as transformation, plasticity, and cloning of eukaryotic cells.

References

1. Hooke R. Of the schematisme or texture of cork, and of the cells and pores of some other such frothy bodies. Micrographia, Observation 18. London: 1665:112-116.
2. Harris H. The Birth of the Cell. New Haven: Yale University Press 1999.
3. Lodish H, Baltimore D, Berk A et al. Molecular Cell Biology, 3rd ed. New York: W.H. Freeman and Company, 1995.
4. Mazzarello P. A unifying concept: the history of cell theory. Nat Cell Biol 1999; 1:E13-E15.
5. Alberts B, Bray D, Hopkin K et al. Essential Cell Biology. 2nd ed. New York: Taylor & Francis Group: Garland Science, 2004.
6. Kleinig H, Sitte P. Zellbiologie. Stuttgart, New York: Gustav Fischer Verlag, 1984.
7. Pollard TD, Earnshaw WC. Cell Biology. Philadelphia, London, New York, St. Louis, Sydney, Toronto: Saunders, Elsevier Science, 2002.
8. Richmond ML. T.H. Huxley's criticism of German Cell Theory: an epigenetic and physiological interpretation of cell structure. J Hist Biol 2000; 33:247-289.
9. Baluska F, Volkmann D, Barlow PW. Eukaryotic cells and their Cell Bodies: Cell Theory revisited. Ann Bot 2004; 94:9-32.
10. Forterre P, Philippe H. Where is the root of the universal tree of life. BioEssays 1999; 21:871-879.
11. Poole A, Jeffares D, Penny D. Early evolution: prokaryotes, the new kids on the block. BioEssays 1999; 21:880-889.
12. Baluska F, Hlavacka A, Volkmann D et al. Getting connected: actin-based cell-to-cell channel in plants and animals. Trends Cell Biol 2004; 14:404-408.
13. Rustom A, Saffrich R, Markovic I et al. Nanotubular highways for intercellular organelle transport. Science 2004; 303:1007-1110.
14. Rivera MC, Lake JA. The ring of life provides evidence for a genome fusion origin of eukaryotes. Nature 2004; 431:152-155.
15. Margulis L. Serial endosymbiotic theory (SET) and composite individuality. Transition from bacterial to eukaryotic genomes. Microbiol Today 2004; 31:172-174.
16. Margulis L, Dolan MF, Guerrero R. The chimeric eukaryote: origin of the nucleus from karyomastigont in amitochondriate protist. Proc Natl Acad Sci USA 2000; 97:6954-6999.
17. Horiike T, Hamada K, Kanaya S et al. Origin of eukaryotic cell nuclei by symbiosis of Archaea in Bacteria is revealed by homology-hit analysis. Nat Cell Biol 2001; 3:210-214
18. Horiike T, Hamada K, Shinozawa T. Origin of eukaryotic cell nuclei by symbiosis of Archaea in Bacteria supported by the newly clarified origin of functional genes. Genes Genet Syst 2002; 77:369-376.
19. Dolan MF, Melnitsky H, Margulis L et al. Motility proteins and the origin of the nucleus. Anat Rec 2002; 268:290-301.
20. Hartman H, Fedorov A. The origin of the eukaryotic cell: a genomic investigation. Proc Natl Acad Sci USA 2002; 99:1420-1425.
21. Baluska F, Volkmann D, Barlow PW. Cell bodies in a cage. Nature 2004; 428:371.
22. Sachs J. Beiträge zur Zellentheorie. Energiden und Zellen. Flora 1892; 75:57-67.
23. Sachs J. Weitere Betrachtungen über Energiden und Zellen. Flora 1892; 81:405-434.
24. Cavalier-Smith T. Genomic reduction and evolution of novel genetic membranes and protein-targeting machinery in eukaryote-eukaryote chimaeras (meta-algae). Philos Trans R Soc Lond B Biol Sci 2003; 358:109-133.
25. Keeling PJ. Diversity and evolutionary history of plastids and their hosts. Am J Bot 2004; 91:1481-1493.

26. Shepherd VA, Beilby MJ, Bisson MA. When is a cell not a cell? A theory relating coenocytic structure to the unusual electrophysiology of Ventricaria ventricosa (Valonia ventricosa). Protoplasma 2004; 223:79-91.
27. Embley TM, van der Giezen M, Horner DS et al. Hydrogenosomes, mitochondria and early eukaryotic evolution. IUBMB Life 2003; 55:387-395.
28. Leon-Avila G, Tovar J. Mitosomes of Entamoeba histolytica are abundant mitochondrion-related remnant organelles that lack a detectable organellar genome. Microbiology 2004; 150:1245-1250.
29. Tovar J, Leon-Avila G, Sanchez LB et al. Mitochondrial remnant organelles of Giardia function in iron-sulphur protein maturation. Nature 2003; 426:172-176.
30. Atsatt PR. Fungus propagules in plastids: the mycosome hypothesis. Int Microbiol 2003; 6:17-26.
31. Maniloff J, Nealson KH, Psenner R et al. Nanobacteria: size limits and evidence. Science 1997; 276:1773-1776
32. Kürner J, Frangakis AS, Baumeister W. Cryo-electron tomography reveals the cytoskeletal structure of Spiroplasma melliferum. Science 2005; 307:436-438.
33. Raoult D, Audic S, Robert C et al. The 1.2-megabase genome sequence of Mimivirus. Science 2004; 306:1344-1350.
34. Benirschke K. Remarkable placenta. Clin Anat 1997; 11:194-205.
35. Baluska F, Liners F, Hlavacka A et al. Cell wall pectins and xyloglucans are internalized into dividing root cells and accumulate within cell plates during cytokinesis. Protoplasma 2005; In press.
36. Dhonuksche P. Visualizing microtubule dynamics and membrane trafficking in live and dividing plant cells. Ph.D. Thesis. University of Amsterdam, 2005.
37. Burgess DR, Chang F. Site selection for the cleavage furrow at cytokinesis. Trends Cell Biol 2005; 15:156-165.
38. Onishi M, Koga T, Morita R et al. Role of phosphatidylinositol 3-phosphate in formation of forespore membrane in Schizosaccharomyces pombe. Yeast 2003; 20:193-206.
39. Shimoda C. Forespore membrane assembly in yeast: coordinating SPBs and membrane trafficking. J Cell Sci 2004; 117:389-396.
40. de Duve C. The birth of complex cells. Scient Amer 1996; 274(4):38-45.
41. de Duve C. The onset of selection. Nature 2005; 433:581-582.
42. Ingber DE. The origin of cellular life. BioEssays 2002; 22:1160-1170.
43. Hartwell LH, Hopfield JJ, Leibler S et al. From molecular to modular cell biology. Nature 1999; 402(suppl):C47-C52.
44. Altman R. Die Elementarorganismen und Ihre Beziehungen zur den Zellen. Leipzig: Verlag von Veit, 1890.
45. Margulis L. Origin of Eukaryotic Cells. New Haven: Yale University Press, 1970.
46. Margulis L. Symbiosis in Cell Evolution. Life and Its Environment on the Early Earth. San Francisco: W. H. Freeman, 1981.
47. Margulis L. Symbiosis in Cell Evolution. San Francisco: W. H. Freeman, 1993.
48. Margulis L, Sagan D. Acquiring Genomes: a Theory of the Origin of Species. New York: Basic Books, 2002.
49. Martin W, Embley TM. Early evolution comes full circle. Science 2004; 431:134-137.
50. López-García P, Moreira D. Metabolic symbiosis at the origin of eukaryotes. Trends Biochem Sci 1999; 24:88-93.
51. Moreira D, López-García P. Symbiosis between methanogenic Archaea and Proteobacteria as the origin of eukaryotes: the syntrophic hypothesis. J Mol Evol 1998; 47:517-530.
52. Pennisi E. The birth of the nucleus. Science 2004; 305:766-768.
53. Hernandez LD, Hoffman LR, Wolfsberg TG et al. Virus-cell and cell-cell fusion. Annu Rev Dev Biol 1996; 12:627-661.
54. Lee J-Y, Yoo B-C, Lucas WJ. Parallels between nuclear-pore and plasmodesmal trafficking of information molecules. Planta 2000; 210:177-187.
55. Carr DJ. Historical perspectives on plasmodesmata. In: Gunning BES, Robards AW, eds. Intercellular Communication in Plants: Studies on Plasmodesmata. Berlin, Heidelberg, New York: Springer Verlag, 1976:291-295.
56. Önfelt B, Nedvetzki S, Yanagi K et al. Membrane nanotubes connect immune cells. J Immunol 2004; 173:1511-1513.
57. Önfelt B, Davis DM. Can membrane nanotubes facilitate communication between immune cells? Biochem Soc Trans 2004; 32:676-678.
58. Vidulescu C, Clejan S, O'Connor KC. Vesicle traffic through intercellular bridges in DU 145 human prostate cancer cells. J Cell Mol Med 2004; 8:388-396.
59. Errington J, Bath J, Wu LJ. DNA transport in bacteria. Nat Rev Mol Cell Biol 2001; 2:538-544.

60. Gauthier A, Thomas NA, Finlay BR. Bacterial injection machines. J Biol Chem 2003; 278:25273-25276.
61. Kumar RB, Das A. Polar location and functional domains of the Agrobacterium tumefaciens DNA transfer protein VirD4. Mol Microbiol 2002; 43:1523-1532.
62. Judd PK, Kumar RB, Das A. The type IV secretion apparatus protein VirB6 of Agrobacterium tumefaciens localizes to a cell pole. Mol Microbiol 2005; 55:115-124.
63. Rohde M, Püls J, Buhrdorf R et al. A novel sheathed surface organelle of the Helicobacter pylori cag type IV secretion system. Mol Microbiol 2003; 49:219-234.
64. Grohmann E, Muth G, Espinosa M. Conjugative plasmid transfer in Gram-positive bacteria. Microbiol Molec Biol Rev 2003; 67:277-301.
65. Kaiser D. Coupling cell movement to multicellular development in myxobacteria. Nat Rev Microbiol 2003; 1:45-54.
66. Skerker JM, Berg HC. Direct observation of extension and retraction of type IV pili. Proc Natl Acad Sci USA 2001; 98:6901-6904.
67. Wall D, Kaiser D. Alignment enhances the cell-to-cell transfer of pilus phenotype. Proc Natl Acad Sci USA 1998; 95:3054-3058.
68. Wolgemuth C, Hoiczyk E, Kaiser D et al. How myxobacteria glide. Curr Biol 2002; 12:369-377.
69. Baluska F, Wojtaszek P, Volkmann D et al. The architecture of polarized cell growth: the unique status of elongating plant cells. BioEssays 2003; 25:569-576.
70. Hoiczyk E, Baumeister W. The junctional pore complex, a prokaryotic secretion organelle, is the molecular motor underlying gliding motility in cyanobacteria. Curr Biol 1998; 8:1161-1168.
71. Heinlein M, Wood MR, Thiel T et al. Targeting and modification of prokaryotic cell-cell junctions by tobacco mosaic virus cell-to-cell movement protein. Plant J 1998; 14:345-351.
72. Menzel D. An interconnected plastidom in Acetabularia: implications for the mechanism of chloroplast motility. Protoplasma 1994; 179:166-171
73. Köhler RH, Cao J, Zipfel WR et al. Exchange of protein molecules through connections between higher plant plastids. Science 1997; 276:2039-2042.
74. Köhler RH, Schwille P, Webb WW et al. Active protein transport through plastid tubules: velocity quantified by fluorescence correlation spectroscopy. J Cell Sci 2000; 113:3921-3930.
75. Kwok EY, Hanson MR. Plastids and stromules interact with the nucleus and cell membranes in vascular strands. Plant Cell Rep 2004; 23:188-195.
76. Kwok EY, Hanson MR. GFP-labeled Rubisco and aspartate aminotransferase are present in plastid stromules and traffic between plastids. J Exp Bot 2004; 55:595-604.
77. Natesan SKA, Sullivan JA, Gray JC. Stromules: a characteristic cell-specific feature of plastid morphology. J Exp Bot 2005; 56:787-797.
78. Gunning BES. Plastid stromules: video microscopy of their outgrowth, retraction, tensioning, anchoring, branching, bridging and tip growth. Protoplasma 2005; 225:33-42.
79. Bereiter-Hahn J, Vöth M. Dynamics of mitochondria in living cells: shape changes, dislocations, fusion, and fission of mitochondria. Microsc Res Tech 1994; 27:198-219.
80. van Gestel K, Verbelen J-P. Giant mitochondria are a response to low oxygen pressure in cells of tobacco (Nicotiana tabacum L.). J Exp Bot 2002; 53:1215-1218.
81. Logan DC. Mitochondrial dynamics. New Phytol 2003; 160:463-478.
82. Westermann B. Merging mitochondria matters. Cellular role and molecular machinery of mitochondrial fusion. EMBO Rep 2002; 3:527-531.
83. Mozdy AD, Shaw JM. A fuzzy mitochondrial fusion apparatus comes into focus. Nat Rev Mol Cell Biol 2003; 4:468478.
84. Boukh-Viner T, Guo T, Alexandrian A et al. Dynamic ergosterol- and ceramide-rich domains in the peroxisomal membrane serve as an organizing platform for peroxisome fusion. J Cell Biol 2005; 168:761-773.
85. Jahn R, Lang T, Südhof TC. Membrane fusion. Cell 2003; 112:519-533.
86. Staehelin LA. The plant ER: a dynamic organelle composed of a large number of discrete functional domains. Plant J 1991; 11:1151-1165.
87. Gamalei YuV. Supercellular plant organization. Russ J Plant Physiol 1997; 44:706-730.
88. Holthuis JC, Levine TP. Lipid traffic: floppy drives and a superhighway. Nat Rev Mol Cell Biol 2005; 6:209-220.
89. Faulkner C, Brandom J, Maule A et al. Plasmodesmata 2004. Surfing the symplasm. Plant Physiol 2005; 137:607-610.
90. Fahrenkrog B, Köser J, Aebi U. The nuclear pore complex: a jack of all trades? Trends Biochem Sci 2005; 29:175-182.
91. Timney BL, Rout MP. Robbing from the pore. Nat Cell Biol 2004; 6:177-179.

92. Sheldrake AR. Effects of osmotic stress on polar auxin transport in Avena mesocotyl sections. Planta 1979; 145:113-117.
93. Baluska F, Samaj J, Hlavacka A et al. Myosin VIII and F-actin enriched plasmodesmata in maize root inner cortex cells accomplish fluid-phase endocytosis via an actomyosin-dependent process. J Exp Bot 2004; 55:463-473.
94. Haupt S, Cowan GH, Ziegler A et al. Two plant-viral movement proteins traffic in the endocytic recycling pathway. Plant Cell 2005; 17:164-181.
95. Oparka KJ. Getting the message across: how do plant cells exchange macromolecular complexes? Trends Plant Sci 2004; 9:33-41.
96. Devos D, Dokudovskaya S, Alber F et al. Components of coated vesicles and nuclear pore complexes share a common molecular architecture. PloS Biol 2004; 2(12):e380.
97. Antonin W, Mattaj IW. Nuclear pore complexes: round the bend? Nat Cell Biol 2005; 7:10-12.
98. Guyader M, Kiyokawa E, Abrami L et al. Role for human immunodeficiency virus type 1 membrane cholesterol in viral internalization. J Virol 2002; 76:10356-10364.
99. Manunta M, Tan PH, Sagoo P et al. Gene delivery by dendrimers operates via a cholesterol dependent pathway. Nucl Acids Res 2004; 32:2730-2739.
100. Vecchi M, Polo S, Poupon V et al. Nucleocytoplasmic shuttling of endocytic proteins. J Cell Biol 2001; 153:1511-1517.
101. Benmerah A, Scott M, Poupon V et al. Nuclear function for plasma membrane-associated proteins? Traffic 2003; 4:503-511.
102. Benmerah A. Endocytosis: signalling from endocytic membranes to the nucleus. Curr Biol 2004; 14:R314-R316.
103. Ramalho-Santos J, Schatten G, Moreno RD. Control of membrane fusion during spermiogenesis and the acrosome reaction. Biol Reprod 2002; 67:1043-1051.
104. Redecker P, Kreutz MR, Bockmann J et al. Brain synaptic junctional proteins at the acrosome of rat testicular germ cells. J Histochem Cytochem 2003; 51:809-819.
105. Giovannetti M, Fortuna P, Citernesi AS et al. The occurrence of anastomosis formation and nuclear exchange in intact arbuscular mycorrhizal networks. New Phytol 2001; 151:717-724.
106. Giovannetti M, Sbrana C, Avio L. Patterns of below-ground plant interconnections established by means of arbuscular mycorrhizal networks. New Phytol 2004; 164:175-181.
107. Glass NL, Kaneko I. Fatal attraction: nonself recognition and heterokaryon incompatibility in filamentous fungi. Eukaryot Cell 2003; 2:1-8.
108. Glass NL, Rasmussen C, Roca MG et al. Hyphal homing, fusion and mycelial interconnectedness. Trends Microbiol 2004; 12:135-141.
109. Xiang X, Fischer R. Nuclear migration and positioning in filamentous fungi. Fung Gen Biol 2004; 41:411-419.
110. Wang XY, Yu CH, Li X et al. Ultrastructural aspects and possible origin of cytoplasmic channels providing intercellular connection in vegetative tissues of anthers. Russ J Plant Physiol 2004; 51:97-106.
111. Guo G-Q, Zheng G-C. Hypotheses for the functions of intercellular bridges in male germ cell development and its cellular mechanisms. J Theor Biol 2004; 229:139-146.
112. Guzicka M, Wozny A. Cytomixis in shoot apex of Norway spruce (Picea abies L. Karst.). Trees 2005; 18:722-724.
113. Zhang WC, Yan WM, Lou CH. Intercellular movement of protoplasm in vivo in developing endosperm of wheat caryopses. Protoplasma 1990; 153:193-203.
114. van Bel A. The phloem, a miracle of ingenuity. Plant Cell Environm 2003; 26:125-149.
115. Telfer WH. Development and physiology of the oocyte-nurse cell syncytium. Adv Insect Physiol 1975; 11:223-319.
116. Spradling A. Germline cysts: communes that work. Cell 1993; 72:649-651.
117. Robinson DN, Cooley L. Stable intercellular bridges in development: the cytoskeleton lining the tunnel. Trends Cell Biol 1996; 6:474-479.
118. Kramerova IA, Kramerov AA. Mucinoprotein is a universal constituent of stable intercellular bridges in Drosophila melanogaster germ line and somatic cells. Dev Dyn 1999; 216:349-360.
119. Haynh J-R, St Johnston D. The origin of asymmetry: early polarisation of the Drosophila germline cyst and oocyte. Curr Biol 2004; 14:R438-R449.
120. Snapp EL, Iida T, Frescas D et al. The fusome mediates intercellular endoplasmic reticulum connectivity in Drosophila ovarian cysts. Mol Biol Cell 2004; 15:4512-4521.
121. Cavalier-Smith T. Economy, speed and size matter: evolutionary forces driving nuclear genome miniaturization and expansion. Ann Bot 2005; 95:147-175.
122. Baluska F, Volkmann D, Barlow PW. Motile plant cell body: a 'bug' within a 'cage'. Trends Plant Sci 2001; 6:104-111.

123. Baluska F, Volkmann D, Barlow PW. Nuclear components with microtubule organizing properties in multicellular eukaryotes: functional and evolutionary considerations. Int Rev Cytol 1997; 175:91-135.
124. Baroux C, Fransz P, Grossniklaus U. Nuclear fusions contribute to polyploidization of the gigantic nuclei in the chalazal endosperm of Arabidopsis. Planta 2004; 220:38-46.
125. Guitton AE, Page DR, Chambrier P et al. Identification of new members of Fertilisation Independent Seed Polycomb Group pathway involved in the control of seed development in Arabidopsis thaliana. Development 2004; 131:2971-2981.
126. Baluska F, Volkmann D, Barlow PW. Actin-based domains of the 'cell periphery complex' and their associations with polarized 'cell bodies' in higher plants. Plant Biol 2000; 2:253-267
127. Ketelaar T, Faivre-Moskalenko C, Esseling JJ et al. Positioning of nuclei in Arabidopsis root hairs: an actin-regulated process of tip growth. Plant Cell 2002; 14:2941-2955.
128. Freitag M, Hickey PC, Raju NB et al. GFP as a tool to analyze the organization, dynamics and function of nuclei and microtubules in Neurospora crassa. Fungal Genet Biol 2004; 41:897-910.
129. Martin R, Walther A, Wendland J. Deletion of the dynein heavy-chain gene DYN1 leads to aberrant nuclear positioning and defective hyphal development in Candida albicans. Eukaryot Cell 2004; 3:1574-1588.
130. Orias JD, Hamilton EP, Orias E. A microtubule meshwork associated with gametic pronucleus transfer across a cell-cell junction. Science 1983; 222:181-184.
131. Janetopoulos C, Cole E, Smothers JF et al. The conjusome: a novel structure in Tetrahymena found only during sexual reorganization. J Cell Sci 1999; 112:1003-1011.
132. Goff LJ, Coleman AW. Transfer of nuclei from a parasite to its host. Proc Natl Acad Sci USA 1984; 81:5420-5424.
133. Saupe SJ. Molecular genetics of heterokaryon incompatibility in filamentous ascomycetes. Microbiol Molec Biol Rev 2000; 64:489-502.
134. Kuhn G, Hijri M, Sanders IR. Evidence for the evolution of multiple genomes in arbuscular mycorrhizal fungi. Nature 2001; 414:745-748.
135. Shiu PKT, Glass NL. Cell and nuclear recognition mechanisms mediated by mating type in filamentous ascomycetes. Curr Opin Microbiol 2000; 3:183-188.
136. Schuurs TA, Dalstra HJP, Scheer LML et al. Positioning of nuclei in the secondary mycelium of Schizophyllum commune in relation to differential gene expression. Fung Genet Biol 1998; 23:150-161.
137. Debuchy R. Internuclear recognition: a possible connection between Euascomycetes and Homobasidiomycetes. Fung Genet Biol 1999; 27:218-223.
138. Thompson-Coffe C, Zickler D. How the cytoskeleton recognizes and sorts nuclei of opposite mating type during the sexual cycle in filamentous ascomycetes. Dev Biol 1994; 165:257-271
139. Wilmut I, Beaujean N, de Sousa PA et al. Somatic cell nuclear transfer. Nature 2002; 419:583-586.
140. Gurdon JB, Byrne JA, Simonsson S. Nuclear reprogramming and stem cell creation. Proc Natl Acad Sci USA 2003; 100:11819-11822.
141. Fujita N, Wade PA. Nuclear transfer: epigenetics pay a visit. Nat Cell Biol 2004; 6:912-922.
142. Woese CR. On the evolution of cells. Proc Natl Acad Sci USA 2002; 99:8742-8747.
143. Woese CR. A new biology for a new century. Microbiol Mol Biol Rev 2004; 68:173-186.
144. Koch AL Development and diversification of the Last Universal Ancestor. J Theor Biol 1994; 168:269-280.
145. Koch AL. The bacterium's way for safe enlargement and division. Appl Environm Microbiol 2000; 66:3657-3663.
146. Simpson AGB, Roger AJ. The real 'kingdoms' of eukaryotes. Curr Biol 2004; 14:R693-R696.
147. Walsh DA, Doolittle WF. The real 'domains' of life. Curr Biol 2005; 15:R237-R240.
148. Sharp MD, Pogliano K. An in vivo membrane fusion assay implicates SpoIIIE in the final stages of engulfment during Bacillus subtilis sporulation. Proc Natl Acad Sci USA 1999; 96:14553-14559.
149. Abanes-De Mello A, Sun Y-L, Aung S et al. A cytoskeleton-like role for the bacterial cell wall during engulfment of the Bacillus subtilis forespore. Genes Dev 2002; 16:3253-3264.
150. Timmis JN, Ayliffe MA, Huang CY et al. Endosymbiotic gene transfer: organelles genomes forge eukaryotic chromosomes. Nat Rev Genet 2004; 5:123-135.
151. Hoiczyk E, Hansel A. Cyanobacterial cell walls: news from an unusual prokaryotic envelope. J Bacteriol 2000; 182:1191-1199.
152. Valentine L. Agrobacterium tumefaciens and the plant: the David and Goliath of modern genetics. Plant Physiol 2003; 133:948-955.
153. Hannon GJ. RNA interference. Nature 2002; 418:244-251.
153. Baulcombe D. RNA silencing in plants. Nature 2004; 431:356-363.
155. McManus MT. Small RNAs and immunity. Immunity 2004; 21:7.

156. Fire A, Xu S, Montgomery MK et al. Potent and specific genetic interference by double-stranded RNA in Caenorhabditis elegans. Nature 1998; 391:806-811.
157. Feinberg EH, Hunter CP. Transport of dsRNA into cells by the transmembrane protein SID-1. Science 2003; 301:1545-1547.
158. Vignery A. Macrophage fusion: are somatic and cancer cells possible partners? Trends Cell Biol 2005; 15:In press
159. Vassilopoulos G, Russell DW. Cell fusion: an alternative to stem cell plasticity and its therapeutic implications. Curr Opin Genet Dev 2003; 13:480-485.

SECTION I
Prokaryotic Cells

CHAPTER 2

Mating Cell-Cell Channels in Conjugating Bacteria

Elisabeth Grohmann*

Abstract

Conjugative plasmid transfer is the most important mechanism for bacteria to deliver and acquire genetic information to cope with rapidly changing environmental conditions. To transfer genetic information intercellularly mating cell-cell channels between donor and recipient bacteria have to be established. For plasmid transfer in Gram-negative bacteria, subassemblies of these mating channels have been discovered, the order in which the transferred DNA contacts the transporter proteins has been determined and crystal structures of key components of the so-called conjugative type IV secretion systems have been solved. In contrast to this, knowledge on conjugative plasmid transfer of sex pheromone-inducible plasmids in *Enterococcus faecalis* is limited to molecular details on the complex regulation processes whereas for broad-host-range plasmids from Gram-positive bacteria investigations on the structure of the conjugative transfer apparatus and the interplay of the secretion components have recently started. The following chapter has the intention to give an overview of the state of the art on conjugative plasmid transfer in Gram-positive and Gram-negative bacteria.

Introduction

Bacterial conjugation is the most important means of gene delivery enabling adaptation of bacteria to changing environmental conditions including spread of antibiotic resistance genes, thereby generating multiply antibiotic resistant pathogens. Multiply resistant pathogens, such as *Pseudomonas aeruginosa*, *Staphylococcus aureus* and *Enterococcus faecalis* represent a serious threat to antibiotic treatment of hospitalized and immuno-suppressed patients. Therefore, much effort has been and is still made towards elucidating the molecular mechanisms of conjugative plasmid transfer.

Bacterial conjugation systems are specialized types of type IV protein secretion systems (T4SS) dedicated to transport proteins (e.g., virulence factors, toxins) from bacterial pathogens to their mammalian hosts. The conjugative T4SS have evolved to transport DNA substrates in addition to proteins intercellularly.

Relevant progress has been made in deciphering the transport pathway of DNA and protein substrates through the Gram-negative (G-) cell envelope (see refs. 1,2). Recently the Christie group[2] provided evidence for the order of transferred (T)-DNA contact with the T4SS proteins of the prototype T4SS of *Agrobacterium tumefaciens* during T-DNA export to the plant nucleus. The best characterized T4SS are the *Agrobacterium* T-DNA transfer system and the conjugative transfer systems of plasmids RP4, F and R388, all originating from G- bacteria. Intense investigations on the transfer mechanisms of plasmids from Gram-positive (G+) bacteria have started only a few years

*Elisabeth Grohmann—University of Technology Berlin, Department of Environmental Microbiology, FR1-2, Franklinstrasse 28/29, D-10587 Berlin, Germany. Email: Elisabeth.grohmann@tu-berlin.de

Cell-Cell Channels, edited by Frantisek Baluska, Dieter Volkmann and Peter W. Barlow.
©2006 Landes Bioscience and Springer Science+Business Media.

ago. An exception represent the well-studied sex-pheromone responsive plasmids of *E. faecalis* whose transfer underlies a complex regulatory mechanism exerted by small secreted signal molecules, the so-called pheromones (for recent comprehensive reviews see refs. 3,4).

The chapter is divided into three parts summarizing the current knowledge of conjugative transfer mechanisms, mating cell-cell channel assembly and structure:

    i. in G- bacteria;

    i. of pheromone-responsive conjugative plasmids in *E. faecalis*; and

    ii. of nonpheromone-responsive plasmids in G+ bacteria.

Conjugative DNA Transfer in Gram-Negative Bacteria

Conjugative DNA transfer systems in G- bacteria, nowadays generally referred to as specialized T4SS, have been extensively studied for more than two decades. T4SS translocate DNA and protein substrates across the bacterial cell envelope. In general, T4SS transport their substrates to recipient cells via direct cell-to-cell contact. But there are also examples of contact-independent protein export and DNA release to and uptake from the extracellular milieu.[1,5,6]

Considerable progress has been made towards the mechanistic understanding of intercellular DNA transport in the plasmid model systems RP4, R388 and F (for recent reviews see refs. 7-9) and in the T-DNA transport system of *A. tumefaciens*.[1,10] Several models for transenvelope DNA/protein transport have been proposed (for a summary see ref. 11) which match considerably well with the experimental data. Recently the order in which T-DNA contacts T4SS proteins on its way through the *A. tumefaciens* cell envelope has been determined by Christies' group.[2,12,13] This discovery is a milestone towards the elucidation of the conjugative DNA/protein secretion mechanism.

On account of these very interesting results the state of the art of T4SS in G- bacteria will be presented on basis of the prototype T4SS, the *A. tumefaciens* T-DNA system (VirB/VirD4 transfer system).

The T4SS Operon Structures

The *A. tumefaciens* VirB/D4 T4SS is encoded by the *virB* and *virD* operons.[13] The *virB* operon codes for 11 genes, *virB1* to *virB11*. The VirB proteins, termed the mating pair formation (mpf) proteins, build a cell envelope-spanning structure required for substrate transfer, and an extracellular filament, the T pilus that mediates attachment to recipient cells.[15] The *virD* operon encodes five genes, *virD1* to *virD5*. *virD1* and *virD2* encode gene products processing the DNA substrate (T-DNA) for transfer. These are named the DNA transfer and replication (Dtr) proteins.[16] *virD3* and *virD5* encode proteins that are not essential for processing or transfer. *virD4* codes for the coupling protein (CP).[17,18] The VirD4 CP is not involved in T-DNA processing or formation of the T-pilus but delivers together with the mpf structure substrates across the cell envelope.[14,15]

T4SS Substrates

The T-Strand-Relaxase Complex

The key enzyme of conjugative plasmid transfer is the DNA relaxase, a transesterase which cleaves a specific phosphodiester bond in the origin of transfer (*oriT*) thereby initiating the conjugative transfer. The relaxase preserves its energy from cleavage of the phosphodiester backbone of the T-strand as a stable phosphotyrosyl intermediate with the T-strand.[19] The complex of processing proteins at *oriT* (relaxase and accessory proteins) is termed the relaxosome. Conjugative DNA transfer proceeds in a 5'-3' direction[16] suggesting that the relaxase, covalently bound to the 5' end of the T-strand, supplies substrate recognition signals and possibly also exerts a piloting function to direct DNA transport through the secretion channel.[20,21]

Protein Substrates

Conjugation systems also export proteins independently of DNA.[22-28] The *A. tumefaciens* T-DNA transfer system translocates the VirE2 protein in a chaperone-independent and the VirE3 and VirF protein in a chaperone-assisted way into the recipient cell. Indirect evidence was also obtained for translocation of the relaxase-T-DNA complex to plant cells.[20] T4SS substrates contain potential secretion signals at their C-termini. The C-termini of *A. tumefaciens* VirB/D4 T4SS substrates carry a conserved Arg-X-Arg motif, whereas many T4SS substrates—including relaxases from various conjugation systems—carry many positively charged residues, mostly Arg within the last 30-50 residues.[10] Recently it was shown that the C-Terminus mediates an interaction between the VirE2 secretion substrate and the VirD4 CP of the VirB/D4 T4SS.[29] It can be suggested that the charged C-termini of T4SS substrates probably contribute to substrate recognition by mediating productive contacts with the CP of the T4SS.[10]

Initiation of T-Strand Transfer: Substrate Recruitment by the Coupling Protein

It is well established that a given T4SS, e.g., encoded by plasmids R388 or RP4 or the *A. tumefaciens* VirB/D4 system, translocates a restricted set of substrates including the cognate plasmid or oncogenic T-DNA, one or a few mobilizable plasmids such as ColE1 or RSF1010, and one or a few proteins.[10,16]

The selectivity of the T4SS is exerted by the respective CPs.[18,30-32] Many experimental data on chimeric T4SS (composed of a CP from one T4SS and a mpf structure from a second T4SS) suggest that the CP links the Dtr processing proteins bound at *oriT*- the relaxosome- to the T4SS, hence the origin of the term "coupling protein".[10] In vitro and in vivo studies have demonstrated several CP-relaxase interactions.[33,34]

Upon recruitment of the relaxosome, how does the CP mediate the next step of transfer? Structural studies have begun to shed light on the answer to this question. Topology studies have shown that CPs consist of an N-proximal region that includes two transmembrane helices and a small periplasmic domain, and a large C-terminal region that resides in the cytoplasm.[35,36] The crystal structure has been solved for the soluble domain (TrwBΔN70) of the TrwB CP of the IncW plasmid R388. The TrwBΔN70 crystal consists of six equivalent protomers that form a spherical particle of overall dimensions of 110 Å in diameter and 90 Å in height. This ring-like structure possesses a central channel of 20 Å in diameter, which traverses the structure, possibly connecting cytoplasm with periplasm.[37,38] Based upon the collected data on CP interactions with mpf subunits the role of the CP can be summarized as follows:

The CP delivers DNA substrates to the secretion apparatus through contacts with the DNA-processing proteins. Then, through contact with VirB10, a putative structural scaffold protein for assembly of the transenvelope T4SS, the CP coordinates passage of the T-strand through the mpf channel. Due to the known structure of the CP one possibility how this translocation occurs would be that the CP acts as a translocase to transport the substrate across the inner membrane.[10]

A Transenvelope Secretion Channel: The mpf Structure

All of the VirB proteins except for the VirB1 lytic transglycosylase are required for substrate export.[39] Thus far, no supramolecular organelles at cellular junctions of mating cells have been identified by high-resolution electron microscopy. However, several experimental findings support the existence of an envelope-spanning secretion apparatus. For example, the presumptive VirB/VirD4 mating channel or subcomplexes thereof have been isolated by membrane solubilization with nonionic detergents. At least two large complexes were detected, one composed of the T-pilus associated proteins VirB2, VirB5, and VirB7, and the second consisting of several other VirB proteins and the VirD4 CP.[40] Independent studies have also reported the isolation of a VirB2/VirB5/VirB7 complex and subcomplexes of other VirB proteins, such as VirB7/VirB9, VirB6/VirB7/VirB9 and VirB7/VirB9/VirB10 by detergent solubilization and immunoprecipitation or GST-pull-down assays (summarized in ref. 10). These results were confirmed by two-hybrid screens and further pairwise interactions were detected in vivo.[41,42]

Taken the above mentioned results together with computer-based predictions and topology studies of individual VirB subunits, a general architecture for the VirB/VirD4 T4SS can be presented: The VirD4 CP and the two mpf ATPases, VirB4 and VirB11, are localized predominantly or exclusively at the cytoplasmic face of the inner membrane. VirB6 is a highly hydrophobic protein predicted to

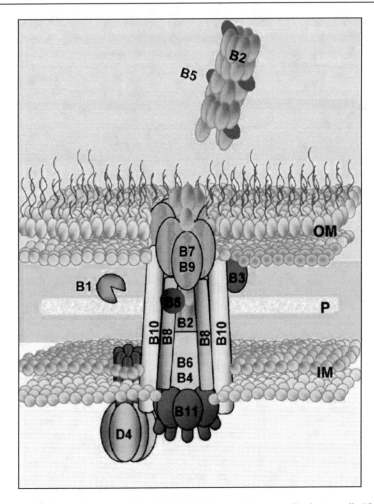

Figure 1. A model showing the *A. tumefaciens* VirB/D4 T4SS as a single, supramolecular organelle. The VirD4 CP is a homomultimeric integral membrane complex required for substrate transfer. The VirB proteins assemble as a secretion channel and an extracellular T pilus. VirD4 and the VirB proteins function together to mediate substrate transfer, and the VirB proteins direct pilus assembly. The adhesive T pilus at the cell surface is postulated to promote aggregation of donor and recipient cells on solid surfaces. IM, inner membrane; P, periplasm; OM, outer membrane. Reprinted with permission from: Christie PJ. Biochim Biophys Acta 2004; 1694:219-234,[10] ©2003 Elsevier.

span the inner membrane several times. VirB8 and VirB10 are bitopic proteins with short N-terminal cytoplasmic domains, a transmembrane helix and large C-terminal periplasmic domains. VirB2 (the major pilin protein), VirB3 and VirB5 are located in the periplasm, VirB2, VirB5, and VirB7 assemble as the extracellular T pilus. VirB7 is a small lipoprotein, which forms together with VirB9 a covalently cross-linked dimer. This dimer or a higher-order VirB7-VirB9 multimer assemble at the outer membrane. VirB9 has nine possible β-sheet outer membrane-spanning segments according to the Schirmer-Cowan algorithm.[43] Therefore, it is the best candidate for forming an oligomeric secretion-like pore mediating secretion of substrates and/or protrusion of the T pilus across the outer membrane.[10] A model for the general architecture of the T4SS is shown in Figure 1.

Energy Supply-VirB4 and VirB11

VirB11 belongs to a large family of ATPases associated with macromolecule secretion systems.[44] Homologues are widely distributed among the G- bacteria, and they are also functional in secretion systems of G+ bacteria and in Archaea.[10] VirB11 is highly insoluble and has been difficult to analyze biochemically. But soluble forms of VirB11 homologues have been characterized enzymatically and structurally. The VirB11-homologue of the *Helicobacter pylori* pathogenicity island, HP0525$_{Cag}$ and several plasmid-encoded VirB11-homologues were shown to assemble as homohexameric rings as demonstrated by electron microscopy.[45] HP0525$_{Cag}$ also presented as a homohexamer by X-ray crystallography. It is a double-stacked ring with a central cavity of about 50 Å in diameter.[46] The overall HP0525$_{Cag}$ structure appears to be highly conserved, even among distantly related NTPases encoded by other transport or fimbrial biogenesis systems.[10] HP0525$_{Cag}$ is also structurally similar to members of the AAA ATPase superfamily.[47] Many AAA ATPases act as energy-dependent unfoldases in substrate remodelling. Considering that the T4SS are export systems, the VirB11-like ATPases might act as chaperones to unfold the protein substrates at the channel entrance.[10] VirB11-like ATPases associate tightly but peripherally at the inner face of the inner membrane.[48,49] Genetic studies showed the coordinated action of VirB11 and the VirD4 CP for substrate transfer.[10] Mutagenesis analyses of VirB11 proved that it participates both in pilus biogenesis and assembly or function of the secretion machine.[50] It is likely that VirD4 and VirB11 hexamers localize next to each other at the inner membrane as depicted in Figure 1. Llosa et al[51] described a model that presents conjugation systems as two separately acting inner membrane translocases: the CP functions as a general recruitment factor for all T4SS substrates which is in agreement with all other prominent T4SS models, as shown in Figure 2. Although it is tempting to assume that the CP delivers the whole substrate to the mpf complex, they propose that the CP translocates the T-strand across the inner membrane while simultaneously transferring the relaxase to VirB11 and the other mpf proteins for secretion. The model of Llosa and coworkers explains very well most experimental data but it still remains difficult to envision how two transport proteins localized next to each other at the inner membrane coordinate their activities to mediate secretion of one substrate, the T-DNA-relaxase complex across the inner membrane.[10]

VirB4 possesses two putative membrane-spanning domains, one close to the N terminus and another located more centrally near the Walker A nucleoside triphosphate binding motif.[48] VirB4 self-interacts as shown by Dang et al[52] and Ward et al.[42] ATP hydrolysis has not been convincingly shown for VirB4-like proteins, but these proteins require intact Walker A motifs to mediate substrate export.[53,54] Bohne et al[55] reported that the presence of a subset of VirB proteins including VirB4 in *Agrobacterium* recipient cells increases the efficiency of plasmid uptake in intraspecies matings significantly. These data led to the proposal that these VirB proteins assemble as a complex that stabilizes mating junctions or perhaps they directly facilitate DNA transport across the recipient cell envelope.[56] However, it was also demonstrated that a Walker A mutation does not decrease the capacity of VirB4 to stimulate plasmid DNA acquisition by recipient cells. Dang et al[52] argued that VirB4 probably contributes structural information required for substrate transfer in either direction across the cell envelope. But an intact ATP-binding motif is necessary for configuring this T4SS specifically for substrate export. Christie concludes that the VirD4 CP, VirB11, and VirB4 must interact in complex and dynamic ways—probably through ATP-powered conformational changes— to energize substrate transfer to and across the inner membrane (ref. 10 and Fig. 1).

Working Models for T4SS

Three different working models describing the possible T4SS architecture and translocation routes are discussed at present (Fig. 2). The first model proposed by Christie[14] suggests that the mpf proteins assemble as a transenvelope channel for substrate export in one step. This model predicts that the T4CP recruits DNA and protein substrates to the translocation apparatus and then coordinates its activity with a VirB11-type ATPase to drive substrate transfer through the mpf channel.[11] The second model is a generalized version of the two-step routing pathway described for the export of the *Bordetella pertussis* toxin.[57] In the first step, an inner membrane translocase delivers substrates

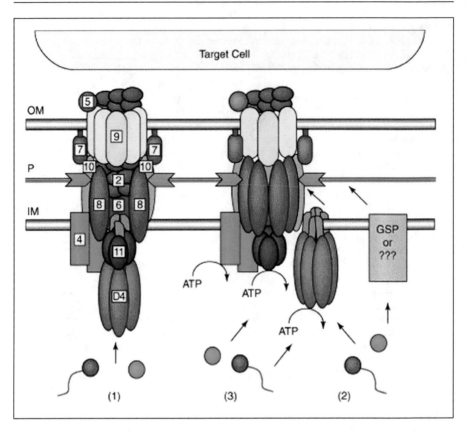

Figure 2. Possible architectures and substrate transloction routes for the T4SS. Numbers refer to VirB/VirD proteins of the *Agrobacterium* T-DNA transport system. Three working models describe the possible machine architectures and translocation routes: (1) a one-step model using a transenvelope channel, (2) a two-step model using the T4CP or alternative translocase for substrate transfer across the inner membrane (IM) and the mpf complex for outer membrane (OM) translocation, and (3) another two-step model, the "shoot and pump model", whereby the T4CP recruits substrates and transports DNA across the IM and delivers protein substrates to the mpf protein export machinery. Blue line, T-strand; red circle, relaxase bound to the T-strand; green circle, protein substrate; P, periplasm. Reprinted with permission from: Ding Z et al. Trends Microbiol 11:527-535,[11] ©2003 Elsevier.

across the inner membrane. In the second step, the T4SS translocase, composed of mpf proteins, transports substrates across the outer membrane.[56] This model predicts that the CP, when present, acts as an inner membrane translocase for both DNA and protein substrates. This activity is exerted completely independent of the mpf proteins.[11] An alternative two-step model, termed the "shoot and pump" model also suggests two inner membrane transporters. However, in this model the T4CP acts as a DNA translocase, whereas the mpf complex translocates protein substrates. In the periplasm, both pathways converge for mpf-dependent transport across the outer membrane.[51] The "shoot and pump" model does not exclude the proposed function of the T4CP as a general recruitment factor. But it postulates that upon recruitment the T4CP translocates DNA and delivers the protein substrate to the mpf channel. The "shoot and pump" model is especially attractive: first, because it accommodates most experimental findings to date and second, because it nicely explains why T4SS evolved to be so highly flexible.[11,51]

Pheromone-Responsive Conjugative Plasmids in *E. faecalis*

Though the enterococcal pheromone-inducible conjugative plasmids such as pCF10, pAD1, and pPD1 represent a unique class of mobile genetic elements spreading virulence traits readily with high transfer efficiency even in aqueous systems, the mechanism of their conjugative DNA secretion system has not been studied intensively. However, the complex regulatory mechanisms underlying specific efficient plasmid exchange has been investigated in great detail. The hereafter presented state of the art will focus on the regulatory machinery that interacts specifically with the pheromone peptides thereby controlling plasmid acquisition of plasmid-free enterococcal recipient cells. The data were derived mainly from a comprehensive review on enterococcal peptide sex pheromones by Chandler and Dunny.[4]

The pheromone plasmids are induced to transfer from donor cells by mating pheromones that are produced by potential recipient cells. The pheromone plasmids have evolved a fascinating and complex regulatory system to ensure their maintenance and stable existence in a population. Their transfer genes are induced by small (7-8 amino acids (aa)) peptides chromosomally encoded by all known enterococcal strains. Each peptide is highly specific for a cognate plasmid or for a family of closely related plasmids. All pheromones analyzed so far are produced by proteolytic processing of the cleaved signal sequences of secreted lipoproteins. The processed peptides are secreted into the growth medium and are utilized by the plasmid-containing donor cells to sense the presence of a nearby recipient cell. The pheromone-responsive plasmids use an interesting combination of host and plasmid encoded proteins to sense exogenous pheromone in order to activate the expression of transfer genes and to avoid self-induction by pheromone that is encoded on the chromosome of the host cell.[4] An overview of the main steps involved in pheromone-induced conjugation in enterococci will be given based upon data obtained from plasmid pCF10, the model plasmid of the Dunny group. pCF10-transfer is induced when the recipient-produced pheromone is detected by the donor cell at the cell surface by the plasmid-encoded lipoprotein PrgZ.[58] PrgZ acts together with the chromosomally encoded oligopeptide permease system (Opp) to import the pheromone into the cytoplasm of the recipient cell.[59] Import of the pheromone is necessary for pheromone response. The interaction process that likely initiates the induction in donor cells is binding of the imported pheromone to the plasmid-encoded cytoplasmic protein PrgX. PrgX is a negative regulator of expression of conjugative functions. Binding of pheromone to PrgX abolishes its repression so that transfer genes are synthesized. Two additional pCF10-encoded proteins, PrgY and iCF10, are required to keep the transfer system off in pCF10-harbouring cells grown in the absence of exogenous pheromone. PrgY and iCF10 are supposed to block self-induction of donor cells by endogenous pheromone.

Exposure of pCF10-containing cells to exogenous pheromone, cCF10, is phenotypically visible by aggregation of the culture resulting from upregulation of the expression of aggregation substance PrgB from the pCF10-encoded *prgB* gene.[60] The aggregation results in close contact between donor- and recipient cells probably enabling effective plasmid transfer, even in liquid medium. Approximately 15 additional genes are encoded 3′ from *prgB*. Data from the Dunny group suggest that many of these genes are upregulated by pheromone.[4]

The PrgZ pheromone binding protein is critical in the first step of pheromone induction: recognition of pheromone and import into the cytoplasm. All pheromone-responsive conjugative plasmids encode a PrgZ-like protein. They are homologues of the peptide-binding OppA proteins found in a wide range of bacterial species.[61] The PrgZ-type proteins are cell surface proteins anchored to a lipid moiety on the outer surface of the cytoplasmic membrane.[58] The PrgZ family of pheromone-binding proteins have been shown to increase the sensitivity of each plasmid system to its cognate pheromone.[58,62]

The pheromones themselves are encoded within the chromosome of *E. faecalis*, processed by host proteins and secreted in extremely small amounts into the culture medium. In the case of pCF10, cCF10 is released at ~10^{-11} M and can induce a donor cell at concentrations of 2×10^{-12} M corresponding to less than five molecules per cell under the conditions tested.[63] Despite multiple pheromones with different spectra of activities secreted by a single cell, each plasmid responds specifically to its cognate pheromone with surprising sensitivity.[64,65] The nucleotide sequence of the *E. faecalis* strain V583[66] showed that the pheromone precursors lie within the N-terminal signal sequences of predicted surface lipoproteins.[67] Proteolytic processing of the pheromone precursor is required prior to release of the mature pheromone into the culture medium. Signal peptidase II cleaves at a specific cysteine residue liberating the signal peptide from the lipoprotein.[68]

How does the plasmid prevent a response to its own host's endogenously produced pheromone? The pheromone plasmids have evolved two independent mechanisms of avoiding this self-induction such that the transfer response is only induced if a nearby recipient is detected. One mechanism is exerted by the synthesis of a plasmid encoded inhibitor peptide, iCF10 in the case of pCF10, which neutralizes endogenously produced pheromone in the culture medium.[4] The inhibitor peptides are proposed to compete with the pheromone for binding to the surface binding protein PrgZ.[69,70] The level of iCF10 in donor cultures has been found to be 10-100-fold above the pheromone level. This molar ratio is just sufficient to neutralize the cCF10 activity released by the same cells. The inhibitor peptides are very similar to each other and to the inducing pheromones. They are 7-8 aa hydrophobic peptides likely being processed from 22 to 23 aa precursors. Despite the apparent homology between the inhibitors and their peptides, their functions are very specific. Data on the pPD1 and the pCF10 plasmid imply that the function of the inhibitor may include more than just competitive inhibition of PrgZ and may instead involve specificity at some other level, since the response of the systems to their cognate pheromones and inhibitors is highly specific.[64,65] This specificity determinant/mechanism remains to be unravelled.

PrgY is the other pCF10-encoded element involved in control of endogenous pheromone. While the inhibitors control endogenous pheromone in the culture medium, the membrane protein PrgY controls endogenous pheromone activity that remains associated with the cell.[4] Buttaro et al[71] showed that a significant amount of pheromone in cCF10 producing cells remains associated to the cell wall. The average plasmid-free recipient cell appears to have twice as much cCF10 in its cell wall than the amount that is secreted into the supernatant. When pCF10 is acquired, the concentration of cCF10 in the supernatant is not affected[70] whereas the cell wall-associated cCF10 is decreased 8-fold from that of plasmid-free recipient cells.[71] PrgY was shown to be involved in this reduction of envelope-associated cCF10 after acquisition of the plasmid, but its mechanism of action is not clear. Other PrgY-like proteins have been identified in recent genome sequencing projects. Curiously, the bacterial species that have been found to encode a PrgY-type protein, do not have characterized peptide-signal systems and are quite distantly related to *E. faecalis*. This finding further deepens the mystery surrounding the role of this protein in the regulation of the pheromone induction process.[4]

So far, no data are available on the DNA transport mechanism of pCF10 and the other pheromone-inducible enterococcal plasmids. However, it seems reasonable to argue that the DNA secretion process of these plasmids might proceed via a T4S-like mechanism as proved for all known G- systems and currently being studied for the broad-host-range conjugative plasmids from G+ bacteria (see below): The following findings are in favour of a T4S-like mechanism: On two pheromone plasmids, namely pAD1 and pAM373, *oriT*s have been found. pAD1 has two *oriT*s, *oriT1* and *oriT2* and encodes a relaxase, TraX, which has been demonstrated to specifically nick in *oriT2*. *oriT*$_{pAM373}$ has been shown to be similar to *oriT2*$_{pAD1}$. Both plasmids are able to mobilize the non conjugative plasmid pAMα1, which encodes two relaxases that are involved in transfer.[72] ORF53 encoded by pAD1 is a protein essential for conjugation, which exhibits structural similarities to TraG-like CPs.[73] Recently nucleotide sequencing of the 67,673-bp pheromone plasmid pCF10 has been completed. pCF10 contains 57 *orf*s, *orf35* encodes a relaxase, *pcfG*[74] with highest homology to LtrB, the relaxase of the *Lactococcus lactis* conjugative plasmid pRS01.[75]

Nonpheromone-Responsive Plasmids in G+ Bacteria

Enterococci harbour also a pheromone-independent conjugative plasmid, namely the 65.1-kb pMG1, that transfers efficiently in broth matings. Interestingly, Southern hybridization of pMG1 DNA showed no homology to pheromone-responsive plasmids and the broad-host-range conjugative plasmids pAMβ1 and pIP501.[76]

Aggregation-mediated plasmid transfer in *Bacillus thuringiensis* and in lactic acid bacteria has been summarized in Grohmann et al[77] and will not be discussed further here. Recently Belhocine et al[78] demonstrated that conjugation is one of the mechanisms by which group II introns, originally discovered on a *L. lactis* conjugative plasmid (pRS01) and within a chromosomally located sex factor in *L. lactis* 712, are broadly disseminated between widely diverged G+ organisms.

Conjugative Transfer of Broad-Host-Range Plasmids

Transfer of broad-host-range G+ plasmids occurs at a variable frequency (generally in the range of 10^{-3} to 10^{-6}) depending on the plasmids and the mating-pair genotype, and mating requires cocultivation of donor and recipient cells on a solid surface.[77] Most conjugative plasmids identified so far in streptococci and enterococci actually show a broad host range (and hence are referred to as broad-host-range plasmids,[79,80] while those found in staphylococci seem to be limited to the genus *Staphylococcus*.

The complete nucleotide sequences of the staphylococcal plasmid pSK41,[81,82] the lactococcal plasmid pMRC01,[83] the enterococcal plasmid pRE25, the streptococcal plasmid pIP501[84,85] and the complete *tra* region of the staphylococcal plasmid pGO1[86] have been determined. Sequence comparisons revealed interesting similarities of the *tra* regions of these self-transmissible plasmids.[77] All of the *tra* regions show a highly modular organization so that the arrangement of the first seven genes is well conserved among the compared *tra* regions, with the exception of an insertion of two genes of unknown function between the putative relaxase gene *traA* and gene *traB* in pMRC01. The pMRC01 *tra* region is the most distantly related and contains seven unique genes. Interestingly, a *traG* gene homologue coding for a putative lytic transglycosylase is present in all plasmids except for pMRC01, while *traK* homologues coding for a putative CP are present in all five plasmids (Fig. 3).

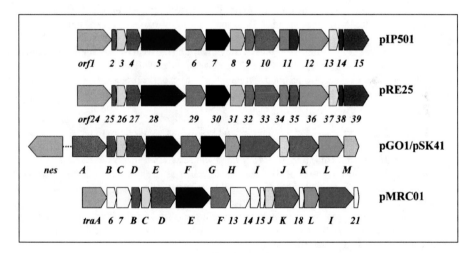

Figure 3. Comparison of the *tra* regions of pIP501, pRE25, pGO1, pSK41, and pMRC01. Similar gene products are shown in the same colour. Cream-colored boxes represent *tra* genes unique to pMRC01. The putative transfer proteins of pRE25 and pIP501 show a high degree of identity (between 80 and 100%), ORF1 to ORF6, ORF8 to ORF9, and ORF14 are 100% identical to the corresponding pRE25 gene products. In pIP501, one big ORF (ORF11) comprises the regions of the corresponding ORF34 and ORF35 in pRE25. The gene products of the *tra* region of pGO1 (*trsA* to *trsM*) and pSK41 (*traA* to *traM*) also exhibit a very high degree of similarity (between 97 and 98% identity). Tra proteins of pMRC01 show 25 to 42% identity to the corresponding proteins of pGO1.[83] pGO1 and pSK41 encode at least one additional *tra* gene, *nes*, located outside the *tra* region. Specific single-strand nicking mediated by Nes at the respective *oriT* site was demonstrated for pGO1.[95] Reprinted with permission from: Grohmann E et al. Microbiol Mol Biol Rev 67:277-301,[77] ©2003 American Society for Microbiology.

Information about the regulatory processes involved in gene transfer of nonpheromone-responsive plasmids in G+ bacteria is scarce: TrsN, a 7.2-kDa protein encoded by pGO1, was shown to repress the synthesis of essential *tra* genes by binding to promoter-like sequences upstream of *trsA*, the first gene of the conjugative gene cluster *trs*.[87]

Tanimoto and Ike[88] detected a gene, *traA*, in the unrelated plasmid pMG1, which is upregulated during conjugation. They found that the *traA* gene product is associated with the formation or stabilization of mating aggregates during broth mating.

Regulation of the pIP501 *tra* Region

The operon organization of the pIP501 *tra* region was elucidated recently by reverse transcription PCR of mRNA isolated from *E. faecalis* harbouring pIP501. All 15 pIP501 *tra* genes, *orf1-orf15*, are transcribed as a single operon of 15.1 kb (Kurenbach and Grohmann, unpublished data). The compact organization of the pIP501 *tra* region makes autoregulation of the *tra* operon by the TraA protein likely. The -10 region of the operon promoter P_{tra} overlaps an inverted repeat structure, proposed to represent the binding site for the TraA relaxase.[89] TraA-binding to P_{tra} has been proved by gel retardation and DNase I footprinting assays. P_{tra}-*lacZ* fusions showed strongly reduced promoter activity when TraA was supplied in trans (Kurenbach and Grohmann, unpublished data).

Environmental stimuli for pIP501 *tra* gene expression have not been detected so far. The *tra* genes are expressed throughout the life cycle of *E. faecalis* and the expression level is independent of the growth phase (Kurenbach and Grohmann, unpublished data). Thus, we conclude that the pIP501 *tra* operon is negatively autoregulated at the transcriptional level by the conjugative DNA relaxase TraA.

Homologies to T4SS Components

Macromolecular transfer systems ancestrally related to the conjugative mpf complexes are called T4SS, as originally proposed by Salmond.[90] T4SS include conjugative transfer apparatus, protein secretion systems of G- pathogens, and natural transformation systems. T4SS are widely distributed among the G- bacterial world (for recent reviews see refs. 1,10,91). We have also found T4 homologues on conjugative elements of G+ bacteria.[77] Exemplarily, the pIP501-encoded T4 homologues will be discussed here. pIP501 encodes one T4 homologue of each of the protein families involved in T-DNA transfer and in G- bacterial plasmid transfer.[92]

Energy Supply

ORF5 (pIP501) encodes a putative VirB4-like ATPase, which could deliver energy for DNA/ protein transport by hydrolysis of ATP. ORF5 shows a score of 71.2 and E value of 3×10^{-13} as a member of the VirB4 family of intracellular trafficking and secretion proteins (COG3451). ORF5 (653 aa) is the largest protein encoded by the pIP501 *tra* region, consistent with the fact that VirB4-homologues of G- bacteria also represent the largest gene product (e.g., VirB4 of the *Agrobacterium* Ti plasmid: 789 aa) of the respective *tra* region. VirB4-type proteins are ubiquitous among the T4SS and are sometimes present in two or more copies. Experimental evidence for a structural contribution of VirB4 to mpf channel formation that is independent of the VirB4 ATPase activity has been provided (for a review see ref. 10).

Mating-Channel Proteins

ORF7 (pIP501) is weakly similar to the family of lytic transglycosylases (pfam01464, score, 36.1; E value 0.007) encoded by bacteriophages and T3 and T4SS. ORF7 contains the soluble bacterial lytic transglycosylase (SLT) domain at its N-terminus. This domain is present in SLTs and in "goose-type" lysozymes (GEWL). It catalyzes the cleavage of the β-1,4-glycosidic bond between acetylmuramic acid and N-acetylglucosamine. At the C-terminus ORF7 possesses high similarity with the COG3942 family of surface antigens (score, 108; E value, $1e^{-24}$) and with the pfam05257 family (score 96.6; E value, $4e^{-21}$) consisting of amidases involved in the cell wall metabolism of bacteria. The pfam05257 family is also known as CHAP (cystein, histidin-dependent amidase/

peptidase) domain family.[93,94] The CHAP domain is often found in association with further protein domains involved in cleavage of peptidoglycan.[93] Proteins containing a N-terminal lytic transglycosylase domain and a C-terminal CHAP domain or vice versa have been characterized. ORF7 also appears to exhibit such a modular structure: a SLT domain at the amino-terminus, between aa 60 and 165, and a CHAP domain between aa 255 and 369. A transmembrane helix has been predicted for the N-terminal moiety of ORF7, approximately between aa 19-35 by several computer programs (HMMTOP, PHDhtm, PROSITE, PSORT etc.). The putative membrane localization is consistent with its proposed role in locally opening the peptidoglycan to facilitate conjugative DNA/protein transport. Interestingly, ORF7 also contains a possible processing site similar to that found in the lytic transglycosylase VirB1. It could be processed to a ORF7* protein (consisting of 210 aa of total 369 aa).

Coupling Proteins

ORF10 (pIP501) belongs to the pfam02534 TraG/TraD family of CPs (score, 291; E value, 1 e[-79]). These proteins contain a P-loop and a Walker B site for nucleotide binding. Putative homologues of CPs have been detected on the chromosomes of many sequenced G+ bacteria (e.g., *E. faecalis* V583) as well as on transposons harboured by them.

Functional Characterization and Protein-Protein Interactions of pIP501-Encoded T4-Homologues

The pIP501-encoded T4 homologues have been expressed with N-terminal tags in *E. coli* and partially purified. ORF7 protein showed hydrolysis activity on peptidoglycan isolated from *E. faecalis* by zymogram analysis (Arends and Grohmann, unpublished data). ATP-binding and ATP-hydrolysis assays for the ORF5 and ORF10 protein are in progress.

The following interactions between putative T4 components have been detected by the yeast two-hybrid assay so far (Abajy and Grohmann, unpublished data). The putative ATPase ORF5 interacts with itself which is consistent with dimerization/oligomerization of the VirB4 protein. ORF5 also binds to ORF7 in the in vivo yeast system. This finding is in agreement with detected VirB4-VirB1 interaction.[42] ORF7 interacts with itself, with ORF5 and with the putative CP ORF10. ORF7 self-association is consistent with VirB1-dimerization shown by yeast two-hybrid assay.[42] Interaction of ORF7 with ORF5 and ORF10 might possibly help incorporate these components in the T4SS structure. ORF10 has been shown to self-associate in vitro (Chmielinska and Grohmann, unpublished data). Further interactions have been detected between T4 homologues and other proteins encoded by the pIP501 *tra* region (Abajy and Grohmann, unpublished data).

Working Model of a T4S-Like Conjugative Mechanism in G+ Bacteria

Based upon our protein-protein interaction data, the predicted localization and function of the (T4SS-like) proteins and preliminary functional characterization of ORF7 we developed a working model for a simplified T4S-like mechanism in G+ bacteria: We suggest that the putative lytic transglycosylase ORF7 should be more important than the VirB1-homologues in G- bacteria, because of the multilayered peptidoglycan in G+ bacteria in contrast to the thinner monolayer of peptidoglycan in G- bacteria: We propose that ORF7 locally opens the peptidoglycan thereby facilitating the establishment of the transenvelope secretion channel. By interaction with ORF5 and ORF10 ORF7 might recruit these proteins and enable their incorporation in the transport apparatus. The putative ATPase ORF5 with high homology to the COG3451 family of intracellular trafficking and secretion proteins could deliver energy for the DNA/protein secretion process by hydrolysis of NTP. The putative CP ORF10 which interacts with the pIP501-encoded relaxase TraA could link the relaxosome consisting of TraA bound at *oriT*$_{pIP501}$, with the transport apparatus. The transport channel presumably consists of ORF5 and non T4SS-like proteins with proposed transmembrane helices such as ORF8, ORF12 and ORF15. Which of these proteins actually act(s) as transporter is currently under investigation.

32 Cell-Cell Channels

References

1. Cascales E, Christie PJ. The versatile bacterial type IV secretion systems. Nat Rev Microbiol 2003; 2:137-149.
2. Cascales E, Christie PJ. Definition of a bacterial type IV secretion pathway for a DNA substrate. Science 2004; 304:1170-1173.
3. Clewell DB, Dunny GM. Conjugation and genetic exchange in enterococci. In: Gilmore MS, ed. The Enterococci: Pathogenesis, Molecular Biology, and Antibiotic Resistance. 1st ed. Washington: ASM press, 2002:265-300.
4. Chandler JR, Dunny GM. Enterococcal peptide sex pheromones: Synthesis and control of biological activity. Peptides 2004; 25:1377-1388.
5. Dillard JP, Seifert HS. A variable genetic island specific for Neisseria gonorrhoeae is involved in providing DNA for natural transformation and is found more often in disseminated infection isolates. Mol Microbiol 2001; 41:263-277.
6. Hofreuter D, Odenbreit S, Haas R. Natural transformation competence in Helicobacter pylori is mediated by the basic components of a type IV secretion system. Mol Microbiol 2001; 41:379-391.
7. Schroeder G, Savvides SN, Waksman G et al. Type IV secretion machinery. In: Waksman G, Caparon M, Hultgren S, eds. Structural Biology of Bacterial Pathogenesis. 1st ed. Washington: ASM press, 2005:179-221.
8. Llosa M, O'Callaghan D. Euroconference on the biology of type IV secretion processes: Bacterial gates into the outer world. Mol Microbiol 2004; 53:1-8.
9. Lawley TD, Klimke WA, Gubbins MJ et al. F factor conjugation is a true type IV secretion system. FEMS Microbiol Lett 2003; 224:1-15.
10. Christie PJ. Type IV secretion: The Agrobacterium VirB/D4 and related conjugation systems. Biochim Biophys Acta 2004; 1694:219-234.
11. Ding Z, Atmakuri K, Christie PJ. The outs and ins of bacterial type IV secretion substrates. Trends Microbiol 2003; 11:527-535.
12. Atmakuri K, Cascales E, Christie PJ. Energetic components VirD4, VirB11 and VirB4 mediate early DNA transfer reactions required for bacterial type IV secretion. Mol Microbiol 2004; 54:1199-1211.
13. Jakubowski SJ, Krishnamoorthy V, Cascales E et al. Agrobacterium tumefaciens VirB6 domains direct the ordered export of a DNA substrate through a type IV secretion system. J Mol Biol 2004; 341:961-977.
14. Christie PJ. The Agrobacterium T-complex transport apparatus: A paradigm for a new family of multifunctional transporters in eubacteria. J Bacteriol 1997; 179:3085-3094.
15. Lai EM, Chesnokova O, Banta LM et al. Genetic and environmental factors affecting T-pilin export and T-pilus biogenesis in relation to flagellation of Agrobacterium tumefaciens. J Bacteriol 2000; 182:3705-3716.
16. Pansegrau W, Lanka E. Enzymology of DNA transfer by conjugative mechanisms. Prog Nucleic Acid Res Mol Biol 1996; 197-251.
17. Zhu J, Oger PM, Schrammeijer B. et al. The bases of crown gall tumorigenesis. J Bacteriol 2000; 182:3885-3895.
18. Hamilton CM, Lee P, Li PL et al. TraG from RP4 and TraG and VirD4 from Ti plasmids confer relaxosome specificity to the conjugal transfer system of pTiC58. J Bacteriol 2000; 182:1541-1548.
19. Byrd DR, Matson MW. Nicking by transesterification: The reaction catalysed by a relaxase. Mol Microbiol 1997; 25:1011-1022.
20. Howard EA, Zupan JR, Citovsky V et al. The VirD2 protein of A. tumefaciens contains a C-terminal bipartite nuclear localization signal: Implications for nuclear uptake of DNA in plant cells. Cell 1992; 68:109-118.
21. Bravo-Angel AM, Gloeckler V, Hohn B et al. Bacterial conjugation protein MobA mediates integration of complex DNA structures into plant cells. J Bacteriol 1999; 181:5758-5765.
22. Rees CED, Wilkins BM. Protein transfer into the recipient cell during bacterial conjugation: Studies with F and RP4. Mol Microbiol 1990; 4:1199-1205.
23. Vergunst AC, Schrammeijer B, den Dulk-Ras A et al. VirB/D4-dependent protein translocation from Agrobacterium into plant cells. Science 2000; 290:979-982.
24. Wilkins BM, Thomas AT. DNA-independent transport of plasmid primase protein between bacteria by the I1 conjugation system. Mol Microbiol 2000; 38:650-657.
25. Schrammeijer B, den Dulk-Ras A, Vergunst AC et al. Analysis of Vir protein translocation from Agrobacterium tumefaciens using Saccharomyces cerevisiae as a model: Evidence for transport of a novel effector protein VirE3. Nucl Acids Res 2003; 31:860-868.

26. Vergunst AC, Van Lier MC, den Dulk-Ras et al. Recognition of the Agrobacterium tumefaciens VirE2 translocation signal by the VirB/D4 transport system does not require VirE1. Plant Physiol 2003; 133:978-988.

27. Simone M, McCullen CA, Stahl LE et al. The carboxy-terminus of VirE2 from Agrobacterium tumefaciens is required for its transport to host cells by the virB-encoded type IV transport system. Mol Microbiol 2001; 41:1283-1293.

28. Luo Z-Q, Isberg RR. Multiple substrates of the Legionella pneumophila Dot/Icm system identified by interbacterial protein transfer. Proc Natl Acad Sci USA 2004; 101:841-846.

29. Atmakuri K, Ding Z, Christie PJ. VirE2, a type IV secretion substrate, interacts with the VirD4 transfer protein at the cell poles of Agrobacterium tumefaciens. Mol Microbiol 2003; 49:1699-1733.

30. Cabezon E, Lanka E, de la Cruz F. Requirements for mobilization of plasmids RSF1010 and ColE1 by the IncW plasmid R388: trwB and RP4 traG are interchangeable. J Bacteriol 1994; 176:4455-4558.

31. Cabezon E, Sastre JI, de la Cruz F. Genetic evidence of a coupling role for the TraG protein family in bacterial conjugation. Mol Gen Genet 1997; 254:400-406.

32. Sastre JI, Cabezon E, de la Cruz F. The carboxyl terminus of protein TraD adds specificity and efficiency to F-plasmid conjugative transfer. J Bacteriol 1998; 180:6039-6042.

33. Schroeder G, Krause S, Zechner EL et al. TraG-like proteins of DNA transfer systems and of the Helicobacter pylori type IV secretion system: Inner membrane gate for exported substrates? J Bacteriol 2002; 184:2767-2779.

34. Szpirer CY, Faelen M, Couturier M. Interaction between the RP4 coupling protein TraG and the pBHR1 mobilization protein Mob. Mol Microbiol 2000; 37:1283-1292.

35. Das A, Xie YH. Construction of transposon Tn3PhoA: Its application in defining the membrane topology of the Agrobacterium tumefaciens DNA transfer proteins. Mol Microbiol 1998; 27:405-414.

36. Lee MH, Kosuk N, Bailey J et al. Analysis of F factor TraD membrane topology by use of gene fusions and trypsin-sensitive insertions. J Bacteriol 1999; 181:6108-6113.

37. Gomis-Ruth FX, Moncalian G, Perez-Luque R et al. The bacterial conjugation protein TrwB resembles ring helicases and F_1-ATPase. Nature 2001; 409:637-641.

38. Gomis-Ruth FX, Moncalian G, de la Cruz F et al. Conjugative plasmid protein TrwB, an integral membrane type IV secretion system coupling protein. Detailed structural features and mapping of the active site cleft. J Biol Chem 2002; 277:7556-7566.

39. Berger BR, Christie PJ. Genetic complementation analysis of the Agrobacterium tumefaciens virB operon: VirB2 through virB11 are essential virulence genes. J Bacteriol 1994; 176:3646-3660.

40. Krall L, Wiedemann U, Unsin G et al. Detergent extraction identifies different VirB protein subassemblies of the type IV secretion machinery in the membranes of Agrobacterium tumefaciens. Proc Natl Acad Sci USA 2002; 99:11405-11410.

41. Das A, Xie Y-H. The Agrobacterium T-DNA transport pore proteins VirB8, VirB9, and VirB10 interact with one another. J Bacteriol 2000; 182:758-763.

42. Ward DV, Draper O, Zupan JR et al. Peptide linkage mapping of the Agrobacterium tumefaciens vir-encoded type IV secretion system reveals protein subassemblies. Proc Natl Acad Sci USA 2002; 99:11493-11500.

43. Cao TB, Saier Jr MH. Conjugal type IV macromolecular transfer systems of Gram-negative bacteria: Organismal distribution, structural constraints and evolutionary conclusions. Microbiol 2001; 147:3201-3214.

44. Planet PJ, Kachlany SC, DeSalle R et al. Phylogeny of genes for secretion NTPases: Identification of the widespread tadA subfamily and development of a diagnostic key for gene classification. Proc Natl Acad Sci USA 2001; 98:2503-2508.

45. Krause S, Barcena M, Pansegrau W et al. Sequence related protein export NTPases encoded by the conjugative transfer region of RP4 and by the cag pathogenicity island of Helicobacter pylori share similar hexameric ring structures. Proc Natl Acad Sci USA 2000; 97:3067-3072.

46. Yeo HJ, Savvides SN, Herr AB et al. Crystal structure of the hexameric traffic ATPase of the Helicobacter pylori type IV system. Mol Cell 2000; 6:1461-1472.

47. Lupas AN, Martin J. AAA proteins. Curr Opin Struct Biol 2002; 12:746-753.

48. Dang TAT, Christie PJ. The VirB4 ATPase of Agrobacterium tumefaciens is a cytoplasmic membrane protein exposed at the periplasmic surface. J Bacteriol 1997; 179:453-462.

49. Rashkova S, Spudich GM, Christie PJ. Mutational analysis of the Agrobacterium tumefaciens VirB11 ATPase: Identification of functional domains and evidence for multimerization. J Bacteriol 1997; 179:583-589.

50. Sagulenko Y, Sagulenko V, Chen J et al. Role of Agrobacterium VirB11 ATPase in T-pilus assembly and substrate selection. J Bacteriol 2001; 183:5813-5825.

51. Llosa M, Gomis-Ruth FX, Coll M et al. Bacterial conjugation: A two-step mechanism for DNA transport. Mol Microbiol 2002; 45:1-8.
52. Dang TAT, Zhou X-R, Graf B et al. Dimerization of the Agrobacterium tumefaciens VirB4 AT-Pase and the effect of ATP-binding cassette mutations on assembly and function of the T-DNA transporter. Mol Microbiol 1999; 32:1239-1253.
53. Rabel C, Grahn AM, Lurz R et al. The VirB4 family of proposed traffic nucleoside triphosphatases: Common motifs in plasmid RP4 TrbE are essential for conjugation and phage adsorption. J Bacteriol 2003; 185:1045-1058.
54. Berger BR, Christie PJ. The Agrobacterium tumefaciens virB4 gene product is an essential virulence protein requiring an intact nucleoside triphosphate-binding domain. J Bacteriol 1993; 175:1723-1734.
55. Bohne J, Yim A, Binns AN. The Ti plasmid increases the efficiency of Agrobacterium tumefaciens as a recipient in virB-mediated conjugal transfer of an IncQ plasmid. Proc Natl Acad Sci USA 1998; 95:7057-7062.
56. Liu Z, Binns AN. Functional subsets of the VirB type IV transport complex proteins involved in the capacity of Agrobacterium tumefaciens to serve as a recipient in virB-mediated conjugal transfer of plasmid RSF1010. J Bacteriol 2003; 185:3259-3269.
57. Burns DL. Type IV transporters of pathogenic bacteria. Curr Opin Microbiol 2003; 6:29-34.
58. Ruhfel RE, Manias DA, Dunny GM. Cloning and characterization of a region of the Enterococcus faecalis conjugative plasmid pCF10, encoding a sex pheromone-binding function. J Bacteriol 1993; 175:5253-5259.
59. Leonard BA, Podbielski A, Hedberg PJ et al. Enterococcus faecalis pheromone binding protein, PrgZ, recruits a chromosomal oligopeptide permease system to import sex pheromone cCF10 for induction of conjugation. Proc Natl Acad Sci USA 1996; 93:260-264.
60. Kao SM, Olmsted SB, Viksnins AS et al. Molecular and genetic analysis of a region of plasmid pCF10 containing positive control genes and structural genes encoding surface proteins involved in pheromone-inducible conjugation in Enterococcus faecalis. J Bacteriol 1991; 173:7650-7664.
61. Tame JR, Murshudov GN, Dodson EJ et al. The structural basis of sequence-independent peptide binding by OppA protein. Science 1994; 264:1578-1581.
62. Nakayama J, Yoshida K, Kobayashi H et al. Cloning and characterization of a region of Enterococcus faecalis plasmid pPD1 encoding pheromone inhibitor (ipd), pheromone sensitivity (traC), and pheromone shutdown (traB) genes. J Bacteriol 1995; 177:5567-5573.
63. Mori M, Sakagami Y, Ishii Y et al. Structure of cCF10, a peptide sex pheromone which induces conjugative transfer of the Streptococcus faecalis tetracycline resistance plasmid, pCF10. J Biol Chem 1988; 263:14574-14578.
64. Dunny GM, Antiporta MH, Hirt H. Peptide pheromone-induced transfer of plasmid pCF10 in Enterococcus faecalis: Probing the genetic and molecular basis for specificity of the pheromone response. Peptides 2001; 22:1529-1539.
65. Ehrenfeld EE, Kessler RE, Clewell DB. Identification of pheromone-induced surface proteins in Streptococcus faecalis and evidence of a role for lipoteichoic acid in formation of mating aggregates. J Bacteriol 1986; 168:6-12.
66. Paulsen IT, Banerjei L, Myers GS et al. Role of mobile DNA in the evolution of vancomycin-resistant Enterococcus faecalis. Science 2003; 299:2071-2074.
67. Clewell DB, An FY, Flannagan SE et al. Enterococcal sex pheromone precursors are part of signal sequences for surface lipoproteins. Mol Microbiol 2000; 35:246-247.
68. Dev IK, Ray PH. Signal peptidases and signal peptide hydrolases. J Bioenerg Biomembr 1990; 22:271-290.
69. Nakayama J, Ono Y, Suzuki A. Isolation and structure of the sex pheromone inhibitor, iAM373, of Enterococcus faecalis. Biosci Biotechnol Biochem 1995; 59:1358-1359.
70. Nakayama J, Ruhfel RE, Dunny GM et al. The prgQ gene of the Enterococcus faecalis tetracycline resistance plasmid pCF10 encodes a peptide inhibitor, iCF10. J Bacteriol 1994; 176:7405-7408.
71. Buttaro BA, Antiporta MH, Dunny GM. Cell-associated pheromone peptide (cCF10) production and pheromone inhibition in Enterococcus faecalis. J Bacteriol 2000; 182:4926-4933.
72. Clewell DB, Francia MV, Flannagan SE et al. Enterococcal plasmid transfer: Sex pheromones, transfer origins, relaxases, and the Staphylococcus aureus issue. Plasmid 2002; 48:193-201.
73. Francia MV, Clewell DB. Transfer origins in the conjugative Enterococcus faecalis plasmids pAD1 and pAM373: Identification of the pAD1 nic site, a specific relaxase and a possible TraG-like protein. Mol Microbiol 2002; 45:375-395.

74. Hirt H, Manias DA, Bryan EM et al. Characterization of the pheromone response of the Enterococcus faecalis conjugative plasmid pCF10: Complete sequence and comparative analysis of the transcriptional and phenotypic response of pCF10-containing cells to pheromone induction. J Bacteriol 2005; 187:1044-1054.

75. Mills DA, McKay LL, Dunny GM. Splicing of a group II intron involved in the conjugative transfer of pRS01 in lactococci. J Bacteriol 1996; 178:3531-3538.

76. Ike Y, Tanimoto K, Tomita H et al. Efficient transfer of the pheromone-independent Enterococcus faecium plasmid pMG1 (Gm^r) (65.1 kilobases) to Enterococcus strains during broth matings. J Bacteriol 1998; 180:4886-4892.

77. Grohmann E, Muth G, Espinosa M. Conjugative plasmid transfer in Gram-positive bacteria. Microbiol Mol Biol Rev 2003; 67:277-301.

78. Belhocine K, Plante I, Cousineau B. Conjugation mediates transfer of the LI.LtrB group II intron between different bacterial species. Mol Microbiol 2004; 51:1459-1469.

79. Clewell DB. Movable genetic elements and antibiotic resistance in enterococci. Eur J Clin Microbiol Infect Dis 1990; 9:90-102.

80. Schaberg DR, Zervos MJ. Intergeneric and interspecies gene exchange in gram-positive cocci. Antimicrob Agents Chemother 1986; 30:817-822.

81. Berg T, Firth N, Apisiridej S et al. Complete nucleotide sequence of pSK41: Evolution of staphylococcal conjugative multiresistance plasmids. J Bacteriol 1998; 180:4350-4359.

82. Firth N, Ridgway KP, Byrne ME et al. Analysis of a transfer region from the staphylococcal conjugative plasmid pSK41. Gene 1993; 136:13-25.

83. Dougherty BA, Hill C, Weidmann JF et al. Sequence and analysis of the 60 kb conjugative, bacteriocin-producing plasmid pMRC01 from Lactococcus lactis DPC3147. Mol Microbiol 1998; 29:1029-1038.

84. Kurenbach B, Bohn C, Prabhu J et al. Intergeneric transfer of the Enterococcus faecalis plasmid pIP501 to Escherichia coli and Streptomyces lividans and sequence analysis of its tra region. Plasmid 2003; 50:86-93.

85. Thompson JK, Collins MA. Completed sequence of plasmid pIP501 and origin of spontaneous deletion derivatives. Plasmid 2003; 50:28-35.

86. Morton TM, Eaton DM, Johnston JL et al. DNA sequence and units of transcription of the conjugative transfer gene complex (trs) of Staphylococcus aureus plasmid pGO1. J Bacteriol 1993; 175:4436-4447.

87. Sharma VK, Johnston JL, Morton TM et al. Transcriptional regulation by TrsN of conjugative transfer genes on staphylococcal plasmid pGO1. J Bacteriol 1994; 176:3445-3454.

88. Tanimoto K, Ike Y. Analysis of the conjugal transfer system of the pheromone-independent highly transferable Enterococcus plasmid pMG1: Identification of a tra gene (traA) up-regulated during conjugation. J Bacteriol 2002; 184:5800-5804.

89. Kurenbach B, Grothe D, Farías ME et al. The tra region of the conjugative plasmid pIP501 is organized in an operon with the first gene encoding the relaxase. J Bacteriol 2002; 184:1801-1805.

90. Salmond GPC. Secretion of extracellular virulence factors by plant pathogenic bacteria. Annu Rev Phytopathol 1994; 32:181-200.

91. Llosa M, de la Cruz F. Bacterial conjugation: A potential tool for genomic engineering. Res Microbiol 2005; 156:1-6.

92. Christie PJ. Type IV secretion: Intercellular transfer of macromolecules by systems ancestrally related to conjugation machines. Mol Microbiol 2001; 40:294-305.

93. Bateman A, Rawlings ND. The CHAP domain: A large family of amidases including GSP amidase and peptidoglycan hydrolases. Trends Biochem Sci 2003; 28:234-237.

94. Rigden DJ, Jedrzejas MJ, Galperin MY. Amidase domains from bacterial and phage autolysins define a family of γ-D,L-glutamate-specific amidohydrolases. Trends Biochem Sci 2003; 28:230-234.

95. Climo MW, Sharma VK, Archer GL. Identification and characterization of the origin of conjugative transfer (oriT) and a gene (nes) encoding a single-stranded endonuclease on the staphylococcal plasmid pGO1. J Bacteriol 1996; 178:4975-4983.

SECTION II
Ciliate Cells

CHAPTER 3

The *Tetrahymena* Conjugation Junction

Eric S. Cole*

Introduction

Life History: Sexual and Asexual Reproduction

Those who study ciliates have struggled over the years to establish a place in the pantheon of model organisms.[1] Eukaryotic cell biologists have turned profitably to yeast models for the powerful genetic tools at their disposal, while developmental biologists have cultivated a gallery of metazoan embryos with contributions from the plant and fungal worlds. Yet ciliates continue to contribute to fundamental aspects of both cell biology and development, often by extreme example, and among the ciliates, *Tetrahymena* has emerged as one of the stars.

Tetrahymena is a freshwater, unicellular organism, approximately 40 to 60 microns in length. They feed on bacteria which they capture by means of four ciliated "membranelles" that form a set of three combs that brush particles into a curved buccal cavity rimmed by a fourth "undulating membrane" (Fig. 1). Ciliates are characterized by three conspicuous features, ciliature that can be specialized for locomotion or food capture, alveolar membranes that lie just beneath the plasma membrane forming a set of flattened sacks, and "nuclear duality".[2] Ciliates posses both a somatic, transcriptionally active macronucleus, and a germinal, transcriptionally silent micronucleus. It has been attractive to borrow the language of germline and soma from the world of metazoan embryos and recent findings suggest that this may reflect more than creative license. Indeed, ciliates may represent the simplest form of life to generate distinct somatic and germinal nuclear lineages.

During vegetative growth, ciliates reproduce by binary fission. In *Tetrahymena*, this involves a remarkable reorganization of the cortical cytoskeleton. The equatorial fission zone separates the anterior division product (proter) containing the functional oral apparatus from the posterior fission product (opisthe) containing the functional water-expulsion organelle, (the contractile vacuole system) along with the cell "anus" or cytoproct (Fig. 1). Each division product would perish without the coordinate synthesis of a new oral structure (Oral Primordium, see Fig. 2) in the posterior opisthe, and a new Contractile Vacuole Pore system in the proter. Events that attend the cortical patterning associated with vegetative cell division have been studied extensively.[3]

When *Tetrahymena* cells are removed from nutrient medium and starved, they have a repertoire of morphogenetic responses that seem adaptive. Starved cells frequently undergo a process of "oral replacement".[4] This involves disassembling the existing oral apparatus and reforming a new one through de novo basal body synthesis. Cells that are maintained under nutritional stress may undergo a second type of transformation to a "dispersal" form, or fast swimmer.[5,6] Finally, nutritional challenge can predispose *Tetrahymena* for sexual reproduction including meiosis, chromosomal recombination and the exchange of meiotic products with another cell. Sex as a strategy for grooming ones' genome has also been richly discussed in ciliates.[7]

*Eric S. Cole—Biology Department, St. Olaf College, 1520 St. Olaf Avenue, Northfield, Minnesota 55057, U.S.A. Email: colee@stolaf.edu

Cell-Cell Channels, edited by Frantisek Baluska, Dieter Volkmann and Peter W. Barlow.
©2006 Landes Bioscience and Springer Science+Business Media.

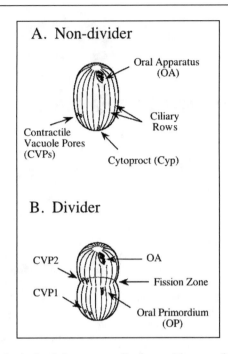

Figure 1. A diagrammatic sketch of cortical structures on *Tetrahymena*. Linear rows (kineties) indicate the longitudinal rows of ciliated basal bodies. A) This represents a nondividing cell showing the oral apparatus located at the anterior end of the cell, and a cytoproct and contractile vacuole pores located at the posterior end of the cell. B) This represents a dividing cell showing the newly formed oral primordium, fission zone and contractile vacuole pores.

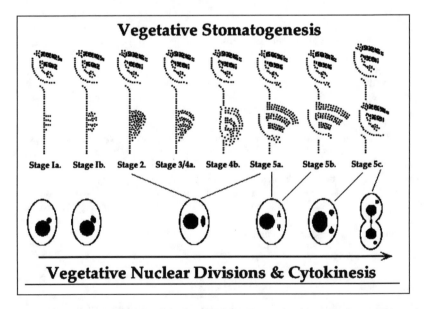

Figure 2. This figure depicts stages in oral development (stomatogenesis) and corresponding stages in nuclear division and cytokinesis (after Lansing et al).[109] Dots represent individual basal bodies.

Nuclear exchange in *Tetrahymena* (and in ciliates in general) is performed in a distinctive way. Metazoan embryos achieve fertilization through whole-cell fusion of sperm and egg. Whole cell fusion also mediates sexual events in haploid, unicellular yeasts (*Saccharomyces*) and algae such as *Chlamydomonas*. *Tetrahymena*, on the other hand, generate a transient region of limited cell fusion.[8,9] Two diploid cells form this fusion junction anterior to their oral apparatus, and emerge from conjugation 12 hours later having exchanged haploid nuclei, while reestablishing cortical integrity.

Developmental Biology

Time Line for Assembly and Disassembly

Events leading to assembly of a nuclear exchange junction have a rich laboratory history. *Tetrahymena* exhibit three requirements for pair formation. First cells must be nutritionally deprived for a minimum of 2 hrs at 30°C. This starvation step has been named "initiation".[10] Second, cells of complementary mating types must be present. *Tetrahymena* possess seven different mating types,[11] determined by the mating-type locus.[12] Any two of these can mate with one another, but under normal circumstances none will mate with themselves. There are, however, cases of "self-mating" cultures associated with clonal senescence.[13,14] Finally, after starvation-induced initiation, cells of complementary mating types must be allowed to collide with one another in a nonshaking culture. This period of cell-cell contact has been referred to as "costimulation".[10,15] The events of initiation and costimulation are illustrated on a time line in Figures 3 and 4.

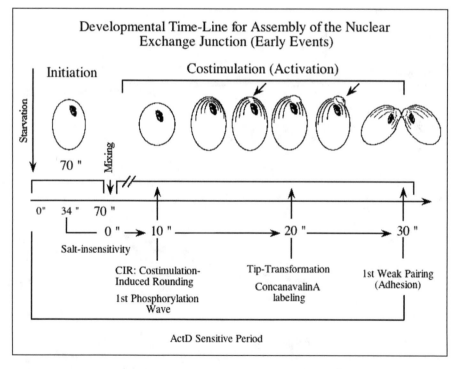

Figure 3. This figure depicts early events associated with pair formation along a developmental time line. Time zero indicates when starvation is initiated. Initiation represents an obligatory period of nutritional deprivation. Costimulation represents the period of obligatory cell-cell collisions. This figure depicts the first 30 minutes of costimulation, also referred to as "activation". By the end of this activation period, loose cell-cell bonds have formed. Arrows indicate the location of membrane remodeling that creates the conjugation junction.

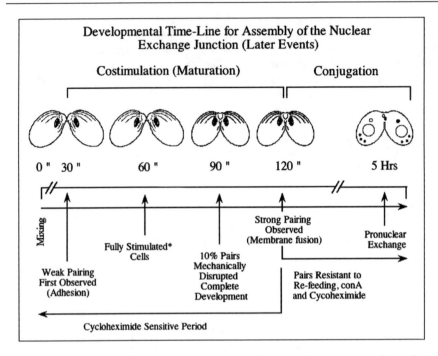

Figure 4. This figure depicts the later events associated with pair formation. After activation, it takes another 90 minutes for the conjugation junction to become fully mature. The nuclear events associated with conjugation are only hinted at (see Cole et al[64] and Cole and Soelter[65] for a complete description of nuclear events).

Starvation-Induced Initiation

The first step in preparing cells for conjugation is starvation. This has been accomplished in one of two ways. *Tetrahymena* can be allowed to deplete a monoxenic culture of bacteria such as *Klebsiella pneumoniae*[13] or, with the development of a defined axenic medium,[16] one can transfer cells from nutrient medium to a nonnutritive medium by centrifugation and resuspension.[17,18] There are obvious advantages to transferring cells from a defined nutrient medium to starvation medium by centrifugation in that one gains control over the synchrony of subsequent developmental events. Initiation can take as little as 70 minutes, (though laboratories routinely starve cells overnight). Initiation can be blocked by exposure to hyperosmotic conditions after 34 minutes of initiation.[19,20] Following initiation, a 15-minute exposure to food can erase most of the program of initiation.[21] This means that starved, mixed cells will require an additional 45 minutes of starvation plus costimulation, (see below) before they are once again ready to pair. Curiously, no single nutrient can block initiation by itself. Only a nutritional upgrade to complete media will inhibit initiation suggesting it is the total nutritional state of the cell that determines readiness for mating, and no single molecular trigger.[22]

Costimulation

Contact-Mediated or Soluble Signal Molecule?

After starvation, cells of complementary mating types can be mixed thereby provoking the mating reaction. This involves a series of developmental events that remodel the anterior cell membrane and its associated cytoskeleton, and results in the formation of a remarkable and dynamic system of cell-cell junctions. This period of obligatory cell-cell interaction lasts between one and two hours.[15,21,23] There has been some controversy as to whether or not costimulation involves soluble signal molecules, or strictly physical cell-cell contact. In reviewing the literature, both appear to play a legitimate role, though cell-cell contact seems the principle trigger.

The first suggestion that there might be a soluble factor associated with mating in *Tetrahymena* came from Takahashi.[24] More extensive evidence came from the Wolfe laboratory.[25,26] These early studies demonstrated that both costimulation and pair formation were profoundly delayed when cells were washed into fresh starvation medium. This delay was eliminated if, after washing, cells were returned to a conditioned starvation medium. The factor that was present in conditioned medium was a macromolecule (retained by dialysis) that was heat-stable (30 minutes in boiling water) and secreted constitutively by cells in either nutrient or starvation medium from sexually mature or immature cells. It had no mating-type specificity, and its synthesis was blocked by cyclo-heximide treatment suggesting that it was an unusually heat-stable, secreted protein. Wolfe named this compound FAC (factor active in conjugation).[27] He went on to demonstrate that FAC acts like neither a hormone (as seen in yeast mating systems and the ciliate *Blepharisma*) nor an agglutinin (as seen in *Chlamydomonas* and *Paramecium*). It does not trigger conspicuous changes in the recipient cell, nor does it cause immediate aggregation of cells. Rather, it seems to somehow facilitate those cell-cell contacts that bring about both costimulation and subsequent cell adhesion. Adair et al[25] suggested we think of this factor as a type of extra-cellular matrix molecule, a provocative idea in that it would suggest a role for ECM proteins in cells that do not form tissues, yet need to stabilize cell-cell interactions for this one event in their life histories. Consistent with this notion is the demonstration that both initiation and costimulation require trace amounts of calcium ion.[22]

Recent studies lend further support to the idea of a soluble mating factor, and raise new questions. Fujishima et al[28] characterized a very early morphological response in cells undergoing costimulation. Starved *Tetrahymena* of complementary mating types exhibit a subtle, yet reproducible and consistent "rounding" of their shape within ten minutes of mixing. Stirring a mixed culture (which can prevent costimulation and subsequent pairing) also prevented cell rounding. When starved cells were washed and mixed into fresh starvation medium, there was a 30-minute delay in both rounding and pairing that was eliminated when cells were washed into conditioned medium. Medium from cells exhibiting a complementary mating type could not bring about rounding in starved cells until both mating types were present. In short, rounding was induced when starved cells of complementary mating types were able to undergo collisions in the presence of some secreted factor (possibly Wolfe's FAC).

Finally, in a very recent study Anafi et al (personal communication) demonstrate that ten minutes after one mixes starved cells of complementary mating types, there is a wave of tyrosine phosphorylation on cytoplasmic proteins as revealed by immuno-fluorescence and western blotting using a monoclonal antibody directed against phospho-tyrosine residues. Furthermore, starved cells of one mating type show elevated tyrosine phosphorylation in response to cell-free, conditioned medium from another mating type. Significantly, conditioned media only stimulates phosphorylation in cells of a different mating type from the cells that were used to condition the medium. This result, if valid, may force us to reopen the question of whether or not there are mating-type specific signal molecules secreted by *Tetrahymena* in addition to soluble factors that promote costimulation and pair formation.

Activation and Maturation

Costimulation can be divided into two stages: activation (first half hour following mixing) and maturation (30 to 120 minutes following mixing,[29] see Figs. 3, 4). During the initial "activation stage," costimulation is sensitive to cycloheximide (10 µg/mL), actinomycin D (20 µg/mL), physical agitation and nutritional enhancement of the medium. After this initial period of activation, loose pairs begin to form that can still be disrupted by either feeding, or mechanical agitation. Curiously, there is a brief period in which loose, "homotypic" pairing can occur.[30] That is, cells of the same mating type form brief associations. These are intrinsically unstable, however, and give way to heterotypic pairing that is reinforced during the subsequent maturation step.

By one hour, mixed cells become "fully stimulated".[10] This condition is revealed when cells are maintained in physical contact with one another for a full hour while weak-pair associations are disrupted through periodic bouts of mechanical agitation. Such cells will form pairs immediately following cessation of the last bout of agitation, but only if they have accumulated one hour of nonshaken contact with one another.

Ninety minutes after mixing, pairs that are violently disrupted (as opposed to gentle disruption) produce cell isolates that can complete the full 12-hour developmental program involving meiosis, mitosis, self-fertilization and differentiation of the macronuclear anlagen.[31,32] This strange scenario suggests that after only one and a half hours of interaction, the conjugation program can be initiated and maintained even in the absence of a mating partner. Two hours after mixing, (1.5 hrs after initial cell pairs are observed), pairs have fully matured. They are now resistant to feeding, cycloheximide, and agitation. This period may be associated with the formation of actual pores connecting the cytoplasm of the mating partners[22] (see below).

Tip Transformation

Membrane Events

In response to costimulation, *Tetrahymena* remodel their anterior tips in a process called tip transformation.[33] Specifically, physical collisions induce a shape change in the anterior region of the cell that begins approximately 30 minutes after mixing. The anterior pole changes from a pointed cap that forms from the convergence of numerous longitudinal ciliary rows and intervening ciliary ridges (Fig. 5A), to a region of smooth, naked membrane (Fig. 5D, asterisk). In costimulated cells, the intervening

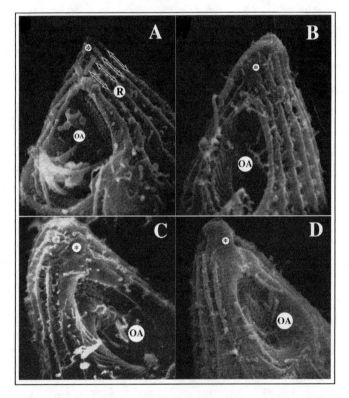

Figure 5. Scanning electron micrographs showing "tip transformation" (reprinted with permission from: Wolfe J, Grimes GW. J Protozool 1979; 26:82-89[33]). OA indicates the oral apparatus at the anterior end of the cell. A) This shows the anterior end of a de-ciliated nonstarved cell in stationary growth phase. Nonciliated ridges "R", can be seen converging at the anterior tip (diagonal arrows). B) This shows a cell after starvation (initiation), and one hour of mixing (costimulation). The first sign of tip transformation is a broadening of the midline seam (*). The membrane now appears smooth in the region where the cortical ridges converged. C) The midline seam (*) has now broadened and the cell's anterior tip appears blunt. D) Two hours after mixing, the tip has been fully modified. This specimen was taken from a culture in which pairs were already observed.

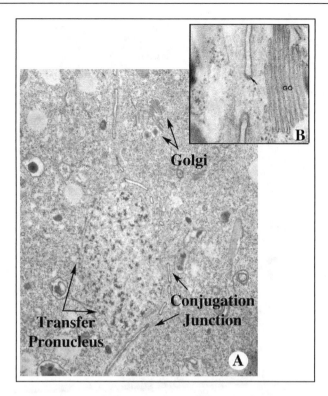

Figure 6. Transmission electron micrographs showing a fully formed conjugation junction. A) This image was captured just prior to nuclear exchange (courtesy of Judy Orias). The transfer pronucleus is pressed up against the conjugation junction that has already formed a series of pores (see arrows coming from right). In the upper part of the photo a golgi apparatus is seen. The insert is from a photo by Elliott (reprinted with permission of the publishers[35]). This clearly shows a set of stacked golgi lamellae (go) adjacent to a conjugation junction pore.

ciliary ridges converge in a seam, or chevron just anterior to the Oral Apparatus. This seam appears to be the site of novel membrane synthesis. In nonmating cells, the endo-membrane system is dispersed predominantly in the form of dictyosomes, or simplified membrane lamellae associated with ciliary basal bodies and the "deep face" of cortical mitochondria.[34] The only time a well-organized, multi-lamellate golgi apparatus has been reported in *Tetrahymena* is during mating when it forms adjacent to the developing conjugation junction (Fig. 6).[35] As tip transformation proceeds, the ridge-seam broadens, producing a blunt, naked, protruding region of membrane at the anterior tip that extends down the ventral face just anterior to the Oral Apparatus (Fig. 5B-D). When a sufficient number of costimulated cells complete tip transformation, loose adhesions form between mating partners (Fig. 7A). As mentioned earlier, transient adhesions can and do form between homotypic pairs (same mating types), but these fail to stabilize and are replaced by heterotypic pairings.[30] As the region of adhesion broadens it forms the "fusion plate"[9] (Fig. 7B). This becomes perforated with hundreds of 0.1-0.2 μm pores that join the cytoplasm of the two mating partners (Fig. 8). Rows of intra-membrane particles form at the perimeter of the fusion plate, as revealed by freeze fracture electron microscopy.[9]

Membrane Proteins and Cell Adhesion: Concanavalin a Receptors

In 1976, Ofer et al[36] demonstrated that the plant lectin concanavalin A (25 micrograms per mL) could inhibit pairing in costimulated *Tetrahymena*, and cause loosely bound pairs to dissociate. ConA is known to bind to mannose-containing glycoproteins. When conA was conjugated to Fluorescein (FITC-conA) and delivered to a mixed, mating culture, it was found to bind strongly to the

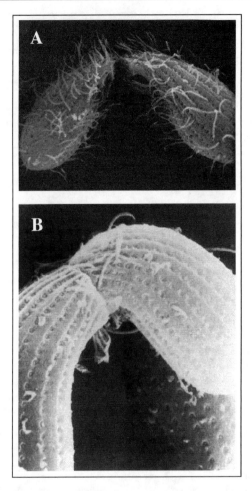

Figure 7. Scanning electron micrographs of mating pairs. A) This figure shows loosely paired cells (reprinted with permission from: Wolfe J, Grimes GW. J Protozool 1979; 26:82-89[33]). B) Mating pairs are shown with a mature conjugation junction (reprinted with permission from: Wolfe J et al. Differentiation 1986; 31:1-9[39]).

developing conjugation junction in *Tetrahymena*.[37] The pattern of conA receptors on *Tetrahymena* during costimulation and pairing has been carefully documented using both FITC-conjugated conA with fluorescence microscopy,[37-41] and ferritin-conjugated conA with transmission electron microscopy.[37,42] FITC-conA decorates the oral region in nonpairing cells, and is also incorporated into food vacuoles. As cells undergo costimulation, labeling of food vacuoles diminishes (ceasing by 20 minutes),[37] suggesting that mating cells cease to feed. Within 15 minutes of mixing, conA-receptors appear in "small clusters near the anterior end, but posterior to the actual tip" (Fig. 9A). Subsequently, punctate conA labeling appears within the region of membrane remodeling located at the anterior tip of the cell (the future conjugation junction), and consolidates into a vertical line (Fig. 9B). Morphologically, this region appears as a region of smooth membrane devoid of the cilia, or membrane ridges that characterize the anterior surface of costimulated, nonmating cells.[9,33] By one hour after mixing, conA labeling is dense and uniform within the, now broadened and transformed tip (Fig. 9C). Pairing follows shortly, and fluorescence becomes restricted to a heart-shaped ring around the perimeter of the conjugation junction[39] (Figs. 9D,10,11).

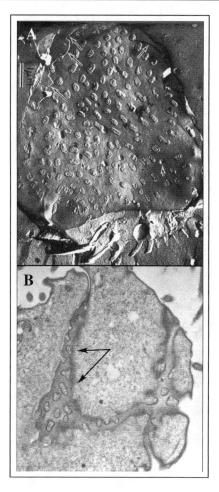

Figure 8. This figure depicts both (A) a freeze fracture image of the conjugation junction and (B) a transmission EM image of a microtome section that grazes the conjugation junction. A) This freeze fracture image shows the surface membrane of one mating partner, from which the adjoining partner's membrane has been sheared. (Reprinted with permission from Wolfe J. J Morphol 172:159-178,[9] ©1982 Wiley-Liss, Inc., a subsidiary of John Wiley & Sons, Inc.) Several hundred 0.1-0.2 micron pores have formed at regular intervals. They appear to be plugged with some kind of button-like material. B) This glancing section through the conjugation junction (arrows) also reveals pores filled with whorls of electron dense material (courtesy of Judy Orias).

It has been pointed out that conA molecules exhibit cross-linking activity, and this could drive the observed "tipping" phenomenon in which conA receptors form aggregates. This has been ruled out by first fixing cells with a variety of agents, and then labeling with conA.[37-39] The study by Watanabe's group is of particular interest in that they demonstrated that when conA labeling is performed prior to observation or fixation, surface labeling is underestimated. Only the most intensely labeled regions appear. When cells are fixed (3 % glutaraldehyde) and then labeled with FITC-conA, a more extensive labeling appears in the anterior end of the cell, though the details of conA tipping and ring formation remain unchanged.

Several studies set out to search for newly synthesized conA-binding glycoproteins as cells underwent costimulation and pairing. Van Bell[43] performed extensive SDS-PAGE analysis on mating and

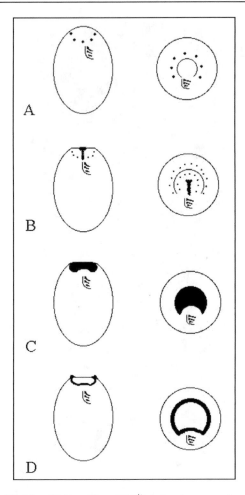

Figure 9. This figure (modified from Wolfe and Feng, 1988[41]) depicts the pattern of appearance of concanavalin A receptors at the anterior tip of a costimulated cell. In stage (A) receptors appear at the oral apparatus and in patches posterior to the cell's un-modified tip. In stage (B) the tip has broadened, and conA receptors appear within this modified surface. In stage (C) the tip is fully modified, and saturated with conA receptors. Finally, after pair formation, conA receptors have cleared from the region of cell-cell contact and appear in a ring circumscribing the conjugation junction.

costimulating cells and probed gels with ^{125}I-conA. Although a number of conspicuous conA-binding proteins were detected, none showed changes in their level of expression during prepairing or conjugation. Wolfe and Feng,[41] isolated cytoskeletal frameworks (triton-extracted cortical residues) and analyzed conA-binding proteins associated with these by Western Blots using conA as a probe. Again, there were no changes associated with pair formation. Other studies identified a number of conA labeling surface proteins (Dentler, 1992). The most recent study by Driscoll and Hufnagel[44] discovered a conA-labeling 28kD protein associated with ciliary membranes that appears to be down regulated in response to starvation, but again, no conA receptors seemed to be synthesized during the period of costimulation. These studies suggest that costimulation induced conA receptor tipping is brought about not by novel synthesis of the conA receptors themselves, but by reorganization of existing surface glycoproteins resulting in recruitment and aggregation of these membrane proteins

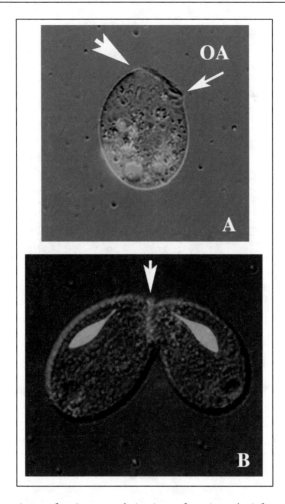

Figure 10. Fluorescence images of conA receptors during tip transformation and pair formation. A) This cell has been costimulated for 1.5 hours. A rhodamine (TRITC)-conjugated conA probe has been employed (red), and a DAPI nuclear counter stain reveals the nucleus in blue. Food vacuoles have incorporated TRITC-conA, and a line of labeling (large arrow) appears anterior to the oral apparatus (small arrow, OA). B) A live mating pair that has been triple-labeled. The conjugation junction (arrow) appears in red (TRITC-conjugated conA), the macronucleus (and some vacuoles) appear blue (DAPI labeling), and the micronuclei have been labeled with a green fluorescent GFP-micronuclear histone (mlh1) as they initiate meiosis I (MLH1::GFP cells courtesy of Douglas Chalker). These images were made from living cells.

at the anterior, conjugation junction. Curiously, general membrane protein synthesis is essential for conA receptor tipping to occur. Pair formation can be blocked by general inhibitors of protein synthesis such as cycloheximide.[22] More specifically, tunicamycin, a reagent that blocks protein glycosylation, blocks pair formation as well.[45] These studies suggest that though conA receptors may already exist diffusely distributed on the surface membrane prior to costimulation, their redistribution and aggregation at the future conjugation junction may depend upon synthesis of other surface membrane glycoproteins. In 1987, Pagliaro and Wolfe[46] demonstrated that binding of conA to membrane receptors causes certain conA binding proteins to become tightly associated with the underlying cytoskeletal frameworks. Specifically a 23 kD and a 25 kD protein become triton-insoluble

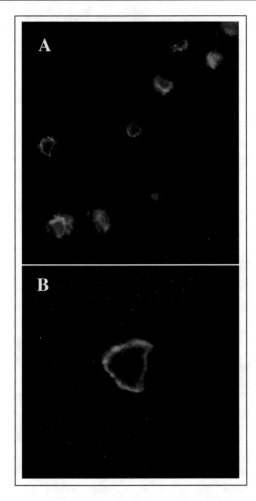

Figure 11. This image depicts isolated conjugation junctions labeled with FITC-conjugated conA. This preparation was made from sonicated mating pairs produced by Johanna Savage. A) 400X micrograph showing heart-shaped conjugation junctions. B) 600X oil immersion image of a single, conA labeled conjugation junction.

when cells are first exposed to conA prior to extraction. This becomes intriguing when coupled with the later discovery that two calcium-binding proteins (TCBP23 and TCBP25) sharing the same molecular weights can be isolated from cortical fractions and appear to decorate the conjugation junction (see below).

One can envision a model in which costimulation induces rapid synthesis of a spatially localized "anchor protein" in the newly formed membrane of the conjugation junction. ConA receptors that were diffusing freely in the plasma membrane would then aggregate at this site and help to mediate cell adhesion while triggering cytoskeletal association of cytoplasmic proteins. In short, conA receptor tipping is one of the first biochemical changes associated with pairing, and it appears to be essential for mediating early cell adhesion events that precede construction of the conjugation junction.

Ultrastructure of the Early Junction and Pore Formation

The finest studies documenting the ultrastructure of the early conjugation junction come from two laboratories, that of Jason Wolfe[9,47] and Judy and Eduardo Orias.[48,49] Normal cortical architecture in

Tetrahymena consists of three membranes: a plasma membrane and alveolar sacs which appear as an outer and inner alveolar membrane in cross section.[50] Beneath the inner-most membrane (the inner alveolar membrane) is a proteinaceous layer referred to as the epiplasm.[51] This layer can be isolated as an intact, cell-ghost structure, and has undergone a great deal of biochemical analysis.[52-58] During tip transformation, the novel membrane that forms the conjugation junction is devoid of cilia, and has none of the typical cortical organelles: basal bodies, mitochondria, mucocysts, and alveolar sacks. The epiplasm layer that lines the cell cortex appears continuous with a denser, tighter layer that lines the membrane of the conjugation junction. This has been referred to as the sub-membrane scaffold (ss),[47] and it too can be isolated as an intact structure (Cole, unpublished, Fig. 11).

As cells become firmly attached, their plasma membranes are separated by a 50 nm gap with strands of "matrix material" extending between them[47] (Fig. 12). On the cytoplasmic side the sub-membrane scaffold appears as a 50 nm thick layer of electron dense material.[47] Transmission

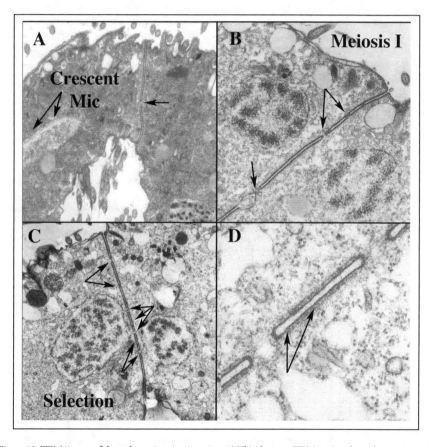

Figure 12. TEM images of the early conjugation junction. A) This depicts a TEM section through a mating pair during early meiosis (crescent stage). The crescent micronucleus appears in the left partner (arrows). A single junctional pore appears at the conjugation junction in this section, (see arrow). (Photo courtesy of Judy Orias.) B) A mating junction after completion of meiosis. Haploid nuclei appear to the left and right of the conjugation junction. In this section, three junctional pores appear (arrows). Photo taken by Anita Ramesh. C) Another, post-meiotic conjugation junction showing seven junctional pores, most concentrated near the apposed nuclei. (Photo courtesy of Judy Orias.) D) A close-up of two junctional pores showing also the 50 nm gap between cells, and the 50 nm layer of electron dense matter lining the cytoplasmic face of the junctional membrane (arrows). (Photo courtesy of Judy Orias.)

electron microscope profiles of the developing conjugation junction reveal that, two hours after mixing, numerous independent membrane fusion events are initiated all across the fusion plate. These result in the formation of hundreds of well-spaced 0.1-0.2 μm diameter pores. Wolfe explains that the scaffolding material becomes thin over regions of future pore formation, the apposed membranes indent towards one another and the space between the membranes fills with some electron dense material.[47] In certain sections, it appears that membrane indentation originates in one cell partner, and somehow evokes response in the apposed membrane.

As pore formation proceeds, the 50nm gap collapses bringing the two apposed membranes into intimate proximity just prior to fusion. Details of the actual membrane fusion event are lacking. After membrane fusion results in pore formation, these appear to be occluded by a "wispy filamentous material".[47] In TEM, these occlusions appear as wheels of electron dense material filling the pore and extending out into the cytoplasm (Fig. 8). In freeze fracture, SEM preparations these appear as plugs of material (Fig. 8A). This material, as well as the "extracellular matrix" material and the sub-membrane scaffolding survive detergent extraction suggesting linkage between sub-membrane scaffolding, integral membrane proteins, and extracellular matrix materials within the conjugation junction. One problem that remains to be explained is how a "fusion plate" perforated by hundreds of 0.1-0.2 micron pores permits transfer of a migrating pronucleus that is over 10 times the diameter of any given pore.

Transformation of the Fusion Plate into the Nuclear Exchange Junction

Membrane Events

At the time of pore formation, the junction becomes stabilized and begins to permit cytoplasmic exchange.[59] The initial pores show fairly consistent dimensions (0.1-0.2 μm in diameter). From casual examination of a number of TEM images, it appears that more pores form over regions of membrane that are adjacent to a nucleus. Over the next 2 hours, pores broaden (possibly fusing with neighboring pores) creating ever larger apertures separated by ever thinner regions of double membrane reticulum. As the transfer pronucleus associates with the membrane of the conjugation junction, that is where we see the most pronounced broadening and fusing of adjacent membrane pores. The largest pores documented by Wolfe,[9] had diameters of nearly 0.45 μm, though more typically 0.32 μm (ten times the area of the smaller, earlier pores). This transformation (depicted in Fig. 16) reaches an extreme when the membrane reticulum separating the apertures is reduced to a network of branching inter-connected tubules with circular cross section of uniform, 90 μm diameter (Fig. 13). This transformation has been beautifully captured by Judy Orias employing both conventional, and high voltage TEM.[48,49] As junction membranes are transformed into this reticulum of branching spaghetti, the electron dense scaffold proteins disappear, and the membranes appear naked (Fig. 13B,D). It is likely that the curtain-organization of the membrane at this point represents an energetically favored state, and does not require protein scaffolding to maintain it (E. Orias, personal communication).

Nuclear-Cytoskeletal Events

As events remodel the membranes of the conjugation junction, the micronucleus of each mating partner undergoes a series of three divisions, meiosis I, meiosis II and a third gametogenic nuclear division that is mitotic in nature.[60-65] During meiosis, there is some form of nuclear assessment going on.[66] When a nucleus is found "defective" by some, as yet ill-understood criterion (aneuploidy, loss of chromosomes, broken chromosomes), meiosis terminates, and all four meiotic products are destroyed. When a nucleus "passes" assessment, one of the four haploid products becomes tethered to the conjugation junction via microtubules[67] and/or 14 nm filaments,[68,69] where it is shielded from signals that trigger destruction of the other three meiotic products. The surviving, nucleus initiates a third, gametogenic nuclear division. This tethering has been named "nuclear selection".[67] The third nuclear division produces what may be accurately thought of as two gametic pronuclei. The resident pronucleus remains in the cytoplasm of its parent cell, while the transfer pronucleus becomes pressed against the conjugation junction. At this point, both resident and transfer pronuclei of a single mating partner show strands of microtubules between them[67] and they (and the

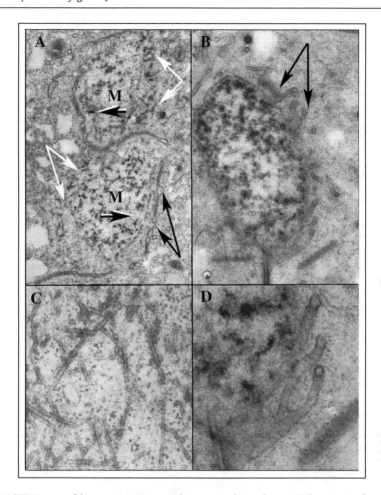

Figure 13. TEM images of the conjugation junction during pronuclear exchange. A) The moment of pronuclear exchange has been captured in this TEM photograph. Black and white arrows lie over the transfer pronuclei as they press into and deform the conjugation junction that has become perforated with pores (black arrows). White arrows indicate the short, criss-crossing microtubules (enlarged in C) that have been implicated in providing the motive force driving the pronuclei across the junction. (Reprinted with permission from: Orias JD et al. Science 222:181-184,[49] ©1983 AAAS.) B) A high voltage, thick section image showing how the conjugation junction membrane has been reduced to a branching curtain of fine, 90 nm diameter tubules. (This image has been enlarged in D). The 50 nm electron dense lining of the conjugation junction (Wolfe's sub-membrane scaffold) is now gone from the membrane. (C,D reprinted with permission from: The Molecular Biology of Ciliated Protozoa. Academic Press, 1986:45-94,[48] ©1986 Elsevier.)

conjugation junction) are decorated with the proteins fenestrin, and TCBP-25 (see below). The transfer pronucleus bears a striking cytological asymmetry in the electron microscope. On the cytoplasmic side, the pronucleus has acquired a thatched cap of short microtubules cris-crossing to form a type of net (Fig. 13C). Some microtubules even appear to form "T"-junctions with one another (Fig. 14). On the junction side, the pronucleus is nearly devoid of microtubules, and closely pressed against the, now-perforated plasma membrane. It is widely assumed that some form of microtubule-based motility is responsible for the impending nuclear exchange event. Vinblastine and nocodazole (drugs that destabilize microtubules) are effective at preventing nuclear exchange.[70-73]

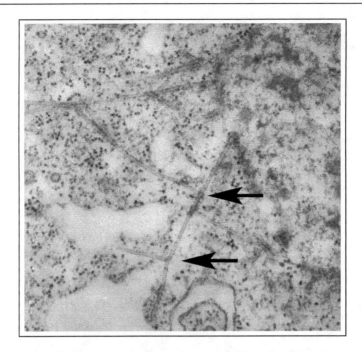

Figure 14. This TEM image (courtesy of Judy Orias) reveals a remarkable series of microtubule "T"-junctions associated with the pronuclear basket.

Both fluorescence microscopy using antisera raised against tubulin,[67] and electron microscopy[49] reveal microtubule involvement. Some type of mechanical pressure appears to be generated in that (A) the transfer pronuclei lose their spherical shape and become flattened or lens-shaped as they press up against the conjugation junction, and (B) the junction itself bows out, deforming as the microtubule driven pronuclei press into it from opposing sides (Fig. 13A). The transfer pronuclei pass, side by side at the same anterior-posterior level within the cell[74] migrate towards the partner's resident pronuclei, and undergo karyogamy (pronuclear fusion). After pronuclear transfer, the conjugation junction is severely disrupted, and the cap of short microtubules and fenestrin that had previously decorated the nucleus remains behind as a "patch" in the gap within the conjugation junction (Fig. 15). How the two plasma membranes are reestablished prior to pair separation has never been studied, though membrane pores persist long after nuclear exchange. A model depicting membrane changes during nuclear exchange appears in Figure 16.

The Conjusome: A p-Granule Nuage Assemblage?

In 1999, Janetopolous et al,[75] discovered a novel, nonmembrane bound, electron-dense structure located adjacent to the nuclear exchange junction. It appeared late in conjugal development at the time that the developing macronucleus was being remodeled. Antibodies to the chromodomain protein pdd1 (implicated in chromatin remodeling during development)[76-78] labeled this structure.[75] This structure, the conjusome, begs further inquiry. It resembles p-granules seen in the embryos of *Caenorhabditis* and *Drosophila*, or the so-called nuage of higher metazoan embryos. These structures have been implicated in distinguishing germ cells from somatic cell lineages early in the development of numerous metazoan embryos. The molecular machinery that is involved in germ-cell determination of *Drosophila* has been found in *Tetrahymena*, where it is involved in the developmental remodeling of the somatic nucleus.[79-81] It is tantalizing to contemplate that the conjusome may represent a ciliate, unicellular version of the nuage assemblage that drives this fundamental developmental decision that distinguishes nuclear fates. It is also interest-

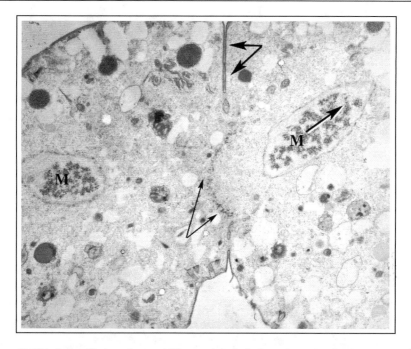

Figure 15. This figure, affectionately named "the morning after", shows the conjugation junction just after migratory pronuclei (M) have been exchanged (photo courtesy of Judy Orias). Black and white arrow indicates the direction of migration of the post-exchange pronucleus. Elongation is due to intranuclear microtubules which appear preparatory to the first post-zygotic mitosis. Thick black arrows indicate the intact portion of the conjugation junction. Thin black arrows indicate the residual microtubule basket left behind as a kind of plug at the post-exchange junction.

ing to note that the conjusome assembles in the anterior cytoplasm that is associated with the conjugation junction, very near where the golgi assembles early in development.

Biochemical Studies

PAGE Analysis

A number of extensive studies have characterized novel protein synthesis in association with initiation, costimulation, and pair formation.[43,82-85,86] Despite these efforts, little progress has been made to assign functions to even the most likely candidates. During starvation –induced initiation, it would seem that there are no new proteins synthesized. Three proteins show a significant reduction in abundance, however: a 43kD, 47kD and 56kD protein.[85] This has lead to speculation that mating readiness may involve removal of inhibitory proteins rather than the acquisition of pair-stimulating proteins. This view is consistent with data from *Paramecium*, suggesting that "immaturin" is expressed in cells that are not yet competent to pair, and vanishes from pairs that are fully mature.[87,88]

During costimulation, numerous authors have seen up-regulation of a protein (or proteins) with the approximate molecular weight of 80kD.[43,82,83] Garfinkel and Wolfe noticed that their 80kD protein was present in isolated nuclear preparations, and speculated on its role as a transcription factor. Van Bell used 2-D PAGE to identify two sets of induced proteins; one set that showed induction early in development (15-60 min after mixing), and a second set that was induced later in development (peaking at 4-6 hrs). In both sets the most prominent protein shares a molecular weight of 80kD, and the author speculates that these may represent post-translational modifications

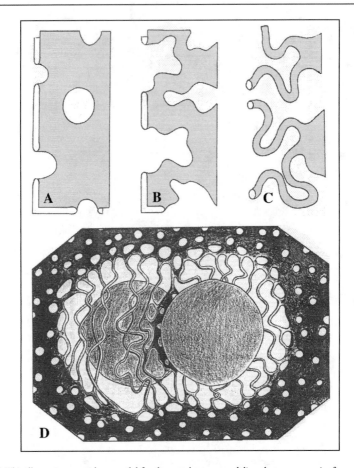

Figure 16. This illustration provides a model for the membrane remodeling that accompanies formation of the conjugation junction (artwork courtesy of Eduardo Orias). In part (A) 0.1-0.2 μm pores have formed between the two cell membranes. In part (B) junctional pores have widened and possibly fused with adjacent pores to create openings of ever larger diameter, and irregular outline. Part C) The process of porewidening has reached its maximum, and adjacent pores are now separated by thin (90 nm) membrane tubules forming a curtain or reticulum. Part (D) shows an en face view of the entire conjugation junction at the time of pronuclear exchange. One can imagine that the pronuclei have just passed each other in their transit across the junction.

of the same molecule. Although this protein has become a familiar marker of mating readiness, no substantive information has been forth-coming regarding its function or identity.

Immunofluorescence

A different quest for proteins of the conjugation junction has involved generation of monoclonal antibodies or polyclonal antisera to novel *Tetrahymena* proteins. Three potentially interesting proteins have been identified.

Fenestrin

This 64kD protein was first identified from a screening of monoclonal antibodies generated against cortical residues of *Tetrahymena pyriformis*.[74] The antibody 3A7 cross-reacted with proteins from both *T. pyriformis* and *T. thermophila*. In vegetative cells, the 3A7 antibody binds to all the "windows" in the cell: the oral apparatus, contractile vacuole pores, the cytoproct and the fission ring. In mating cells this antibody decorates the conjugation junction on the cytoplasmic face of the membrane in much the same pattern as that seen by FITC-conA on the extracellular face of the membrane, though it remains in the central disk and is not excluded to the perimeter as seen with conA labeling. 3A7 also decorates both the migratory and stationary pronuclei, though not before the third gametogenic nuclear division. Fenestrin appears to adhere to the cytoplasmic face of the sub-membrane scaffolding (ss) first recognized by Wolfe.[9] It also appears to decorate filaments that extend out from the conjugation junction and appear to anchor the pronuclei. Recently, a group claims to have sequenced the fenestrin gene by mass spectrometry of the protein fragments.[89]

Calmodulin Homologs

Takemasa et al[90,91] cloned the genes for two *Tetrahymena* calmodulin homologs: TCBP-25 and TCBP-23 respectively. These genes encode EF – hand, Ca^{2+} binding proteins. Polyclonal antisera were generated directly against these proteins that had been expressed in *E. coli*.[92,93] Both proteins appear to associate with the epiplasm. Furthermore, the antisera to TCBP-25 labels migratory and stationary pronuclei in mating cells, and faintly decorates the exchange junction as well.[94] This pattern is quite similar to the fenestrin labeling of the conjugation junction, though other "windows" into the cell were not labeled. A similar protein was found in *Paramecium*, (PCBP 25α) where it was shown to be a target for phosphorylation.[95] These findings are significant in that they may implicate calcium ions in the dynamic processes associated with nuclear exchange during conjugation.

49 kD Filament-Forming Protein

This curious protein was isolated from *Tetrahymena thermophila* by in vitro assembly /disassembly.[96] It appears to form 14 nm filaments in the presence of calcium and ATP. A polyclonal antiserum was generated against it[97,98] and its gene was later cloned.[99] The sequence revealed it to have homology to citrate synthase and it has been touted as a dual-function protein.[100] Immuno-fluorescence revealed that this protein forms fine cables that radiate from the conjugation junction at the time of nuclear selection, and appear to associate with the gametic pronuclei.[68,69] This protein is of interest for two reasons. First, this protein may be involved in capturing one of the meiotic nuclei, shielding it from programmed nuclear elimination, and tethering it to the conjugation junction for a subsequent gametogenic mitosis. Second, if its assembly is mediated by calcium, then we may see one reason for calmodulin family proteins to also be assembled at the conjugation junction.

Genetic Studies

Two types of mutant analyses have been conducted in *Tetrahymena*. First, clonal senescence produces cell lines (referred to as "star" cells) which show varying degrees of micronuclear aneuploidy, and are infertile. Star-cells reveal interesting defects during conjugation. Second, chemical mutagenesis using nitrosoguanidine, has produced a growing gallery of conjugation mutants in a series of mutant screens.[64,65]

Aneuploid Cells and Their Developmental Phenotype

When an aneuploid, "star" cell partner is mated to a diploid partner, conjugation is aberrant. Pairing occurs, and micronuclei in both partners undergo meiosis. In the diploid partner, one of the meiotic products is "selected", becomes tethered to the conjugation junction, and undergoes the third gametogenic division. The other three haploid nuclei are destroyed. In the aneuploid, "star" partner, none of the meiotic products is selected, and all four are destroyed. The diploid partner transfers its gametic pronucleus across to the star partner, but receives nothing in exchange. After unilateral nuclear transfer, development arrests, and pairs separate prematurely (9 hours rather than 12 hours at 30 C). The exconjugants are capable of remating immediately, and the second round of pairing proceeds normally establishing whole genome homozygotes. This alternative developmental program has been called "genomic exclusion".[101]

Two points are worth emphasizing about genomic exclusion pairs. First, such pairs are mechanically weakened, and easily disrupted.[102] One could predict, that unilateral nuclear association would result in the formation of fewer nuclear pores on the unoccupied site, and that this would result in a less robust conjugation junction. This could be tested by performing TEM sectioning on Star X Star matings. Second, it has been shown that star-matings produce a transferable factor that inhibits the developmental program in diploid X diploid matings.[66] Furthermore, it has been argued that some form of micronuclear assessment occurs during first meiosis (Cole et al, in preparation). When mating involves an aneuploid "star" cell, micronuclear assessment results in the generation of a dominant, conjugal arrest activity that prevents the pair from entering the postzygotic developmental program.

bcd Mutants Affect Nuclear Selection and Fusion

Nitrosoguanidine mutagenesis produced bcd, (broadend cortical domains), in a 1987 screen for pattern mutants. These cells exhibit supernumerary cortical organelles: multiple oral apparatuses, extra contractile vacuoles and extra cytoprocts.[103] Subsequent studies revealed that they were also sterile, and exhibited an interesting, pleiotropic conjugal phenotype.[104] First, in matings between bcd homozygotes, nuclear selection frequently captured two meiotic nuclei rather than the typical one. These each divided and one saw pairs with two pronuclei on each side of the conjugation junction. Second, such pairs arrested shortly after nuclear exchange, and multiple pronuclei aggregated at the conjugation junction without undergoing nuclear fusion. One interpretation, is that the bcd mutant results in a broadening of the nuclear-selection field as well as all the other cortical domains. It has not escaped our attention that every cortical domain affected by the bcd mutation is also a target for fenestrin involvement. It is attractive to consider that the bcd mutation may affect fenestrin distribution, either directly or indirectly.

Pair Separation Failures

A second pattern mutant that exhibits a conjugation phenotype is janus A. This mutant produces a mirror image duplication of its ventral cortical pattern.[105] Mating is substantially normal up to 12 hours, when pair separation normally occurs. At this point, pairs remain attached at the conjugation junction.[106] This proves fatal, in that pairs are unable to rebuild their oral apparatus and enter the vegetative feeding pathway. Other mutants have been shown to produce the same phenotype: cnj6 and mra.[64,107] Actinomycin D or cycloheximide treatments delivered after macronuclear anlagen formation also result in a pair-separation failure.[108] This phenotype is difficult to understand without more facts. The janA connection to this (a cortical phenotype derangement) remains mysterious, though potentially, janA pairs produce an expanded region of pair fusion, resulting in two-cell parabiosis.

cnj10 Pleiotropy

This mutant exhibits a conjugal phenotype that is quite intriguing.[64] cnj10 X cnj10 matings show three instances where there is a failure of nuclear-cortical interaction. First, nuclear selection frequently fails. Second, even in pairs that complete nuclear selection and generate gametic pronuclei, nuclear transfer fails. Nuclear fusion also fails in most partners. Finally, even though postzygotic development proceeds relatively normally in some pairs, none of the post-zygotic nuclei become anchored at the posterior cortex (the normal fate of two postzygotic nuclear products destined to become micronuclei). Hence, all the nuclei wander freely in the cytoplasm and differentiate into a plethora of macronuclear anlagen. It is compelling to suggest that the gene identified by the cnj10 mutation is necessary for nuclear-cortical interactions and nuclear-nuclear interactions during fertilization.

Epilogue

The *Tetrahymena* conjugation junction presents a remarkable biological system promising insight into a number of fundamental problems of cell biology and development. It represents a rare case in which one sees specific, developmental induction of a golgi complex. Moreover, it also represents a unique event that somehow coordinates hundreds of independent membrane fusion events within a confined membrane domain. It reveals a potentially novel microtubule-based motility system in the appearance of "T-junctions" within the pronuclear transfer basket. It may participate in the developmental assembly of a p-granule/ nuage-like organelle that directs differentiation of a somatic nucleus from a germ-line nucleus, and finally, it represents a rich, dynamic dialog between nucleus and cell cortex. This dynamic structure will bear watching in the years to come.

Acknowledgements

I wish to thank Joseph Frankel for leading me into the spectacular world of ciliates, Jason Wolfe for his encouragement in writing this chapter on a subject near and dear to him, and a special thanks to Judy and Ed Orias whose keen observations and boundless generosity fill these pages. This work was supported by an NSF grant # MCB 0444700 "The Gene Stream".

References

1. Sapp J. Beyond the Gene: Cytoplasmic inheritance and the Struggle for Authority in Genetics. New York: Oxford University Press, 1987.
2. Lee JJ, Leedale GF, Bradbury PC. An Illustrated Guide to the Protozoa: Organisms Traditionally Referred to as Protozoa, or Newly Discovered Groups. 2nd ed. Lawrence: Society of Protozoologists, 2000.
3. Frankel J. Pattern Formation: Ciliate Studies and Models. New York: Oxford University Press, 1989.
4. Frankel J. Participation of the undulating membrane in the formation of oral replacement primordia in Tetrahymena pyriformis. J Protozool 1969; 16:26-35.
5. Nelsen EM. Transformation in Tetrahymena pyriformis: Development of an inducible phenotype. Dev Biol 1978; 66:17-31.
6. Nelsen EM, DeBault LE. Transformation in Tetrahymena pyriformis—Description of an inducible phenotype. J Protozool 1978; 25:113-119.
7. Bell G. Sex and death in protozoa: The history of an obsession. Cambridge: Cambridge University Press, 1988.
8. Wolfe J. Cytoskeletal reorganization and plasma membrane fusion in conjugation Tetrahymena. J Cell Sci 1985; 73:69-85.
9. Wolfe J. The conjugation junction of Tetrahymena: Its structure and development. J Morphol 1982; 172:159-178.
10. Bruns PJ, Brussard TB. Pair formation in Tetrahymena pyriformis, an inducible developmental system. J Exp Zool 1974; 188:337-344.
11. Nanney DL, Caughey PA. Mating type determination in Tetrahymena pyriformis. Proc Natl Acad Sci USA 1953; 39:1057-1053.
12. Lynch TJ, Brickner J, Nakano KJ et al. Genetic map of randomly amplified DNA polymorphisms closely linked to the mating type locus of Tetrahymena thermophila. Genetics 1995; 141:1315-1325.
13. Elliot AM, Nanney DL. Conjugation in Tetrahymena. Science 1952; 116:33-34.
14. Nanney DL, Caughey PA. An unstable nuclear condition in Tetrahymena pyriformis. Genetics 1955; 40:388-398.
15. Bruns PJ, Palestine RF. Costimulation in Tetrahymena pyriformis: A developmental interaction between specially prepared cells. Dev Biol 1975; 42:75-83.
16. Thompson GA. Studies of membrane formation in Tetrahymena pyriformis. I. Rates of phospholipid biosynthesis. Biochemistry 1967; 6:2015-2022.
17. Dryl S. Antigenic transformation in Paramecium aurelia after autologous antiserum treatment during autogamy and conjugation. J Protozool 1959; 6(Suppl):25.
18. Bruns PJ, Brussard TB. Positive selection for mating with functional heterokaryons in Tetrahymena pyriformis. Genetics 1974; 81:831-841.
19. Bruns PJ, Cassidy-Hanley D. Biolistic transformation of macro- and micronuclei. Meth Cell Biol 2000; 62:501-512.
20. Wellnitz WR, Bruns PJ. The prepairing events in Tetrahymena thermophila. Analysis of blocks imposed by high concentrations of Tris-HCl. Exp Cell Res 1979; 119:175-180.
21. Wellnitz WR, Bruns PJ. The prepairing events in Tetrahymena. II. Partial loss of developmental information upon refeeding starved cells. Exp Cell Res 1982; 137:317-328.
22. Allewell JM, Oles J, Wolfe J. A physicochemical analysis of conjugation in Tetrahymena pyriformis. Exp Cell Res 1976; 97:394-405.
23. Finley MJ, Bruns PJ. Costimulation in Tetrahymena. II. A nonspecific response to heterotypic cell-cell interactions. Dev Biol 1980; 79:81-94.
24. Takahashi M. Does fluid have any function for mating in Tetrahymena? Scient Rep Tohoku Univ 1973; 36:223-229.
25. Adair WS, Barker R, Turner Jr RS et al. Demonstration of a cell-free factor involved in cell interactions during mating in Tetrahymena. Nature 1978; 274:54-55.

26. Wolfe J, Turner R, Barker R et al. The need for an extracellular component for cell pairing in Tetrahymena. Exp Cell Res 1979; 121:27-30.
27. Wolfe J, Meal KJ, Soiffer R. Limiting conditions for conjugation in Tetrahymena: Cellular development and factor active in conjugation. J Exp Zool 1980; 212:37-46.
28. Fujushima S, Tsuda M, Mikami Y et al. Costimulation-induced rounding in Tetrahymena thermophila: Early cell-shape transformation induced by sexual cell-to-cell collisions between complementary mating types. Dev Biol 1993; 155:198-205.
29. Allewell NM, Wolfe J. A kinetic analysis of the memory of a developmental interaction: Mating interactions in Tetrahymena pyriformis. Exp Cell Res 1977; 109:15-24.
30. Kitamura A, Sugai T, Kitamura Y. Homotypic pair formation during conjugation in Tetrahymena thermophila. J Cell Sci 1986; 82:223-234.
31. Virtue MA, Cole ES. A cytogenetic study of devolopment in mechanically disrupted pairs of Tetrahymena thermophila. J Euk Microbiol 1999; 46:597-605.
32. Kiersnowska M, Kaczanowski A, Morga J. Macronuclear development in conjugants of Tetrahymena thermophila, which were artificially separated at meiotic prophase. J Euk Microbiol 2000; 47:139-147.
33. Wolfe J, Grimes GW. Tip transformation in Tetrahymena: A morphogenetic response to interactions between mating types. J Protozool 1979; 26:82-89.
34. Kurz S, Tiedtke A. The Golgi apparatus of Tetrahymena thermophila. J Euk Microbiol 1993; 40:10-13.
35. Elliott AM. Biology of Tetrahymena. Dowden Hutchinson and Ross Inc, 1973.
36. Ofer L, Levkovits H, Loyter A. Conjugation in Tetrahymena pyriformis. The effect of polylysine, Con A, bivalent metals on the conjugation process. J Cell Biol 1976; 7:287-293.
37. Frisch A, Loyter A. Inhibition of conjugation in Tetrahymena pyriformis by Con A. Localization of Con A binding sites. Exp Cell Res 1977; 110:337-346.
38. Watanabe S, Toyohara A, Suzaki T et al. The relation of concanavalin A receptor distribution to the conjugation process in Tetrahymena thermophila. J Protozool 1981; 28:171-175.
39. Wolfe J, Pagliaro L, Fortune H. Coordination of concanavalin-A-receptor distribution and surface differentiation in Tetrahymena. Differentiation 1986; 31:1-9.
40. Pagliaro L, Wolfe J. Concanavalin A binding induces association of possible mating-type receptors with the cytoskeleton in Tetrahymena. Exp Cell Res 1987; 168:138-152.
41. Wolfe J, Feng S. Concanavalin A receptor "tipping" in Tetrahymena and its relationship to cell adhesion during conjugation. Development 1988; 102:699-708.
42. Suganuma Y, Yamamoto H. Conjugation in Tetrahymena: Its relation to concanavalin A receptor distribution on the cell surface. Zool Sci 1988; 5:323-330.
43. van Bell CT. An analysis of protein synthesis, membrane proteins, and concanavalin A-binding proteins during conjugation in Tetrahymena thermophila. Dev Biol 1983; 98:173-181.
44. Driscoll C, Hufnagel LA. Affinity-purification of concanavalin A-binding ciliary glycoconjugates of starved and feeding Tetrahymena thermophila. J Euk Microbiol 1999; 46:142-146.
45. Frisch A, Levkowitz H, Loyter A. Inhibition of conjugation and cell division in Tetrahymena pyriformis by tunicamycin: A possible requirement of glycoprotein synthesis for induction of conjugation. Biochem Biophys Res Commun 1976; 72:138-145.
46. Pagliaro L, Wolfe J. Concanavalin-A binding induces association of possible mating type receptors with the cytoskeleton of Tetrahymena. Exp Cell Res 1987; 168:138-152.
47. Wolfe J. Cytoskeletal reorganization and plasma membrane fusion in conjugating Tetrahymena. J Cell Sci 1985; 73:69-85.
48. Orias E. Ciliate conjugation. In: Gall LG, ed. The Molecular Biology of Ciliated Protozoa. Orlando: Academic Press, 1986:45-94.
49. Orias JD, Hamilton EP, Orias E. A microtubular meshwork associated with gametic pronuclear transfer across a cell-cell junction. Science 1983; 222:181-184.
50. Allen RD. Fine structure of membranous and microfibrillar systems in the cortex of Paramecium caudatum. J Cell Biol 1971; 49:1-20.
51. Sibley JT, Hanson ED. Identity and function of a subcortical cytoskeleton in Paramecium. Arch Protistenk 1974; 116:221-235.
52. Vaudaux P. Isolation and identification of specific cortical proteins in Tetrahymena pyriformis GL. J Protozool 1976; 23:458-464.
53. Williams NE, Honts JE, Jaeckel-Williams RF. Regional differentiation of the membrane skeleton in Tetrahymena. J Cell Sci 1987; 87:457-463.
54. Williams NE, Honts JE. The assembly and positioning of cytoskeletal elements in Tetrahymena. Development 1987; 100:23-30.

55. Williams NE, Honts JE. Isolation and fractionation of the Tetrahymena cytoskeleton and oral apparatus. Meth Cell Biol 1995; 47:301-306.

56. Williams NE, Honts JE, Dress VM et al. Monoclonal antibodies reveal complex structure in the membrane skeleton of Tetrahymena. J Eukar Microbiol 1995; 42:422-427.

57. Honts JE, Williams NE. novel cytoskeletal proteins in the cortex of Tetrahymena. J Eukar Microbiol 2003; 50:9-14.

58. Williams NE. The epiplasm gene EPC1 influences cell shape and cortical pattern in Tetrahymena thermophila. J Eukar Microbiol 2004; 51:201-206.

59. McDonald BB. The exchange of RNA and protein during conjugation in Tetrahymena. J Protozool 1966; 13:277-285.

60. Nanney DL. Nucleocytoplasmic interactions during conjugation in Tetrahymena. Biol Bull 1953; 105:133-148.

61. Elliot AM, Hayes RE. Mating types in Tetrahymena. Biol Bull 1953; 105:269-284.

62. Ray CJ. Meiosis and nuclear behavior in Tetrahymena pyriformis. J Protozool 1956; 3:604-610.

63. Martindale DW, Allis CD, Bruns PJ. Conjugation in Tetrahymena thermophila: A temporal analysis of cytological stages. Exp Cell Res 1982; 140:227-236.

64. Cole ES, Soelter TA. A mutational analysis of conjugation in Tetrahymena themophila 2. Phenotypes affecting middle and late development: Third prezygotic division, pronuclear exchange, pronuclear fusion and postzygotic development. Dev Biol 1997; 189:233-245.

65. Cole ES, Cassidy-Hanley D, Hemish J et al. A mutational analysis of conjugation in Tetrahymena themophila 1. Phenotypes affecting early development: Meiosis to nuclear selection. Dev Biol 1997; 189:215-232.

66. Cole ES, Virtue MA, Stuart KR. Development in electrofused conjugants of Tetrahymena thermophila. J Eukar Microb 2001; 48:266-279.

67. Gaertig J, Fleury A. Spatiotemporal reorganization of intracytoplasmic microtubules is associated with nuclear selection and differentiation during developmental process in the ciliate Tetrahymena thermophila. Protoplasma 1992; 167:74-87.

68. Numata O, Sugai T, Watanabe Y. Control of germ cell nuclear behavior at fertilization by Tetrahymena intermediate filament protein. Nature 1985; 314:192-193.

69. Takagi I, Numata O, Watanabe Y. Involvement of 14-nm filament-forming protein and tubulin in gametic pronuclear behavior during conjugation in Tetrahymena. J Eukar Microbiol 1991; 38:345-351.

70. Hamilton EP. Dissection of fertilization and development in Tetrahymena using anti-microtubule drugs [Ph.D. dissertation]. Santa Barbara: Division of Biological Sciences, University of California Santa Barbara, 1984.

71. Hamilton EP, Suhr-Jessen PB. Autoradiographic evidence for self-fertilization in Tetrahymena thermophila. Exp Cell Res 1980; 126:391-396.

72. Hamilton EP, Suhr-Jessen PB, Orias E. Pronuclear fusion failure: An alternate conjugational pathway in Tetrahymena thermophila induced by vinblastine. Genetics 1988; 118:627-636.

73. Kaczanowski A, Gaertig J, Kubiak J. Effect of the antitubulin drug nocodazole on meiosis and postmeiotic development in Tetrahymena thermophila. Induction of achiasmatic meiosis. Exp Cell Res 1985; 158:244-256.

74. Nelsen EM, Williams NE, Yi H et al. "Fenestrin" and conjugation in Tetrahymena thermophila. J Euk Microbiol 1994; 41:483-495.

75. Janetopoulos C, Cole E, Smothers JF et al. The conjusome: A novel structure in Tetrahymena found only during sexual reorganization. J Cell Sci 1999; 112:1003-1111.

76. Smothers JF, Mizzen CA, Tubbert MM et al. Pdd1p associates with germline-restricted chromatin and a second novel anlagen-enriched protein in developmentally programmed DNA elimination structures. Development 1997; 124:4537-4545.

77. Smothers JF, Madireddi MT, Warner FD et al. Programmed DNA degradation and nucleolar biogenesis occur in distinct organelles during macronuclear development in Tetrahymena. J Euk Microbiol 1997; 44:79-88.

78. Madireddi MT, Coyne RC, Smothers JF et al. Pdd1p, A novel chromodomain-containing protein, links heterochromatin assembly and DNA elimination in Tetrahymena. Cell 1996; 87:75-84.

79. Mochizuki K, Gorovsky MA. A Dicer-like protein in Tetrahymena has distinct functions in genome rearrangement, chromosome segregation, and meiotic prophase. Genes Dev 2005; 19:77-89.

80. Mochizuki K, Gorovsky MA. Conjugation-specific small RNAs in Tetrahymena have predicted properties of scan (scn) RNAs involved in genome rearrangement. Genes Dev 2004; 18:2068-2073.

81. Mochizuki K, Fine NA, Fujisawa T et al. Analysis of a piwi-related gene implicates small RNAs in genome rearrangement in Tetrahymena. Cell 2002; 110:689-699.

82. Ron A, Suhr-Jessen PB. Protein synthesis patterns in conjugating Tetrahymena thermophila. Exp Cell Res 1981; 133:325-330.
83. Garfinkel MD, Wolfe J. Alterations in gene expression induced by a specific cell interaction during mating in Tetrahymena thermophila. Exp Cell Res 1981; 133:317-324.
84. Suhr-Jessen PB. Stage-specific changes in protein synthesis during conjugation in Tetrahymena thermophila. Exp Cell Res 1984; 151:374-383.
85. van Bell CT, Williams NE. Membrane protein differences correlated with the development of mating competence in Tetrahymena thermophila. J Euk Microbiol 1984; 31:112-116.
86. Suhr-Jessen P, Salling L, Larsen HC. Polypeptides during early conjugation in Tetrahymena thermophila. Exp Cell Res 1986; 163:549-557.
87. Haga N. Transformation of sexual maturity to immaturity by microinjecting immature genomic DNA in Paramecium. Zool Sci 1991; 8:11-24.
88. Miwa I. Destruction of immaturin activity in early mature mutants of Paramecium caudatum. J Cell Sci 1984; 72:111-120.
89. Joachimiak E, Sikora J, Kaczanowska J et al. Characterization of the fenestrin, a cytoskeletal protein involved in Tetrahymena cell polarity and of its coding gene. J Euk Microbiol 2005; 52:7S.
90. Takemasa T, Ohnishi K, Kobayashi T et al. Cloning and sequencing of the gene for Tetrahymena calcium-binding 25-kDa protein (TCBP-25). J Biol Chem 1989; 264:19293-19301.
91. Takemasa T, Takagi T, Kobayashi T et al. The third calmodulin family protein in Tetrahymena. Cloning of the cDNA for Tetrahymena calcium-binding protein of 23 kDa (TCBP-23). J Biol Chem 1990; 265:2514-2517.
92. Hanyu K, Takemasa T, Numata O et al. Immunofluorescence localization of a 25-kDa Tetrahymena EF-hand Ca^{2+}-binding protein, TCBP-25, in the cell cortex and possible involvement in conjugation. Exp Cell Res 1995; 219:487-493.
93. Hanyu K, Numata O, Takahashi M et al. Immunofluorescence localization of a 23-kDa Tetrahymena calcium-binding protein, TCBP-23, in the cell cortex. J Biochem (Tokyo) 1996; 119:914-919.
94. Numata O, Hanyu K, Takeda T et al. Tetrahymena calcium-binding proteins, TCBP-25 and TCBP-23. Meth Cell Biol 2000; 62:455-465.
95. Kim K, Son M, Peterson JB et al. Ca^{2+}-binding proteins of cilia and infraciliary lattice of Paramecium tetraurelia: Their phosphorylation by purified endogenous $Ca^{(2+)}$-dependent protein kinases. J Cell Sci 2002; 115:1973-1984.
96. Numata O, Watanabe Y. In vitro assembly and disassembly of 14-nm filament from Tetrahymena pyriformis. The protein component of 14-nm filament is a 49,000-Dalton protein. J Biochem (Tokyo) 1982; 91:1563-1573.
97. Numata O, Hirono M, Watanabe Y. Involvement of Tetrahymena intermediate filament protein, a 49K protein, in the oral morphogenesis. Exp Cell Res 1983; 148:207-220.
98. Numata O, Tomiyoshi T, Kurasawa Y et al. Antibodies against Tetrahymena 14-nm filament-forming protein recognize the replication band in Euplotes. Exp Cell Res 1991; 193:183-189.
99. Numata O, Takemasa T, Takagi I et al. Tetrahymena 14-NM filament-forming protein has citrate synthase activity. Biochem Biophys Res Commun 1991; 174:1023-1034.
100. Numata O. Multifunctional proteins in Tetrahymena: 14nm filament protein/citrate synthase and translation elongation factor-1 alpha. Int Rev Cytol 1996; 164:1-35.
101. Allen SL, File SK, Koch SL. Genomic exclusion in Tetrahymena. Genetics 1967; 55:823-837.
102. Gaertig J, Kaczanowski A. Correlation between the shortened period of cell pairing during genomic exclusion and the block in post-transfer nuclear development in Tetrahymena thermophila. Dev Growth Diff 1987; 29:553-562.
103. Cole ES, Frankel J, Jenkins LM. bcd: A mutation affecting the width of organelle domains in the cortex of Tetrahymena thermophila. Wilhelm Roux's Arch Dev Biol 1987; 196:421-433.
104. Cole ES. Conjugal blocks in Tetrahymena pattern mutants and their cytoplasmic rescue. I. Broadened cortical domains (bcd). Dev Biol 1991; 148:403-419.
105. Frankel J, Nelsen EM. Positional reorganization in compound janus cells of Tetrahymena thermophila. Development 1987; 99:51-68.
106. Cole ES, Frankel J. Conjugal blocks in Tetrahymena pattern mutants and their cytoplasmic rescue. II. Janus A. Dev Biol 1991; 148:420-428.
107. Kaczanowski A. Mutation affecting cell separation and macronuclear resorption during conjugation in Tetrahymena thermophila: Early expression of the zygotic genotype. Dev Genet 1992; 13:58-65.
108. Ward JG, Herrick G. Effects of the transcription inhibitor actinomycin D on postzygotic development of Tetrahymena thermophila conjugants. Dev Biol 1994; 173:174-184.
109. Lansing TJ, Frankel J, Jenkins LM. Oral ultrastructure and oral development in the misaligned undulating membrane mutant of Tetrahymena thermophila. J Protozool 1985; 32:126-139.

Section III
Algal Cells

CHAPTER 4

Cytoplasmic Bridges in *Volvox* and Its Relatives

Harold J. Hoops, Ichiro Nishii and David L. Kirk*

...we believe that there are enough reasons from the developmental and structural aspects as shown in this work to distinguish the cytoplasmic connecting strands of Volvox from the plasmodesmata of higher plants. In this regard it would be more suitable to refer to the connecting strands described in this work simply as "cytoplasmic bridges".[1]

Abstract

The volvocaceans are a closely related group of green flagellates that range in size and complexity from colonial forms that contain a small number of identical cells, to *Volvox* in which there is a division of labor between several thousand terminally differentiated somatic cells and a small number of asexual reproductive cells called gonidia. Similar cytoplasmic bridges link the cells of all volvocean embryos, but the formation, structure and function of such bridges have been studied most extensively during *Volvox carteri* embryogenesis.

Each mature *V. carteri* gonidium produces an embryo that executes 11-12 rounds of rapid, stereotyped, synchronous cleavage divisions in the absence of any growth. Large numbers of cytoplasmic bridges are formed in each cleavage furrow as a result of incomplete cytokinesis. The bridges are tightly packed in bands that are concentric with the inner surface of the hollow embryo, and the bridge bands of all cells are linked to form a single, coherent cytoplasmic-bridge system that runs through the entire embryo and holds it together.

A fully cleaved embryo contains all of the cells that will be present in an adult of the next generation, but it is inside-out with respect to the adult configuration: its gonidia are on the outside and the flagellar ends of its somatic cells are on the inside. This awkward arrangement is corrected by inversion, in which the embryo rapidly turns itself right-side-out through a combination of cell shape changes and movements. The cytoplasmic-bridge system plays a central role in inversion by providing the physical framework against which the cells exert force to reverse the curvature of the cellular sheet. Recent studies show that a kinesin located in the cytoplasmic bridges appears to provide the critical driving force of inversion. Inversion is a defining characteristic of the volvocaceans, and all of the volvocaceans that have been tested have a gene product that is homologous to the kinesin that drives inversion in *V. carteri*. So we think that the mechanism of inversion is probably similar in all of them.

In all of the smaller volvocaceans, and in about half of all *Volvox* species, the cytoplasmic bridges disappear soon after inversion has been completed and a newly formed ECM has taken over the job of holding the organism together. However, three lineages of *Volvox* have independently evolved the trait of retaining cytoplasmic bridges throughout the life of the adult. And in each case this trait has coevolved with a novel form of embryonic development in which embryos begin to divide while they are still quite small, and then grow between successive divisions—all the while remaining connected to the adult somatic cells. Here we provide new information about the formation and the structure of persistent cytoplasmic bridges in one such species of *Volvox*, and review the evidence that leads us to believe that adult cytoplasmic bridges serve as conduits to convey nutrients from somatic cells to developing embryos.

*Corresponding Author: David L. Kirk—Department of Biology, Campus Box 1229, Washington University, 1 Brookings Drive, St. Louis, Missouri 63130, U.S.A. Email: kirk@biology.wustl.edu

Cell-Cell Channels, edited by Frantisek Baluska, Dieter Volkmann and Peter W. Barlow.
©2006 Landes Bioscience and Springer Science+Business Media.

Introduction

Late in the 17th century Antoni van Leeuwenhoek watched with fascination as the little, round green "animalcules" to which Linnaeus would later give the name "*Volvox*" rolled about restlessly before his simple lens.[2] More than 130 years would pass before microscopes were improved sufficiently that another great microbiologist, Christian Ehrenburg, would realize that each cell within such a *Volvox* spheroid was connected to its neighbors by "filamentous green tubes".[3] Such intercellular connections were seen in free-swimming adults of both species of *Volvox* then known (*V. globator* and *V. aureus*), and in 1896 Arthur Meyer used plasmolysis to demonstrate that these intercellular connectors were actually cytoplasmic extensions of the cells, indicating that the adult organisms in both of these species were syncytial.[4] In that same paper, however, Meyer described a third species of *Volvox* (which he imaginatively named *V. tertius*) in which fine connections existed between cells of the developing embryo, but then those bridges disappeared as the embryo developed into an adult.[4]

It would turn out that the latter pattern—cytoplasmic bridges present in the embryo but not the adult—is ancestral. It is the pattern seen in all of the smaller and simpler relatives of *Volvox* in the family Volvocaceae (genera such as *Pandorina*, *Eudorina* and *Pleodorina*) as well as in about half of all species of *Volvox*. Persistence of bridges in the adult is a derived trait that is seen only in some species of *Volvox*. We will return to a discussion of that derived trait toward the end of this chapter.

A Brief Overview of *Volvox carteri* Development

In her 1933 paper describing embryonic development in two species of *Volvox* found in her native South Africa, Mary Agard Pocock first suggested that cytoplasmic bridges probably were formed as a result of incomplete cytokinesis.[5] Over the years, that suggestion was repeated by investigators studying development in various members of the family Volvocaceae,[1,6-18] but formation of the cytoplasmic bridges has been studied systematically only in *Volvox carteri* embryos.[19,20] For that reason, a brief overview of embryonic development in *V. carteri* is called for as background for the subsequent discussion of cytoplasmic-bridge formation and function (details of *V. carteri* development are available elsewhere.[21]

A young adult spheroid of *V. carteri* consists of several thousand small, biflagellate somatic cells at the surface of a transparent sphere of glycoprotein-rich extracellular matrix and a few large asexual reproductive cells, called "gonidia", that lie just beneath the somatic cell monolayer (Fig. 1). When

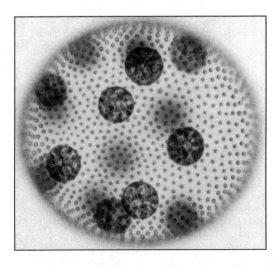

Figure 1. *Volvox carteri*. A young adult asexual spheroid, as shown here, contains several thousand *Chlamydomonas*-like, biflagellate somatic cells at the surface, and 16 large gonidia (asexual-reproductive cells) just beneath the surface, of a transparent sphere of extracellular matrix.

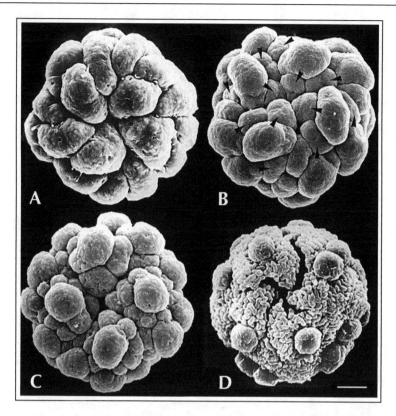

Figure 2. Scanning electron micrographs (SEMs) of cleaving *V. carteri* embryos in polar view (cleavage normally takes place inside the adult spheroid.). Each mature gonidium undergoes a stereotyped set of rapid, synchronous cleavage divisions. A) The first five divisions are symmetrical, producing a 32-cell embryo in which all cells are about the same size and shape. B) The cells in one half of the 32-cell embryo divide asymmetrically to produce large–small sister-cell pairs (here shown linked by arrowheads). The large cells are gonidial initials that will produce one gonidium each, whereas the small cells are somatic initials, each of which will divide symmetrically to produce a clone of somatic cells. C) Gonidial initials divide asymmetrically a total of three or four times (producing additional somatic initials at each division) and then withdraw from the division cycle while the somatic initials continue to divide two or three more times. D) By the end of cleavage the embryo contains all of the cells of both types that will be in an adult spheroid of the next generation, but the embryo is inside-out with respect to the adult configuration: the gonidia are on the outside, and the flagellar ends of the somatic cells are on the inside. The swastika-shaped slit at one pole of the embryo is called the phialopore; it was formed in early cleavage, and it is the opening through which the embryo will now turn itself right-side-out. Bar: 10 μm.

a gonidium is mature, it behaves like a stem cell and initiates a series of rapid, synchronous cleavage divisions that produce a new juvenile individual containing a new cohort of somatic and gonidial cells (Fig. 2). The prospective gonidial and somatic cells of the new generation are set apart during cleavage by a stereotyped set of asymmetric cleavage divisions (Fig. 2B). Each large cell produced by asymmetric division produces one gonidium while its smaller sister cell produces a clone of somatic cells.

By the end of cleavage the embryo is a hollow sphere in which all of the cells of both types that will be present in an adult of the next generation are already present, but the embryo is inside-out with respect to the adult configuration: its gonidia are on the outside, and the flagellar ends of its somatic cells face the interior. That awkward situation is rapidly corrected as the embryo turns itself

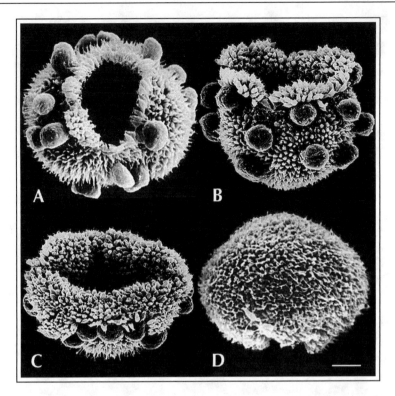

Figure 3. Inversion of a *V. carteri* embryo; A) polar view; B-D) equatorial views. A,B) Shortly after the end of cleavage the four lips of cells flanking the phialopore curl outward and backward over the rest of the embryo. C) More and more of the cellular monolayer is turned right-side-out as the region of maximum curvature moves progressively toward the pole of the embryo opposite the phialopore. D) By about 40 minutes after inversion began, the entire monolayer has been turned right-side-out, with its flagella now facing the exterior. Bar: 10 μm.

right-side-out in a process called inversion (Fig. 3). Inversion is a diagnostic feature of the family Volvocaceae; it is seen in all members of the family, and will be discussed in more detail shortly.

The Formation of a Cytoplasmic-Bridge System in *Volvox carteri* Embryos

Volvocacean embryos are held together during cleavage and inversion by a network of cytoplasmic bridges. Some early papers suggesting that such bridges were produced by incomplete cytokinesis[1,5-18] might have left the impression that a bridge or two was formed in each cleavage furrow. However, scanning electron microscope (SEM) examination of *V. carteri* embryos[19] revealed that bridges are actually formed by the hundreds in each of the first few cleavage furrows (Fig. 4A). Then at each division all preexisting bridges are preserved and divided between the daughter cells while many additional bridges are being formed in the furrow between them. Thus, for example, each cell of an 8-cell embryo is linked to its neighbors by ~250 bridges, half of which were inherited, and half of which were newly formed during the third division. Bridge transmission and formation continues throughout cleavage, resulting in a progressive increase in the total number of bridges in the embryo—although the number of bridges per cell decreases, as the cells become smaller. By the end of cleavage the embryo contains about 100,000 bridges, and an average cell is connected to its neighbors by about 25 bridges.[19]

It is important to note that these cytoplasmic bridges are not located randomly on the cells. They are organized into closely packed, curved bands of bridges (Fig. 4B) that are concentric with the

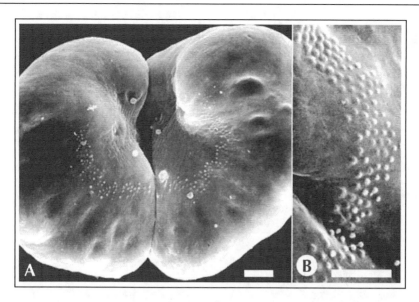

Figure 4. Scanning electron micrographs (SEMs) of fragments of *V. carteri* embryos that were dissected after they had been fixed and dried. A) One half of a 4-cell embryo. Broken cytoplasmic bridges appear as bumps (or sometimes shallow pits) on the surface of the cell, and are arranged in curved bands that are concentric with the inner surface of the embryo. B) A bridge band on a 4-cell embryo at higher magnification; note the close packing and regular spacing of the broken bridges. Bars: 5 μm. Reproduced from: Green KJ, Kirk DL. J Cell Biol 1981; 91:743-755,[19] by copyright permission of the Rockefeller University Press.

inner surface of the hollow embryo, closer to the end of the cell with the nucleus and basal body than to the end of the cell containing the chloroplast (Fig. 4A). Transmission electron microscopy (TEM) of thin sections reveals that throughout cleavage, the bridge bands are present at a level that is occupied on the cellular interior by a bowl-shaped zone of Golgi-rich cytoplasm that cups the nucleus and separates it from the chloroplast.[20] Most importantly, the bridge bands of all cells are aligned and connected, forming a single, coherent cytoplasmic-bridge system that is continuous throughout the embryo, and that holds it together.[19,20]

The position of the cytoplasmic-bridge bands on the dividing cells is linked to the mechanism by which they are formed during cytokinesis.[20] As alluded to in the preceding paragraph, all cells of the cleaving embryo have a very regular, polarized organization: the basal bodies (BBs) are located at the inner tip of each cell, the nucleus is positioned just below the BBs and is surrounded by cytoplasm containing numerous Golgi bodies and vesicles. The large chloroplast fills the outer portion of the cell. Each cell is girdled by an extensive array of cortical microtubules (MTs) that originate near the basal bodies and extend toward the chloroplast end of the cell just under the plasma membrane. As sister nuclei are formed during mitotic telophase, parallel sheets of additional MTs emanating from the vicinity of the (now-separated) sister BBs extend through the cytoplasm in the inter-nuclear region. These parallel MT arrays are called the cleavage MTs, and they are part of the "phycoplast", a characteristic cytoskeletal array that delineates the plane of cell division in most green algal cells. Division of the BB/nuclear end of the cell occurs by an ingressive furrow that is similar to the furrow often seen in metazoan cytokinesis. However, cytokinesis in the sub-nuclear region of the cell occurs by a different mechanism: by MT-guided alignment and fusion of Golgi-derived vesicles—much like the formation of a cell plate in higher-plant cytokinesis.[20] The cytoplasmic bridges are apparently formed at the regularly spaced places where these vesicles fail to fuse, and the bridges remain in this sub-nuclear region, closely associated with the nearby MT arrays, throughout the rest of the cleavage period.

Structural Features of the *V. carteri* Cytoplasmic-Bridge System

The cytoplasmic bridges of the *V. carteri* embryo are extremely regular in appearance. They are ~200 nm in diameter, spaced 500 nm center-to-center on average (Fig. 4B), and are associated with at least two types of conspicuous membrane specializations. The inner face of the plasma membrane throughout the bridge region is coated with electron-dense material that appears to be thickest in the center of each bridge, so that the bridges appear as dark rings when sectioned transversely (Fig. 5A,D). Intimately associated with (or embedded in) that electron-dense coating is an extensive array of concentric or spiral cortical striations with a spacing of ~25 nm that not only girdle each bridge throughout its length (Fig. 5B), they continue out under the adjacent plasma membrane of the cell body as concentric rings that abut the rings surrounding neighboring bridges (Fig. 5A). In bridges that have been sectioned longitudinally, these cortical striations often appear as a series of short, uniformly spaced parallel line segments, near the cell surface (Fig. 5C). Extremely similar cortical striations in the bridge region were first observed by Jeremy Pickett-Heaps in *V. tertius* embryos,[11] and were later seen in the cytoplasmic bridges of *Eudorina elegans* embryos.[13] The molecular nature

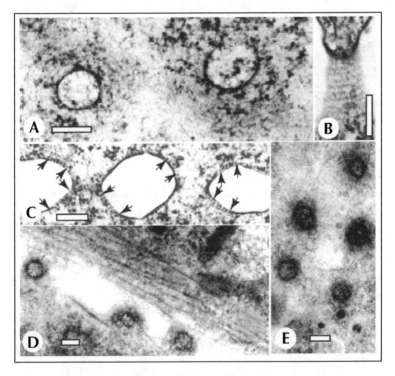

Figure 5. Transmission electron micrographs (TEMs) of cytoplasmic bridges in cleaving *V. carteri* embryos. A) In a thick section parallel to a cleavage furrow, each cytoplasmic bridge is cut transversely and appears as a dark ring. Note the concentric cortical striations surrounding each bridge. B) The electron dense coating on the inner surface of the plasma membrane is apparent at the upper end of this glancing section along the length of a cytoplasmic bridge, and the regular series of cortical striations is apparent at the lower end. C) A longitudinal thin section through two bridges. Note the electron-dense coating on the inner face of the plasma membrane throughout the bridge region. The arrows mark regions where the plasma membrane is cut slightly obliquely and the cortical striations appear as a series of regularly spaced short line segments. D) A thick section through the bridge region reveals the numerous cortical MTs that are found in the vicinity of all bridges. E) In a thick section through the bridge region of an inverting embryo it can be seen that the bridges appear to be linked to one another by a fibrous network. Bars: 200 nm. Reproduced from: Green KJ. J Cell Biol 1981; 91:756-769,[20] by copyright permission of the Rockefeller University Press.

of the cortical striations is unknown, but there is some evidence supporting the notion that they serve to strengthen the membrane in the bridge region.[20]

Of particular significance with respect to the role that the cytoplasmic bridges will play during inversion is the fact (mentioned earlier) that throughout embryogenesis the bridges are associated with numerous cortical MTs that originate in the BB region and run to the opposite end of the cell (Fig. 5D). During cleavage, additional MTs are often seen running into or through cytoplasmic bridges, at an angle to the main MT arrays, but such MTs are no longer seen during inversion. However, a second type of cytoskeletal element is seen in association with the bridges during inversion—a set of fibers of unknown composition that connect neighboring bridges (Fig. 5E). It has been postulated that these fibers account for the observation that the center-to-center spacing of the bridges is essentially constant during inversion, despite the considerable stresses that the bridge system appears to experience.[20] Although strands of ER have been seen within the cytoplasmic bridges of other species of *Volvox*,[1,7,9,11] none have been seen in the cytoplasmic bridges of *V. carteri* embryos.[21]

The Function of the Cytoplasmic-Bridge System in Cleaving Embryos

The most obvious function of the cytoplasmic-bridge system in a volvocacean embryo is to hold the embryo together. The best evidence that they actually do this is as follows: the cytoplasmic bridges of *V. carteri* embryos routinely break down shortly after inversion—but normally only after enough extracellular matrix (ECM) has been assembled to hold the cells together once the bridges are gone. However, if assembly of the ECM is blocked with a specific inhibitor, the bridges break down on schedule, and the juvenile then falls apart into a suspension of single cells.[22]

It has been postulated that the cytoplasmic-bridge system also provides the communication channels for signals that synchronize mitotic cycling in the cleaving embryo. Mitotic activities are so well synchronized in *V. carteri* that, for example, when one cell is in anaphase (which lasts only five minutes), all cells are in anaphase.[19] However, as appealing as it is to propose that such perfect mitotic synchrony is mediated by chemical signals passing through the cytoplasmic bridges, there is no direct evidence that this is the case.

The Function of the Cytoplasmic-Bridge System in Inverting Embryos

The role of the cytoplasmic-bridge system in synchronizing mitosis may be in doubt, but there can be little doubt that the bridge system plays a central role in inversion.[15,17,20] The bridge system that is formed during cleavage, and in which all the bridges of all cells are aligned and linked into a single, coherent network, remains intact during inversion. It constitutes the only structural framework against which the cells can exert force in order to turn the embryo inside out. And they do indeed exert force against it.

By the end of cleavage the cytoplasmic-bridge system links the cells at their widest points, just below the level of their nuclei (Fig. 6A). The only place where the neighboring cells are not linked by the cytoplasmic-bridge system in this way is across the phialopore, an opening at the anterior end of the embryo that is formed in very early cleavage, and that is the site where inversion begins. The outward curling of the cellular monolayer begins when cells near the phialopore execute two coordinated activities: (i) they change shape, forming long, narrow, MT-reinforced "stalks" at their outer (chloroplast) ends and (ii) they move inward individually relative to the cytoplasmic-bridge system, going as far as they can go in that direction. In combination, these two changes cause the cells to go from being linked at their widest points in the sub-nuclear region to being linked at their slender outermost tips. And that, in turn, forces the cellular monolayer to curl outward (Fig. 6B).

Subsequently the region of maximum curvature moves toward the posterior pole of the embryo as cells that have passed through the region of curvature withdraw their stalks and take on the appearance of a simple columnar epithelium, while other cells that were located further from the phialopore move into the region of curvature by changing shape and moving relative to the cytoplasmic-bridge system—just as the cells closer to the phialopore did earlier.

By this process, the embryo turns itself right-side-out in about 40 minutes, the BB/flagellar ends of all cells are brought to the outer surface, and all of the cells are now linked by the cytoplasmic-bridge

Figure 6, viewed on next page. Diagrammatic representation of the cellular mechanism of inversion as deduced from EM analysis of inverting embryos. A) Before inversion the cytoplasmic-bridge system (red line) links the cells (green outlines) at their widest points, just below the nuclei (blue). The opening at the top of the embryo is the phialopore, which was formed in early cleavage. B) Inversion begins when cells bordering the phialopore undergo two changes: (1) they extend long, thin, MT-reinforced stalks at their outer ends (outward-pointing arrows), and (2) they move inward individually, past the cytoplasmic-bridge system (inward-pointing arrows). As a result, these cells go from being linked at their widest points to being linked at their narrowest, outermost tips, and that forces the cell sheet to bend outward. Inversion will continue as cells further from the phialopore change shape and move, entering the bend region, while cells that have passed through the bend region change to a simple columnar shape.

Figure 7, viewed on next page. Inversion arrest in an *invA*⁻ mutant embryo. A) The mutant appears to begin inverting quite normally, but it then stops abruptly after the lips of cells bordering the phialopore have just begun to curl backward. B) In a closer view of the cells circled in (A) the reason for the inversion arrest is obvious: the cells have changed shape, but they have failed to move relative to the bridges that link them to their neighbors (arrowheads). Reprinted from: Nishii I et al. Cell 113:743-753,[26] ©2003 with permission from Elsevier.

Figure 8, viewed on next page. Immunofluorescent localization of the InvA kinesin. Tubulin is labeled green, InvA red, and DNA blue. Each arrowhead marks the inner end of a cell whose outermost tip is marked with a small arrow. A) In a preinversion embryo, InvA is located in a narrow band near the equator of each cell, the region where earlier EM studies have shown that the cytoplasmic-bridge system is located before inversion (compare with Fig. 6A). B) During early inversion the InvA label is found progressively closer to the outer tips of the cells in cells that are progressively closer to the phialopore, and it is located at the outermost tips of cells in the bend region. Again, this corresponds with the location of the cytoplasmic bridge system in early inversion (compare with Fig. 6B). Reprinted from: Nishii I et al. Cell 113:743-753,[26] ©2003 with permission from Elsevier.

system at their (now-innermost) chloroplast ends.[15,17,20] Shortly after inversion a primary layer of ECM is formed that encircles the embryo and holds all of the cells in a fixed orientation,[22] and then the cytoplasmic bridges disappear by a process that has yet to be analyzed.

Inversion is a hallmark of the family Volvocaceae. But how general is the cellular mechanism of inversion that has been described for *V. carteri*? The first EM-level examination of inversion was performed by Jeremy Pickett-Heaps, studying *V. tertius*.[11] He reported that although cells in cleaving embryos had cytoplasmic bridges at the sub-nuclear level, by the end of inversion they were connected only at the chloroplast ends. He inferred (but provided no evidence) that the cells were initially connected at both the sub-nuclear and chloroplast regions, but that during inversion the sub-nuclear bridges were broken, leaving the bridges at the chloroplast ends of the cells to serve as "hinges" that would allow the cells to swing outward. Subsequently, investigators studying *Pandorina morum*[14] and *Eudorina elegans*[13,16] made observations and drew conclusions very similar to those of Pickett-Heaps. However, none of the micrographs in any of those four reports shows cells linked simultaneously in the sub-nuclear and chloroplast regions: the only bridges shown in cleaving embryos are sub-nuclear, and bridges at the chloroplast ends of the cells are shown only in post-inversion cells.[11,13,14,16] Thus, all published micrographs are consistent with the concept that in *P. morum, E. elegans* and *V. tertius*—as in *V. carteri*—each cell possesses a single band of cytoplasmic bridges that is located in the sub-nuclear region during cleavage, but that ends up at the chloroplast end of the cells by the end of inversion. This, in turn, is consistent with the idea that the structure of the cytoplasmic-bridge system, and its function during inversion, is similar in all volvocacean embryos.

An Inversion Motor in the Cytoplasmic Bridges

The development of a transposon-tagging system for *V. carteri*[23] provided an effective way of cloning developmentally important genes.[24,25] The first inversion-specific gene tagged with this transposon (which was named *Jordan* because it jumped so well) was *invA*.[26] In mutants in which *invA* has been inactivated by a *Jordan* insertion, inversion appears to start in the usual way, but then it stops abruptly as soon as the lips of cells flanking the phialopore have curled outward

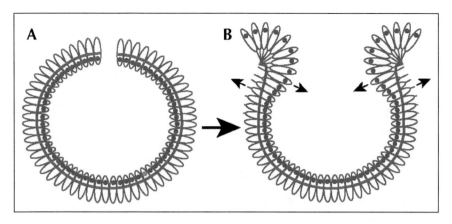

Figure 6. Please see figure legend on previous page.

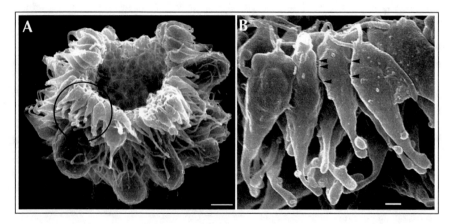

Figure 7. Please see figure legend on previous page.

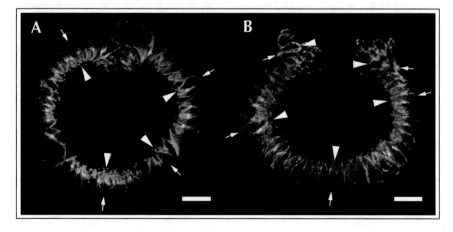

Figure 8. Please see figure legend on previous page.

partway (Fig. 7A). Although the mutant cells change shape as wild-type cells do during inversion, they fail to move with respect to the cytoplasmic-bridge system (Fig. 7B). This mobility defect and the resulting inversion block can both be cured in either of two ways: (i) by excision of *Jordan* from the *invA* locus, or (ii) by transformation of the mutant strain with a wild-type *invA* transgene.[26] These findings led to the conclusion that embryonic cells require InvA (the protein product of the *invA* gene) to move with respect to the cytoplasmic-bridge system, in order to complete the inversion process.

Sequencing revealed that InvA was a novel type of kinesin. Although the amino acid sequence of its motor domain is rather similar to that of the KIF4/chromokinesin subfamily of kinesins, outside of the motor domain it has no significant similarity to any of the hundreds of kinesins from other organisms that have been characterized to date (with the exception of an InvA orthologue that we have cloned from a closely related alga, as discussed below).

Immunolocalization studies revealed that InvA is located in the cytoplasmic-bridge region throughout embryogenesis (Figs. 8,9). Its sequence indicated that it is most likely a plus-end-directed microtubule motor,[26] and the numerous cortical MTs that run past each cytoplasmic bridge (Fig. 5D) are known to be oriented with their minus ends proximal to the basal bodies and their plus ends toward the opposite end of the cell. These facts led to the following simple model for the way that InvA functions during inversion: When InvA is activated during inversion it attempts to "walk" along the adjacent cortical MTs toward their distal ends. But InvA is not free to move, because it is anchored in the cytoplasmic bridges. Therefore, the force it exerts on the MTs results in the MTs—and with them the entire cell—moving in the opposite direction, past the cytoplasmic-bridge system (see Fig. 10), thereby generating the curvature that turns the cell sheet inside out (as shown in Fig, 6B).

Although no kinesins orthologous to InvA have been found outside of the green algae, an *invA* orthologue was cloned from *Chlamydomonas reinhardtii*, the closest unicellular relative of *V. carteri*.[26] It encodes a kinesin that is 90% identical to InvA in the motor domain, and 82% identical to it overall. More recently we have used an affinity-purified anti-InvA antibody to probe Western blots, and have detected a cross-reacting protein of the same size as InvA in protein extracts of *Gonium pectorale, Pandorina morum, Eudorina elegans* and *Pleodorina californica,* all of which are volvocaceans that are intermediate between *Chlamydomonas* and *Volvox* in size and complexity (I. Nishii, to be published). Our working hypothesis is that these InvA homologues all play a similar role in the inversion of the corresponding embryos. Attempts to determine the function of the InvA orthologue in unicellular *Chlamydomonas* are in progress.

The Evolutionary Origins of Cytoplasmic Bridges That Persist in the Adult

As noted earlier, in about half of the species of *Volvox*—and in all of the smaller and simpler volvocaceans—the cytoplasmic bridges that are present during embryonic stages break down after inversion, and are absent from adults. That clearly is the ancestral pattern. However, in certain species of *Volvox* the cytoplasmic bridges persist throughout the life of the adult. Smith believed that such bridges were so significant that in his 1944 taxonomic treatment of the genus *Volvox* he proposed that the genus be subdivided into four sections that differed with respect to the nature of the cytoplasmic bridges in the adult.[27] In the section Merillosphaera he placed eight *Volvox* species (including *V. carteri*) that lack any bridges in the adult. The section Copelandosphaera was reserved for *V. dissipatrix*, in which adult cells are connected by "delicate cytoplasmic strands smaller in diameter than flagella". The section Janetosphaera was reserved for *V. aureus*, in which adult cells are connected by "cytoplasmic strands approximately the same diameter as flagella". The remaining eight species, in which the cytoplasmic bridges in adults are so broad that the cells appear star-shaped when viewed from their flagellar ends, were placed by Smith in the section Euvolvox.

Smith's decision to sort *Volvox* species with persistent cytoplasmic bridges into three separate sections of the genus has been validated recently by a molecular phylogenetic analysis of 59 volvocine algae,[28] which indicates that the retention of cytoplasmic bridges in the adult has evolved independently in each of those three sections of the genus (Fig. 11).

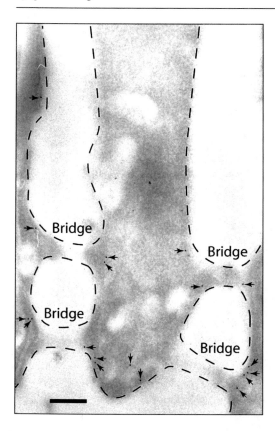

Figure 9. Immunogold localization of the InvA kinesin at the EM level. A section through the outermost tips of three cells that are connected by four cytoplasmic bridges (outlined with dashed lines) in the bend region of an immunogold-labeled embryo. Arrows point to individual immunogold particles. Significant numbers of gold particles were found only in the immediate vicinity of the cytoplasmic bridges. Reprinted from: Nishii I et al. Cell 113:743-753,[26] ©2003 with permission from Elsevier.

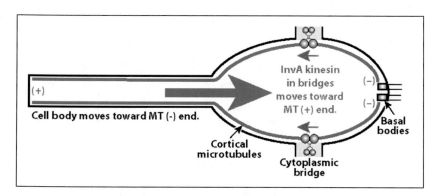

Figure 10. A model indicating how InvA is thought to drive inversion. InvA is a plus-end-directed MT motor that is firmly attached to the cytoplasmic bridges, which are all connected in turn to the rest of the cytoplasmic-bridge system. When InvA is activated during inversion, it attempts to walk to the plus ends of the cortical MTs that are at the chloroplast end of the cell. But neither InvA nor the bridges in which it is activated are free to move away from the rest of the cytoplasmic-bridge system. Therefore, the force exerted on the MTs by InvA causes the MTs—and thus the cell in which they are embedded—to move inward, past the cytoplasmic-bridge system. This, in turn generates the bend region required to turn the cell sheet inside out, as diagrammed in Figure 7. Reprinted from: Nishii I et al. Cell 113:743-753,[26] ©2003 with permission from Elsevier.

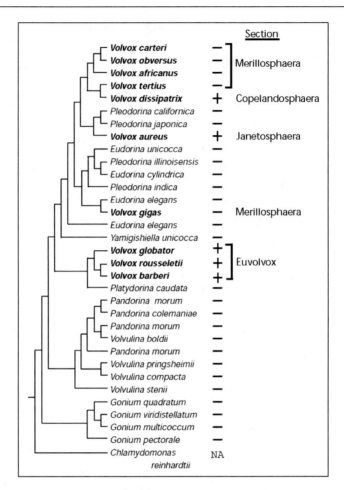

Figure 11. Phylogenetic relationships among 33 taxa of volvocine algae. Adapted (with simplifications approved by the author) from the minimum-evolution tree derived by Nozaki[30] from the sequences of five chloroplast genes. Symbols to the right of the species names: +, cytoplasmic bridges present in adults; –, no cytoplasmic bridges present in adults; NA, not applicable. Right-hand column: the sections of the genus *Volvox* in which Smith[29] placed the various *Volvox* species that are included in this analysis. Note that retention of cytoplasmic bridges in the adult is a trait that appears to have evolved independently in three lineages that correspond exactly to the three sections of the genus *Volvox* into which Smith distributed such species 60 years ago.

Structure and Formation of Persistent Cytoplasmic Bridges in *Volvox*, Section Euvolvox

Nearly a century and a half after Eherenburg discovered the cytoplasmic bridges of *V. globator*, Ikushima and Maruyama first examined them with the EM,* and found that the cytoplasmic bridges of embryos and adults were extremely different in number, size and structure.[9] In cleaving embryos, the bridges were about 100 nm in diameter and so numerous that multiple bridges were seen in every thin section. In contrast, the adult bridges were much greater in diameter, but much less

---

*They called the alga that they found in a local pond a "*globator* type," because they were not exactly certain what species it belonged to, but it clearly was a member of Smith's section Euvolvox, and for simplicity we refer to it as *V. globator*.

numerous. Each adult cell had only four to six bridges connecting it to its neighbors, but those bridges were so broad that they made the cells appear star-shaped. Of particular interest was the structure of the "granule" that had long been known to be present in the cytoplasmic bridges of all members of the section Euvolvox, mid-way between the cells connected by that bridge.[4,5] The granule, they discovered, was actually a disk-like structure that was different from anything seen in the embryonic bridges of any other volvocacean species or in the adult cytoplasmic bridges of *V. aureus* (section Janetosphaera—the only other species of *Volvox* with adult cytoplasmic bridges that had been examined).[1,7] They named this structure the "medial body". Each medial body was ~200 nm thick, and consisted of two 40-nm-thick electron-dense discs separated by a less-dense region ~120 nm thick. Although the medial bodies in their published micrographs ranged in diameter from about 300 to 1,000 nm, each of them fully spanned the region of the bridge in which it was located. In the surprisingly common cases in which two or three medial bodies were located edge-to-edge, the cytoplasmic bridge was split into that number of branches on each side of the medial body region, with one medial body spanning each branch of the bridge. At first glance the medial bodies appeared to occlude the bridges, but on closer examination it was found that they were traversed by numerous "canaliculi"[9] (probably ER tubules).

This study raised—but did not answer—the question of how the many thin cytoplasmic bridges of the embryo might be related developmentally to the much smaller number of broad bridges that connect adult somatic cells. Nor did it provide any information about how the medial bodies are formed, or about the structure of the cytoplasmic bridges that are known to connect adult somatic cells to developing embryos in the section Euvolvox.[5] One of us (Hoops) made some observations relative to those issues during an ultrastructural analysis of the flagellar apparatus in another species in the section Euvolvox, *V. rousseletii*,[29] which had previously been the object of several developmental studies.[5,30,31] His previously unpublished observations regarding *V. rousseletii* cytoplasmic bridges will be reviewed next.

There are at least three types of cytoplasmic bridges in *V. rousseletii*: "E-E" bridges linking embryonic cells to each other, "S-S" bridges linking adult somatic cells to each other, and "S-E" bridges linking adult somatic cells to developing embryos (Fig. 12).

In certain respects, the E-E bridges of *V. rousseletii* resemble those of *V. carteri* embryos that were discussed above: they are about 170 nm in diameter, occur in regularly spaced clusters (Figs. 12, 13A,B) that are located at the sub-nuclear level in cleaving embryos (Fig. 12, upper set of E-E bridges), but at the chloroplast ends of the cells by the time inversion has been completed (Fig. 13A). They differ from the *V. carteri* bridges in one important respect, however: an ER tubule traverses each *V. rousseletii* E-E bridge (Fig. 13A,B), whereas ER has not yet been detected in any of the cytoplasmic bridges of *V. carteri* embryos.[20] (We note with great interest, however, that ER tubules also traverse the cytoplasmic bridges in *V. aureus*,[1,7] a member of the section Janetosphaera, in which persistent cytoplasmic bridges apparently evolved independently).

During embryonic stages, ribosomes are generally absent from much of the ER tubule that is located within a bridge (Fig. 13A, arrow), even though these same tubules are studded heavily with ribosomes in the cell body. Moreover, a little later—by the time the spheroid is beginning to accumulate extracellular matrix and expand—ribosomes are entirely absent from the part of the ER tubule located within the bridge (Fig. 13C). At this time the ER tubule within the bridge also becomes markedly swollen, and electron-dense material that is different from anything seen in the cell body begins to accumulate between the ER membrane and the bridge plasma membrane (Fig. 13D). We believe that this accumulation of electron-dense material in the bridge represents an early step in medial body formation, but so far we have not observed other intermediate stages of medial body construction. Nor have we detected any differences between S-S bridges and prospective S-E bridges during these early stages of medial body formation.

The medial bodies in S-S bridges between two adult somatic cells are symmetrical, with two flat ~40 nm-thick electron dense disks separated by a less dense disc of about the same thickness (Fig. 14A). They range fairly widely on either side of the average diameter of ~250 nm, but (as in *V. globator*) whatever the diameter of a medial body, it appears to completely span the part of the cytoplasmic bridge where it is located. ER elements continue to traverse the bridges after the medial bodies have

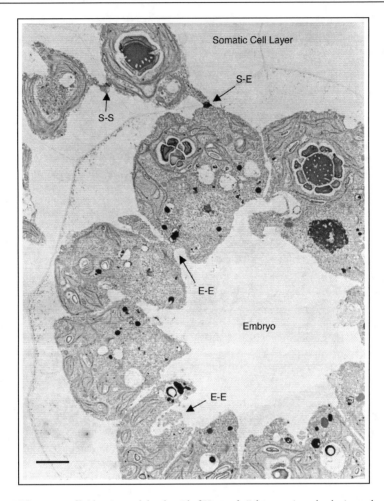

Figure 12. Three types of bridges in an adult spheroid of *V. rousseletii* that contains a developing embryo. S-S, a bridge connecting two somatic cells of the adult; S-E, a bridge connecting an adult somatic cell to a cell of the embryo; E-E, bridges connecting cells of the developing embryo to one another (in the upper example, "E-E" identifies two bridges that have been cut longitudinally, whereas in the lower one it identifies a cluster of bridges that have been cut transversely.). Bar: 2 μm.

appeared (Fig. 14A, arrows). In a longitudinal section through the edge of a bridge, some of those ER elements are seen to be branched and convoluted (Fig. 14B, arrow), and in transverse sections (not shown) the ER elements are often seen to be arranged in a ring near the bridge periphery. Whereas the ER elements within the bridges appeared to be dilated during the early-post-embryonic stage, by this time the portions within the bridge are narrower, appearing to be pinched by the medial body.

The medial bodies of S-E bridges also have two electron dense layers surrounding a more electron lucent layer (Fig. 14C,D), but they are hourglass-shaped rather than disk-shaped as are the medial bodies between somatic cells. The minimum diameters of the S-E medial bodies are only slightly larger than the S-S medial bodies. However, the plate on the side towards the somatic cell is convex and both it and the electron lucent layer are slightly thicker than the comparable structures between somatic cells. More importantly, the dense structure on the side toward the developing embryo is about five times as thick as a disk in an S-S bridge (and about 4 times

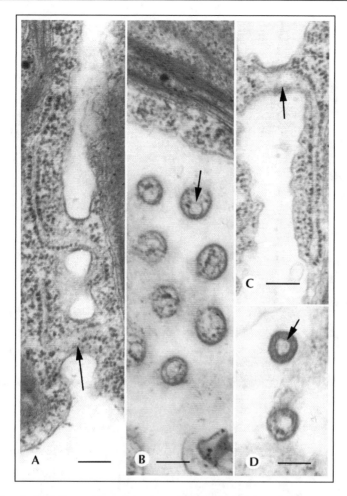

Figure 13. Cytoplasmic bridges in *V. rousseletii* embryos and juveniles. A) Longitudinal section through bridges connecting cells of a recently inverted embryo. Note the ribosome-studded ER (arrow) traversing at least two of these bridges. B) Transverse sections of bridges similar to those in (A). Each bridge contains an ER tubule (arrow). C) Longitudinal section of a bridge in a juvenile in which the cells have begun to move apart, owing to the deposition of ECM between them. The ER tubule that is traversing this bridge is studded with ribosomes in the cell to the right, but is swollen and free of ribosomes within the bridge. Note the electron-dense material that has begun to accumulate in the space between this ER tubule and the plasma membrane (arrow). D) Bridges at the same stage as those in (C), but sectioned transversely. In the upper bridge, which is sectioned through its midpoint, electron-dense material completely fills the space between the ER and the plasma membrane (arrow). Bars: 200 nm.

thicker than the disk on the somatic cell side of the same medial body). It has a highly crenulated outer edge (Fig. 14C,D). Thus the S-E medial body is highly asymmetric, unlike the S-S medial body. Similar numbers of ER elements are seen on both sides of the medial body in an S-E bridge (Fig. 14C,D, arrows), and we believe that some or all of these ER strands extend through it. However, owing to the thickness and opacity of the medial body, we have not been able to establish with certainty that they do.

When either S-S or S-E medial bodies are initially formed, and the cell bodies are still close together, the cytoplasmic bridges are widest next to the cell bodies and taper from both sides

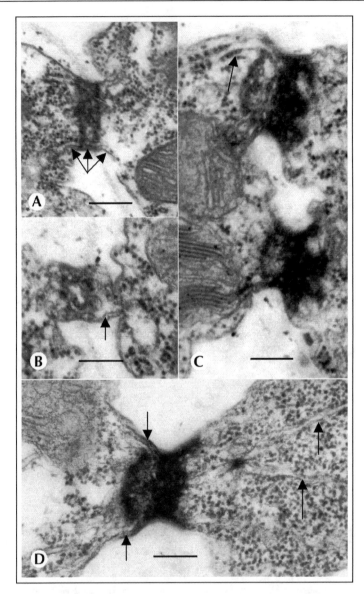

Figure 14. Cytoplasmic bridges in *V. rousseletii* young adults (early in the period of ECM deposition and spheroid expansion). A) an S-S bridge connecting two young adult somatic cells. A symmetrical medial body, consisting of two similar electron-dense layers separated by an less-dense intermediate layer, is present in the middle of the bridge. Note ER tubules that traverse the bridge and the medial body (arrows). B) A section near the edge of a medial body like the one in (A) reveals a convoluted and branching ER element traversing the bridge (arrow). C) Two S-E bridges connecting the somatic cell on the left to a cell of an embryo on the right. Note that in distinction to the medial body in the S-S bridge, these medial bodies are very asymmetric, with the electron-dense structure on the side toward the embryo much thicker and more irregular than the part on the somatic side. D) Another S-E bridge joining a somatic cell on the left with a cell of the embryo on the right . Note the ER elements that appear to enter the medial body from the somatic cell and lead away from the medial body on the embryonic side (arrows). Bars: 200 nm.

toward the medial body, which lies in the narrowest part of the bridge at that stage (Fig. 14). However, as adjacent cell bodies are moved apart by the process of ECM deposition and spheroid expansion, the cytoplasmic bridges become drawn out into increasingly narrow strands, leaving the medial body region to become the widest part of the bridge.[5,30] As in *V. globator*, whenever two *V. rousseletii* medial bodies are found side-by-side, the bridge is bifurcated in that region (Fig. 14C).

What Is the Relationship between Embryonic and Adult Cytoplasmic Bridges?

The above studies do not address the question of how the four to six broad bridges that are present in *V. rousseletii* adult somatic cells are related developmentally to the many, thin cytoplasmic bridges present in an embryo. Because there is no evidence that new cytoplasmic bridges can be formed in any volvocacean after cleavage and inversion, we believe that the medial body-containing bridges of the adult almost certainly develop from the more numerous, but smaller bridges found in the embryo. How might this happen? Two possibilities suggest themselves: either a few of the embryonic bridges might enlarge and become transformed into adult bridges while the rest of the bridges break down, or each adult bridge might be formed by fusion of several embryonic bridges (followed by additional changes in structure, of course).

We strongly prefer the latter possibility. Among our reasons are the following: (i) Whereas each bridge in the embryo is traversed by a single ER tubule (Fig. 13), each adult bridge is traversed by multiple ER tubules (Fig. 14). We find it easier to visualize how multiple ER elements traversing each bridge might result from side-to-side fusion of several small bridges that had one tubule each than by the de novo development of multiple tubules within an existing bridge. (ii) Pocock[5] was the first to observe that in *V. rousseletii* adults many cytoplasmic bridges are seen that branch in the middle and connect one cell to two others. Such Y-shaped bridges have since been documented in all published light micrographs that show an adult Euvolvox spheroid at a magnification and focal plane such that numerous adult bridges are visible.[4,5,9,30,32] A Y-shaped bridge could not be formed by enlargement of a single embryonic bridge, because each embryonic bridge is unbranched and always connects only two cells—sister cells. However, a branching bridge could be formed by fusion of two bridges that were located near one another on one cell, a, but that connected cell a to two different sister cells, b, and c (because those bridges had been formed in two different division cycles). We believe that the regular presence of Y-shaped bridges in Euvolvox adults provides the strongest evidence supporting the hypothesis that adult cytoplasmic bridges are formed by fusion of embryonic bridges.

A very different kind of support for the notion that such a fusion of bridges might be possible comes from the observation that treatment of a fully cleaved *V. carteri* embryo with concanavalin A can cause adjacent embryonic bridges to fuse.[21]

The Functions of Persistent Cytoplasmic Bridges

In the 1830s, Ehrenburg considered *Volvox globator* to be a "social animal" composed of thousands of sentient individuals that cooperate to establish and achieve common goals—such as swimming toward the light.[3] When he discovered that the individual "animals" (now known as somatic cells) of a *V. globator* spheroid were all interconnected by "filamentous tubes" (a.k.a., cytoplasmic bridges) he had no hesitation about assigning those tubes a function. To him it seemed obvious that they must be the channels by which the individual animals communicated with each other to set and achieve their common goals. The idea that the cytoplasmic bridges probably functioned for intercellular communication and coordination of phototactic behavior persisted for decades after *V. globator* had come to be thought of as a multicellular alga, rather than as a social animal. But Meyer cast serious doubt on this idea when he reported in 1896 that *Volvox tertius* adults lack any intercellular connections, but are just as well coordinated, and swim toward the light just as well as *V. globator* adults do.[4] A decade later Mast used microsurgery to show that even in species such as *V. globator* in which the cytoplasmic bridges persist in the adult, those bridges are not required for phototactic coordination.[33] No evidence inconsistent with Mast's conclusion has ever been published. Nevertheless, the notion that the bridges might serve for communication and phototactic coordination continues to raise its head from time to time (e.g., see ref. 9).

Another idea that has surfaced occasionally is that persistent bridges are required to hold adult Euvolvox cells in fixed positions. Pocock provided evidence that this is not the case: she reported that the bridges in *V. rousseletii* adults tended to break down under adverse environmental conditions, but that a spheroid remained intact even after all cytoplasmic bridges had disappeared from an entire quadrant of it.[5] Clearly, in the section Euvolvox, as in the section Merillosphaera, once an ECM has been assembled it is able to hold the cells in place in the absence of cytoplasmic bridges.

The third function for persistent bridges that has been proposed—and the one that we find most attractive—is that they serve to channel materials from the somatic cells to the developing embryos. This idea was first proposed by Ferdinand Cohn in 1875 in one of the most important papers on *Volvox* development of the 19th century.[34] Much more recently it has reappeared as one form of the source–sink hypothesis that Bell proposed to account for the evolution of sterile somatic cells in *Volvox*.[35] If such an asymmetric flow of materials from somatic to germ cells were proven to take place through the cytoplasmic bridges in members of the section Euvolvox, it would be analogous (but clearly not homologous) to the asymmetric flow of materials through plasmodesmata in higher plants (see ref. 36 for a review), or the flow of materials from nurse cells to developing oocytes in *Drosophila* and other dipterans (see ref. 37 for a review).

In evaluating this material-flow hypothesis, it is important to consider the following important fact: In all three of the *Volvox* lineages in which persistent cytoplasmic bridges have evolved independently (Fig. 11), they have coevolved with a novel form of asexual reproduction. In all species of *Volvox* that lack bridges in the adult—plus all of the smaller volvocaceans, and hundreds of other species of green algae—asexual reproduction involves "multiple fission", rather than the more familiar binary fission. In multiple fission, an asexual reproductive cell (or "gonidium") first grows 2^n-fold and then executes n rapid division in the absence of any further growth, thereby forming 2^n daughter cells.[21] This clearly is the ancestral pattern of cell division in the volvocine algae. However, in all three sections of the genus *Volvox* in which cytoplasmic bridges are present in the adult, a different pattern is observed: Gonidia in species of *Volvox* with persistent bridges begin to divide while they are not a great deal larger than somatic cells, and then they grow before they divide again, nearly doubling in mass between successive divisions. Although this resembles the pattern of growth and development characteristic of the land plants and animals with which most people are more familiar, it clearly is a novel, evolutionarily derived trait in the volvocine algae. And—it bears repeating—this same deviation from the ancestral pattern appears to have coevolved with persistent cytoplasmic bridges in three independent lineages. We interpret this to mean that the two phenomena are linked in terms of their selective value: that is, that the new pattern of embryonic growth provides a selective advantage only in the presence of persistent cytoplasmic bridges through which materials can flow, and vice versa.

The asymmetric structure of the medial body in an S-E bridge (Fig. 14C,D) is consistent with the hypothesis that the bridges and/or their medial bodies may function asymmetrically, permitting unidirectional flow of nutrients.

The rate of growth of a Euvolvox embryo can be prodigious; it can nearly double its mass, and then divide, every hour for 12-15 hours. This is a growth rate considerably greater than even lab strains of yeast can achieve, and it may well be a eukaryotic record worthy of Guinness-Book recognition. Considering this rather astonishing growth rate, Ferdinand Cohn stated matter-of-factly, with respect to *V. globator*, that "…it becomes obvious that the substances produced in the sum of the somatic cells during their life (carbohydrates, protoplasm, chlorophyll) aid the eight reproductive cells, so that the young colonies are not exclusively nurtured by the mother cells, but also by the united efforts of the whole cell family".[34] Mary Agard Pocock, in contrast, expressed the opposite opinion—that in *V. rousseletii*, "…the bulk of the necessary food, which must be considerable, is manufactured by the daughter [the embryo] itself". However—astute observer and objective scientist that she was—she went on to provide what remains about the best evidence indicating that the somatic cells do export materials to the developing embryos through the cytoplasmic bridges. She noted that that the somatic cells immediately surrounding a developing embryo usually are "nearly devoid of starch", which she took to indicate that at least some of the nutrient required for growth of the embryo came to it from somatic cells, through the cytoplasmic bridges. She also pointed out that *V. rousseletii* somatic cells always grow until the gonidia begin to divide, at which time they stop growing and then later decrease in size, "particularly in the posterior region" of the spheroid where the developing embryos are located.[5] She also noted that a *V. rousseletii* somatic cell that had suffered accidental severance of all of its cytoplasmic bridges was larger than its neighbors, and suggested that this probably was because it was unable to export materials through the cytoplasmic-bridge system—thereby implying once more that somatic cells normally do export such materials.

Altogether, then, there is a fair amount of indirect evidence compatible with the hypothesis that in *Volvox* species that have persistent cytoplasmic bridges, material may flow through the bridges from somatic cells ("the source") to gonidia and/or embryos ("the sink").[35] But as yet there is no direct evidence that this is the case. However, with all of the methods that have been developed for labeling various cytoplasmic components and monitoring their subsequent distribution, direct tests of this hypothesis should now be feasible.

Assuming that material does flow from somatic cells to embryos, what path would it be most likely to take through the cytoplasmic bridges? We assume that the ER would be involved, because adult cytoplasmic bridges that evolved independently in two sections of the genus *Volvox*—the sections Janetosphaera and Euvolvox—are both traversed by ER (unfortunately, EM studies of the cytoplasmic bridges in the section Copelandosphaera have not been reported to date.). One possibility is that materials are moved in the space between the ER and the plasma membrane by ER-associated molecular motors, as in the movement of certain cargos through plasmodesmata (reviewed in ref. 36). In the micrographs in Figure 14, the space between the ER and the plasma membrane is filled with electron-dense material that would appear to be a barrier to transport; but we have no way to rule out the possibility that this material is more permeable than it appears. It may even be that its apparent density is at least partially a fixation artifact (much as the often-observed "neck constriction" in plasmodesmata is thought to be at least partially an artifact of tissue preparation and fixation).[38] So it is possible that the medial body is more open in living cells than it appears to be in electron micrographs. Nevertheless, we believe that the morphology of the medial bodies favors the notion that material flow occurs via the ER lumen. Higher plants provide a precedent for such transport, because although much of the transport through plasmodesmata is believed to occur outside the ER (as mentioned above), there are some cases where the movement of materials appears to be through the "dilated desmotubules" (modified ER) instead.[36]

If the ER lumen were to provide the conduits for material flow, what role might the medial bodies play? We note the lack of precedent for motor proteins within the ER cisternae and wonder if the medial bodies might supply contractile activity from the exterior, causing something resembling peristaltic waves in the ER? As we noted earlier, in fixed specimens the ER tubules of adult bridges look as though they are being pinched by the medial bodies; but in life might such pinching be a dynamic process, promoting unidirectional flow through the ER lumen? Immunocytology of medial bodies with antibodies directed against a variety of cytoskeletal and motor proteins could provide a step toward learning whether something of that sort might be possible.

References

1. Bisalputra T, Stein JR. The development of cytoplasmic bridges in Volvox aureus. Can J Bot 1966; 44:1697-1702.
2. van Leeuwenhoek A. Concerning the worms in sheeps, livers, gnats and animalicula in the excrement of frogs. Phil Trans R Soc Lond 1700; 22:509-518.
3. Ehrenberg CG. Zur Erkenntniss der Organisation in der Richtung des kleinsten Raumes. Abhandl Kgl Akad Wiss Berlin 1832; 1-154.
4. Meyer A. Die Plasmaverbindung und die Membranen von Volvox. Bot Zeitg 1896; 54:187-217.
5. Pocock MA. Volvox in South Africa. Ann South Afr Mus 1933; 16:523-625.
6. Metzner J. A morphological and cytological study of a new form of Volvox-I. Bull Torrey Bot Club 1945; 72:86-113.

7. Dolzmann R, Dolzmann P. Untersuchungen über die Feinstruktur und die Funktion der Plasmodesmen von Volvox aureus. Planta 1964; 61:332-345.
8. Stein JR. On cytoplasmic strands in Gonium pectorale (Volvocales). J Phycol 1965; 1:1-5.
9. Ikushima N, Maruyama S. Protoplasmic connections in Volvox. J Protozool 1968; 15:136-140.
10. Deason TR, Darden Jr WH, Ely S. The development of the sperm packets of the M5 strain of Volvox aureus. J Ultrastruct Res 1969; 26:85-94.
11. Pickett-Heaps JD. Some ultrastructural features of Volvox, with particular reference to the phenomenon of inversion. Planta 1970; 90:174-190.
12. Marchant HJ. Plasmodesmata in algae and fungi. In: Gunning BES, Robards AW, eds. Intercellular Communication in Plants. Berlin: Springer–Verlag,1976:59-80.
13. Marchant HJ. Colony formation and inversion in the green alga Eudorina elegans. Protoplasma 1977; 93:325-339.
14. Fulton AB. Colony development in Pandorina morum. II. Colony morphogenesis and formation of the extracellular matrix. Dev Biol 1978; 64:236-251.
15. Viamontes GI, Kirk DL. Cell shape changes and the mechanism of inversion in Volvox. J Cell Biol 1977; 75:719-730.
16. Gottlieb B, Goldstein ME. Colony development in Eudorina elegans (Chlorophyta, Volvocales). J Phycol 1977; 13:358-364.
17. Viamontes GI, Fochtmann LJ, Kirk DL. Morphogenesis in Volvox: Analysis of critical variables. Cell 1979; 17:537-550.
18. Birchem R, Kochert G. Mitosis and cytokinesis in androgonidia of Volvox carteri f. weismannia. Protoplasma 1979; 100:1-12.
19. Green KJ, Kirk DL. Cleavage patterns, cell lineages, and development of a cytoplasmic bridge system in Volvox embryos. J Cell Biol 1981; 91:743-755.
20. Green KJ, Viamontes GI, Kirk DL. Mechanism of formation, ultrastructure and function of the cytoplasmic bridge system during morphogenesis in Volvox. J Cell Biol 1981; 91:756-769.
21. Kirk DL. Volvox: Molecular-genetic origins of multicellularity and cellular differentiation. Cambridge and New York: Cambridge University Press, 1998.
22. Hallman A, Kirk DL. The developmentally regulated ECM glycoprotein ISG plays an essential role in organizing the ECM and orienting the cells of Volvox. J Cell Sci 2000; 113:4605-4617.
23. Miller SM, Schmitt R, Kirk DL. Jordan, an active Volvox transposable element similar to higher plant transposons. Plant Cell 1993; 5:1125-1138.
24. Kirk MM, Stark K, Miller SM et al. regA, a Volvox gene that plays a central role in germ-soma differentiation, encodes a novel regulatory protein. Development 1999; 126:639-647.
25. Miller SM, Kirk DL. glsA, a Volvox gene required for asymmetric division and germ cell specification, encodes a chaperone-like protein. Development 126:649-658.
26. Nishii I, Ogihara S, Kirk DL. A kinesin, InvA, plays an essential role in Volvox morphogenesis. Cell 2003; 113:743-753.
27. Smith GM. A comparative study of the species of Volvox. Trans Amer Microsc Soc 1944; 63:265-310.
28. Nozaki H. Origin and evolution of the genera Pleodorina and Volvox (Volvocales). Biologia (Bratislava) 2003; 58:425-431.
29. Hoops HJ. Somatic cell flagellar apparatuses in two species of Volvox (Chlorophyceae). J Phycol 1984; 20:20-27.
30. McCracken MD, Starr RC. Induction and development of reproductive cells in the K-32 strains of Volvox rousseletii. Arch Protistenkd 1970; 112:262-282.
31. McCracken MD, Barcellona WJ. Ultrastructure of sheath synthesis in Volvox rousseletii. Cytobios 1981; 32:179-187.
32. Rich F, Pocock MA. Observations on the genus Volvox in Africa. Ann South Afr Mus 1933; 16:427-471.
33. Mast SO. Light reactions in lower organisms. II. Volvox globator. J Comp Neurol Psychol 1907; 17:99-180.
34. Cohn F. Die Entwickelungsgeschichte der Gattung Volvox. Beitr Biol Pfl 1875; 1:93-115.
35. Bell G. The origin and early evolution of germ cells as illustrated by the Volvocales. In: Halvorson HO, Monroy A, eds. The Origin and Evolution of Sex. New York: Alan R Liss, 1985:221-256.
36. Heinlein M, Epel BL. Macromolecular transport and signaling through plasmodesmata. Int Rev Cytol 2004; 235:93-164.
37. Mahajan-Miklos S, Cooley L. Intercellular cytoplasm transport during Drosophila oogenesis. Dev Biol 1994; 165:336-351.
38. Radford JE, Vesk M, Overall RL. Callose deposition at plasmodesmata. Protoplasma 1998; 210:30-37.

SECTION IV
Fungal Cells

CHAPTER 5

Vegetative Hyphal Fusion in Filamentous Fungi

Nick D. Read* and M. Gabriela Roca

Abstract

The formation of channels between fungal hyphae by self fusion is a defining feature of filamentous fungi and results in the fungal colony being a complex interconnected network of hyphae. During the vegetative phase hyphal fusions are commonly formed during colony establishment by specialized conidial anastomosis tubes (CATs) and then later by specialized fusion hyphae in the mature colony. CAT induction, homing and fusion in *Neurospora crassa* provides an excellent model in which to study the process of vegetative hyphal fusion because it is simple and experimentally very amenable. Various mutants compromised in hyphal fusion have been isolated and characterized. Although the self-signalling ligand(s) involved in CAT induction and homing has/have not been identified, MAP kinase signalling is downstream of the initial ligand-receptor interaction(s), and has features in common with MAP kinase signalling during mating cell interactions in the budding yeast and during fungal infection structure (appressorium) formation. Hyphal fusion also resembles yeast cell mating and appressorium formation in other ways. Vegetative hyphal fusion between hyphae of different genotypes (nonself fusion) usually results in a form of programmed cell death which normally prevents heterokaryons from developing further. This process in *N. crassa* is controlled by heterokaryon incompatibility (*het*) loci. Understanding hyphal fusion in the model fungus, *N. crassa*, provides a paradigm for self-signalling mechanisms in eukaryotic microbes and might also provide a model for understanding somatic cell fusion in other eukaryotic species.

Introduction

Cell fusion occurs at various stages and serves numerous functions during the vegetative and sexual phases of the filamentous fungal life cycle.[1] This review is concerned with vegetative cell fusion which can occur between hyphae of the same genotype (i.e., self fusion) or different genotypes (i.e., nonself fusion). The latter normally results in an incompatible response and a form of programmed cell death.[2-3]

Buller was the first to provide a detailed description of the process of vegetative hyphal fusion in 1931,[4-5] and show how it gives rise to the fungal colony as a complex interconnected hyphal network (Fig. 1).[6] Remarkably, it is only recently that a systematic analysis of the cell biology and genetics of vegetative hyphal fusion has been initiated,[1,7-8] and most of this work has been done with the model filamentous fungus, *Neurospora crassa*.[9-10] Here we review: (1) the different types of vegetative hyphal fusion that occur in filamentous fungi, (2) their possible roles, (3) features that vegetative hyphal fusion has in common with yeast cell mating and infection structure differentiation in

*Corresponding Author: Nick D. Read—Fungal Cell Biology Group, Institute of Cell Biology, University of Edinburgh, Rutherford Building, Edinburgh EH9 3JH, U.K. Email: Nick@fungalcell.org

Cell-Cell Channels, edited by Frantisek Baluska, Dieter Volkmann and Peter W. Barlow.
©2006 Landes Bioscience and Springer Science+Business Media.

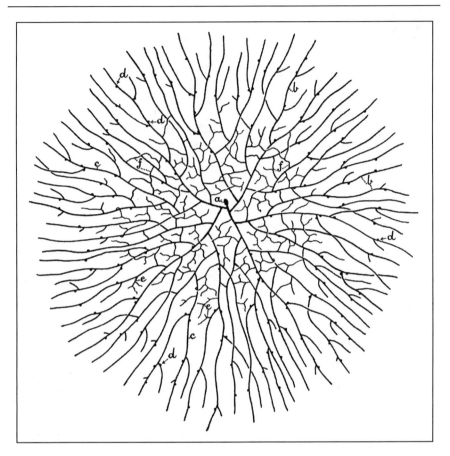

Figure 1. Buller's classic figure showing that hyphal fusion produces a fungal colony (in this case a colony of *Coprinus sterquilinus*) with an interconnected network of hyphae. a) a basidiospore which has germinated; b) and c) leading hyphae at the colony periphery which are avoiding each other; d) clamp connections; e) fusion hyphae prior to fusion; f) fusion hyphae which have anastomosed. (Reproduced from ref. 4.)

fungal pathogens, and (4) the heterokaryon incompatibility response that results from nonself fusions. Particular attention is given to recent work on *N. crassa*.

Fusion between Spores and Spore Germlings

During the early stages of colony establishment, fungal spores (or the germ tubes which arise from them) commonly fuse. This process has been best described between conidia and conidial germlings in ascomycete and mitosporic fungi (Fig. 2A).[7,11-12] However, fusions between ascospore and urediospore germlings have also been shown (Fig. 2B).[4,13-15]

Fusion between conidia or conidial germlings involves the formation and interaction of specialized hyphae called *conidial anastomosis tubes* (CATs) which are morphologically and physiologically distinct from germ tubes,[7,11-12] and under separate genetic control.[11] CATs are short, thin, usually unbranched hyphae that arise from conidia or conidial germ tubes.[7,11-12] What had not been appreciated until recently is that CAT fusion is an extremely common process that has been described in at least 73 species of 21 genera, including many plant pathogens, and thus is probably a common feature of colony establishment in ascomycete and mitosporic fungi.[7] How widespread fusion is between other types of spores and spore germlings remains to be determined.

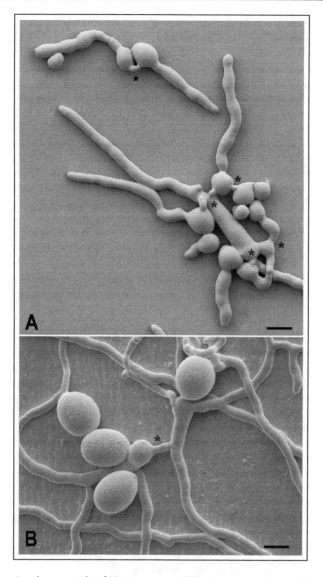

Figure 2. A) Germinated macroconidia of *Neurospora crassa* with long germ germ tubes avoiding each other and conidial anastomosis tubes which have homed and fused (asterisks). (Reproduced from ref. 11 with permission.) B) Germinated ascospores of *Sordaria macrospora* showing fusion between germlings (asterisk). (Reproduced with permission from: Read ND, Beckett A. Mycol Res 100:1281-1314, ©1996 British Mycological Society.[15]) The samples were prepared using low-temperature scanning electron microscopy. Bar = 10 μm.

In *Neurospora crassa*, CAT formation is conidium density dependent and thus may involve quorum sensing.[11] The nature of the CAT inducer is unknown but it activates a mitogen-activated protein (MAP) kinase cascade which has orthologs in the MAP kinase pathway involved in: (1) pheromone signalling in *Saccharomyces cerevisiae*,[16-17] and (2) appressorium formation in fungal plant pathogens.[17-20] Mutations in the genes encoding the MAP kinase kinase kinase, NRC-1, and the MAP kinase, MAK-2, are unable to form CATs (Fig. 3).[11] Increased phosphorylation of MAK-2

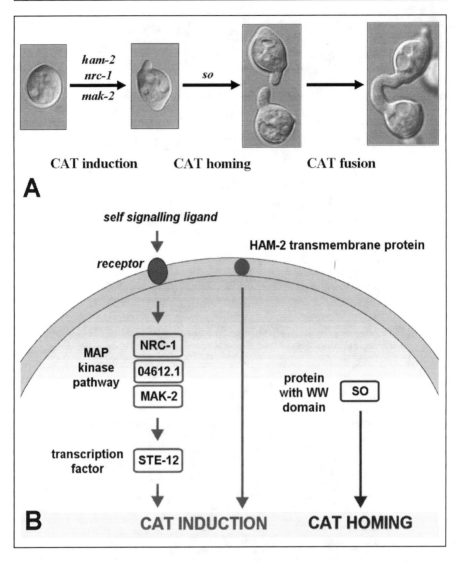

Figure 3. Conidial anastomosis tube induction, homing and fusion in *Neurospora crassa*, and signalling which occurs during CAT induction and homing. A) Mutants blocked in CAT induction and homing. B) Model of the signalling pathways involved in CAT induction and homing. 04612.1 is the *N. crassa* NCU number (http:www.broad.mit.edu/annotation/fungi/neurospora/) for the predicted ortholog of *STE7* in *Saccharomyces cerevisiae*. (Reproduced with permission from: Roca MG et al. FEMS Microbiol Lett 2005; 249:191-198.[7])

was shown to correlate with the onset of fusion between conidial germlings during colony establishment.[17] The downstream target of these MAP kinases seems to be the transcription factor, *pp-1* which is an ortholog of the budding yeast *ste-12*.[7-8,21] HAM-2, a putative transmembrane protein of unknown function,[22] has also been implicated in CAT induction because it is unable to form CATs.[11]

CATs home towards each other. CAT homing has been shown to involve the secretion and reception of a chemoattractant at CAT tips of *Neurospora* by using optical tweezers as a novel experimental tool to move spores or germlings relative to each other.[11]

The identities of the CAT inducer and CAT chemoattractant are unknown, but conceivably could be the same self-signalling ligand. However, it is known that neither signalling molecule is cAMP in *Neurospora* because CATs form and home in a mutant that lacks cAMP.[11]

Another *Neurospora* gene, *so* encodes a protein which plays a role in the biochemical machinery involved in the synthesis and/or secretion of the CAT chemoattractant, or in the signalling apparatus involved in the perception and/or transduction of the chemoattractant signal. This mutant forms CATs but they are unable to home towards or fuse with other CATs of the *so* mutant or wild type (Fig. 3).[23]

Once two CAT tips make contact, they adhere, a fusion pore forms between them, and organelle and cytoplasmic mixing occurs. The movement of organelles between fusion partners is very slow.[7]

Fusion between Hyphae in the Mature Colony

A detailed description of vegetative hyphal fusion in mature colonies of *N. crassa* has been provided by Hickey et al using confocal microscopy to perform time lapse imaging of living hyphae.[24] Three basic phases in the process leading up to hyphal fusion have been defined: (1) pre-contact, (2) post-contact, and (3) post-fusion.[1,3,24] The different stages involved in these three phases are summarized in Figure 4. Hickey et al[24] demonstrated that during the precontact phase specialized fusion hyphae that are fusion-competent exhibit positive tropisms by growing (homing) towards each other, and their close vicinity to other hyphae can induce the formation of further fusion hyphae (Fig. 5). During the post-contact phase, these hyphae make contact, adhere, commonly swell at their tips, and their intervening cell walls break down. Finally, during the post-fusion phase the plasma membranes of the two hyphae fuse providing cytoplasmic continuity and allowing the movement of organelles between them. The bulk flow of organelles and cytoplasm from one hypha to another through the connecting channel can often be very rapid resulting in their immediate mixing. The fusion pore usually increases in size following fusion (Fig. 5) and septa often form in the vicinity of the site of fusion.[24]

The formation and homing of fusion hyphae were shown to be intimately associated with the dynamic behaviour of the Spitzenkörper, a cluster of secretory vesicles, cytoskeletal elements, and other proteins which plays a crucial role in hyphal extension.[25] However, in contrast to vegetative hyphae, the Spitzenkörper in fusion hyphae persists after they have made contact with each other, stopped growing and a fusion pore formed. This suggests that the Spitzenkörper in fusion hyphae continues to provide secretory vesicles for wall synthesis during cell swelling, adhesive molecules for adhesion, and lytic enzymes to allow dissolution of cell wall material.[24]

All of the mutants (*nrc-1*, *mak-2*, *ham-2* and *so*) which are compromised in CAT induction and homing (Fig. 3) are also unable to undergo hyphal fusion in mature colonies. However, we do not know the precise stages at which the fusion process is blocked in mature colonies.

Nematophagous fungi undergo a specialized type of self-fusion which results in the formation of simple rings (Fig. 6A), complex rings that can form net-like structures (Fig. 6B), or constricting rings. These structures are all designed to trap nematodes either by ensnaring them passively (with simple and complex rings), or actively (with constricting rings).[26] The development of these structures can be elicited by the presence of nematodes. Pramer and Stoll[27] coined the term *nemin* to describe the inducer. Nordbring-Hertz[28] later showed that di- and tri-peptides containing valine were very effective in inducing traps under nutrient poor conditions.

CAT Fusion as a Model for Studying Vegetative Hyphal Fusion

Initial studies to systematically analyse the cell biology and genetics of vegetative hyphal fusion in *N. crassa* have concentrated on fusion between hyphae in mature colonies.[17,23-24] Research is now shifting to focusing more on fusion between conidia and conidial germlings[7,11,17,23] because they provide a much simpler and more experimentally amenable system in which to study the process of vegetative hyphal fusion.[7] CAT fusion is thus being used as a model to study basic fundamental aspects of vegetative hyphal fusion. However, although there are many features of CAT fusion which are common to the fusion of hyphae in the mature colony, there are also some differences. Firstly, CATs are short (Fig. 2)[7,11] whilst fusion hyphae vary from being short peg-like structures (Fig. 5) to

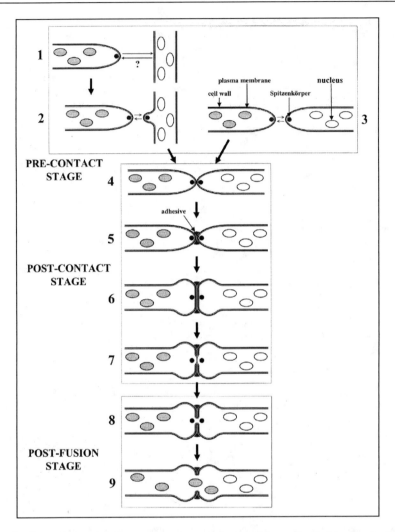

Figure 4. Diagram showing the precontact, post-contact and post-fusion stages involved in vegetative hyphal fusion in the mature colony (diagram modified from that shown in ref. 1). Two types of precontact behaviour are shown: (a) a tip of a fusion hypha induces a branch of a new fusion hypha with which it subsequently fuses (shown in stages 1 and 2); and (b) two tips of fusion hyphae grow towards each other and subsequently fuse (shown in stage 3). A third type of precontact behaviour (not shown here) involves tip-to-side fusion.[24] Nuclei are coloured grey and white to indicate that they belong to different hyphae and not necessarily because they are genetically different. Stage 1: a tip of a fusion hyphae secretes an unknown, diffusible, extracellular signal (small arrow) which induces Spitzenkörper formation; it is not known whether the other secretes a chemotropic signal at this stage. Stages 2 and 3: tips of fusion hyphae each secrete diffusible, extracellular chemotropic signals (small arrows) that regulate Spitzenkörper behaviour; hyphal tips grow towards each other. Stage 4: cell walls of hyphal tips make contact; hyphal tip extension ceases and the Spitzenkörper persists. Stage 5: secretion of adhesive material at the hyphal tips. Stage 6: switch from polarized to more isotropic growth, resulting in swelling of adherent hyphal tips. Stage 7: dissolution of cell wall and adhesive material; plasma membranes of two hyphal tips make contact. Stage 8: plasma membranes of hyphal tips fuse and the fusion pore forms; the Spitzenkörper remains associated with the fusion pore as it begins to widen and cytoplasm starts flow between hyphae. Stage 9: the fusion pore widens and the Spitzenkörper disappears; organelles (such as nuclei, vacuoles and mitochondria) commonly exhibit rapid bulk flow between the fused hyphae, possibly due to differences in turgor pressure.

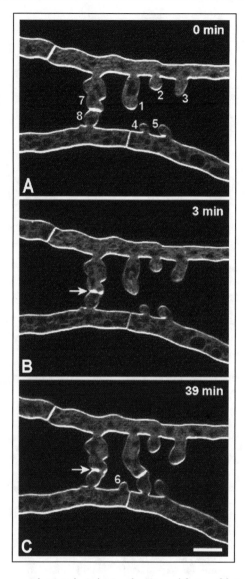

Figure 5. Confocal microscopy showing the induction, homing and fusion of fusion hyphae in the mature colony of *Neurospora crassa* after staining with FM1-43. Note the growth of three branches (1-3) from one hypha towards two short branches on the opposite hypha (4 and 5). After 39 min, a further branch (6) has been initiated on the lower hypha. Branches 7 and 8 fused prior to time 0 min. Different stages in fusion pore formation (arrows) can be observed after 3 and 39 min. Bar = 10 μm. (Reproduced with permission from: Hickey PC et al. Fungal Genet Biol 2002; 37:109-119.[24])

being reasonably long (Fig. 5), and often dichotomously branched hyphae.[24] The isolation and characterization of mutants which form fusion hyphae but not CATs (or vica versa) should be useful to understand the significance of the morphological variation between these different types of hyphae. The second difference relates to fusion hyphae and CATs being in a different physiological state. Cytoplasmic and organelle mixing is usually very rapid between hyphae in the mature colony but is

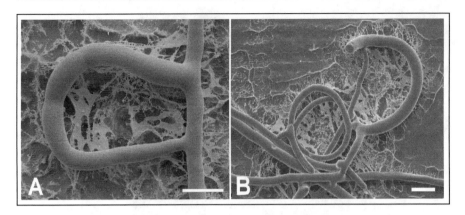

Figure 6. Low-temperature scanning electron microscopy of ring traps produced by the nematophagous fungus *Arthrobotrys oligospora*. A) Simple ring trap. B) Net-like ring traps. Bars = 10 μm.

very slow between fused CATs. We have speculated that there might be slight differences in the turgor pressures of fusion hyphae which results in the rapid bulk flow of cytoplasm and organelles.[24] It may be that the turgor pressure differential between fusing CATs is much less or nonexistent.

Functions of Vegetative Hyphal Fusion

We have suggested that CAT fusion may have two roles.[7] Firstly, CAT fusion may function in improving the chances of colony establishment by allowing heterogeneously distributed nutrients or water within the environment to be shared between different conidial germlings. Secondly, CAT fusion may play an important role in providing a mechanism for gene transfer as a prelude to nonmeiotic gene recombination following nonself fusion.[7,29-30] Our preliminary results indicate that the incompatible response following heterokaryon formation, that normally results in a form of programmed cell death (see below), is suppressed for an extended period during the early stages of colony establishment.[7] CAT fusion during this period may provide a 'window of opportunity' for nonmeiotic recombination to occur. If true, then this may provide a mechanism to explain how much of the genetic variation arises in species in which sexual forms are rare or nonexistent in nature.[7]

Within the mature colony, vegetative hyphal fusion seems to serve a number of different functions. By interconnecting hyphae within the fungal colony (Fig. 1), hyphal fusion significantly contributes to its supracellular state.[31] The supracellular fungal colony is a combination of being a syncytium (in which hyphae have fused together) and a coenocyte (in which individual mitoses are not associated with individual cell divisions). Indeed, adjacent hyphal compartments are separated by septa possessing pores which are commonly open and allow movement of cytoplasm and organelles between hyphal compartments. The supracellular nature of the colony of a filamentous fungus allows it to act as a cooperative, and confers novel yet little understood mechanisms for long distance communication, translocation of water, and transport of nutrients within the typically heterogeneous environments in which they reside.[31] This can be particularly important for the flow of nutrients and water to fruitbodies and other multicellular structures.[5] Furthermore, self fusion between multiple colonies can allow them to act cooperatively in supporting one or more large fruitbodies such as toadstools.[4] In general terms, hyphal fusion must contribute significantly to the overall homeostasis of a fungal colony.

Features of Hyphal Fusion in Common with Yeast Cell Mating and Appressorium Formation

Our initial studies on the intracellular signal transduction pathways involved in vegetative hyphal fusion indicate that this process of self fusion exhibits similarities (e.g., with regard to MAP kinase signalling) to that of the nonself fusion involved in mating cell fusion in the budding yeast. The latter involves cells of each mating type producing a different peptide pheromone (α- or a-factor).[32] However, it seems likely that the process of hyphal self fusion involves the production of the same self-signalling ligand by the CATs or fusion hyphae which are going to fuse. How a self-signalling ligand can provide recognition between two hyphae of the same genotype to orchestrate their induction and homing prior to fusion, is at present a mystery. However, it seems probable that this CAT inducer/CAT chemoattractant, which may or may not be the same molecular species, is a peptide because this would provide specificity[11] (the homing and fusion of CATs has been shown to be strongly species specific).[33] Other features which hyphal fusion and yeast cell mating have in common are that they both involve chemotropic growth of cells towards each other, cell adhesion and cell wall digestion.[34]

Hyphal fusion also exhibits a number of similarities with the formation of an appresorium (a type of infection structure) and its penetration of a plant host cell. Firstly, mutations in *mak-2* homologues in various plant pathogens inhibit appressorium induction indicating that MAP kinase signalling of the type involved CAT induction and in hyphal fusion in the mature colony, is involved in this process too.[17-20] Secondly, both hyphal fusion and appressorium formation typically involve cell adhesion, swelling, and cell wall digestion.[24,35]

It will be interesting to determine how much of the genetic and biochemical machinery involved in hyphal fusion is also involved in yeast cell mating and in appressorium formation.

Vegetative Hyphal Fusion and Heterokaryon Incompatibility

Vegetative hyphal fusion between different colonies carries risks of transmitting infectious cytoplasmic elements such as mycoviruses[36] and senescence plasmids,[37] as well as risks of nuclear parasitism.[38] As a cellular defence mechanism, filamentous fungi have evolved means for recognizing fusions between hyphae of different genotypes and this typically results in an incompatible reaction which prevents heterokaryons from developing further.[2,3,39]

In *N. crassa*, heterokaryon incompatibility results from allelic differences at heterokaryon incompatibility (*het*) loci. Fusion between *het*-incompatible hyphae typically results in rapid hyphal compartmentation and death of the hyphal fusion cell (Fig. 7). This involves the plugging of septal pores to compartmentalize and physically isolate hyphal segments, increased vacuolarisation of the cytoplasm, increased permeability of the plasma membrane and organelle membranes, the release of hydrolytic enzymes into the cytoplasm, and organelle degradation.[2-3] The destruction of the heterokaryotic cell (and often some surrounding cells) can be complete within 30 min following hyphal fusion.[40]

In *N. crassa* there are 11 *het* loci with two or three alleles at each locus.[41] Five of these *het* genes (*mat A-1*, *mat a-1*, *het-c*, *het-6* and *un-24*) have been cloned and the products of these genes have been found to be very diverse indicating that each *het* locus involves a different mechanism of nonself recognition.[2] Nonself recognition at the *het-c* locus has been shown to be mediated by a heterocomplex of polypeptides encoded by *het-c* alleles of alternative specificity and to be localized at the plasma membrane.[42]

tol and *vib-1* have been found to be downstream effectors of heterokaryon incompatibility in *N. crassa*: *tol* suppresses *mat* heterokaryon incompatibility[43-46] whilst *vib-1* (which encodes a putative transcription factor) suppresses both *mat* and *het-c* incompatibility.[47-48]

Concluding Comments

Although much is known about nonself fusion between nonidentical cells (e.g., sperm and egg in animals; mating cells in yeast), little is known about self fusion between genetically identical cells (e.g., between myoblasts during muscle formation), and particularly with regard to the mechanistic basis of self-signalling. Self fusion, in the form of hyphal fusion, is a defining feature of colony morphogenesis in filamentous fungi. Understanding hyphal fusion in the model genetic system *N. crassa*, provides a paradigm for self-signalling mechanisms in eukaryotic microbes and might also provide a model for understanding somatic cell fusion in other eukaryotic species.[1]

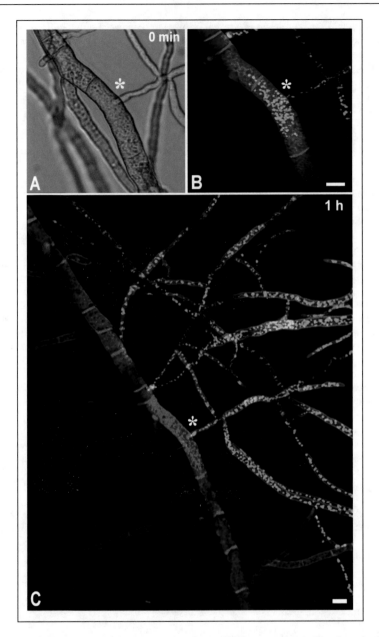

Figure 7. Heterokaryon incompatibility in a mature colony of *Neurospora* crassa resulting from the fusion of *het-c^{OR}* and *het-c^{PA}* strains. Both strains are stained with the membrane-selective dye FM4-64 and the nuclei in the *het-c^{OR}* strain are labelled with H1-GFP.[48] A) Brightfield image demonstrating that a thin fusion hypha has fused with the underside of a thick hypha (asterisk). B) Confocal image of the same field of view as illustrated in (A) showing the migration of nuclei (fluorescing green) from the *het-c^{OR}* strain into the *het-c^{PA}* strain lacking fluorescently labelled nuclei. Bar = 20 μm. C) Incompatible response 1 h after fusion. Note firstly the intense staining of cytoplasm of the fusion cell due to the increased permeability of the plasma membrane to the FM4-64 dye, and secondly, the loss in GFP fluorescence from nuclei due to their breakdown. Bar = 20 μm. (Roca MG, Glass NL and Read ND, unpublished).

References

1. Glass NL, Rasmussen C, Roca MG et al. Hyphal homing, fusion and mycelial interconnectedness. Trends Microbiol 2004; 12:135-141.
2. Glass NL, Kaneko I. Fatal attraction: Nonself recognition and heterokaryon incompatibility in filamentous fungi. Eukaryot Cell 2003; 2:1-8.
3. Glass NL, Jacobson DJ, Shiu PKT. The genetics of hyphal fusion and vegetative incompatibility in filamentous ascomycete fungi. Annu Rev Genet 2000; 34:165-186.
4. Buller AHR. Researches on Fungi, vol. 4. London: Longmans, 1931.
5. Buller AHR. Researches on Fungi, vol. 5. London: Longmans, 1933.
6. Gregory PH. The fungal mycelium: An historical perspective. Trans Brit Mycol Soc 1984; 82:1-11.
7. Roca MG, Read ND, Wheals AE. Conidial anastomosis tubes in filamentous fungi. FEMS Microbiol Lett 2005; 249:191-198.
8. Glass NL, Fleißner A. Rewiring the network: Understanding the mechanism of function of anastomosis in filamentous ascomycete fungi. In: Kues U, Fisher R, eds. The Mycota: Growth, Differentiation and Sexuality, Vol. 1. 2nd ed. Berlin: Springer-Verlag, 2006, (in press).
9. Davis R. Neurospora: Contributions of a Model Organism. Oxford: Oxford University Press, 2000.
10. Davis RH, Perkins DD. Neurospora: A model of model microbes. Nat Rev Genet 2002; 3:397-403.
11. Roca MG, Arlt J, Jeffree CE et al. Cell biology of conidial anastomosis tubes in Neurospora crassa. Eukaryot Cell 2005; 4:911-919.
12. Roca MG, Davide LC, Mendes-Costa MC et al. Conidial anastomosis tubes in Colletotrichum. Fungal Genet Biol 2003; 40:138-145.
13. Bary de A, Woronin M. Sphaeria Lemaneae, Sordaria coprophila, fimiseda, Arthrobotrys oligospora. In: de Bary A, Woronin M, eds. Beitrage zür Morphologie und Physiologie der Pilze. Winter: Frankfurt am main: Verlag von C, 1870:1-89.
14. Manners JG, Bampton SS. Fusion of uredospore germ tubes in Puccinia graminis. Nature 1957; 179:483-484.
15. Read ND, Beckett A. Ascus and ascospore morphogenesis. Mycol Res 1996; 100:1281-1314.
16. Gustin MC, Albertyn J, Alexander M et al. MAP kinase pathways in the yeast Saccharomyces cerevisiae. Microbiol Mol Biol Rev 1998; 62:1264-1300.
17. Pandey A, Roca MG, Read ND et al. Role of a mitogen-activated protein kinase pathway during conidial germination and hyphal fusion in Neurospora crassa. Eukaryot Cell 2004; 3:348-358.
18. Lev S, Sharon A, Hadar R et al. A mitogen-activated protein kinase of the corn leaf pathogen Cochliobolus heterostrophus is involved in conidiation, appressorium formation, and pathogenicity: Diverse roles for mitogen-activated protein kinase homologs in foliar pathogens. Proc Natl Acad Sci USA 1999; 96:13542-13547.
19. Takano Y, Kikuchi T, Kubo Y et al. The Colletotrichum lagenarium MAP kinase gene CMK1 regulates diverse aspects of fungal pathogenesis. Mol Plant Microbe Interact 2000; 13:374-383.
20. Xu JR, Hamer JE. MAP kinase and cAMP signaling regulate infection structure formation and pathogenic growth in the rice blast fungus Magnaporthe grisea. Genes Dev 1996; 10:2696-2706.
21. Li D, Bobrowicz P, Wilkinson HH et al. A mitogen-activated protein kinase pathway essential for mating and contributing to vegetative growth in Neurospora crassa. Genetics 2005; 170:1091-1104.
22. Xiang Q, Rasmussen C, Glass NL. The ham-2 locus, encoding a putative transmembrane protein, is required for hyphal fusion in Neurospora crassa. Genetics 2002; 160:169-180.
23. Fleißner A, Sarkar S, Jacobson DJ et al. The so locus is required for vegetative cell fusion and postfertilization events in Neurospora crassa. Eukaryot Cell 2005; 4:920-930.
24. Hickey PC, Jacobson D, Read ND et al. Live-cell imaging of vegetative hyphal fusion in Neurospora crassa. Fungal Genet Biol 2002; 37:109-119.
25. Harris SD, Read ND, Roberson RW et al. Polarisome meets Spitzenkörper: Microscopy, genetics, and genomics converge. Eukaryot Cell 2005; 4:225-229.
26. Barron GL. The Nematode-Destroying Fungi. Guelph, Ontario: Canadian Biological Publications, 1977.
27. Pramer D, Stoll NR. Nemin: A morphogenic substance causing trap formation by predaceous fungi. Science 1959; 129:966-969.
28. Nordbring-Hertz B, Friman E, Veenhuis M. Hyphal fusion during initial stages of trap formation in Arthrobotrys oligospora. Antonie van Leeuwenhoek 1989; 55:237-244.
29. Pontecorvo G. The parasexual cycle in fungi. Annu Rev Microbiol 1956; 10:393-400.
30. Rosewich UL, Kistler HC. Role of horizontal gene transfer in the evolution of fungi. Annu Rev Phytopath 2000; 38:325-363.
31. Read ND. Environmental sensing and the filamentous fungal lifestyle. In: Gadd, ed. Fungi in their Environment. Cambridge: Cambridge University Press, 2006; in press.

32. Kurjan J. The pheromone response pathway in Saccharomyces cerevisiae. Annu Rev Biochem 1992; 61:1097-1129.
33. Köhler E. Zur Kenntnis Der Vegetativen Anastomosen Der Pilze (II. Mitteilung). Planta 1930; 10:495-522.
34. Madden K, Snyder M. Cell polarity and morphogenesis in budding yeast. Annu Rev Microbiol 1998; 52:687-744.
35. Tucker SL, Talbot NJ. Surface attachment and prepenetration stage development by plant pathogenic fungi. Annu Rev Phytopathol 2001; 39:385-417.
36. Cortesi P, McCulloch CE, Song H et al. Genetic control of horizontal virus transmission in the chestnut blight fungus, Cryphonectria parasitica. Genetics 2001; 159:107-118.
37. Debets F, Yang X, Griffiths AJ. Vegetative incompatibility in Neurospora: Its effect on horizontal transfer of mitochondrial plasmids and senescence in natural populations. Curr Genet 1994; 26:113-119.
38. Debets AJM, Griffiths AJ. Polymorphism of het-genes prevents resource plundering in Neurospora crassa. Mycol Res 1998; 102:1343-1349.
39. Jacobson DJ, Beurkens K, Klomparens KL. Microscopic and ultrastructural examination of vegetative incompatibility in partial diploids heterozygous at het loci in Neurospora crassa. Fungal Genet Biol 1998; 23:45-56.
40. Saupe SJ. Molecular genetics of heterokaryon incompatibility in filamentous ascomycetes. Microbiol Mol Biol Rev 2000; 64:489-502.
41. Perkins DD, Radford A, Sachs MS. The Neurospora Compendium. San Diego: Academic Press, 2001.
42. Sarkar S, Iyer G, Wu J et al. Nonself recognition is mediated by HET-C heterocomplex formation during vegetative incompatibility. EMBO J 2002; 21:4841-4850.
43. Newmeyer D. A suppressor of the heterokaryon-incompatibility associated with mating type in Neurospora crassa. Can J Genet Cytol 1970; 12:914-926.
44. Jacobson DJ. Control of mating type heterokaryon incompatibility by the tol gene in Neurospora crassa and N. tetrasperma. Genome 1992; 35:347-353.
45. Vellani TS, Griffiths AJ, Glass NL. New mutations that suppress mating-type vegetative incompatibility in Neurospora crassa. Genome 1994; 37:249-255.
46. Shiu PK, Glass NL. Molecular characterization of tol, a mediator of mating-type-associated vegetative incompatibility in Neurospora crassa. Genetics 1999; 151:545-555.
47. Xiang Q, Glass NL. Identification of vib-1, a locus involved in vegetative incompatibility mediated by het-c in Neurospora crassa. Genetics 2002; 162:89-101.
48. Xiang Q, Glass NL. Chromosome rearrangements in isolates that escape from het-c heterokaryon incompatibility in Neurospora crassa. Curr Genet 2004; 44:329-338.
49. Freitag M, Hickey PC, Raju NB et al. GFP as a tool to analyze the organization, dynamics and function of nuclei and microtubules in Neurospora crassa. Fungal Genet Biol 2004; 41:897-910.

SECTION V
Plant Cells

CHAPTER 6

Plasmodesmata:
Cell-Cell Channels in Plants

Yoselin Benitez Alfonso, Laurence Cantrill and David Jackson*

Abstract

Plasmodesmata (PD) permit diffusion of small metabolites and proteins, as well as active trafficking of specific RNAs and proteins. Their structure and distribution vary according to species, cell type, physiological function and stage of development.

Two mechanisms for PD trafficking have been described: targeted and nontargeted. Transgenic plants expressing cytoplasmic GFP (Green Fluorescent Protein), illustrate nontargeted transport, as GFP moves freely between immature cells by diffusion. The transport through PD of viral proteins is the most studied example of targeted movement.

PD are also involved in the regulation of cell fate and development, as PD transport of transcription factors and RNAs, which act as short and long distance signals, has also been reported.

Cell-cell communication is highly regulated by developmental and environmental factors. Blocking or selective occlusion of PD has been reported during stomatal development, regeneration of plant tissues, embryogenesis, phloem development and reproductive development.

Here we review current models of PD structure and function, though many of their molecular components are still unknown.

Introduction

Communication between cells regulates development in both animals and plants.[1-3] In plants this involves two pathways: the apoplastic pathway, which is often mediated by receptor-ligand interactions, and the symplastic pathway. The symplastic pathway is mediated by plasmodesmata (PD): intercellular channels that function to exchange electrical signals,[4] small metabolites,[5] and nucleic acids and proteins.[6-11]

Trafficking through plasmodesmata can be classified in two kinds: targeted trafficking, which is selective and actively regulated, and nontargeted trafficking, which is nonselective and presumably passive.[10,12] Nontargeted trafficking has been shown for small cytoplasmic proteins, which likely move between cells by simple diffusion. The size limit for molecules that can move passively through PD is called the size exclusion limit (SEL). In animals, passive transport can also occur through gap junctions, but usually only of smaller ions and metabolites[3,13] (reviewed in the chapter "Gap Junctions: Cell-Cell Channels in Animals" in this volume).

Targeted trafficking is illustrated by proteins that interact with PD to change the SEL, allowing their own movement.[8,12] Viruses and viroids use this as a mechanism to spread their infection,[8,14,15] for example by using virally encoded Movement Proteins (MPs) (reviewed in the chapter "Viral Movement Proteins Target Cell-Cell Channels in Prokaryotes and Plants" in this volume).

*Corresponding Author: David Jackson—Cold Spring Harbor Laboratory, 1 Bungtown Rd., Cold Spring Harbor, New York 11724, U.S.A. Email: jacksond@cshl.edu

Cell-Cell Channels, edited by Frantisek Baluska, Dieter Volkmann and Peter W. Barlow. ©2006 Landes Bioscience and Springer Science+Business Media.

Macromolecular transport through symplastic cell-cell connections is not restricted to the plant kingdom; as similar structures have been recently discovered in animal systems.[1] These "tunneling nanotubes" (TNTs) are channels established between animal cells that allow transport of endosomes. Formation and function of these channels, as for PDs, appears to be dependent on the actin-myosin system (see the chapter "Tunneling Nannotubes as Cell-Cell Channels in Animal Cells" in this volume).

This chapter will review the current knowledge about the structure of plasmodesmata channels, and the mechanisms proposed for protein and RNA trafficking.

Structure of Plasmodesmata

The structure of plasmodesmata is a topic that is far from being resolved. Due to the limits of electron microscopy,[16] the difficulty in separating component proteins of PD from plant cell walls[17,18] and the likely lethality of most plasmodesmal structural mutants[2] the model we have of PD is frustratingly incomplete. Our current vision of the plasmodesma is based on EM images and has developed over many decades from different processing and staining techniques applied to a wide range of tissue types. The standard model describes a simple unbranched channel composed of a plasma membrane-lined tube traversing the cell wall between the cytoplasm of adjacent cells (Fig. 1). The membrane-lined tube is continuous with the plasma membrane of each cell and it encloses a cytoplasmic sleeve or annulus, which in turn surrounds an axial desmotubule that is continuous with the endomembranes (endoplasmic reticulum or ER) of the adjacent cells.

Primary PD are formed during cytokinesis, at sites where ER tubules become trapped by wall material deposited by Golgi vesicles at the cell plate.[19-21] In contrast, secondary PD are formed through existing cell walls.[22,23] However they cannot be distinguished structurally from primary PD, other than by their existence between previously separated cells or their location in non division walls.[24-26] Both primary and secondary PD can exist in simple or branched forms. Simple single channels are thought to be the initial PD structure, which is later modified into complex branched structures with central cavities, as observed during the source-sink transition in leaves.[20] The mechanism of secondary PD formation has been reviewed extensively by Kollmann and Glockmann.[21]

Interpretation of the internal substructure of PD has been the subject of intense debate about the components of the desmotubule and cytoplasmic annulus. In transverse section, the gap between the plasma membrane and the desmotubule is seen to consist of electron dense regions delineating electron lucent particles. The latter have been interpreted as unstained protein particles of around 4.5 nm in diameter, separated by cytoplasmic gaps of between 2.6 and 3.6 nm.[2,27] These dimensions correspond to the Stokes radii of fluorescent molecules known to move passively from cell to cell.[5,28] In other studies, the electron dense regions of the cytoplasmic annulus have been interpreted as spokes linking the appressed ER to the plasma membrane, with the unstained regions in the annulus providing gaps for intercellular movement.[29] The desmotubule itself has been characterized as a solid proteinaceous rod[30] or as a membrane based cylinder with an internal lumen,[31] a view that is currently the most favoured. Despite such differing opinions, most studies propose a helical arrangement of protein particles in the central part of the plasmodesma. These partially block the cytoplasmic annulus and alternate with spaces that together provide a series of small channels through which molecules might "spiral" from cell to cell in a size-restricted way.[32]

Surrounding each plasmodesma in the cell wall matrix is an electron lucent region. Modifications observed within this region include external spirals,[27,33] electron dense strands between the plasma membrane and the stained part of the wall,[34] and particulate sphincters[33] as well as callose collars in the neck of the channel.[35-37] Callose constrictions in many cases might be artifacts caused by wound induced deposition of callose during fixation, as they are absent from material fixed in the presence of a callose synthase inhibitor.[38] This treatment also showed PD with a more open appearance, providing an explanation for underestimations of SEL in earlier work. Finally, at a higher order of organization, PD are frequently clustered together in pit fields, and the cell wall in these regions appears to be thinner than usual.[39]

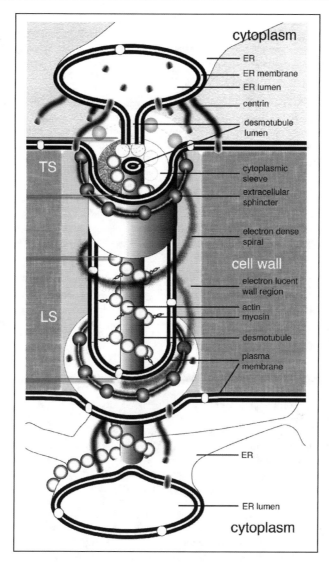

Figure 1. Model of a simple unbranched plasmodesma based on electron micrographs and immunohistochemistry. In the transverse section (TS), the plasmodesma appears as a circular structure delineated by an outer ring of plasma membrane surrounding the cytoplasmic sleeve and the central desmotubule. Within the cytoplasmic sleeve, electron lucent regions are surrounded by an electron dense matrix. In the longitudinal section (LS), particles are arranged helically around the desmotubule and these might be complexes of actin and myosin. These complexes could connect the plasma membrane to the desmotubule and regulate the size exclusion limit within the cytoplasmic sleeve or provide a motile system for trafficking macromolecules from cell to cell. The axial desmotubule (LS view) is composed of appressed endoplasmic reticulum (ER) with a narrow central lumen (TS view) that might provide an additional route for intercellular movement of substances. Contractile proteins such as centrin possibly link the plasma membrane to the ER in the neck region or interact within regions of plasma membrane to control the plasmodesmal aperture. Particulate extracellular sphincters and spirals of electron dense material could regulate the outer dimensions of the plasmodesma and provide structural stability. Model adapted with permission from Overall and Blackman (1996).[32]

It is widely accepted that the cytoplasmic annulus or sleeve provides the main passage for intercellular movement. Fluorescent probes that have been microinjected[56,61] or loaded passively across cell membranes or via the phloem[62,63] are thought to diffuse intercellularly along this pathway. Macromolecules in the form of RNAs and proteins are also thought to pass along the cytoplasmic sleeve during active trafficking[11] or passive diffusion.[64] The idea of an SEL for passive movement was introduced early in the history of microinjection studies,[56] though with advances in technology and exploration of a wider variety of cell types the estimates of SEL have expanded from around 1kDa[56] up to nearly 70 kDa in specific cell types.[57,58] Despite this apparent lack of control, it is clear that many tissues maintain a tight regulation on their SEL, with even small molecules being excluded from intercellular movement at certain stages of development.[59-61]

The cytoplasmic sleeve might not be the only route for intercellular movement through PDs. Growing evidence suggests the desmotubule provides an additional pathway for the movement of substances between cells. Dyes microinjected into the cortical ER can move from cell to cell,[65] and fluorescent lipid analogues diffuse from cell to cell via the endomembrane network, even though similar labelling of the plasma membrane results in no label movement. PD with open desmotubules have been described in tobacco trichomes[66] and Lazzaro and Thomson in 1996[67] even demonstrated vacuole continuity through PD in secretory trichomes of chickpea. They speculated that vacuole membranes rather than ER might be present in the PD of many mature plant cells. They also proposed the vacuolar-tubular continuum as a pathway for solute diffusion, a concept elaborated by Gamalei et al in 1994[68] and Velikanov et al in 2005[69] into a vacuolar/ER/desmotubule symplast that runs in parallel with the cytoplasmic symplast. Roberts and Oparka in 2003[12] have also pointed out the potential for ER-mediated transport between companion cells and sieve elements, as the latter contain little cytoplasm but retain parietal ER. Furthermore there is tantalising evidence that plant viruses might use aspects of the ER network to target both the secretory pathway and PD.[46,70]

To date, no PD specific genes have been identified, but a great deal of effort has been dedicated to this problem and a range of methods are identifying proteins closely associated with PDs. An initial approach involved rhodamine phalloidin staining and antibodies against actin to demonstrate the presence of this key cytoskeletal protein in PD.[40-42] Following this, a whole suite of well-known proteins were found to be present in PD, including myosin,[42,43] centrin,[44] calreticulin,[45,46] a pathogenesis-related protein[47] and ubiquitin.[48] These findings have contributed greatly to the conception of the current structural model.[32,49] Early attempts at biochemical characterization of PD proteins from maize mesocotyl identified PAP26, which was found to cross-react with anti connexin antibodies (connexins are gap junction proteins) however these findings have not been reproduced.[50] Later, antibodies against a 41 kDa protein isolated from a plasmodesmal extract were shown to localize to PD.[51] A proteomics approach by Blackman et al in 1998[111] identified four putative PD associated proteins in the alga *Chara corallina,* and antibodies raised against a 45 kDa protein localized to PD. More recently Faulkner et al in 2005[18] have extended this work by identifying several new peptide sequences, and have shown that two previously identified proteins, heat shock cognate 70[52] and cp-wap13[51] are also associated with PD in this species. Using different proteomics strategies a reversibly glycosylated polypeptide has also been identified by several groups as a PD associated protein.[17,53,54] Finally, in a recent innovative genomics approach, 12 GFP fusion proteins from a *Nicotiana benthamiana* random cDNA overexpression library have been localized to PD.[55] Whether these proteins localize to PDs in their native form, and if they are structural or regulatory is yet to be determined.

Mechanisms of Protein and RNA Transport through PDs

Models for macromolecular transport through PD need to explain the different modes of transport. Two general routes for protein transport have been proposed. The nontargeted diffusion of small molecular weight proteins through the cytoplasmic sleeve may be important to provide signaling coordination between large groups of cells.[10,12,71] In addition, PDs apparently interact with some larger proteins and RNAs, allowing their transport by a targeted pathway.[7,10,12]

Nontargeted Trafficking

Nontargeted movement depends on the abundance of the protein in the cytoplasm, and on its molecular weight.[71] Studies that support this theory were mostly carried out using the heterologous protein, GFP, and it remains to be shown how the findings relate to native plant proteins. Cytoplasmically localized GFP can move freely from cell to cell in many plants and cell types.[71-73,112] However, GFP carrying a nuclear localization signal (NLS) is unable to move from cell to cell, or moves less extensively than cytoplasmically localized proteins of a similar size. This restriction may be based purely on the size of the nuclear localized protein, or on other factors, such as interaction with chromatin.[2,10,71] Targeting to other sub-cellular locations can also limit nontargeted protein movement. For example, addition of cytoskeletal localization sequences or of ER targeting and retention sequences to GFP blocks its ability to move.[6,57,74]

Nontargeted trafficking of a single plant protein, LEAFY (LFY), has been reported, and similar to GFP, the movement of different LFY fusion proteins is correlated with their size and subcellular localization.[75]

Targeted Trafficking

The spreading of plant viral infection was the earliest characterized system involving targeted protein trafficking.[6,76] Viral movement proteins (MPs) interact with PD to increase the SEL and allow trafficking of the MP-associated viral genome, and even of viral particles.[14] Many structure-function studies of MPs have been published; one recent example of the red clover necrotic mosaic virus MP suggested a correlation between movement and targeting to the plant cell wall.[77] Mutations in this MP showed that different domains are involved in its movement and in its intracellular localization, supporting the idea that movement in this case is targeted.

Recently, a number of plant proteins and sub-cellular structures that may be involved in viral movement have been described. For example, heat shock cognate 70 related chaperones,[52] the actin-myosin cytoskeleton, the endoplasmic reticulum,[78-80] calcium-dependent and independent protein kinases,[81-83] pectin methyl esterase,[70,84] and several proteins of unknown function are associated with PDs and in some cases interact with viral MPs and/or play a role in PD gating. In the following paragraphs some of these will be described.

A tobacco PD-associated protein kinase (PAPK) was recently identified by Lee and coworkers.[84] PAPK belongs to the casein kinase I family and localizes to PD. It phosphorylates tobacco mosaic virus movement protein (TMV MP) in vitro, and this phosphorylation is known to affect MP function.[82] Another kinase from tobacco (NtRIO) that is homologous to the yeast protein kinase Rio1p (RIght Open reading frame 1p) was isolated by a far-Western screening using a recombinant tomato mosaic virus movement protein (ToMV MP).[83] The association appears to be phosphorylation-dependent suggesting that NtRIO might have a role in the negative regulation of MP trafficking.

Calreticulin, a lectin-like chaperone involved in the regulation of intracellular Ca^{2+} homeostasis and in the folding of newly-synthesized glycoproteins, has been reported to interfere with the spreading of viral infection.[46] This protein accumulates in PD, and interacts directly with the viral MP. Plants overexpressing calreticulin redirect the TMV MP from PD to microtubules, restricting cell-cell movement of the MP.[46]

Last but not least, a plant protein that interacts with a plant paralog of a viral MP has been identified. This protein, NON-CELL AUTONOMOUS PATHWAY PROTEIN1 (NCAPP1), is a putative receptor in the PD trafficking pathway.[85] NCAPP1 is related to GP40, a protein localized at the nuclear periphery that participates in nuclear protein import.[10] This observation supports the idea that trafficking through PD may be mechanistically related to nuclear import and export through nuclear pores.[86] Alteration of NCAPP1 expression blocks viral MP and PP16 (MP related Phloem Protein 16) transport, but does not affect trafficking of KNOTTED-1, the first plant protein found to traffic cell to cell (see next section).[85]

Plasmodesmata as a Route for Trafficking of Developmental Signals

Remarkably, some plant transcription factors might function as short distance signals by trafficking between cells.[64] Evidence suggests that most of them, like KNOTTED-1 (KN1)[11,87] and SHORT ROOT,[88] move selectively through PD. These transcription factors play a role in specification of cell fate. The mechanism of transcription factor movement and its implications in development are discussed in the chapter "Cell-Cell Transport of Transcription Factors in Plants" by Alexis Maizel in this volume.

The trafficking of mRNA has also been suggested to provide intercellular as well as long distance signals throughout the plant.[87,89,90] For instance, microinjection experiments showed that KN1 mRNA can traffic between cells, but only in the presence of the KN1 protein.[87] This result was confirmed using a trichome rescue assay system.[90] This assay is based on the transformation of *glabrous1* (*gl1*) mutant plants, which lack trichomes, with a vector that carries a fusion of KN1 to GL1 under the control of the *Arabidopsis* rubisco small subunit 2b (*RbcS2b*) promoter. This promoter drives expression in mesophyll cells; therefore expression of the cell autonomous GL1 protein alone does not rescue trichomes. When GL1 is fused to KN1, the fusion protein traffics into the epidermal cell layer and rescues trichomes. A variation of this system in which a stop codon was inserted between GL1 and KN1 was used to show that the KN1 mRNA could also be transported to the epidermis, but again only in the presence of the KN1 protein.[90]

Grafting experiments in tomato support the hypothesis that long distance transport of regulatory mRNAs also occurs in plants.[89] The *Mouse ears* (*Me*) mutation results from the expression of a gene fusion between a pyrophosphate-dependent phosphofructokinase (PFP) gene and the *LeT6* KNOX (KN1-related homeobox) gene, controlled by the native PFP promoter. This fusion causes overexpression of the PFP-LeT6 fusion protein. When a wild type scion was grafted onto a *Mouse ears* (*Me*) mutant rootstock, the *Me* phenotype developed in the wild type shoot. This phenotypic transmission was correlated with trafficking of the *LeT6* fusion mRNA from the stock to the developing scion shoot through the phloem.[89]

Haywood and coworkers[91] have also described the transport of GIBBERELLIC ACID-INSENSITIVE (GAI) transcripts from the phloem sap into sink tissues. *GAI* encodes a transcriptional regulator that functions in gibberellic acid signaling. When dominant gain-of-function constructs of *GAI* genes from pumpkin and *Arabidopsis* were expressed in the phloem sap of tomato plants, the transgenic RNA was able to exit the phloem and move into the shoot apex, resulting in a leaf phenotype.[91] Both the LeT6 and GAI studies show that ectopic expression of regulatory mRNAs can result in phenotypic changes, but neither prove that the RNA itself is the signal.

The trafficking of RNAs has also been characterized by the study of viruses and viroids.[76] Viral RNAs move as ribonucleoprotein complexes with MPs, which bind RNAs nonspecifically.[6] More recently, evidence for specificity in RNA transport was described. The discovery of specific sequence motifs necessary for the directional trafficking of the potato spindle tuber viroid RNA from bundle sheath into mesophyll cells suggests a developmental regulatory strategy for RNA transport.[92]

The isolation of the phloem proteins PP1, PP2, PP16 and some lectins, which show RNA-binding activity and the capacity to increase PD SEL, provides additional evidence that plant RNAs also traffic as ribonucleoprotein complexes in the phloem.[93-95]

Phloem RNA trafficking may also be involved in the systemic propagation of post transcriptional gene silencing / RNAi.[96] For example, single-stranded RNAs of around 25 nucleotides in length have been found in the phloem sap of pumpkin.[97] The isolation of the phloem protein CmPSRP1 (*C. maxima* Phloem Small RNA Binding Protein 1), which binds selectively to 25-nucleotide single stranded RNAs, suggests that small RNAs also require RNA binding proteins to mediate their cell-cell trafficking.[97]

Developmental and Environmental Regulation of PD SEL

Biotic and abiotic factors such as turgor, hypoxia, secondary messengers, or treatment with sodium azide, cytoskeletal antagonists or divalent cations all have an influence on the function of PD (reviewed by Gillespie and Oparka in 2005).[98] Transport between cells is likely to be controlled via rapid conformation changes in plasmodesmal proteins, such as actin[41] or myosin.[38,99] PD transport may also be regulated over longer time periods by gross structural modification of PD. For example, wounding induces callose deposition[100] and this polymer is also implicated in the inhibition of virus movement from cell to cell.[101] Local callose synthesis is also central to the regulation of symplasmic organisation during dormancy cycling in shoot apices.[102]

On a broader level, regulation of intercellular transport can be observed in the formation and occlusion of PD. de novo formation of secondary PD seems to accompany cell wall elongation and cell expansion[103,104] and allows the establishment of new regions of intercellular communication.[25] Aniline blue staining shows that callose accumulates in PD during the rapid phase of cotton fiber elongation, leading to PD closure. The duration of closure positively correlates with fiber length and the expression of a fiber-specific beta-1,3-glucanase (which is involved in degradation of callose) appears to be related with PD reopening.[105] Permanent occlusion of PD has been demonstrated most clearly during stomatal development, where new wall material is deposited across the channels leaving the cells completely isolated from the surrounding epidermis.[106,107] During regeneration in tissue cultures, selective occlusion of PD and even the complete removal of PD from some walls have been observed, and this process is thought to define symplasmic domains in embryogenic tissue cultures as well as in reproductive tissues in vivo (and references cited).[19]

The analysis of fluorescent tracer uptake during embryogenesis has provided a system to investigate the relation between PD SEL and differentiation.[58,108] Kim et al[63] in 2002 have shown that the trafficking of fluorescent tracers of around 10 kDa is permitted during early stages of embryogenesis but that there is a downregulation of PD SEL after the torpedo stage. Trafficking during embryo development was more carefully analyzed using transgenic lines that express a cell autonomous GFP targeted to the endoplasmic reticulum (erGFP) and either cytoplasmic GFP (sGFP) or a fusion of two copies of cytoplasmic GFP (2XsGFP) controlled by a tandem array of GAL4 upstream activation sequences (UAS). These lines were generated by transformation of the *Arabidopsis* enhancer trap line J2341 (which expresses erGFP) with constructs that express sGFP or 2XsGFP under the control of the UASs. In the J2341 line, a shoot meristem region specific enhancer drove GAL4 expression, and induced the expression of erGFP in this region.[58] In contrast, sGFP expressed using this enhancer line spread throughout the entire embryo at all developmental stages analyzed (heart, late heart and mid-torpedo). Furthermore, plants expressing 2XsGFP revealed a decrease in PD permeability during development of the embryo.[58]

Another interesting example of developmental regulation of PD SEL can be observed in the formation of the vascular system. A comprehensive study to characterize the SEL of PDs in the different domains formed during phloem development has been published by Stadler et al[57] in 2005. After differentiation, the cells of the phloem system show a higher PD SEL than other mature tissues. Quantitative analysis has been performed by studying unloading of a range of GFP-fusion proteins expressed under the control of the phloem companion cell (CC) specific *AtSUC2* (*Arabidopsis thaliana* sucrose transporter 2) promoter in transgenic *Arabidopsis* plants. This study suggested that the SEL of PDs between sieve elements (SEs) and CCs is around 67 kDa, and revealed new symplasmic domains around the root protophloem.[57] These studies provide new tools to study symplasmic domains during development.

Independent studies analyzing cell-cell transport during reproductive development also support the idea of developmental regulation of PD permeability. For example, intercellular communication in the shoot apical meristem decreases during the initiation of floral development, and it is hypothesized that this may be functionally significant in limiting the trafficking of floral repressors.[60]

As summarized above, increasing evidence indicates a dynamic regulation of PD SEL during development and in response to biotic and abiotic factors. For this reason, quantifying changes in PD permeability remains an important goal. A new method which measures the coefficient of conductivity for cell-to-cell movement of GFP in transgenic *Nicotiana benthamiana* plants has been reported.[110] This value evaluates the capacity for diffusion through PD in different environmental conditions, including during pathogen infection.

Future Perspectives

We are starting to learn some of the molecular components of PD, though we are still lacking any PD specific proteins or a detailed structure function model of PD transport. Clearly there is great diversity in PD form and function in different cell types and developmental stages, and the challenge is to understand which components of PD are modified to generate this developmental flexibility, and what are the core components of these intriguing channels.

We described novel genomic and proteomic approaches that promise to identify new functional components of the PD pathway. Other strategies to characterize PD are emerging, as well as a number of new fluorescent techniques to analyze PD trafficking.[53] One of the main problems in applying proteomics approaches to the study of PDs is the difficulty in their isolation from the plant cell wall. In a recent attempt to obtain material for PD proteomics, the frequency and features of PDs in *Arabidopsis* suspension cells has been explored.[17]

A genetic approach to isolate genes specifically involved in the regulation of transport through PD is another promising approach. A screen based on the movement of fluorescent tracers in embryo defective mutants was performed,[63] and two mutants with increased size exclusion limit (*ise*) of the embryo PD were identified. *ISE1* encodes a DEAD-box RNA helicase putatively involved in an intercellular signaling pathway that regulates embryo development.[53] By EMS mutagenesis of transgenic *Arabidopsis* plants that express GFP under the control of the *SUC2* promoter, several candidate protein trafficking mutants have also been obtained (Benitez-Alfonso and Jackson, unpublished). These mutants show a restriction in GFP unloading from the phloem to the meristematic cells of the root tip. Mapping the genes responsible for these phenotypes is currently in progress and hopefully will identify new components involved in PD function or structure.

Acknowledgements

Research on PDs in the Jackson lab is funded by the National Science Foundation, grant number IBN-0213025.

References

1. Baluska F, Hlavacka A, Volkmann D et al. Getting connected: Actin-based cell-to-cell channels in plants and animals. Trends Cell Biol 2004; 14:404-448.
2. Zambryski P. Cell-to-cell transport of proteins and fluorescent tracers via plasmodesmata during plant development. J Cell Biol 2004; 164:165-168.
3. Gallagher KL, Benfey PN. Not just another hole in the wall: Understanding intercellular protein trafficking. Genes Dev 2005; 19:189-189.
4. Holdaway-Clarke TL, Walker NA, Reid RJ et al. Cytoplasmic acidification with butyric acid does not alter the ionic conductivity of plasmodesmata. Protoplasma 2001; 215:184-190.
5. Terry BR, Robards AW. Hydrodynamic radius alone governs the mobility of molecules through plasmodesmata. Planta 1987; 171:145-157.
6. Heinlein M. Plasmodesmata: Dynamic regulation and role in macromolecular cell-to-cell signaling. Curr Opin Plant Biol 2002; 5:543-552.
7. Cilia ML, Jackson D. Plasmodesmata form and function. Trends Cell Biol 2004; 16:500-506.
8. Heinlein M, Epel BL. Macromolecular transport and signaling through plasmodesmata. Int Rev Cytol 2004; 235:93-164.
9. Lucas WJ, Lee JY. Plasmodesmata as a supracellular control network in plants. Nature Rev Mol Cell Biol 2004; 5:712-726.
10. Kim JY. Regulation of short-distance transport of RNA and protein. Curr Opin Plant Biol 2005; 8:45-52.
11. Kim JY, Yuan Z, Cilia M et al. Intercellular trafficking of a KNOTTED1 green fluorescent protein fusion in the leaf and shoot meristem of Arabidopsis. Proc Natl Acad Sci USA 2002; 99:4103-4108.
12. Roberts AG, Oparka KJ. Plasmodesmata and the control of the symplasmic transport. Plant Cell Envirom 2003; 26:103-124.
13. Goldberg GS, Valiunas V, Brink PR. Selective permeability of gap junction channels. Biochim Biophys Acta 2004; 1662:96-101.
14. Beachy RN, Heinlein M. Role of P30 in replication and spread of TMV. Traffic 2000; 1:540-544.
15. Citovsky V, Zambryski P. Systemic transport of RNA in plants. Trends Plant Sci 2000; 5:52-54.

16. Overall RL. Substructure of plasmodesmata. In: van Kesteren P, van Bel AJE, eds. Plasmodesmata, Structure, Function, Role in Cell Communication. Berlin, Heidelberg, New York: Springer, 1999:130-148.
17. Bayer E, Thomas CL, Maule AJ. Plasmodesmata in Arabidopsis thaliana suspension cells. Protoplasma 2004; 223:93-102.
18. Faulkner CR, Blackman LM, Cordwell SJ et al. Proteomic identification of putative plasmodesmatal proteins from Chara corallina. Proteomics 2005; 5:2866-2875.
19. Ehlers K, van Bel AJE. The physiological and developmental consequences of plasmodesmal connectivity. In: van Kesteren P, van Bel AJE, eds. Plasmodesmata, Structure, Function, Role in Cell Communication. Berlin, Heidelberg, New York: Springer, 1999:243-260.
20. Roberts IM, Boevink P, Roberts AG et al. Dynamic changes in the frequency and architecture of plasmodesmata during the sink-source transition in tobacco leaves. Protoplasma 2001; 218:31-44.
21. Staehelin LA, Hepler PK. Cytokinesis in higher plants. Cell 1996; 84:821-824.
22. Kollmann R, Glockmann C. Multimorphology and nomenclature of plasmodesmata in higher plants. In: van Kesteren P, van Bel AJE, eds. Plasmodesmata, Structure, Function, Role in Cell Communication. Berlin, Heidelberg, New York: Springer, 1999:149-172.
23. Ehlers K, Kollmann R. Primary and secondary plasmodesmata: Structure, origin and functioning. Protoplasma 2001; 216:1-30.
24. Steinberg G, Kollmann R. A quantitative analysis of the interspecific plasmodesmata in the nondivision walls of the plant chimera Laburnocytisus adamii (Poit.) Schneid. Planta 1994; 192:75-83.
25. van der Schoot C, Dietrich MA, Storms M et al. Establishment of a cell-to-cell pathway between separate carpels during gynoecium development. Planta 1995; 195:450-455.
26. Ehlers K, Kollmann R. Synchronization of mitotic activity in protoplast-derived Solanum nigrum L. microcalluses is correlated with plasmodesmal connectivity. Formation of branched plasmodesmata in regenerating Solanum nigrum protoplasts. Planta 2000; 210:269-278.
27. Overall RL, Wolfe J, Gunning BES. Intercellular communication in Azolla roots: I. Ultrastructure of plasmodesmata. Protoplasma 1982; 111:134-150.
28. Fisher DB. The estimated pore diameter for plasmodesmal channels in the Abutilon nectary trichome should be about 4nm, rather than 3nm. Planta 1999; 208:299-300.
29. Ding B, Turgeon R, Parthasarathy MV. Substructure of freeze-substituted plasmodesmata. Protoplasma 1992; 169:28-41.
30. Tilney LG, Cooke TJ, Connelly PS et al. The structure of plasmodesmata as revealed by plasmolysis, detergent extraction, and protease digestion. J Cell Biol 1991; 112:739-747.
31. Gunning BES, Overall RL. Plasmodesmata and cell-to-cell transport in plants. Bioscience 1983; 33:260-265.
32. Overall RL, Blackman LM. A models of the macromolecular structure of plasmodesmata. Trends Plant Sci 1996; 1:307-311.
33. Badelt K, White RG, Overall RL et al. Ultrastructural specializations of the cell wall sleeve around plasmodesmata. Am J Bot 1994; 81:1422-1427.
34. Schulz A. Plasmodesmal widening accompanies the short-term increase in symplasmic phloem unloading in pea root tips under osmotic stress. Protoplasma 1995; 188:22-37.
35. Olesen P, Robards AW. The neck region of plasmodesmata: General architecture and some functional aspects. In: Robards AW, Lucas WJ, Pitts JD et al, eds. Parallels in Cell to Cell Junctions in Plants and Animals. Berlin, Germany: Springer-Verlag, 1990:145-170.
36. Northcote DH, Davey R, Lay J. Use of antisera to localize callose, xylan and arabinogalactan in the cell-plate, primary and secondary walls of plant cells. Planta 1989; 178:353-366.
37. Delmer DP, Volokita M, Solomon M et al. A monoclonal antibody recognizes a 65 kDa higher plant membrane polypeptide which undergoes cation dependent association with callose synthase in vitro and colocalizes with sites of high callose deposition in vivo. Protoplasma 1993; 76:33-42.
38. Radford JE, White RG. Localisation of a myosin-like protein to plasmodesmata. Plant J 1998; 14:743-750.
39. Juniper BE. Some speculations on the possible roles of the plasmodesmata in the control of differentiation. J Theor Biol 1977; 66:583-592.
40. White RG, Badelt K, Overall RL et al. Actin associated with plasmodesmata. Protoplasma 1994; 180:169-184.
41. Ding B, Kwon MO, Warnberg L. Evidence that actin filaments are involved in controlling the permeability of plasmodesmata in tobacco mesophyll. Plant J 1996; 10:157-164.
42. Blackman LM, Overall RL. Immunolocalisation of the cytoskeleton to plasmodesmata of Chara corallina. Plant J 1998; 14:733-742.

43. Baluska F, Cvrckova F, Kendrick-Jones J et al. Sink plasmodesmata as gateways for phloem unloading. Myosin VIII and calreticulin as molecular determinants of sink strength? Plant Physiol 2001; 126:39-46.
44. Blackman LM, Harper JDI, Overall RL. Localization of a centrin-like protein to higher plant plasmodesmata. Eur J Cell Biol 1999; 78:297-304.
45. Baluska F, Samaj J, Napier R et al. Maize calreticulin localizes preferentially to plasmodesmata in root apex. Plant J 1999; 19:481-488.
46. Chen MH, Tian GW, Gafni Y et al. Effects of calreticulin on viral cell-to-cell movement. Plant Physiol 2005; 138:1866-1876.
47. Murillo I, Cavallarin L, San Segundo B. The maize pathogenesis-related PRms protein localizes to plasmodesmata in maize radicles. Plant Cell 1997; 9:145-156.
48. Ehlers K, Schulz M, Kollmann R. Subcellular localisation of ubiquitin in plant protoplasts and the function of ubiquitin in selective degradation of outer-wall plasmodesmata in regenerating protoplasts. Planta 1996; 199:139-151.
49. Blackman LM, Overall RL. Structure and function of plasmodesmata. Aust J Plant Physiol 2001; 28:709-727.
50. Yahalom A, Warmbrodt RD, Laird DW et al. Maize mesocotyl plasmodesmata proteins cross-react with connexin gap junction protein antibodies. Plant Cell 1991; 3:407-417.
51. Epel BL, van Lent JWM, Cohen L et al. A 41 kDa protein isolated from maize mesocotyl cell walls immunolocalizes to plasmodesmata. Protoplasma 1996; 191:70-78.
52. Aoki K, Kragler F, Xoconostle-Cazares B et al. A subclass of plant heat shock cognate 70 chaperones carries a motif that facilitates trafficking through plasmodesmata. Proc Natl Acad Sci USA 2002; 99:16342-16347.
53. Faulkner C, Brandom J, Maule A et al. Plasmodesmata 2004. Surfing the symplasm. Plant Physiol 2005; 137:607-610.
54. Sagi G, Katz A, Guenoune-Gelbart D et al. Class 1 reversibly glycosylated polypeptides are plasmodesmal-associated proteins delivered to plasmodesmata via the Golgi Apparatus. Plant Cell 2005; 17:1788-1800.
55. Escobar NM, Haupt S, Thow G et al. High-throughput viral expression of cDNA-green fluorescent protein fusions reveals novel subcellular addresses and identifies unique proteins that interact with plasmodesmata. Plant Cell 2003; 15:1507-1523.
56. Goodwin PB. Molecular size limit for movement in the symplast of the Elodea leaf. Planta 1983; 157:124-130.
57. Stadler R, Wright KM, Lauterbach C et al. Expression of GFP-fusions in Arabidopsis companion cells reveals nonspecific protein trafficking into sieve elements and identifies a novel post-phloem domain in roots. Plant J 2005; 41:319-331.
58. Kim I, Cho E, Crawford K et al. Cell-to-cell movement of GFP during embryogenesis and early seedling development in Arabidopsis. Proc Natl Acad Sci USA 2005; 102:2227-2231.
59. Ormenese S, Havelange A, Bernier G et al. The shoot apical meristem of Sinapis alba L. expands its central symplasmic field during the floral transition. Planta 2002; 215:67-78.
60. Gisel A, Hempel FD, Barella S et al. Leaf-to-shoot apex movement of symplastic tracer is restricted coincident with flowering in Arabidopsis. Proc Natl Acad Sci USA 2002; 99:1713-1717.
61. Cantrill LC, Overall RL, Goodwin PB. Changes in macromolecular movement accompany organogenesis in thin cell layers of Torenia fournieri. Planta 2005; 222:933-946.
62. Duckett CM, Oparka KJ, Prior DAM et al. Dye-coupling in the root epidermis of Arabidopsis is progressively reduced during development. Development 1994; 120:3247-3255.
63. Kim I, Hempel FD, Sha K et al. Identification of a developmental transition in plasmodesmatal function during embryogenesis in Arabidopsis thaliana. Development 2002; 129:1261-1272.
64. Wu X, Dinneny JR, Crawford KM et al. Modes of intercellular transcription factor movement in the Arabidopsis apex. Development 2003; 130:3735-3745.
65. Cantrill LC, Overall RL, Goodwin PB. Cell-to-cell communication via plant endomembranes. Cell Biol Int 1999; 23:653-661.
66. Waigmann E, Turner A, Peart J et al. Ultrastructural analysis of leaf trichome plasmodesmata reveals major differences from mesophyll plasmodesmata. Planta 1997; 203:75-84.
67. Lazzaro MD, Thomson WW. The vacuolar-tubular continuum in living trichomes of chickpea (Cicer arietinum) provides a rapid means of solute delivery from base to tip. Protoplasma 1996; 193:181-190.
68. Gamalei YV, van Bel AJE, Pakhomova MV et al. Effects of temperature on the conformation of the endoplasmic reticulum and on starch accumulation in leaves with the symplasmic minor-vein configuration. Planta 1994; 194:443-453.

69. Velikanov GA, Volobueva OV, Belova LP et al. Vacuolar symplast as a regulated pathway for water flows in plants. Russ J Plant Physiol 2005; 52:326-331.
70. Chen MH, Sheng J, Hind G et al. Interaction between the tobacco mosaic virus movement protein and host cell pectin methylesterases is required for viral cell-to-cell movement. EMBO J 2000; 19:913-920.
71. Crawford KM, Zambryski PC. Subcellular localization determines the availability of nontargeted proteins to plasmodesmatal transport. Curr Biol 2000; 10:1032-1040.
72. Imlau A, Truernit E, Sauer N. Cell-to-cell and long distance trafficking of the green fluorescent protein in the phloem and symplastic unloading of the protein into sink tissues. Plant Cell 1999; 11:309-322.
73. Ishiwatari Y, Fujiwara T, McFarland KC et al. Rice phloem thioredoxin h has the capacity to mediate its own cell-to-cell transport through plasmodesmata. Planta 1998; 205:12-22.
74. Di Sansebastiano GP, Paris N, Marc-Martin S et al. Specific accumulation of GFP in a nonacidic vacuolar compartment via a C-terminal propeptide-mediated sorting pathway. Plant J 1998; 15:449-457.
75. Sessions A, Yanofsky MF, Weigel D. Cell-cell signaling and movement by the floral transcription factors LEAFY and APETALA1. Science 2000; 289:779-782.
76. Flores R, Hernandez C, Alba AE et al. Viroids and viroid-host interactions. Annu Rev Phytopathol 2005; 43:117-139.
77. Tremblay D, Vaewhongs AA, Turner KA et al. Cell wall localization of Red clover necrotic mosaic virus movement protein is required for cell-to-cell movement. Virology 2005; 333:10-21.
78. Heinlein M, Epel BL, Padgett HS et al. Interaction of tobamovirus movement proteins with the plant cytoskeleton. Science 1995; 270:1983-1985.
79. McLean BG, Zupan J, Zambryski PC. Tobacco mosaic virus movement protein associates with the cytoskeleton in tobacco cells. Plant Cell 1995; 7:2101-2114.
80. Reichel C, Mas P, Beachy RN. The role of the ER and cytoskeleton in plant viral trafficking. Trends Plant Sci 1999; 4:458-462.
81. Trutnyeva K, Bachmaier R, Waigmann E. Mimicking carboxyterminal phosphorylation differentially effects subcellular distribution and cell-to-cell movement of Tobacco mosaic virus movement protein. Virology 2005; 332:563-577.
82. Lee JY, Taoka KI, Yoo BC et al. Plasmodesmal-associated protein kinase in tobacco and Arabidopsis recognizes a subset of noncell-autonomous proteins. Plant Cell 2005; 17:2817-2831.
83. Yoshioka K, Matsushita Y, Kasahara M et al. Interaction of tomato mosaic virus movement protein with tobacco RIO kinase. Mol Cells 2004; 17:223-229.
84. Dorokhov YL, Makinen K, Frolova OY et al. A novel function for a ubiquitous plant enzyme pectin methylesterase: The host-cell receptor for the tobacco mosaic virus movement protein. FEBS Letts 1999; 461:223-428.
85. Lee JY, Yoo BC, Rojas MR et al. Selective trafficking of noncell-autonomous proteins mediated by NtNCAPP1. Science 2003; 299:392-396.
86. Lee JY, Yoo BC, Lucas WJ. Parallels between nuclear-pore and plasmodesmal trafficking of information molecules. Planta 2000; 210:177-187.
87. Lucas WJ, Bouche-Pillon S, Jackson DP et al. Selective trafficking of KNOTTED1 homeodomain protein and its mRNA through plasmodesmata. Science 1995; 270:1980-1983.
88. Nakajima K, Sena G, Nawy T et al. Intercellular movement of the putative transcription factor SHR in root patterning. Nature 2001; 413:307-311.
89. Kim M, Canio W, Kessler S et al. Developmental changes due to long-distance movement of a homeobox fusion transcript in tomato. Science 2001; 293:287-289.
90. Kim JY, Rim Y, Wang J et al. A novel cell-to-cell trafficking assay indicates that the KNOX homeodomain is necessary and sufficient for intercellular protein and mRNA trafficking. Genes Dev 2005; 19:788-793.
91. Haywood V, Yu TS, Huang NC et al. Phloem long-distance trafficking of GIBBERELLIC ACID-INSENSITIVE RNA regulates leaf development. Plant J 2005; 42:49-68.
92. Qi Y, Pelissier T, Itaya A et al. Direct role of a viroid RNA motif in mediating directional RNA trafficking across a specific cellular boundary. Plant Cell 2004; 16:1741-1752.
93. Gomez G, Torres H, Pallas V. Identification of translocatable RNA-binding phloem proteins from melon, potential components of the long-distance RNA transport system. Plant J 2005; 41:107-116.
94. Golecki B, Schulz A, Thompson GA. Translocation of structural P proteins in the phloem. Plant Cell 1999; 11:127-140.
95. Xoconostle-Cazares B, Yu X, Ruiz-Medrano R et al. Plant paralog to viral movement protein that potentiates transport of mRNA into the phloem. Science 1999; 283:94-98.

96. Himber C, Dunoyer P, Moissiard G et al. Transitivity-dependent and -independent cell-to-cell movement of RNA silencing. EMBO J 2003; 22:4523-4533.
97. Yoo BC, Kragler F, Varkonyi-Gasic E et al. A systemic small RNA signaling system in plants. Plant Cell 2004; 16:1979-2000.
98. Gillespie T, Oparka KJ. Plasmodesmata-gateways for intercellular communication in plants. In: Flemming AJ, ed. Intercellular Communication in Plants. Oxford: Blackwell Publishing, 2005:109-146.
99. Samaj J, Peters M, Volkman D et al. Effects of myosin ATPase inhibitor 2,3-butanedione 2 monoxime on distributions of myosin, F-actin, microtubules, and cortical endoplasmic reticulum in maize root apices. Plant Cell Physiol 2000; 41:571-582.
100. Hughes JE, Gunning BES. Glutaraldehyde-induced deposition of callose. Can J Bot 1980; 58:250-258.
101. Rinne PLH, Boogaard R, van den Mensink MGJ et al. Tobacco plants respond to the constitutive expression of the tospovirus movement protein NSM with a heat-reversible sealing of plasmodesmata that impairs development. Plant J 2005; 43:688-707.
102. Rinne PL, Kaikuranta PM, van der Schoot C. The shoot apical meristem restores its symplasmic organization during chilling-induced release from dormancy. Plant J 2001; 26:249-264.
103. Seagull RW. Differences in the frequency and disposition of plasmodesmata resulting from root cell elongation. Planta 1983; 159:497-504.
104. Schnepf E, Sych A. Distribution of plasmodesmata in developing Sphagnum leaflets. Protoplasma 1983; 116:51-56.
105. Palevitz BA, Hepler PK. Changes in dye coupling of stomatal cells of Allium and Commelina demonstrated by microinjection of Lucifer yellow. Planta 1985; 164:473-479.
106. Ruan YL, Xu SM, White R et al. Genotypic and developmental evidence for the role of plasmodesmatal regulation in cotton fiber elongation mediated by callose turnover. Plant Physiol 2004; 136:4104-4113.
107. Erwee MG, Goodwin PB, van Bel AJE. Cell-cell communication in the leaves of Commelina cyanea and other plants. Plant Cell Environm 1985; 8:173-178.
108. Ueki S, Citovsky V. Control improves with age: Intercellular transport in plant embryos and adults. Proc Natl Acad Sci USA 2005; 102:1817-1818.
109. Liarzi O, Epel BL. Development of a quantitative tool for measuring changes in the coefficient of conductivity of plasmodesmata induced by developmental, biotic, and abiotic signals. Protoplasma 2005; 225:67-76.
110. Blackman LM, Gunning BES, Overall RL. A 45 kDa protein isolated from the nodal walls of Chara corallina is localised to plasmodesmata. Plant J 1998; 15:401-411.
111. Itaya A, Liang GQ, Woo YM et al. Nonspecific intercellular protein trafficking probed by green-fluorescent protein in plants. Protoplasma 2000; 213:165-175.

CHAPTER 7

Sieve-Pore Plugging Mechanisms

Aart J.E. van Bel*

Abstract

S ieve tubes seem to dispose of diverse mechanisms which are able to seal or plug sieve pores—very specialized and evolutionarily transformed cell-cell channels—in response to injury. Two major devices for their closure are distinguished: a putative rapid (several seconds) plugging by phloem-specific proteins and a slower (several minutes) sealing by callose. Sieve-pore plugging is probably executed by several classes of phloem-specific proteins such as water-soluble proteins, proteins which occur as a fine meshwork in the sieve-element lumen, parietal proteins attached to the sieve-element plasma membrane and proteins ordered in so-called forisomes. A striking feature of the forisomes is the calcium-mediated reversibility of dispersion and contraction, even in vitro. The calcium-dependence of forisome conformation and callose production suggests a general calcium regulation of sieve-pore closure.

Cell Biology of the Sieve Element/Companion Cell Complex

Sieve tubes in angiosperms are longitudinally arranged modules of sieve element/companion cell complexes (SE/CCs) which extend by successive differentiation at their meristematic ends.[1] During ontogeny, the SE/CC-precursor divides longitudinally giving rise to two completely disparate cells.[2-4] One may or may not divide transversely, producing metabolically active companion cells (CCs) with a dense cytoplasm. The other cell goes through a "semi-apoptosis" during which most of the cellular outfit is sacrificed to the transport function by partial autolysis leading to the differentiation of a sieve element (SE).

A mature SE is delineated by a plasma membrane and only contains a few mitochondria, a reduced ER and SE plastids, all in a parietal position.[1,5,6] In addition to that, SEs in dicotyledons are lined with parietal proteins. Between SEs and CCs, special plasmodesmata (PPUs) occur with numerous branches at CC-side and one single corridor at the SE side.[7] Plasmodesmata are absent at other SE-interfaces.[8] Standard plasmodesmata (PDs) occur at the interface between CCs and phloem parenchyma cells (PPCs). As a third PD-type of the SE/CC, meristematic PDs between SE mother cells are transformed into sieve pores during differentiation of the SE/CC in order to increase mass flow capacity through the sieve tubes.[9]

The absence of a nucleus makes an SE virtually fully dependent on the genetic and metabolic activities of its CC(s). Consequently, turnover of macromolecular SE components must take place in the CCs and trafficked through the PPUs.[10] Evidence for the production of SE proteins in CCs and protein turnover in SEs has been provided by experiments using ^{35}S-methionine.[11,12] The genes for the phloem lectin PP2[13] and the filamentous phloem-specific protein PP1[14] have shown to be expressed exclusively in the CCs of cucurbits. The rate of expression is related to the vascular differentiation.[15] Carrier proteins engaged in sucrose uptake by SEs were also found to be produced in

*Aart J.E. van Bel—Plant Cell Biology Group, Institute of General Botany, Senckenbergstrasse 17, Justus-Liebig-University, D-35390 Giessen, Germany. Email: aart.v.bel@bot1.bio.uni-giessen.de

Cell-Cell Channels, edited by Frantisek Baluska, Dieter Volkmann and Peter W. Barlow. ©2006 Landes Bioscience and Springer Science+Business Media.

CCs.[16] It is therefore likely that all phloem-specific proteins are produced in CCs and transported to SEs via the PPUs.[11,12,17-21]

The nature of the intercellular CC-to-SE transfer is still a mystery,[22-25] the more as the proteins trafficked have diverse molecular weights and do not show common motifs.[18] The current belief is that the proteins are actively carried by cytoskeleton components through the PPUs.[26]

Mass Flow and Phloem-Specific Proteins

Mass flow is created by a hydrostatic difference between sources and sinks. Continuous collection of photassimilates by the source ends of the sieve tubes results in a high local turgor pressure in the leaf vein ends. By contrast, substances are released at the sink ends of the sieve tubes, resulting in a local drop of turgor potential. The difference in turgor potential drives a mass flow of sieve tube sap which contains a multitude of solutes and soluble macromolecules. The phloem-specific proteins belong the longer known classes of sieve-tube-specific macromolecules.[5,27] They occur in SEs in mostly soluble and several nonsoluble forms. Predictions on their number go up to approx. 500, although only some 200 have been shown with certainty.[21,28]

Presumably most of the phloem-specific proteins are in a soluble form, a few are present as insoluble deposits along the SE plasma membrane of dicotyledons,[5,9] whereas again others are dispersed as a fine meshwork throughout the SE lumina.[29] Apart from these proteins free-lying in the sieve tube lumen, proteins in a more enclosed form have been described. In many plant families, proteinaceous bodies are enclosed in so-called SE plastids. These plastids occur in SEs of all angiosperms and are of a smaller size ($<1\mu$) than plastids in other cell types.[5,30] In SEs of legumes, spindle-like protein bodies up to a size of 100 micron in length have been found.[31-53]

Their species-specificity has been employed to show that phloem-specific proteins are phloem-mobile. Stock-specific proteins were found in phloem exudates collected from scions of intergeneric grafts of cucurbits.[36] These proteins were immunolocalized in SEs of scions.[37] RNA blot analysis and reverse transcription polymerase chain reaction indicate that proteins rather than their transcripts are translocated across the graft border.[36] Even structural proteins can cross intergeneric graft borders of cucurbit species.[36,37] The filamentous PP1 of *Cucurbita maxima* can undergo transformational changes and is translocated in a 88 kDa globular form.[38]

Sieve-Plate Plugging by Phloem-Specific Proteins

The function of phloem-specific proteins is largely unknown with one obvious exception: phloem-specific proteins are involved in plugging sieve plates after wounding.[39] It explains why such diverse proteinaceous forms have been found in sieve tubes in electron microscopic (EM) pictures published in the seventies.[5] Thick clots of proteins deposited on the sieve plates in these EM pictures[40,41] elicited a fierce debate between the proponents of mass flow and those who advocated other mechanisms for phloem translocation.[42] However, as the sieve pores in EM pictures were virtually empty after careful preparation,[29,41,43-45] the massive protein deposits on sieve plates are now considered to be artefacts induced by wounding. Apparently, preparative cutting for EM triggered plugging reactions induced by wounding.

Mass flow of solutes through the sieve tubes has been visualized by CLSM.[39] Experiments in which CFDA (carboxyflorescein diacetate) was introduced into the sieve tubes of intact plants showed rapid CF translocation unhindered by sieve plates.[39] CLSM pictures also evidence that mass flow is not impeded by protein deposition onto sieve plates of *Vicia faba*.[29] Some degree of protein plugging of sieve plates leaving corridors for mass flow has also been found in EM pictures of *Vicia faba* phloem tissue exposed to "gentle" fixation.[29] The occurrence of mass flow explains the need for extensive anchoring of SE plastids, mitochondria and ER between each other and the SE plasma membrane. The resultant myctoplasmic sheath along the SE plasma membrane is entangled with the SE plasma membrane by anchors of unknown molecular origin.[29]

Sealing of sieve plates is presented in most textbooks as a means against leakage of the precious phloem sap under pressure. More than any other cell type SEs are endangered to act as a massive leak for plant saps. The internal pressure is high, the sieve pores between the SEs are large and down-regulation of turgor pressure is difficult due to an inadequate cellular machinery. However, the function of plugging may be less obvious than it may appear. When leaves or twigs are snatched away by wind or herbivores, the driving force for phloem mass flow is dissipated due to removal of the source; phloem sap is not exuding under these circumstances. Therefore, maintenance of the pressure conditions in the plant or a defence against phytopathogens are more likely functions of SE plugging. Severed sieve tubes give easy access to phytopathogens. The lectin character of PP2 in *Cucurbita* phloem sap[46] may be meaningful in this frame. It should be noted that the above leakage argument may still hold for tree stem browsing by herbivores or beetles.

Mechanisms of Sieve-Plate Sealing

Plugging of sieve pores may be a closure response to cell damage evolutionarily inherited from standard PDs. Abrupt damage-related changes in turgor may be responsible for PD closure, as sudden changes in turgor pressure brought about PD closure in tobacco leaf hairs.[47] Yet, it remains unclear whether the closure mechanisms of PDs and sieve pores are actually similar. PD closure has often been related to production of callose in the walls around the PD neck region.[48-53] In this way, callose does not cause plugging but rather "strangling" of PD orifices. The same would hold for sieve pore closure provided that the callose production mechanisms around PDs and sieve pores are similar. Production of callose around the sieve pores may be too slow to meet the demands for rapid sealing of sieve pores. Therefore, proteins seem to be more likely candidates responsible for a sudden sieve plate plugging presenting a provisional line of defence, whereas sealing by callose may have a more definitive nature.[1]

In *Vicia faba* sieve tubes, parietal phloem-specific proteins as well as the spindle-like proteins seem to be responsible for sieve plate sealing. Both are instantly deposited onto the sieve plate in response to damage. It is not excluded that also water-soluble proteins are also involved in sealing, but I am not aware of any convincing report on this point. Among the best investigated phloem-specific proteins are the so-called forisomes in legume sieve tubes which used to be designated as crystalline protein bodies. These "crystalline protein bodies " have only been observed in members of the *Fabaceae*.[31-35]

Dispersion and Contraction of Forisomes

The striped appearance of forisomes in EM pictures led to the belief that they had crystalline properties. Recent investigation have shown that forisomes do not possess a crystalline structure (Peters et al unpublished). Forisomes presumably occur exclusively in the SEs of *Fabaceae* and their size often corresponds with the SE diameter. They mostly posses a spindle-like form with or without tails,[33,34] are up to 100 μm in length as in *Canavallia* (Hanakam et al, unpublished) and are composed of numerous subunits (Knoblauch et al unpublished). Quite recently, sequencing of forisome proteins has been accomplished, but their spatial structure remains unresolved as yet (Noll et al, unpublished).

Forisomes disperse in response to damage or to turgor disturbance and seem to recontract spontaneously in sieve tubes.[54] The expansion which can amount to 300% of the contracted volume has been linked to calcium influx caused by damage.[55] Both in vivo and in vitro (isolated forisomes), forisomes disperse in reaction to calcium supply and contract upon addition of the calcium chelator EDTA.[54-55] None of the other (divalent) ions has the same effect on forisomes.[54] As for the turgor effect, the involvement of mechanoreceptors has been postulated: either mechano-receptive calcium channels or mechanoreceptors linked with calcium channels.[54]

Physiological Triggers of the Sieve-Plate Plugging

The source of calcium set free into the SE lumen may be in the cell wall or in the SE ER. Fluorochrome-recorded Ca^{2+}-influx into SEs is mostly seen as a narrow fluorescent lining along the SE plasma membrane. This points to the narrow parietal ER as the storage compartment for calcium[56] set free in response to injury. On the other hand, forisome dispersion in reaction to externally supplied calcium[54] and to electrical potential waves (Hafke et al, unpublished) speaks for the cell wall as the source of calcium. The latter events have been linked to the activity of calcium channels residing in the plasma membrane, although calcium channels have been identified in SEs only once.[57] In vivo spontaneous recontraction of the forisomes after a turgor shock or an electrical stimulus have been ascribed tentatively to the activity of calcium pumps. Ongoing experiments using SE protoplasts are expected to elucidate presence and nature of calcium channels and pumps in the SE plasma membrane.

How dispersion and contraction of forisomes operate in the absence of ATP or other organic energy suppliers is unknown. Perhaps, sudden binding of calcium at vital amino-acid cross-links modifies the molecular protein structure allowing a drastic influx of water. How the escape of water also provides energy[55] and how a forisome attains its original structure after withdrawal of calcium remains a mystery.

In addition to Ca^{2+}, oxygen may be responsible for coagulation of phloem-specific proteins. Usually, oxygen concentration in the phloem is close to 1%.[58] By consequence, wounding is accompanied by a sharp rise in oxygen which may result in clogging of some proteins. A high reactivity of cucurbit phloem sap to oxygen has been ascribed to the coagulation of certain proteins.[59] It may explain why phloem sap in aphid stylets starts gelling after microcauterisation and why collection of phloem sap by microcapillaries has hardly been successful thus far. However, clogging of sieve element sap inside cut aphid stylets may also be due to the absence of saliva components which may prevent protein clogging in several putative ways.[60]

References

1. van Bel AJE. The phloem, a miracle of ingenuity. Plant Cell Environ 2003; 26:125-150.
2. Esau K. The Phloem. Encyclopedia of Plant Anatomy, Vol 5. Berlin: Bornträger, 1969.
3. Thorsch J, Esau K. Changes in the endoplasmic reticulum during differentiation of a sieve element in Gossypium hirsutum. J Ultrastr Res 1981; 74:183-194.
4. Thorsch J, Esau K. Nuclear degeneration and the association of endoplasmic reticulum with the nuclear envelope and microtubules in maturing sieve elements of Gossypium hirsutum. J Ultrastr Res 1981; 74:195-204.
5. Behnke H-D, Sjolund RD. Sieve elements. Comparative Structure, Induction and Development. Berlin: Springer, 1990.
6. Sjolund RD. The phloem sieve element. A river flows through it. Plant Cell 1997; 9:1137-1146.
7. Ding B, Turgeon R, Parthasaraty MV. Substructure of freeze-substituted plasmodesmata. Protoplasma 1992; 169:28-41.
8. Kempers R, Ammerlaan A, van Bel AJE. Symplasmic constriction and ultrastructural features of the sieve element/companion cell complex in the transport phloem of apoplasmically and symplasmically phloem-loading species. Plant Physiol 1998; 116:271-278.
9. van Bel AJE, Knoblauch M. Sieve element and companion cell: The story of the comatose patient and the hyperactive nurse. Austr J Plant Physiol 2000; 27:477-487.
10. van Bel AJE, Ehlers K, Knoblauch M. Sieve elements caught in the act. Trends Plant Sci 2002; 7:126-132.
11. Fisher DB, Wu K, Ku MSB. Turnover of soluble proteins in the wheat sieve tube. Plant Physiol 1992; 100:1433-1441.
12. Sakuth T, Schobert C, Pecsvaradi A et al. Specific proteins in the sieve-tube exudate of Ricinus communis L. seedlings: Separation, characterization and in-vivo labelling. Planta 1993; 191:207-213.
13. Bostwick DE, Dannenhoffer JM, Skaggs MI et al. Pumpkin phloem lectin genes are specifically expressed in companion cells. Plant Cell 1992; 4:1539-1548.
14. Clark AM, Jacobsen KR, Dannenhoffer JM et al. Molecular characterization of a phloem-specific gene encoding for the filament protein, phloem protein 1 (PP1) from Cucurbita maxima. Plant J 1997; 12:49-61.
15. Dannenhoffer JM, Schulz A, Skaggs MI et al. Expression of the phloem lectin is developmentally linked to vascular differentiation in cucurbits. Planta 1997; 201:405-414.
16. Kühn C, Franceschi VR, Schulz A et al. Macromolecular trafficking indicated by localization and turnover of sucrose transporters in enucleate sieve elements. Science 1997; 275:1298-1300.
17. Nakamura S, Hayashi H, Mori S et al. Protein phosphorylation in the sieve tubes of rice plants. Plant Cell Physiol 1993; 34:927-933.
18. Schobert C, Großmann P, Gottschalk M et al. Sieve-tube exudate from Ricinus communis L seedlings contains ubiquitin and chaperones. Planta 1995; 196:205-210.

19. Schobert C, Baker L, Szederkenyi J et al. Identification of immunologically related proteins in sieve-tube exudate collected from monocotyledonous and dicotyledonous plants. Planta 1998; 206:245-252.
20. Thompson GA, Schulz A. Long-distance transport of macromolecules. Trends Plant Sci 1999; 4:354-360.
21. Hayashi H, Fukuda A, Suzui N et al. Proteins in the sieve tube-companion cell complexes: Their detection, localization and possible functions. Austr J Plant Physiol 2000; 27:489-496.
22. Kragler F, Monzer J, Shash K et al. Cell-to-cell transport of proteins: Requirement of unfolding and characterisation of binding to a putative plasmodesmatal receptor Plant J 1998; 15:367-381.
23. Oparka KJ, Turgeon R. Sieve elements and companion cells – Traffic control centers of the phloem. Plant Cell 1999; 11:739-750.
24. Oparka KJ, Santa Cruz S. The great escape: Phloem transport and unloading of macromolecules. Annu Rev Plant Physiol Plant Mol Biol 2000; 51:323-347.
25. Lucas WJ, Yoo BC, Kragler F. RNA as a long-distance information macromolecule in plants. Nature Rev Mol Cell Biol 2001; 2:849-857.
26. Oparka KJ. Getting the message across how do plant cells exchange macromolecular complexes. Trends Plant Sci 2004; 9:33-41.
27. Cronshaw J, Sabnis DD. Phloem proteins. In: Behnke H-D, Sjolund RD, eds. Sieve Elements Comparative Structure, Induction and Development. Berlin: Springer, 1990:275-283.
28. Haebel S, Kehr J. Matrix-assisted desorption/ionization time of flight mass spectrometry peptide mass fingerprints and post source decay: A tool for the identification and analysis of phloem proteins from Cucurbita maxima Duch. separated by two-dimensional polyacrylamide gel electrophoresis. Planta 2001; 213:586-593.
29. Ehlers K, Knoblauch M, van Bel AJE. Ultrastructural features of well-preserved and injured sieve elements: Minute clamps keep the phloem transport conduits free for mass flow. Protoplasma 2000; 214:80-92.
30. Behnke H-D. Distribution and evolution of forms and types of sieve-element plastids in the dicotyledons. ALISO 1991; 3:167-182.
31. Palevitz BA, Newcomb EH. The ultrastructure and development of tubular and crystalline P protein in the sieve elements of certain papilionaceous legumes. Protoplasma 1971; 72:399-427.
32. Wergin WP, Palevitz BA, Newcomb EH. Structure and development of P-protein in phloem parenchyma and companion cells of legumes. Tissue and Cell 1975; 7:227-242.
33. Lawton DM. P-protein crystals do not disperse in uninjured sieve elements in roots of runner bean (Phaseolus multiflorus) fixed with glutaraldehyde. Ann Bot 1978; 42:353-361.
34. Lawton DM. Ultrastructural comparison of the tailed and tailless P-proteins respectively of runner bean (Phaseolus vulgaris) and garden pea (Pisum sativum) with tilting stage electron microscopy. Protoplasma 1978; 97:1-11.
35. Behnke H-D. Non dispersive protein bodies in sieve elements: A survey and review of their origin, distribution and taxonomic significance. IAWA Bulletin 1991; 12:143-175.
36. Golecki B, Schulz A, Thompson GA. Translocation of structural P proteins in the phloem. Plant Cell 1999; 11:127-140.
37. Golecki B, Schulz A, Carstens-Behrens U et al. Evidence for graft transmission of structural phloem proteins or their precursors in heterografts of Cucurbitaceae. Planta 1998; 206:630-640.
38. Leineweber K, Schulz A, Thompson GA. Dynamic transitions in the translocated phloem filament protein. Austr J Plant Physiol 2000; 27:733-741.
39. Knoblauch M, van Bel AJE. Sieve tubes in action. Plant Cell 1998; 10:35-50.
40. Robidoux J, Sandborn EB, Fensom DS et al. Plasmatic filaments and particles in mature sieve elements of Heracleum sphondylium under the electron microscope. J Exp Bot 1973; 79:349-359.
41. Johnson RPC, Freundlich A, Barclay GF. Transcellular strands in sieve tubes: What are they? J Exp Bot 1976; 101:1127-1136.
42. Spanner DC. Sieve plates, open or occluded? A critical review. Plant Cell Environ 1978; 1:7-20.
43. Evert RF, Tucker CM, Davis JD et al. Light microscope investigation of sieve element ontogeny and structure in Ulmus americana. Amer J Bot 1969; 56:999-1017.
44. Kollmann R. Cytologie des Phloems. In: Ruska H, Sitte P, eds. Grundlagen der Cytologie. Jena: Gustav Fisher, 1973:479-505.
45. Sjolund RD, Shih CY. Freeze-fracture analysis of phloem structure in plant tissue cultures. I. The sieve element reticulum. J Ultrastruct Res 1983; 82:111-121.
46. Read SM, Northcote DH. Chemical and immunological similarities between the phloem proteins of three genera of Cucurbitaceae. Planta 1983; 158:119-127.
47. Oparka KJ, Prior DAM. Direct evidence for pressure generated closure of plasmodesmata. Plant J 1992; 2:741-750.

48. Drake GA, Carr DJ, Anderson WP. Plasmolysis, plasmodesmata, and the elctrical coupling of oat coleoptile cells. J Exp Bot 1978; 29:1205-1214.
49. Wolf S, Deom CM, Beachy R et al. Plasmodesmatal function is probed using transgenic tobacco plants that express a virus MP. Plant Cell 1991; 3:593-604.
50. Radford J, Vesk M, Overall RL. Callose deposition at plasmodesmata. Protoplasma 1998; 201:30-37.
51. Botha CEJ, Cross RHM. Towards reconciliation of structure with function in plasmodesmata – Who is the gatekeeper? Micron 2000; 21:713-721.
52. Iglesias VA, Meins F. Movement of plant viruses is delayed in a β-1,3-glucanase-deficient mutant showing a reduced plasmodesmatal SEL and enhanced callose deposition. Plant J 2000; 21:157-166.
53. Roberts A, Oparka KJ. Plasmodesmata and the control of symplastic transport. Plant Cell Environ 2003; 26:103-124.
54. Knoblauch M, Peters WS, Ehlers K et al. Reversible calcium-regulated stopcocks in legume sieve tubes. Plant Cell 2001; 13:1221-1230.
55. Knoblauch M, Noll GA, Müller T et al. ATP-independent contractile proteins from plants. Nature Materials 2003; 2:600-603.
56. Arsanto JP. Ca^{2+} binding sites and phosphatase activities in sieve element reticulum and P-protein of chickpea phloem: A cytochemical and X-ray microanalysis survey. Protoplasma 1986; 132:160-171.
57. Volk G, Franceschi V. Localization of a calcium channel-like protein in the sieve element plasma membrane. Aust J Plant Physiol 2000; 27:779-786.
58. van Dongen JT, Schurr U, Pfister M et al. Phloem metabolism and function have to cope with low internal oxygen. Plant Physiol 2003; 131:1529-1543.
59. Alosi MC, Melroy DI, Park RB. The regulation of gelation of exudate from Cucurbita by dilution, glutathione and glutathione reductase. Plant Physiol 1988;86: 1089-1094.
60. Will T, van Bel AJE. Physical and chemical interactions between aphids and plants. J Exp Bot 2006; in press.

CHAPTER 8

Actin and Myosin VIII in Plant Cell-Cell Channels

Jozef Samaj,* Nigel Chaffey, Uday Tirlapur, Jan Jasik, Andrej Hlavacka, Zhan Feng Cui, Dieter Volkmann, Diedrik Menzel and Frantisek Baluska

Abstract

Plasmodesmata (PD) are cell-cell channels interconnecting all the cells of the plant body into a huge syncytium, which makes plants 'supracellular' organisms. Recent studies have clearly revealed that both the targeting and gating of PD is highly regulated. Importantly, it is known that molecules below the size exclusion limits, such as auxin and calcium, cannot pass freely through plasmodesmata, strongly implicating a very effective sieve-like structure within PD. Even though the first PD-resident proteins emerge from recent molecular studies, the majority of these elusive proteins remain to be unveiled. Convincing evidence suggests that F-actin meshworks, supported by myosins of class VIII and perhaps also by ARP2/3 proteins and other actin-binding proteins, are components of gateable PD and potentially involved in sieve-like nature of these cell-cell channels. Interestingly, there are several structural and functional similarities between PD and nuclear pores. For example, homeodomain transcription factors are able to pass through both nuclear pores as well as PD via similar mechanisms. Some cells are also symplasmically isolated such as stomata and trichoblasts initiating root hairs, and root cap statocytes. The root cap statocytes and their PD are depleted in both F-actin and myosin VIII, implying that these cytoskeletal molecules are essential for transport through PD. Gravistimulation triggers the opening of a special subset of PD in the root cap statocytes suggesting that a physical force, such as gravity, is capable of targeting molecules/ processes which regulate the gating of PD.

Introduction

Plasmodesmata (PD) are cell-cell channels which traverse the cell walls of neighbouring plant cells providing a symplasmic continuum between these cells organized into complex tissues and/ or organs. In this way almost all plant cells within an organ are interconnected and the plant body can be considered as a large syncytium. PD not only provide physical cell-cell connectivity but also, and more importantly, play a crucial role in intercellular transport of nutrients, RNAs, signalling molecules, transcription factors and the spread of diverse plant viruses. A broad spectrum of molecules destined for intercellular transport is first targeted to PD and then transported across them in order to reach the next cell. Mechanisms controlling this transport have just started to be unveiled during recent years. Accordingly, knowledge of the structural components of PD has also improved (reviewed by ref. 1). Here we are focusing on the actomyosin cytoskeleton,

*Corresponding Author: Jozef Samaj—Institute of Cellular and Molecular Botany, University of Bonn, Kirschallee 1, 53115 Bonn, Germany, Email: jozef.samaj@uni-bonn.de. Also, Institute of Plant Genetics and Biotechnology, Slovak Academy of Sciences, Akademicka 2, SK-950 07 Nitra, Slovak Republic.

Cell-Cell Channels, edited by Frantisek Baluska, Dieter Volkmann and Peter W. Barlow.
©2006 Landes Bioscience and Springer Science+Business Media.

which represents an important structural and functional component of PD. In particular, we would like to highlight the potential functions of the actomyosin cytoskeleton in loading of PD and transport of cargo through the PD apparatus.

How Are Proteins and Other Molecules Targeted to PD?

Two principal mechanisms of protein targeting to PD are plausible; involving either nontargeted movement via diffusion through cytoplasm, or targeted movement based on association of the transported protein to other endomembrane proteins and/or to cytoskeletal proteins with subsequent transport along cytoskeletal arrays.

The most comprehensive information comes from screens for interaction partners of virus movement proteins (see other chapters in this volume) as well as studies focused on targeting of these proteins to PD. These movement proteins are absolutely essential for virus spread and associate with other proteins, such as SNAREs and small GTPases of the Rab family, which are involved in fusion and movement of intracellular vesicles. For example, the movement protein of cauliflower mosaic virus (CMV) directly interacts with protein PRA1 (prenylated Rab acceptor) which further associates both with Rab and SNAP (v-SNARE), while the movement protein of grape vine fan leaf virus interacts with cytokinesis-specific syntaxin KNOLLE, a plant t-SNARE protein involved in membrane fusion events.[2,3] Additionally, GFP-tagged N-terminal part of Rab11 localizes to PD when expressed in *Nicotiana benthamiana* leaf cells,[4] while expression of antisense construct of LeRab11a in tomato plants causes a pleiotropic phenotype encompassing impaired growth, reduced apical dominance, abnormal floral structure with branched inflorescences, and ectopic shoot growth on leaves.[5] This phenotype might be related to the specific function of LeRab11a in the movement mechanism of proteins through PD. Altogether, these data suggest a possible role of vesicular trafficking in subcellular targeting of movement proteins to PD.

On the other hand, the movement protein of tobacco mosaic virus as well as NtNCAPP1 and CmPP16, two proteins that participate in its movement, all localize to the ER.[6,7] Since the ER, along with the plasma membrane, is also continuous from cell-to-cell through PD, in the form of the PD desmotubule,[1] and closely associates with the cytoskeleton, this might represent an alternative mechanism of protein delivery to PD. Although the ER nature of the desmotubule is widely accepted and almost unquestioned, there are new data which suggest that tubular vacuoles can also traverse PD. For instance, pulsed field gradient NMR technique revealed vacuolar transport of water from cell-to-cell in root apices.[8-10] Moreover, in vivo studies revealed that tubular and dynamic vacuoles invade the phragmoplast[11,12] and obviously traverse the forming cell plates in dividing BY-2 cells.[12]

Proteins destined for targeted cell-cell traffic through PD are usually described as noncell-autonomous proteins. Interestingly, homeodomain transcription factors such as KNOTTED1 (KN1), LEAFY (LFY) and SHORT-ROOT (SHR) are able to move actively from cell-to-cell by their effective passage through plamodesmata. Moreover, these noncell-autonomous proteins can get enriched in target cells, switching and controlling their developmental fate.[13-15] This discovery was surprising since these proteins are much bigger than the size exclusion limit (SEL) of the basal PD pore, which is normally 800-1200Da. However, PD behave as very dynamic structures which are able to dilate their pores allowing the passage of molecules that are bigger than 40 kDa. This unique feature of PD enables RNAs and proteins to move between cells. While LFY is moved by diffusion from the L1 to L2 and L3 cell layers in the floral primordium,[14] SHR is moved by targeted movement from the stele, where it is originally expressed, outwards to the endodermis cell layer in roots and gets enriched in nuclei of these target cells.[15] For this cell-cell movement, SHR must be present in the cytoplasm.[17] Passing of SHR through the endodermis to the cortex is prevented by a strong nuclear localization signal.[15]

How Are Molecules Transported through PD?

Gating of PD is highly selective. Although small cytoplasmic molecules below the SEL can pass PD by diffusion, and this is also true to a limited extent and only under certain circumstances for some bigger molecules, there are some examples for the active exclusion of these small molecules. For instance small signalling molecules such as calcium and auxin can not pass freely across PD. Other molecules need the active transport machinery for their transport through the PD channel. As described above, some trafficking proteins likely form multi-protein complexes with specialized proteins associated with endomembranes including the membranes of vesicles, endosomes and/or ER. These target them efficiently to PD with the involvement of the cytoskeleton.

After reaching the PD, there are three possible ways that vesicle-associated proteins can traverse them. First, after selective vesicle degradation, the enclosed proteins are released and accomplish chaperone-assisted transport through PD. Secondly, proteins associated with the surface of vesicles are released for their cell-cell transport after selective fusion of vesicles with the PM near the PD. The third possibility encompasses a recently discovered mode of cell-cell transport both in plants[18] and in cultured animal cells,[19] where entire vesicles containing, for instance, intact replication complexes[18] and/or entire endosomes[19] are transported either through PD or in the case of animal cells through tunnelling nanotubes (TNTs) representing cytoplasmic bridges between adjacent cells[19] which have some resemblance to PD.[19,20] Interestingly, these TNTs contain filamentous actin organized obviously in a sieve-like meshwork preventing passage of 'unlicensed' molecules as well as providing a cytoskeletal track along which endosomes and other vesicles move. Disruption of the actin cytoskeleton abolishes intercellular movement.[19] Recently, similar F-actin and myosin V enriched TNTs have been described in neuronal astrocytes, in which these cell-cell channels are induced via reactive oxygen species (ROS) and the MAPK signalling pathways.[21]

For several years, it has been known that PD are enriched with actin and myosin in algal and plant cells.[22-29] Additionally, manipulating the integrity of the actin cytoskeleton by cytoskeletal drugs that disrupt the F-actin-based cytoskeleton, such as cytochalasin D and latrunculin B, or by profilin microinjection, alter PD architecture and also interfere with both the gating of PD and transport of soluble molecules.[23] Structural models describe actin filaments and associated myosin motors traversing the cytoplasmic sleeve and providing cytoskeletal tracks for noncell autonomous proteins through the pore, implicating a role for actomyosin in the gating of PD.[23,24,27] Indeed, actin filaments traverse myosin VIII-enriched PD in *Arabidopsis* and maize root apices (Fig. 1A,B). Similar transcellular structures have already been described in detail (Fig. 1C,D) by Bohumil Nemec in 1901.[30] He also provided convincing data revealing that these axial structures are involved in transmission of excitability along the cell files as well as throughout the whole root apices.

Recently, it was shown that microinjection of myosin VIII antibody into epidermis cells of roots and leaves of *Arabidopsis* resulted in an enhanced movement of Lucifer Yellow and FITC-labelled dextrans between cells.[31] This enhanced cell-to-cell movement of fluorescent tracers was likely connected to an increased SEL of PD, suggesting that myosin VIII is required to limit transit through the PD pores in actively transporting PD. Thus, in addition to a role in the gating mechanism, myosin VIII and F-actin may also play roles in molecular transport through the PD pores.

Actin and Myosin VIII in PD of Algae and Herbaceous Plants

Actin and myosin were localized to PD of *Chara corallina*[23,24] and several herbaceous species including maize, cress, barley, onion and *Arabidopsis* using immunofluorescence, confocal laser scanning microscopy and immunogold EM localization techniques with heterologous and/or ATM1-specific antibodies (class VIII plant myosins).[22,25,26,28] Interestingly, BDM treatment, interfering with the ATPase activity of myosins, caused a strong constriction of the neck region of PD[25] further supporting a role of myosins in controlling PD structure and/or gating. Recently, it was shown by immunofluorescence that myosin VIII is recruited to PD and pit fields of plasmolysed root cap cells upon osmotic stress.[32] The same myosin is also recruited to the surface of plastids in gravi-sensing columella cells. Moreover, actin meshworks associated with PD[28] can be controlled via actin-branching proteins of the ARP2/3 family, since these proteins have been localized to PD in maize and tobacco as well.[33] These dynamic meshworks not only exert mechanical work but can even store memory.[34,35]

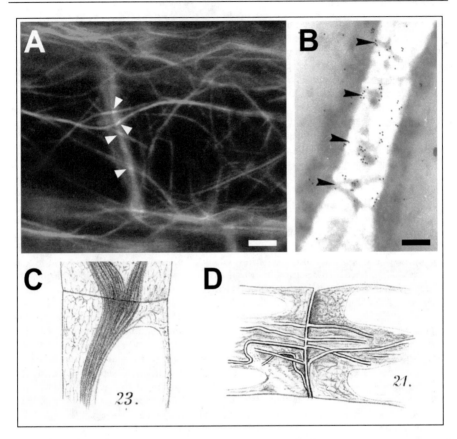

Figure 1. A) In vivo localization of actin in plasmodesmata (indicated by arrowheads) in living cortical cells of *Arabidopsis* root treated with mild osmoticum (100 mM NaCl). Actin is visualized in green colour with GFP-FABD2 construct[82] and cell walls are counterstained red with propidium iodide. B) In situ localization of myosin VIII in pit-field (indicated by arrowheads) of cortical root cells of maize using immunogold EM labelling (for more detail information see ref. 26). C,D) Actin cables resembling structures traversing cross-walls of root apices of *Allium cepa* (C) and *Cucurbita melopepo* (D). Reprinted with permission from the relatives of Bohumil Nemec.[30] Scale bars = 2 μm (A), 200 nm (B).

Root Cap Statocytes Are Symplasmically Isolated and Depleted in Myosin VIII

Nonlinear multiphoton laser scanning microscopy, combined with noninvasive near-infrared femtosecond laser-assisted targeted loading/permeation of propidium iodide into selected cells of the root tips is a novel technique that facilitates vital imaging of plasmodesmal gating.[36,37] Using this technique it has been demonstrated that during root hair initiation the trichoblasts are symplasmically isolated from the adjacent cells, and also that some cells within the root apex have 'one-way' PD (facilitating unidirectional transport of molecules) precluding their movement into adjacent cells.[37] The existence of such unidirectional 'one-way' PD are also know from studies on the shoot apex.[38] Moreover, it is now well know that the PD of the differentiated stomatal guard cells are permanently closed, unlike those of the root trichoblasts which are temporarily closed.

We have recently taken advantage of the novel non-linear multiphoton laser scanning microscopy in combination with noninvasive near-infrared femtosecond laser-assisted dye loading protocol to show experimental data/evidence that root cap statocytes in *Arabidopsis* have

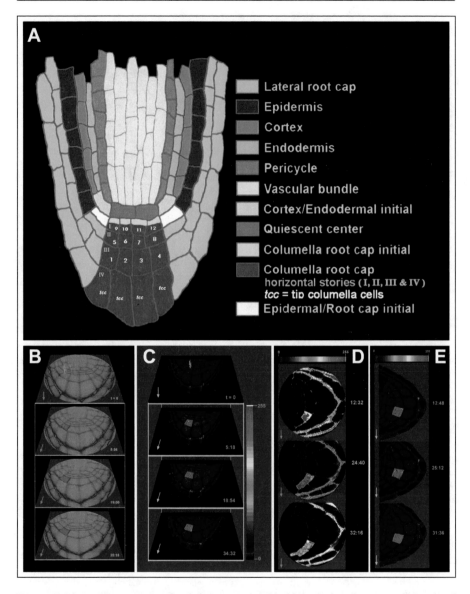

Figure 2. A) Schematic representation of *Arabidopsis* root apex with highlighted columella root cap cells (numbered 1-12) which were used for plasmodesmal gating experiments. B-E) Noninvasive near-infrared femtosecond laser-assisted loading/ permeation of targeted root cap cells with propidium iodide and concomitant visualisation of dye-coupling with multiphoton laser scanning microscopy in vertically growing (B,C) and gravistimulated *Arabidopsis* root tips. Note that only after gravistimulation the root cap cell in position 5 (D) shows selective unidirectional gating of plasmodesmata towards the lower side of the root whereas the neighbouring cell, in position 6, (E) has closed plasmodesmata (for more details see refs. 36,37).

temporarily closed PD (Fig. 2A-C). Surprisingly, this closed PD state can be rapidly relieved in gravistimulated root caps, at least in some of the outermost statocytes (Fig. 2D,E). This finding suggests that plant cells can selectively—and rapidly—regulate the gateability of their PD in

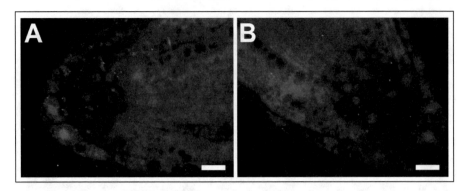

Figure 3. A,B) Immunofluorescence labelling of *Arabidopsis* root apices with myosin VIII antibody. Note that central root cap statocytes are depleted in myosin VIII but show faintly labeled nuclei. Scale bars = 18 μm (A,B).

response to physical phenomena such as gravity. As gating of PD was also reported to be sensitive to light,[39] it is conceivable that they act as sensors of the external physical environment.

The root cap is built of unique tissue as its cells are progressively heading towards the loss of their multicellular, or, more precisely, supracellular state, since all cells will eventually be sloughed off from the plant body at the root cap periphery.[40] These so-called border cells proved to be viable, despite predictions to the contrary, and in fact are very important for the root rhizosphere as they protect roots as well as communicate with diverse microorganisms associated with the root surface.[41] This process reveals that, despite their supracellular state, plant cells are able to revert back to a unicellular state. As myosin VIII is known to be associated with the cell periphery[42] and emerges as the elusive linker molecule between the cytoskeleton and cell wall components,[32,43] it is perhaps not surprising that root cap cells express only small amounts of myosin VIII, which is not associated with root cap PD (Fig. 3A,B; see also refs. 32,42).

PD which are active in cell-cell transport have large amounts of myosin VIII and F-actin.[27] Additionally, immunodepletion of endogenous class VIII myosins using microinjection of myosin VIII antibody into living cells results in unspecific opening of PD.[31] Root cap cells rapidly up-regulate myosin VIII and target it to pit-fields if they are challenged with mechanical stresses associated with protoplast formation, osmotic stress, or cell wall weakening due to the enzymatic degradation of cell wall components.[32] These latter findings suggest that myosin VIII expression, localization, and in due course also its function, are closely related to mechanical stresses. Thus, myosin VIII perhaps represents a molecular component of the elusive mechanosensor of plant cells.[32,42] As already pointed out, this myosin VIII-based function might be closely associated with PD gating mechanisms as these are also known to be mechanosensitive.[44]

PD and Nuclear Pores: Common Structural and Functional Aspects

PD and nuclear pores share several features. First of all, their structural architecture shows striking similarities.[45] Both are membraneous channels supported by a central element which is mechanically supported via spokes interlinking the central elements with rim structures. The most obvious structural difference is that PD are more complex due to the fact that the central element is membranous in nature (typically considered to be a compressed ER element) and are also lined by the plasma membranes at their peripheries. The functional similarities are even more striking as both these channels are highly selective, having typical exclusion limits, but still not allowing passage of some critical signalling molecules below this limit. Moreover, both the PD and nuclear pores can be targeted by the same homeodomain proteins.[46-48]

There are also evolutionary parallels between the nuclear pores and PD. As nuclei emerge as descendants of endosymbiotic cells which entered the host cells before mitochondria and plastids,[49-51] one can consider nuclear pores as highly specialized cell-cell channels.[51] Moreover, cell-cell channels are obvious at all levels of cellular complexity, starting with viruses and ending with complex eukaryotic cells.[51] Therefore, it is not surprising that homeodomain transcription factors can traverse both nuclear pores and PD.[46-48] Furthermore, there are several reports indicating that these transcription factors can also move from cell to cell in animals, although the nature of this intercellular transport is still unclear.[46,48] As for nuclear pores, PD are complex channels which are highly gateable. They too allow passage of very large molecules—whilst still being extremely selective and preventing passage of small molecules which, in theory, should be freely permeable. This feature implies the existence of a complex apparatus that can recognize the identity of approaching molecules and take 'decisions' as to their mode of passage through these cell-cell channels.

We have localized calreticulin to plant PD and hypothesized that it might control the gating of PD via regulation of calcium levels.[27,52] This accumulation of calreticulin to PD was confirmed recently by other groups, both in vitro and in vivo,[53,54] and calreticulin is now known to interact directly with tobacco mosaic virus movement protein within PD.[54] Calreticulin over-expression interferes with targeting of this movement protein to PD suggesting that the interaction between the movement protein and calreticulin is essential for the cellular export of viruses across PD.[54] Intriguingly in this respect, calreticulin was also identified as a protein regulating nucleocytoplasmic transport, acting as a receptor for the calcium-dependent nuclear export in mammalian cells.[55] In addition, plant and animal homeodomain proteins use similar mechanisms for their cell-cell transport which closely resemble their transport across nuclear pores.[51] Specific mutations in KNOTTED1 and in other homedomain proteins compromise their cell-cell transfer and transport across nuclear pores, both in plant and animal cells. Even more surprisingly, nuclear transport proved to be relevant for cell-cell transfer of these proteins.

Last but not least, the nuclear pore complex might turn out to be regulated via contractile proteins such as actin and myosins, as is the case for PD. Nuclear actin and myosins were proposed as critical molecules regulating nucleocytoplasmic transport[56] and myosin heavy chain-like proteins were localized to nuclear pores.[57] Recently, these ideas were supported further[58] and actin as well as filaments composed of 4.1 protein have been located to the nuclear pore complexes.[59] Moreover, large coil-coil proteins located on the nucleoplasmic side of pore complexes have similarity to myosins and are involved in the nuclear export of heterogeneous ribonucleoproteins.[60] Finally, very recent paper revealed that, as it is the case with the PD, also nuclear pore trafficking is based on the actin cytoskeleton.[61]

PD and Cell Plates: Endocytic Connections?

New PD arise first during cell plate formation, and there are further parallels between nuclear pores, PD, and cytokinetic cell plates in plant cells. WPP-domain proteins associate with the outer nuclear membrane in the vicinity of nuclear pores and are also targeted to the growing cell plate.[62,63] The WPP domain is necessary and sufficient for protein targeting both to the cell plate and nuclear pores.[62] What might be the link between cell plate and nuclear pores? Recent data suggest that an endocytic machinery drives cell plate formation,[64,65] and also supports molecular trafficking between cells via PD.[66,67] Moreover, endocytic proteins such as clathrin and adaptin share a common molecular architecture with seven proteins of the yNup84/vNup107-160 subcomplex of the nuclear pore complex.[68] Myosin VIII antibody faintly labels nuclei in root cap statocytes (Fig. 3A,B) and, in accordance with this, bioinformatics predicts localization of myosin VIII to nuclei (Andrej Hlavacka, unpublished data). In addition, while myosin VIII does not localize to phragmoplast during the early stages of cell plate formation (Fig. 4A-H), it rapidly redistributes to the cell plate after disassembly of phragmoplast microtubules and actin filaments (Fig. 4E-H). New PD are formed within callosic cell plates which are also enriched in calreticulin (Fig. 4I). Thus, both myosin VIII and calreticulin localize to the cell plate and PD. Interestingly, treatment with caffeine or 2,6-dichlorobenzonitrile, interfering with the cell plate assembly, results in increased amounts of myosin VIII localized to the cell periphery, aberrant callose-enriched cell plates, and increased callose at PD (Fig. 4J,K,M). Surprisingly, this resembles the situation observed in root cells recovering from F-actin-depletion after latrunculin B treatment (Fig. 4L). It might be that the mechanically

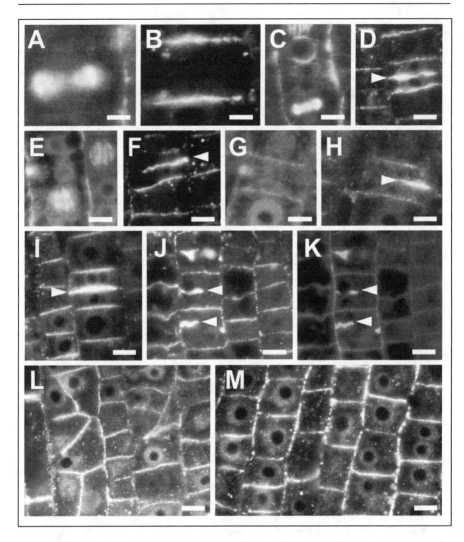

Figure 4. Immunofluorescence localization of tubulin (A,C,E), actin (G), myosin VIII (B,D,F,H,J,L,M), calreticulin (I), and fluorescence localization of callose (K) in meristematic cells of maize root apices. Note that myosin VIII does not localize to phragmoplast (B-H) but rapidly redistributes to the early cell plate after depolymerization of both microtubules and actin filaments (D,F,H, indicated by arrowheads). Calreticulin also localizes abundantly to the cell plate (I, indicated by arrowhead). Inhibition of cellulose biosynthesis with DCB results in aberrant cell plates (indicated by arrowheads) which cannot complete cytokinesis and are enriched in both myosin VIII (J) and callose (K). Chaotic division planes in latrunculin B-treated cells (L) and absence of cytokinesis in caffeine-treated cells (M) are associated with enhanced myosin VIII presence at the cell peripheries/plasmodesmata and absence of myosin VIII-enriched cell plates. Scale bars = 5 μm (A, B), 8 μm (C-H), 10 μm (I-M).

stressed cells lacking F-actin respond similarly to osmotically stressed cells which are recruiting myosin VIII and callose to cell periphery domains.[32]

Calreticulin is an interacting partner not only for the virus movement protein but also for KNOLLE, the SNARE which resides in endosomes[69] and drives cell plate formation.[53] Moreover, virus movement protein also accumulates at the cell plate during cytokinesis, similar to KNOLLE and calreticulin.[53] All this suggest very close links between endocytic machinery, cell plate assembly and PD formation, as well as cell-cell transport across mature PD.

Cytoskeleton, Endocytosis and PD

Fluid-phase endocytosis of the marker dye Lucifer Yellow (LY) was detected preferentially at F-actin- and myosin VIII-enriched PD and pit-field domains in the inner cortex cells of maize roots.[28] Tubulo-vesicular protrusions of the PM at PD and pit-fields (Fig. 5), similar to plasmatubules,[70] are involved in endocytic uptake of LY which requires both F-actin and myosin-based forces.[28] This internalisation is completely blocked or inhibited by the cytoskeletal drugs latrunculin and 2,3 butanedione monoxime (BDM) which disrupt the actin cytoskeleton and myosin function, respectively. Previously, we have shown that the myosin inhibitor BDM

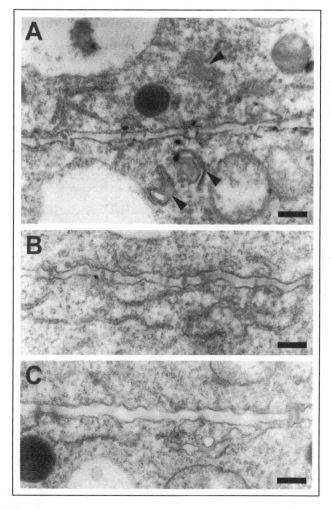

Figure 5. A-C) Transmission electronmicrographs showing plasmatubules, tubulo-vesicular protrusions and multivesicular bodies associated with primary plasmodesmata of transverse cell walls in inner cortex of maize roots. Note that multilamellar bodies are also present in the vicinity of plasmodesmata (A). Arrowheads highlight multilamellar endosomes. Scale bars = 700 nm (A-C).

causes cytoskeletal and endomembrane changes in PD domains, e.g., it promoted the accumulation of cortical microtubules around PD which can be involved in blocking endocytotic uptake at these domains.[71] Microtubules are not required for fluid-phase endocytosis since the microtubular drugs show positive effects on LY uptake, strengthening the hypothesis about steric hindrance of endocytosis by cortical microtubules located close to the PM.[28] Actin and myosin VIII are inherent not only for PD of primary plant tissues but also for secondary tissues generated by cambium activities during wood formation. This topic is in focus of the second part of our chapter.

Actin and Myosin VIII in PD of Woody Tissues in Trees

Preamble

The plant-related contributions in this volume primarily relate to the primary plant body—that formed by the action of primary meristems of the root and shoot systems—reflecting our greater knowledge of those tissues. However, as we become more confident in the study of plant biology, attention is rightly being directed to the secondary plant body—tissues made by the action of secondary meristems, e.g., the vascular cambium. Study of these aspects of plant anatomy is still largely in its infancy, but, given the importance of such vascular cambial derivatives as wood—as a fuel source in many local economies, on the one hand, and as a modulator of global climate on the other—is long overdue.[72]

Although this section will concentrate on cytoskeletal aspects of PD in the secondary vascular system (SVS—the vascular cambium and its derivatives—secondary phloem and secondary xylem) of woody plants, it should be read in conjunction with Barnett's chapter in this volume on cell-cell channels in wood.

Necessary Anatomical Background to Woody Tissues in Trees

The SVS of both angiosperm and gymnosperm trees comprises the vascular cambium—a sheath of meristematic cells that encircles the shoot/root—and its derivatives—secondary phloem ('inner bark', produced to the outside of the cambium) and secondary xylem ('wood', produced to the inside of the cambium) (Fig. 6A). Although cell types differ between the two tree types mentioned, in angiosperms the cambium consists of two cell types—long, thin fusiform cells (which give rise to the vertical elements of the secondary vascular tissues—fibres, tracheids, vessel and sieve elements, companion cells, and axial parenchyma), and more cuboid ray cells that generate the ray cells of both vascular tissues (and which extend horizontally through the SVS from phloem through the cambium and into the xylem). At maturity, xylem and phloem fibres, vessel elements and tracheids are dead and devoid of cell contents; all other cell types—particularly the ray cells of both vascular tissues—remain alive for periods of up to several years.

What Do We Know?

To date, studies on actin and myosin associated with PD in woody plants are largely those of Chaffey and Barlow,[29,73] using indirect immunofluorescence procedures with actin and myosin VIII antibodies on thick, chemically-fixed sections, imaged with epifluorescence or confocal laser scanning microscopy (CLSM). At the fluorescence microscope level individual PD are not visible; instead, sites of plasmodesma(ta) are inferred by the distribution of callose immunolocalisation.[74] Such sites are compared with myosin immunolocalisations,[29,73] and further corroborated by transmission electron microscope (TEM) studies of PDI location/distribution in angiosperm trees.[75]

Due to space constraints, we will primarily consider cytoskeletal aspects of PD within the long-lived ray cells of the secondary xylem tissues of the two angiosperm trees studied—*Aesculus hippocastanum* L. (horse-chestnut) and *Populus tremula* L. x *P. tremuloides* Michx. (hybrid aspen).

Within the cell walls of neighbouring mature xylem ray cells, both myosin and actin are localised to the pit fields—regions of extremely thin primary wall with high densities of PD—of tangential walls, between neighbouring cells within the same radial file, and transverse walls, between vertically adjacent ray cells (Fig. 6B,C).

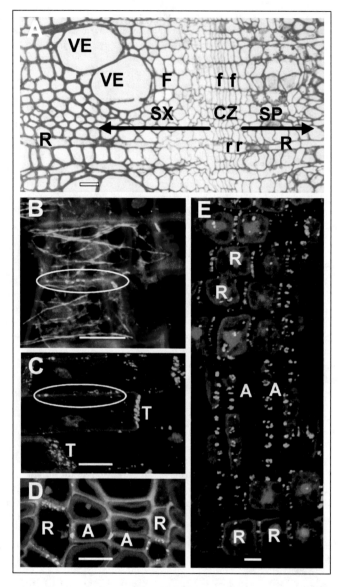

Figure 6. Aspects of the cytoskeleton in the secondary vascular system (SVS) of woody angiosperm trees: A is a bright field light micrograph, B-E are two-colour confocal laser scanning micrographs of FITC (fluorescein isothiocyanate)-tagged indirect immunolocalisations of actin (B) and myosin VIII (C-E) in radial longitudinal (B,C,E) and transverse (A,D) sections of stem (A-C,E) and root (D) tissues of hybrid aspen (A-D) and horse-chestnut (E). A) SVS showing the major tissues (CZ, cambial zone, SX, secondary xylem, SP, secondary phloem) and some important cell types (F, fibres, VE, vessel elements, R, ray cells, f, fusiform cambial cells, r, ray cambial cells). B) Two vertically adjacent xylem ray cells—note the axially arranged actin microfilament strands extending from one end of the cell to the other and actin at the pit fields in the transverse walls (ellipse). C) Similar to B, but showing anti-myosin localisation at pit fields in transverse (ellipse) and tangential (T) walls. D) Myosin localisation at pit fields in radial walls between adjacent axial parenchyma (A) and ray (R) cells. E. Myosin at pit fields between axial (A) and ray (R) cells of the secondary phloem. C,E) Modified from reference 73. B,D) Modified from reference 29. Scale bars = 40 μm (A), 20 μm (C,E), 10 μm (B,D).

Xylem rays in these angiosperm trees are predominantly one-cell wide, hence there is limited potential for radial (circumferential) symplasmic continuity within the rays. However, axial parenchyma cells are produced towards the end of a year's growth—at the growth ring boundary. Like the xylem ray cells, axial parenchyma cells are alive at maturity and long-lived. Myosin is immunolocalised at the pit fields between radially adjacent axial parenchyma cells, between vertically adjoining axial parenchyma cells, and between axial parenchyma cells and radially adjacent xylem ray cells (Fig. 6D). Presence of this second living cell type within the wood not only allows the possibility of symplasmic transport between adjacent, but otherwise symplasmically isolated, xylem rays at any given level in the stem or root (albeit probably only at growth ring boundaries), but also provides an opportunity for symplasmic transport between xylem rays at different levels within the organs.

The complexity of the potential symplasmic network within the tree SVS is increased by the more extensive cell-cell transport channels that permeate the living cells of the secondary phloem. Apart from the fibres, all of the other cells of this secondary vascular tissue are alive at maturity and live for up to a few years before they are crushed as a result of outwards expansion of the cambial cylinder. Although presence of actin has not been specifically examined at the PD here, myosin is immunolocalised at the pit fields (Fig. 6E) and it is reasonable to infer that actin should be found here, too.

Although the existence of a 3-dimensional symplasmic continuum in the wood could probably have been inferred from TEM studies of pit field/PD distributions in and between different secondary xylem cell types, the recent immunolocalisation work extends its significance by indicating the potential for such a transport pathway to be controlled by PD 'gating' using an actomyosin complex within both the wood and the secondary phloem. Any such gating could permit the selective transfer of 'information' (whether molecular or otherwise) from one domain of the organism to another, or assist in (re)directing photosynthate/storage reserves.

What Don't We Know?

Work in this area is limited—primarily due to the technical challenges of the material,[76] and is thus of a rather preliminary nature. However, what is known indicates interesting possibilities for manipulation of tree physiology,[29] particularly in efforts to alter the ratio of the two secondary vascular tissues that are made, or influence biomass production.

However, whilst it is attractive to suggest existence of a previously unsuspected 3-D transport superhighway ramifying throughout the tree invoking cytoskeletal gating of PD between cells, possibly associated with microtubule-microfilament-facilitated intracellular transport,[29] it is much better to speculate from a position of knowledge, and several important questions remain to be answered. For example: Does the type of myosin present at the PD act in concert with actin? Can any actomyosin complex envisaged contract/expand as anticipated, and hence control the diameter of the PD transport route? How are the various cytoskeletal proteins arranged at/in the PD? Such information is probably crucial to any attempt to realistically infer function from structure at these important sites. What effect(s)—if any—would manipulating the symplasmic transport pathway within the SVS have on subsequent development of the woody tissues in trees?

Conclusions and Outlook: From the Actomyosin-Based PD in Plants to the Actomyosin-Based Cell-Cell Channels in Animals

The actomyosin cytoskeleton represents an integral structural component of PD which is likely involved in both gating and transport through PD. Future functional studies will unveil specific interactions among actin, myosins, transport machinery proteins, transported cargo and molecules (e.g., proteins and RNAs), as well as mechanistic details of how this PD transport is regulated in diverse plant tissues. Surprisingly, these plant data might turn out to be highly relevant also for animal biology and medicine as myosin- and F-actin-enriched cell-cell channels are present also in animal and human cells (see also this volume).[19,21,77-81] Moreover, the cell-cell spread of homeodomain proteins is accomplished via very similar mechanisms and plant homeodomain proteins can accomplish cell-cell transport in animal cells, further minimizing the differences between plant and animal cells.

Last but not least, it is known that plant viral movement proteins which gate PD in plant cells can induce F-actin-and myosin-enriched cell-cell channels in insect cells (Taiyun Wei, personal communication). Future studies will shed more light on these exciting topics, which are relevant not only for basic cell biology but also for applied biology, especially in biomedicine.

Acknowledgements

This work was supported by research fellowships from Alexander von Humboldt Foundation and DAAD (Bonn, Germany) to J. Samaj and F. Baluska as well as by grants obtained from Deutsches Zentrum für Luft- und Raumfahrt (DLR, Bonn, Germany); from Grant Agency APVT (grant no. APVT-51-002302) and Vega (Grant Nr. 2/5085/25), Bratislava, Slovakia. N. Chaffey's work on the cytoskeleton of trees was funded by the Commission of the European Communities, Agriculture and Fisheries (FAIR) specific RTD programme, CT98-3972, 'Wood formation processes: the key to improvement of the raw material'. U.K.T. and Z.F.C. are grateful to the financial support from Biotechnology and Biological Sciences Research Council, U.K. (BBSRC PowderJect LINK 43/ LKE17445).

References

1. Lucas WJ, Lee J-Y. Plasmodesmata as a supracellular control network in plants. Nat Rev Mol Cell Biol 2004; 5:712-726.
2. Huang Z, Andrianov VM, Han Y et al. Identification of Arabidopsis proteins that interact with cauliflower mosaic virus (CaMV) movement protein. Plant Mol Biol 2001; 47:663-675.
3. Laporte C, Vetter G, Loudes AM et al. Involvement of the secretory pathway and the cytoskeleton in intracellular targeting and tubule assembly of Grapevine fan-leaf virus movement protein in tobacco BY-2 cells. Plant Cell 2003; 15:2058-2075.
4. Escobar NM, Haupt S, Thow G et al. High-throughput viral expression of cDNA-green fluorescent protein fusions reveals novel subcellular addresses and identifies unique proteins that interacts with plasmodesmata. Plant Cell 2003; 15:1507-1523.
5. Lu C, Zainal Z, Tucker GA et al. Developmental abnormalities and reduced fruit softening in tomato plants expressing an antisense Rab11 GTPase gene. Plant Cell 2001; 13:1819-1833.
6. Más P, Beachy RN. Replication of tobacco mosaic virus on endoplasmic reticulum and role of the cytoskeleton and virus movement protein in intracellular distribution of viral RNA. J Cell Biol 1999; 147:945-958.
7. Lee JY, Yoo BC, Rojas MR et al. Selective trafficking of noncell-autonomous proteins mediated by NtNCAPP1. Science 2003; 299:392-396.
8. Velikanov GA, Volobueva OV, Khokhlova LP. The study of the hydraulic conductivity of the plasmodesmal transport channels by the pulse NMR method. Russ J Plant Phys 2001; 48:375-383.
9. Volobueva OV, Velikanov GA, Baluska F. Regulation of intercellular water exchange in various zones of maize root under stresses. Russ J Plant Phys 2004; 51:676-683.
10. Velikanov GA, Volobueva OV, Belova LP et al. Vacuolar symplast as a regulated pathway for water flows in plants. Russ J Plant Phys 2004; 52:326-331.
11. Kutsuna N, Hasezawa S. Dynamic organization of vacuolar and microtubule structures during cell cycle progression in synchronized tobacco BY-2 cells. Plant Cell Physiol 2002; 43:965-973.
12. Kutsuna N, Kumagai F, Sato MH et al. Three-dimensional reconstruction of tubular structure of vacuolar membrane throughout mitosis in living tobacco cells. Plant Cell Physiol 2003; 44:1045-1054.
13. Lucas WJ, Bouche-Pillon S, Jackson DP et al. Selective trafficking of KNOTTED1 homeodomain protein and its mRNA through plasmodesmata. Science 1995; 270:1980-1983.
14. Wu X, Dinneny JR, Crawford KM et al. Modes of intercellular transcription factor movement in the Arabidopsis apex. Development 2003; 130:3735-3745.
15. Gallagher KL, Paquette AJ, Nakajima K et al. Mechanisms regulating SHORT-ROOT intercellular movement. Curr Biol 2004; 14:1847-1851.
16. Nakajima K, Sena G, Nawy T et al. Intercellular movement of the putative transcription factor SHR in root patterning. Nature 2001; 413:307-311.
17. Sena G, Jung JW, Benfey PN. A broad competence to respond to SHORT-ROOT revealed by tissue-specific ectopic expression. Development 2004; 131:2817-2826.
18. Kawakami S, Watanabe Y, Beachy RN. Tobacco mosaic virus infection spreads cell to cell as intact replication complexes. Proc Natl Acad Sci USA 2004; 101:6291-6296.

19. Rustom A, Saffrich R, Markovic I et al. Nanotubular highways for intercellular organelle transport. Science 2004; 303:1007-1010.
20. Baluska F, Hlavacka A, Volkmann D et al. Getting connected: Actin-based cell-to-cell channel in plants and animals. Trends Cell Biol 2004; 14:404-408.
21. Zhu D, Tan KS, Zhang X et al. Hydrogen peroxide alters membrane and cytoskeleton properties and increases intercellular connections in astrocytes. J Cell Sci 2005; 118:3695-3703.
22. White RG, Badelt K, Overall RL et al. Actin associated with plasmodesmata. Protoplasma 1994; 180:169-184.
23. Overall RL, Blackman LM. A model of the macromolecular structure of plasmodesmata. Trends Plant Sci 1996; 1:307-311.
24. Blackman LM, Overall RL. Immunolocalisation of the cytoskeleton to plasmodesmata of Chara corallina. Plant J 1998; 14:733-741.
25. Radford JE, White RG. Localization of myosin-like protein to plasmodesmata. Plant J 1998; 14:743-750.
26. Reichelt S, Knight AE, Hodge TP et al. Characterisation of the unconventional myosin VIII in plant cells and its localization at the post-cytokinetic cell wall. Plant J 1999; 19:555-568.
27. Baluska F, Cvrckova F, Kendrick-Jones J et al. Sink plasmodesmata as gateways for phloem unloading: Myosin VIII and calreticilin as molecular determinants of sink strenght. Plant Physiol 2001; 126:39-47.
28. Baluska F, Samaj J, Hlavacka A et al. Myosin VIII and F-actin enriched plasmodesmata in maize root inner cortex cells accomplish fluid-phase endocytosis via an actomyosin-dependent process. J Exp Bot 2004; 55:463-473.
29. Chaffey NJ, Barlow PW. The cytoskeleton facilitates a three-dimensional symplasmic continuum in the long-lived ray and axial parenchyma cells of angiosperm trees. Planta 2001; 213:811-823, (Erratum: Planta 2001; 214, 330-331).
30. Nemec B. Die Reizleitung und die Reizleitenden Strukturen bei den Pflanzen. Jena: Verlag von Gustaf Fischer, 1901.
31. Volkmann D, Mori T, Tirlapur UK et al. Unconventional myosins of the plant-specific class VIII: Endocytosis, cytokinesis, plasmodesmata/pit-fields, and cell-to-cell coupling. Cell Biol Int 2003; 27:289-291.
32. Wojtaszek P, Anielska-Mazur A, Gabrys H et al. Recruitment of myosin VIII towards plastid surfaces is root-cap specific and provides the evidence for actomyosin involvement in root osmosensing. Funct Plant Biol 2005; 32:1-16.
33. Van Gestel K, Slegers H, von Witsch M et al. Immunological evidence for the presence of plant homologues of the actin related protein Arp3 in tobacco and maize: Subcellular localization to actin-enriched pit fields an emerging root hairs. Protoplasma 2003; 222:45-52.
34. Tseng Y, Kole TP, Le JSH et al. How actin crosslinking and bundling proteins cooperate to generate an enhanced cell mechanical response. Biochem Biophys Res Comm 2005; 334:183-192.
35. Weaver AM. Cytoskeletal interactions with the outside world. Dev Cell 2005; 9:35-38.
36. Tirlapur UK, König K. Technical advance: Near-infrared femtosecond laser pulses as a novel noninvasive means for dye-permeation and 3D imaging of localised dye-coupling in the Arabidopsis root meristem. Plant J 1999; 20:363-370.
37. Tirlapur UK, König K. Femtosecond near-infrared laser pulses as a versatile noninvasive tool for intra-tissue nanoprocessing in plants without compromising viability. Plant J 2002; 31:365-374.
38. Ueki S, Citovsky V. Control improves with age: Intercellular transport in plant embryos and adults. Proc Natl Acad Sci USA 2005; 102:1817-1818.
39. Epel BL, Erlanger MA. Light regulates symplastic communication in etiolated corn seedlings. Physiol Plant 1991; 83:149-153.
40. Barlow PW. The root cap: Cell dynamics, cell differentiation and cap function. J Plant Growth Regul 2003; 21:261-286.
41. Hawes MC, Brigham LA, Wen F et al. Function of root border cells in plant health: Pioneers in the rhizosphere. Annu Rev Phytopathol 1998; 36:311-327.
42. Baluska F, Barlow PW, Volkmann D. Actin and myosin VIII in developing root cells. In: Staiger CJ, Baluska F, Volkmann D, Barlow PW, eds. Actin: A Dynamic Framework for Multiple Plant Cell Functions. Dordrecht, The Netherlands: Kluwer Academic Publishers, 2000:457-476.
43. Baluska F, Samaj J, Wojtaszek P et al. Cytoskeleton – plasma membrane – cell wall continuum in plants: Emerging links revisited. Plant Physiol 2003; 133:482-491.
44. Oparka KJ, Prior DAM. Direct evidence for pressuregenerated closure of plasmodesmata. Plant J 1992; 2:741-750.
45. Lee JY, Yoo BC, Lucas WJ. Parallels between nuclear-pore and plasmodesmal trafficking of information molecules. Planta 2000; 210:177-178.

46. Prochiantz A, Joliot A. Can transcription factors function as cell-cell signalling molecules? Nat Rev Mol Cell Biol 2003; 4:814-819.
47. Maizel A. Cell-cell movements of transcription factors in plants. In: Baluska F, Volkmann D, Barlow PW, eds. Cell-Cell Channels. Georgetown: Landes Bioscience, 2006:176-182.
48. Tassetto M, Maizel A, Osorio J et al. Plant and animal homeodomains use convergent mechanisms for intercellular transfer. EMBO Rep 2005; 6:885-890.
49. Baluska F, Volkmann D, Barlow PW. Cell bodies in a cage. Nature 2004b; 428:371.
50. Baluska F, Volkmann D, Barlow PW. Eukaryotic cells and their cell bodies: Cell theory revisited. Ann Bot 2004c; 94:9-32.
51. Baluska F, Volkmann D, Barlow PW. Cell-cell channels and their implications for Cell Theory. In: Baluska F, Volkmann D, Barlow PW, eds. Cell-Cell Channels. Georgetown: Landes Bioscience, 2006:1-18.
52. Baluska F, Samaj J, Napier R et al. Maize calreticulin localizes preferentially to plasmodesmata in root apices. Plant J 1999; 19:481-488.
53. Laporte C, Vetter G, Loudes AM et al. Involvement of the secretory pathway and the cytoskeleton in intracellular targeting and tubule assembly of Grapevine fanleaf virus movement protein in tobacco BY-2 cells. Plant Cell 2003; 15:2058-2075.
54. Chen M-H, Tian G-W, Gafni Y et al. Effects of calreticulin on viral cell-to-cell movement. Plant Physiol 2005; 138:1866-1876.
55. Holaska JM, Black BE, Rastinejad F et al. Ca^{2+}-dependent nuclear export mediated by calreticulin. Mol Cell Biol 2002; 22:6286-6297.
56. Schindler M, Jiang L-W. Nuclear actin and myosin as control elements in nucleocytoplasmic transport. J Cell Biol 1986; 102:859-862.
57. Berrios M, Fisher PA, Matz EC. Localization of a myosin heavy chain-like polypeptide to Drosophila nuclear pore complexes. Proc Natl Acad Sci USA 1991; 88:219-233.
58. Tonini R, Grohovaz F, Laporta CAM et al. Gating mechanism of the nuclear pore complex channel in isolated neonatal and adult mouse liver nuclei. FASEB J 1999; 13:1395-1403.
59. Kiseleva E, Drummond SP, Goldberg MW et al. Actin- and protein-4.1-containing filaments link nuclear pore complexes to subnuclear organelles in Xenopus oocyte nuclei. J Cell Sci 2004; 117:2481-2490.
60. Green DM, Johnson CP, Hagan H et al. The C-terminal domain of myosin-like protein 1 (Mlp1p) is a docking site for heterogeneous nuclear ribonucleoproteins that are required for mRNA export. Proc Natl Acad Sci USA 2003; 100:1010-1015.
61. Minakhina S, Myers R, Druzhinina M et al. Crosstalk between the actin cytoskeleton and Ran-mediated nuclear transport. BMC Cell Biol 2005; 6:32.
62. Patel S, Rose A, Meulia T et al. Arabidopsis WPP-domain proteins are developmentally associated with the nuclear envelope and promote cell division. Plant Cell 2004; 16:3260-3273.
63. Jeong SY, Rose A, Joseph J et al. Plant-specific mitotic targeting of RanGAP requires a functional WPP domain. Plant J 2005; 42:270-282.
64. Baluska F, Liners F, Hlavacka A et al. Cell wall pectins and xyloglucans are internalized into dividing root cells and accumulate within cell plates during cytokinesis. Protoplasma 2005; 225:141-155.
65. Samaj J, Read ND, Volkmann D et al. The endocytic network in plants. Trends Cell Biol 2005; 15:425-433.
66. Haupt S, Cowan GH, Ziegler A et al. Two plant-viral movement proteins traffic in the endocytic recycling pathway. Plant Cell 2005; 17:164-181.
67. Oparka KJ. Getting the message across: How do plant cells exchange macromolecular complexes? Trends Plant Sci 2004; 9:33-41.
68. Devos D, Dokudovskaya S, Alber F et al. Components of coated vesicles and nuclear pore complexes share a common molecular architecture. PloS Biol 2004; 2:2085-2093.
69. Uemura T, Ueda T, Ohniwa RL et al. Systematic analysis of SNARE molecules in Arabidopsis: Dissection of the post-Golgi network in plant cells. Cell Struct Funct 2004; 29:49-65.
70. Chaffey NJ, Harris N. Plasmatubules: Fact or artefact? Planta 1985; 165:185-190.
71. Samaj J, Peters M, Volkmann D et al. Effects of myosin ATPase inhibitor 2,3-butanedione monoxime on distributions of myosins, F-actin, microtubules, and cortical endoplasmic reticulum in maize root apices. Plant Cell Physiol 2000; 41:571-582.
72. Chaffey NJ. Introduction. In: Chaffey NJ, ed. Wood Formation in Trees: Cell and Molecular Biology Techniques. London: Taylor and Francis, 2002a:1-8.
73. Chaffey NJ, Barlow PW. Myosin, microtubules, and microfilaments: Cooperation between cytoskeletal components during cambial cell division and secondary vascular differentiation in trees. Planta 2002; 214:526-436.

74. Radford JE, Vesk M, Overall RL. Callose deposition at plasmodesmata. Protoplasma 1998; 201:30-37.
75. Sauter JJ, Kloth S. Plasmodesmatal frequency and radial translocation rates in ray cells of poplar (Populus x canadensis Moench 'robusta'). Planta 1986; 168:377-380.
76. Chaffey NJ. An introduction to the problems of working with trees. In: Chaffey NJ, ed. Wood Formation in Trees: Cell and Molecular Biology Techniques. London: Taylor and Francis, 2002b:9-16.
77. Onfelt B, Nedvetzki S, Yanagi K et al. Cutting Edge: Membrane nanotubes connect immune cells. J Immunol 2004; 173:1511-1513.
78. Vidulescu C, Clejan S, O'Connor KC. Vesicle traffic through intercellular bridges in DU 145 human prostate cancer cells. J Cell Mol Med 2004; 8:388-396.
79. Koyanagi M, Brandes RP, Haendeler J et al. Cell-to-cell connection of endothelial progenitor cells with cardiac myocytes by nanotubes: A novel mechanism for cell fate changes? Circ Res 2005; 96:1039-1041.
80. Watkins SC, Salter RD. Functional connectivity between immune cells mediated by tunneling nanotubules. Immunity 2005; 23:309-318.
81. Hsiung F, Ramirez-Weber F-A, Iwaki DD et al. Dependence of Drosophila wing imaginal disc cytonemes on Decapentaplegic. Nature 2005; 437:560-563.
82. Voigt B, Timmers T, Samaj J et al. GFP-FABD2 fusion construct allows in vivo visualization of the dynamic actin cytoskeleton in all cells of Arabidopsis seedlings. Eur J Cell Biol 2005; 84:595-608.

CHAPTER 9

Cell-Cell Communication in Wood

John R. Barnett*

Abstract

The sapwood of trees contains living cells in the form of ray and axial parenchyma. Communication between these cells is important in controlling the differentiation of xylem cells, movement of nutrients and water, defence against pathogens and damage repair. These parenchyma cells are therefore linked symplasmically by plasmodesmata enabling them to coordinate their activities. This chapter reviews the distribution of plasmodesmata in mature sapwood and the cambium and differentiation zone in an attempt to explain their role in the wood. The extent to which symplasmic connections are important in unloading nutrients into the cambium and zone of differentiating wood cells is confused by the fact that many of the differentiating cells appear to be isolated from the symplasm. This suggests that these cells rely on the apoplasmic pathway for input of nutrients, and that secondary formation of plasmodesmata may be one factor in the control of the fate of differentiating wood cells.

Introduction

It is a common misconception that wood in trees is a dead tissue whose sole functions are to provide support for the crown and a system of channels for water flow from roots to leaves. If this were indeed the case then there would be no need for cell-cell communication in wood. However, the sapwood of all trees comprises a substantial proportion of living parenchyma (Fig. 1A-C) involved in nutrient storage and mobilization, water movement and hydraulic safety, defence against pathogens, and damage repair.[1] In older trees, the conversion of sapwood to heartwood involves active metabolic processes requiring living parenchyma capable of remaining metabolically active for many years after being formed by the vascular cambium. All of these processes must take place in a coordinated way which requires that the parenchyma cells communicate with each other. The parenchyma die only when the sapwood in which they are embedded has been converted to heartwood, which is truly dead tissue.

Wood is usually considered by anatomists to comprise two cell systems, the axial and the ray systems, but it might equally be considered to consist of a living parenchyma network embedded in a mass of dead supportive and conductive tissue. If it were possible to remove one system leaving the other in place, either would be seen to form a continuous network throughout the stem. On the other hand, the two systems are not isolated physiologically. Rays remove water and mineral nutrients from the conducting system in the xylem, and photosynthates from the phloem and distribute them to the vascular cambium and its differentiating derivatives.

The process of wood formation by the vascular cambium is also a highly coordinated process, requiring input of water, minerals and nitrogen from the mature xylem and photosynthates from the phloem.[2] The fact that each species has its own characteristic anatomical arrangement of cell types implies communication between meristematic cells of the cambial zone to ensure

*John R. Barnett—School of Plant Sciences, The University of Reading, Whiteknights, P.O. Box 221, Reading, RG6 6AS, U.K. Email: j.r.barnett@reading.ac.uk

Cell-Cell Channels, edited by Frantisek Baluska, Dieter Volkmann and Peter W. Barlow.
©2006 Landes Bioscience and Springer Science+Business Media.

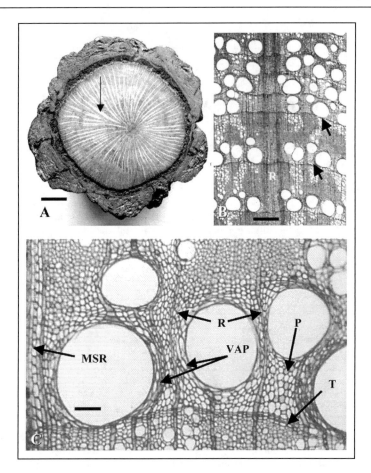

Figure 1. Parenchyma in wood of oak. A) Cross section through a stem of *Quercus suber*. About 50% of the wood (the lighter-shaded radial streaks (arrow)) is made up of broad multiseriate rays comprised of living parenchyma. Scale bar = 1 cm. B) Low magnification light micrograph of a transverse section through sapwood of *Quercus robur*. A line of parenchyma cells (terminal parenchyma) marks the growth ring boundaries (arrows). A broad parenchymatous multiseriate ray runs from top to bottom of the micrograph (R). Scale bar = 1 mm. C) Greater magnification of *Quercus robur* sapwood showing part of a multiseriate ray at the left of the micrograph (MSR), a number of parenchymatous uniseriate rays (R), the terminal parenchyma of the growth ring boundary (T), vessel-associated parenchyma (VAP) and noncontact axial parenchyma, identifiable by the presence of cell contents (P). Scale bar = 0.1 mm.

coordinated development. For example, in a species with solitary vessels, a fusiform initial which has just given rise to a vessel element must "know" that it should not produce another one immediately, but must instead produce a different wood cell type. This means that information must be entering the cell about the nature of the differentiation pathway being followed by its own previously-formed derivatives and those of cells produced by adjacent cambial cells. This information is essential to ensure switching of the gene expression required for formation of each wood cell in the pattern.[3]

The primary functions of cell-cell communication in wood therefore involve:
- Transport of nutrients and water within the parenchyma system, and to the cambium and developing xylem

- Transfer of signals about the physiological state of the wood, producing responses such as release of nutrients or stimulation of defence reactions in parenchyma
- Movement of growth regulators and molecules controlling gene expression within the cambium and zone of differentiating wood cells

Communication may be symplasmic, with protoplasts of adjacent cells being linked by plasmodesmata through pit fields, or apoplasmic involving movement of material through unlignified cell walls or intercellular spaces associated with rays. Metabolites may move between the symplasm and apoplasm by crossing the plasmalemma. Plasmodesmata may form primarily in the cell plate at the time of cytokinesis or secondarily in a preexisting primary wall. The apoplasmic route is probably the one taken by water and solutes[2] and it seems likely that the symplasmic route is used where more careful control over movement of materials such as signalling molecules is required. These aspects are considered in more detail below.

Xylem Parenchyma

Parenchyma cells in a woody stem may be categorised in two ways. The first is purely anatomical, may be applied to both angiosperms and gymnosperms, and classifies them as either ray parenchyma or axial parenchyma.

Ray parenchyma are organised into radiating arms which vary in size and structure between species (Fig. 1A-C). Although rayless species, such as are found in the genus *Hebe* do occur, the majority of woody taxa which have been investigated have rays occupying between 20 and 40% of the volume of the wood, occasionally reaching 75% as in *Dillenia indica*.[4] The individual cells are classified as procumbent (longer radially than vertically) or upright (longer vertically than radially). According to species, the rays may be entirely of one type (homogeneous) or made up of a combination of more than one type (heterogeneous). The physiological basis for these differences is not understood. The cells have numerous plasmodesmata linking their protoplasts with those of other ray parenchyma. Where there has been secondary thickening of the wall, the plasmodesmata may be confined to pit fields comprising areas of unthickened primary wall, although plasmodesmata that have been overgrown by secondary wall deposition are common (Fig. 2).

Ray cells have also been sub-divided into contact cells, which have a pitted wall in common with a vessel, and isolation cells having no pitting to vessels (Fig. 3). The contact pits between the parenchyma cell and the vessel have a loosely woven microfibrillar membrane.[5] There are no functional plasmodesmata in the membrane, so uptake of water and solutes from the vessel into the contact parenchyma is apoplasmic. It was found that the contact parenchyma cells have a loose-textured, porous layer lining the wall, the so-called protective layer (see also Figs. 3, 4).[6] It has been hypothesised that this increases the surface area of the plasmalemma of the cell in contact with the apoplast, increasing the rate at which water and solutes from the vessel can be taken up by the contact cell protoplast.[7] However, Chafe and Chauret recorded the presence of a protective layer in isolation parenchyma,[8] so it appears that this layer may have a role in apoplasmic transport beyond the contact cells.[9,10] This view has been supported by Wiesniewski et al who found that xylem parenchyma cells were impermeable to a solution of lanthanum nitrate used as an apoplasmic tracer, except around their pit membranes and in the protective layer.[11] Van Bel and van der Schoot also suggested that the loose pectocellulosic structure of the protective layer would permit easy and massive passage of solutes at the apoplasm/symplasm interface.[12]

Sauter and Kloth found that in rays, contact cells had fewer plasmodesmata per unit area of tangential wall than isolation cells, suggesting that they were less involved in radial transport.[13] They calculated that the flow rate of sugars through isolation cell plasmodesmata, based on figures presented for plasmodesmata in other cell types by Gunning and Robards,[14] was sufficient to explain observed movements in the tree. It was proposed that isolation cells were the main pathway for symplasmic movement of sugars. Contact cells of rays and axial parenchyma were suggested as unloading organic nitrogen compounds synthesised in the roots from the xylem vessels apoplasmically, then taking them into the symplasm before passing them via plasmodesmata into the isolation cells.[2] This behaviour is analogous to that of companion cells or albuminous cells in the phloem.[15]

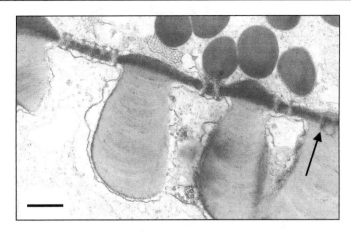

Figure 2. Plasmodesmata passing through pit fields in the tangential wall between a maturing ray cell (bottom) and a ray cell of the cambial zone (top). Note the wall thickenings in the maturing cell have overgrown some plasmodesmata (arrow). Scale bar = 1 μm. From reference 29, with kind permission of Springer Science+Business Media, ©1982 Springer-Verlag.

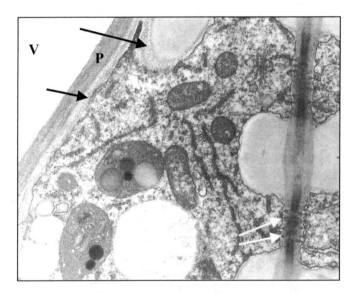

Figure 3. Transmission electron micrograph of a contact parenchyma cell with pit (P) to a vessel element (V) in *Quercus robur*. There are no plasmodesmata in the pit membrane. To the right is an isolation cell connected symplasmically with the contact cell by plasmodesmata through the common pit membrane (white arrows). Note the loose-textured protective-layer lining the pit membrane and the cell wall (black arrows). Scale bar = 1 μm.

Axial parenchyma cells are produced by several transverse divisions of a cambial derivative to produce a vertical file of cells known as a parenchyma strand. The arrangement of parenchyma strands in wood as seen in transverse section varies according to species. In some, they appear isolated (solitary) and embedded among other cell types. In others, they may be associated with vessels individually or form an almost continuous layer or layers surrounding the vessel. In still others they may form bands alternating with bands of vessels and fibres. The variety of arrangements is almost

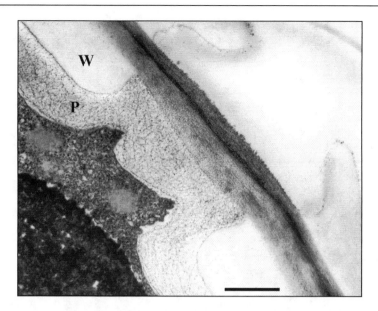

Figure 4. Transmission electron micrograph of a pit between a vessel and a contact parenchyma cell in *Sorbus aucupara*. Note the loose-textured protective layer (P) lining the secondary wall of the parenchyma cell (W). Scale bar = 1 μm.

endless and is a major tool in wood identification. Whatever the arrangement, however, protoplasts of axial parenchyma cells of a strand are connected symplasmically to each other and to cells of other parenchyma strands or rays by plasmodesmata (Figs. 5,6).

Another classification of xylem parenchyma is based on function and divides them into storage or specialized cells.[5] Storage cells can be either ray or axial parenchyma cells and, in addition to the usual storage substances (starch, lipid, protein), they can contain tannins and crystals of, for example, calcium oxalate. Specialized cells are always contiguous with a vessel and hence only occur in angiosperms. Czaninski further sub-divided specialized cells into contact cells and vessel-associated cells.[5] Sauter et al considered contact cells to have a role in the control of water movement through metabolically controlled breakdown of stored starch and release of sucrose into the adjacent vessel through the pit membrane.[16] This would explain the observation, first made by Hartig,[17] that the xylem sap in vessels of some deciduous trees contains sugars in solution in winter and early spring. Species showing this phenomenon include members of the genera *Acer, Betula, Carpinus, Alnus and Populus*.[16] Braun proposed that this activity enables temperate deciduous trees to take up and move water prior to dormancy breakage and leaf emergence in the spring, and tropical trees to move water at times when relative humidity is too high for transpiration to occur.[18]

According to Czaninski, vessel-associated cells have distinct and characteristic cytological and cytochemical properties.[5] She found that they had a dense and active cytoplasm containing no significant quantities of starch at any time of year. The primary cell wall in the pit fields between the cell and its adjacent vessel is composed of comprised mainly pectin and cellulose and had no plasmodesmata, whereas the membrane between neighbouring storage parenchyma was sometimes lignified and had plasmodesmata. Not all the parenchyma cells contiguous with a vessel were of the same type, however. The primary function of these vessel-associated cells was believed to be in tylose formation or secretion of defensive compounds such as phenolics into the vessel. It was also suggested that they might provide an alternative route for water movement around a damaged region of a vessel.

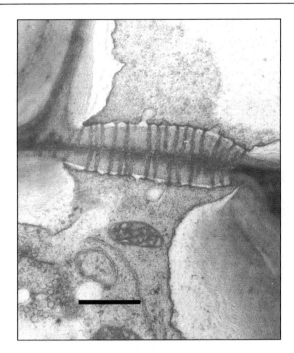

Figure 5. Transmission electron micrograph of a pit membrane in *Querbus robur* containing numerous plasmodesmata between laterally adjacent axial parenchyma cells. Scale bar = 1 μm. From reference 29, with kind permission of Springer Science+Business Media, ©1982 Springer-Verlag.

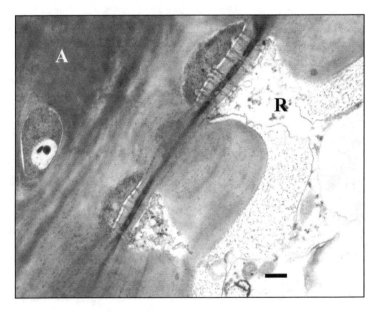

Figure 6. Pit membranes between an axial parenchyma cell (A) and a ray parenchyma cell (R) in *Quercus robur*. Scale bar = 1 μm.

Origin and Function of Rays

The first rays are extensions of the medullary rays in the primary stem, but new ones are initiated throughout the life of the tree. Bannan suggested that fusiform cambial cells in coniferous trees need to be in contact with a ray in order to survive.[19] Those which are not either differentiate terminally, or divide to produce a new set of ray initials. In this way the frequency of rays in the wood and hence the frequency of ray contacts with the fusiform initials are maintained. Ziegler suggested that since the cambium is supplied by the rays this enables a feedback system regulating equidistant spacing between rays.[20] Philipson et al also suggested that conversion of fusiform initials with poor ray contacts into ray initials ensured that the ratio of ray to fusiform initials was maintained.[21]

Ethylene is known to stimulate ray production in secondary xylem.[22] Eklund proposed that rays drain ethylene (or a precursor) from the xylem where it is formed during differentiation of xylem elements,[23] leading Lev Yadun and Aloni to suggest that a cambial cell which was not close to or in contact with a ray would experience a higher concentration of ethylene and be stimulated to form ray initials.[22]

All rays are continuous from the point in the wood at which they were initiated, through the cambial zone to the phloem (Fig. 7). Here they become loaded with photosynthates from the sieve elements of the phloem via phloem companion cells in angiosperms or the albuminous cells of phloem rays in gymnosperms. The ray thus provides a transport system for moving nutrients from the active phloem, and water with solutes from the active xylem to the vascular cambium and differentiating

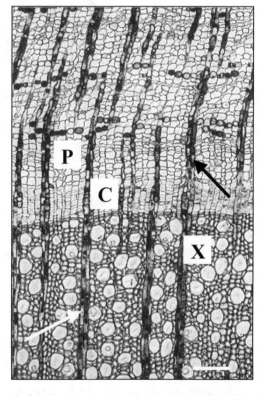

Figure 7. Light micrograph of a transverse section through phloem (P), cambium (C) and xylem (X) of a mature tree of *Pyrus communis*. The rays (arrow) are continuous through all three tissues except where they have moved out of the plane of section. In this specimen the cambium was dormant and the rays contain large amounts of storage materials. Scale bar = 0.1 mm.

secondary xylem and phloem cells.[2] Materials not required for the moment by the cambium and its differentiating derivatives are stored in the rays or the axial parenchyma with which they are in contact (Fig. 7). This storage process has been reported to take place towards the end of the growing season when cambial activity is declining.[24-26] The stored carbohydrate, lipid and protein can be mobilised as required, for example in early spring, when cambial activity and differentiation of secondary tissues begins and the leaves have not yet fully formed or become net exporters.

As the vascular cambium moves outwards away from the wood cells it has produced, older parenchyma cells are left behind and become more remote from the source of photosynthate and from the metabolite sink formed by the cambium and differentiating xylem cells. The continuity of the symplasm, however, ensures that ray cells and axial parenchyma deep within sapwood can be supplied with sufficient raw materials to enable them to maintain their metabolic activity until they are required to participate in heartwood formation. If the axial parenchyma were isolated and surrounded only by dead fibres, tracheids and vessel elements, they would be unlikely to be able to sustain themselves using the nutrients with which they had been endowed on their formation from the cambium. They would also be isolated from signals controlling their physiological roles in nutrient storage and distribution, water movement, wound repair and heartwood formation and would not have the resources to fulfil these roles. Cell-cell communication is therefore essential in ensuring maintenance of living tissue in wood.

Symplasmic movement of materials through the rays and axial parenchyma system has been discussed by Chaffey and Barlow.[27] Immunofluorescence studies of the distribution of actin, myosin and tubulin in the ray and axial parenchyma system of *Aesculus hippocastanum* rootwood led them to support the suggestion of Overall et al, that a contractile acto-myosin complex is involved in pumping materials through plasmodesmata.[28] They suggested that the more randomly organised cytoskeleton in the cambial zone allows release of transportate for sequestration by the cambial zone and zone of cell differentiation. Such a model provides a mechanism for, and is in keeping with the suggestion of Sauter and Kloth, that transport through the ray system is mainly symplasmic.[13]

Distribution of Plasmodesmata in Nonparenchymatous Wood Cells

In addition to parenchyma, wood is made up of some or all of three different cell types: tracheids, vessel elements, and fibres. Their presence or absence, arrangement and proportions varies according to species. They are dead at maturity and have neither protoplast nor active plasmodesmata. The mature pits in their walls act as channels for water and solute movement from one cell to the next. The pits between fibres and between fibres and parenchyma cells often have vestigial plasmodesmata in their membranes, relics of the active plasmodesmata present in the differentiating cell's wall.[29] In some species, in which tracheary elements have a thickened torus-like structure in the mature pit membrane, vestigial plasmodesmata may also be present. These form secondarily in pit fields in the radial walls of fusiform cambium cells and become concentrated in the pit membrane thickening during the phase of secondary wall formation (Figs. 8,9).[30,31]

A cell type intermediate in form between fibres and tracheids called the fibre tracheid is present in many species although there is still disagreement about how these should be defined. According to the International Association of Wood Anatomists,[32] a fibre tracheid is thick-walled with lenticular- to slit-shaped bordered pit apertures. This distinguishes it from a tracheid which has circular pit apertures and a fibre which has simple pits. The situation with regard to fibre tracheids is confused by the fact that there is a continuum of structural variation with regard to the size, shape and number of bordered pits from thick-walled fibres through to thinner-walled tracheids. In this respect, analysis of plasmodesmatal distribution in the pit membranes of these cells during their development can be of help. Barnett found evidence to suggest that the developing pit membranes of fibres possess plasmodesmata, while those of tracheids do not.[29] Thus, presence or absence of plasmodesmata during the development of a cell may indicate whether it leans more to the fibre or the tracheid side of the fibre/tracheid continuum. It was also suggested that the evolutionary origins of fibre tracheids from either fibres or tracheids might be indicated by the presence or absence of plasmodesmata.

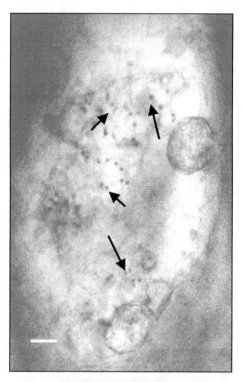

Figure 8. High voltage transmission electron micrograph of a thick (1 μm) section in the plane of the radial wall of a fusiform cambial cell of *Sorbus aucuparia*. The lighter region is the primary pit field containing randomly dispersed plasmodesmata which appear as black spots (arrows). These have arisen secondarily. Scale bar = 1 μm. Reproduced with permission from reference 30, ©1987 The Annals of Botany Company.

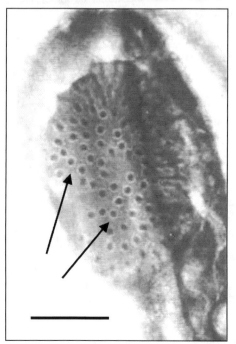

Figure 9. High voltage transmission electron micrograph of a thick (1 μm) section in the plane of the radial wall between developing fibres in *Sorbus aucuparia*. The oval structure is a swollen pit membrane in which the dispersed plasmodesmata present in the pit field of the cambial cell (Fig. 8) have aggregated and become closely packed (arrows). Scale bar = 1 μm. Reproduced with permission from reference 30, ©1987 The Annals of Botany Company.

Plasmodesmata in Differentiating Xylem

Plasmodesmata are increasingly being recognised as the channels through which pass a range of molecules involved in the control of plant tissue differentiation.[33-37] They might therefore be expected to play a major role in the differentiation of wood cells, controlling the development of the anatomical arrangement of cell types. In addition, the fusiform cells of the cambium and their differentiating xylem derivatives represent the major sink for transportates from the rays. Movement of materials into meristematic and differentiating cells must be through the radial walls, although it is not known for certain whether this movement is symplasmic, apoplasmic or a combination of both. The distribution and abundance of plasmodesmata might be expected, therefore, to provide some clues to their role in both processes.

At the moment, however, evidence for the presence of plasmodesmata in the radial walls between ray parenchyma and differentiating fusiform cells is equivocal. In some cases such as *Pinus radiata*, and primitive woody angiosperm dicotyledonous plants such as members of the Winteraceae and Trochodendraceae, which have only tracheids, there are none.[38] In the more advanced woody angiosperms, however, they are present in greater or lesser numbers according to the differentiating fusiform cell type.[29]

The distribution of plasmodesmata in the cell walls of vascular cambium and differentiating xylem has received relatively little attention and few species have been examined with this in mind. Cronshaw and Wardrop carried out the first electron microscope investigation of sections through differentiating secondary xylem elements and noted that plasmodesmata were absent from the sites of bordered pit development in *Pinus radiata*.[39] This was confirmed for the same species by Barnett and Harris who also observed that plasmodesmata were apparently absent from any part of the wall of the developing tracheids, including the tangential wall formed at cell division.[38] Examination of fusiform cells in the cambial zone also failed to find plasmodesmata in either their radial or tangential walls. On the other hand, plasmodesmata were abundant in the radial and horizontal walls between ray cells at all stages of development. This means that the fusiform cambial cells and their derivatives communicate with their neighbouring cells only via the apoplasm; an interesting observation in that it paralleled the observed isolation from the symplasm of cells in tissue culture by loss of plasmodesmata prior to their differentiation into tracheary elements.[40] An alternative control mechanism for the formation of vessel elements has been recently proposed by Barlow, who suggests that a new vessel element is determined as a result of contact between a new ray and a fusiform cell.[41] A stimulus (possibly auxin) produced by the newly developing vessel element then travels vertically downwards causing fusiform cells below to also develop into vessel elements. It is also suggested that the time taken for this process could account for vessels crossing growth ring boundaries. If this idea is correct, then the anatomical arrangement of vessels should be related to the distribution pattern of rays. Vessel multiples would arise from almost simultaneous formation of new rays in close vertical alignment at different heights in the stem. A careful analysis of ray distribution and initiation could help to test this interesting idea.

Czaninski reported that plasmodesmata were absent from the pit membranes of vessel elements,[5] an observation confirmed by Barnett in a study of differentiating secondary xylem in angiosperm woods.[29] Cells destined to become fibres and axial parenchyma had plasmodesmata in their developing pit membranes whereas those differentiating into vessel elements and tracheids had none. The pit membranes of developing tracheids in the primitive vessel-less angiosperms *Drimys winteri* and *Trochodendron aralioides* also lacked plasmodesmata.[29] The situation is not clear-cut, however. A recent quantitative study of numbers of plasmodesmata in developing xylem of one-year-old and 20-year-old hybrid aspen (*Populus tremula* x *tremuloides*), however, showed plasmodesmata in the walls between enlarging vessel elements and enlarging fibres and between enlarging vessel elements and rays, albeit in smaller numbers than in the walls of parenchyma or developing fibres (unpublished observations).

It may be significant that the fusiform cambial cells in the majority of coniferous trees produce only tracheids. There is thus no need for information concerning the fate of neighbouring cells to enter the cambial cells or differentiating tracheids via the symplasm. The metabolites required by the developing fusiform cells presumably enter by the apoplasmic pathway having moved out of the symplasm of the rays. Exceptions to tracheid formation occur when resin canals are formed towards the end of the growing season, or sparsely distributed axial parenchyma are formed in species such as larch and redwood. Whether plasmodesmatal connections are present in the walls of these cells during their development has still to be investigated, but it is possible that insertion of secondary plasmodesmata may be a controlling factor in their formation.

The situation with regard to the origin of plasmodesmata in developing wood of angiosperms is complicated by the fact that during the cell enlargement stage of xylem cell differentiation, the circumference of a putative vessel and the length of a putative fibre may each increase by several hundred percent. There is thus a constant jostling and shuffling of cells resulting in the formation of many new cell contacts and the loss of others. In the case of vessel element enlargement in *Quercus*, for example, the original cambial derivative from which the vessel element differentiates would have been in lateral contact with a maximum of six other cambial zone cells; after enlargement as many as twenty or more cells may be found to share a common wall with the vessel (Fig. 1C). Fibres increase in length by tip growth and may force their way between the derivatives of several vertical files of cambial cells. Plasmodesmata between cells which have come into contact during growth must have formed secondarily, while it is probable that many connections present in the cambium are broken.

Formation of Plasmodesmata in Cambium and Developing Wood Cells

While plasmodesmata in tangential walls between ray parenchyma are almost certainly formed primarily through this wall at the cell plate stage of development, there can be no doubt that plasmodesmata in developing fusiform xylem cells are formed secondarily. A study of a transverse section of a gymnosperm wood reveals long radial files of tracheids, produced by an uninterrupted sequence of periclinal divisions of single fusiform initials (Fig. 10). These files have long common ancestral radial walls whose thickness and integrity have been maintained by addition of

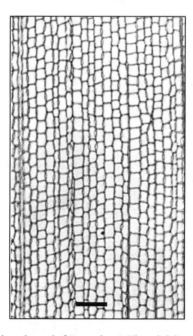

Figure 10. Transverse section through wood of *Pinus sylvestris*. The radial direction in the wood is from top to bottom of the micrograph. Note the long uninterrupted rows of cells with common radial ancestral walls. Scale bar = 0.1 mm.

new wall material at each division. There is no opportunity for the primary formation of plasmodesmata in these walls; yet in angiosperms, it is these radial walls which contain most plasmodesmata, usually in the form of clusters in developing pit membranes in fibres or parenchyma cells.[30,31] On the other hand, and contrary to expectation, plasmodesmata are rare in the tangential walls formed at periclinal division. This suggests that where such connections are lacking, they have never been formed, rather than being lost during differentiation. That plasmodesmata are rare in the walls of developing vessel elements and tracheids, but common in the walls of developing fibres and parenchyma, suggests further that active production of secondary plasmodesmata may be used as one element in controlling the fate of a cambial derivative. Isolation of cells into symplasmic domains by closing of plasmodesmatal connections is now thought to allow groups of cells to differentiate into distinct types.[42] An alternative method of creating domains may function in wood by leaving cells destined to become tracheary elements at the meristematic stage isolated from the symplasm.

In conclusion, cell-cell channels in the form of plasmodesmata are important in wood in conveying materials from both xylem and phloem to the meristematic zone producing wood (and secondary phloem). The large numbers of plasmodesmata in the tangential walls between ray parenchyma support this view. However, the relatively small numbers, or even a total lack of plasmodesmata between these ray parenchyma and cells of the developing xylem suggest that materials leave the rays to enter the cambial cells and their developing derivatives apoplasmically. Such plasmodesmata as exist in the walls of the latter may function as channels for messengers controlling the direction of cell differentiation in symplasmic domains.

References

1. Barnett JR. Xylem Physiology. In: Burley J, Evans J, Youngquist J, eds. Encyclopedia of Forest Sciences. London, New York: Academic Press, 2004:1583-1590.
2. Bel AJE van. Xylem-phloem exchange in the rays: The undervalued route of transport. J Exp Bot 1990; 41:631-644.
3. Savidge RA. Xylogenesis, genetic and environmental regulation – a review. IAWA J 1996; 17:269-311.
4. Iqbal M, Ghouse AKM. Cambial concept and organisation. In: Iqbal M, ed. The Vascular Cambium. Taunton, England, New York: Research Studies Press Ltd, John Wiley and Sons Inc., 1990:1-36.
5. Czaninski Y. Vessel associated cells. IAWA Bull 1977; 3:51-55.
6. Schmid R. Fine structure of pits in hardwoods. In: Côté WA, ed. Cellular Ultrastructure of Woody Plants. New York: Syracuse University Press, 1965:291-304.
7. Barnett JR, Cooper P, Bonner LJ. The protective layer as an extension of the apoplast. IAWA J 1993; 14:163-171.
8. Chafe SC, Chauret G. Cell wall structure in the xylem parenchyma of trembling aspen. Protoplasma 1974; 80:129-147.
9. Wooding FBP, Northcote DH. An anomalous wall thickening and its possible role in the uptake of stem-fed tritiated glucose by Pinus picea. J Ultr Res 1965; 12:463-472.
10. Chafe SC. Cell wall formation and protective layer development in the xylem parenchyma of trembling aspen. Protoplasma 1974; 89:335-354.
11. Wisniewski W, Ashworth E, Schaffer K. Use of lanthanum to characterize cell wall permeability to deep supercooling and extracellular freezing in woody plants. I. Intergeneric comparisons between Prunus, Cornus and Salix. Protoplasma 1987; 139:105-116.
12. Bel AJE van, Schoot C. Primary function of the protective layer in contact cells: Buffer against oscillations in the hydrostatic pressure in the vessels? IAWA Bull 1988; 9:285-288.
13. Sauter JJ, Kloth S. Plasmodesmatal frequency and radial translocation rates in ray cells of poplar (Populus x canadensis Moench "robusta"). Planta 1986; 168:377-380.
14. Gunning BES, Robards AW. Plasmodesmata: Current knowledge and outstanding problems. In: Gunning BES, Robards AW, eds. Intercellular Communication in Plants: Studies on Plasmodesmata. Springer-Verlag, 1976:297-311.
15. Outer RW den. Histological investigations of the secondary phloem in gymnosperms. Mededelingen Landbouwhogschool Wageningen 1967; 67(7):119.

16. Sauter JJ, Iten W, Zimmermann MH. Studies on the release of sugar into the vessels of sugar maple (Acer saccharum). Can J Bot 1973; 51:1-8.
17. Hartig T. Beitrage für physiologischen Forstbotanik. Allgemeine Forst Jagdzeitung 1860; 36:257-261.
18. Braun HJ. The significance of the accessory tissues of the hydrosystem for osmotic water shifting as the second principle of water ascent, with some thoughts concerning the evolution of trees. IAWA Bull 1984; 5:275-294.
19. Bannan MW. The annual cycle of size changes in the fusiform cambial cells of Chamaecyparis and Thuja. Can J Bot 1951; 29:421-437.
20. Ziegler H. Storage, mobilization and distribution of reserve material in trees. In: Zimmermann MH, ed. The Formation of Wood in Forest Trees. New York: Academic Press, 1964:303-320.
21. Philipson WR, Ward JM, Butterfield BG. The vascular cambium: Its development and activity. London: Chapman and Hall, 1971.
22. Lev-Yadun S, Aloni R. Differentiation of the ray system in woody plants. Bot Rev 1995; 61:45-84.
23. Eklund L. Endogenous levels of oxygen, carbon dioxide and ethylene in stems of Norway spruce trees during one growing season. Trees: Struct Funct 1990; 4:150-154.
24. Sauter J, van Cleve B. Biochemical, immunochemical and ultrastructural studies of protein storage in poplar (Populus x canadensis "robusta") wood. Planta 1990; 183:92-100.
25. Höll W. Distribution, fluctuation and metabolism of food reserves in the wood of trees. In: Savidge RA, Barnett JR, Napier R, eds. Cell and Molecular Biology of Wood Formation. Oxford: BIOS Scientific Publishers, 2000:347-362.
26. Sauter J. Photosynthate allocation to the vascular cambium: Facts and problems. In: Savidge RA, Barnett JR, Napier R, eds. Cell and Molecular Biology of Wood Formation. Oxford: BIOS Scientific Publishers, 2000:71-83.
27. Chaffey N, Barlow PW. The cytoskeleton facilitates as three-dimensional symplasmic continuum in the long-lived ray and axial parenchyma cells of angiosperm trees. Planta 2001; 213:811-823.
28. Overall RL, White RG, Blackman LM et al. Actin and myosin in plasmodesmata. In: Steiger C, Baluska F, Volkmann D, Barlow PW, eds. Actin: A Dynamic Framework for Multiple Plant Cell Functions. Dordrecht: Kluwer, 2000:497-515.
29. Barnett JR. Plasmodesmata and pit development in secondary xylem elements. Planta 1982; 155:251-260.
30. Barnett JR. Changes in the distribution of plasmodesmata in developing fibretracheid pit membranes of Sorbus aucuparia L. Ann Bot 1987; 59:269-279.
31. Barnett JR. The development of fibretracheid pit membranes in Pyrus communis L. IAWA Bull 1987; 8:134-142.
32. IAWA. Multilingual glossary of terms used in wood anatomy. Committee on Nomenclature. International Association of Wood Anatomists. Verlangsanstalt Buchdruckerei Konkordia Winterthur 1964.
33. Heinlein M. Plasmodesmata: Dynamic regulation and role in macromolecular cell-cell signalling. Curr Opin Plant Biol 2002; 5:543-552.
34. Ruiz-Medrano R, Xoconostle-Cazares B, Kragler F. The plasmodesmatal transport pathway for homoeotic proteins, silencing signals and viruses. Curr Opin Plant Biol 2004; 17:641-650.
35. Lucas WJ, Lee JY. Plant cell biology: Plasmodesmata as a supracellular control network in plants. Nat Rev Mol Cell Biol 2004; 5:712-726.
36. Zambryski P. Cell-to-cell transport of proteins and fluorescent tracers via plasmodesmata during plant development. J Cell Biol 2004; 164:165-168.
37. Gallagher KL, Benfey PN. Not just another hole in the wall: Understanding intercellular protein trafficking. Genes Dev 2005; 19:189-195.
38. Barnett JR, Harris JM. Early stages of bordered pit formation in radiate pine. Wood Sci Technol 1975; 9:233-241.
39. Cronshaw J, Wardrop AB. The organisation of the cytoplasm in differentiating xylem. Aust J Bot 1963; 12:15-23.
40. Carr DJ. Plasmodesmata in growth and development. In: Gunning BES, Robards AW, eds. Intercellular Communication in Plants: Studies on Plasmodesmata. Berlin, Heidelberg, New York: Springer-Verlag, 1976:243-290.
41. Barlow PW. From cambium to early cell differentiation within the secondary vascular system. In: Holbrook NM, Zwieniecki M, Melcher PJ, eds. Vascular Transport in Plants. Oxford: Elsevier/Academic Press, 2005:in press.
42. Itaya A, Ma FS, Qi YJ. Plasmodesma-mediated selective protein traffic between "symplasmically isolated" cells probed by a viral movement protein. Plant Cell 2002; 14:2071-2083.

CHAPTER 10

TMV Movement Protein Targets Cell-Cell Channels in Plants and Prokaryotes:
Possible Roles of Tubulin- and FtsZ-Based Cytoskeletons

Manfred Heinlein*

Abstract

Cell-to-cell communication in plants occurs via plasmodesmata (Pd), dynamic membrane-lined pores in the plant cell wall that provide symplasmic continuity between adjacent cells. Communication through Pd involves the trafficking of protein and RNA macromolecules, and also includes the trafficking of viruses. Virus movement depends on virus-encoded movement proteins (MP), of which the MP of *Tobacco mosaic virus* (TMV) has been most thoroughly studied. The many cellular activities of this protein include the interaction with Pd, which results in the deposition of fibrous material in the Pd cavity as well as in an increased size-exclusion limit of the channel. The protein also interacts with plant microtubules and actin filaments, and the alignment of the protein to microtubules has been correlated with the function of this protein. Intriguingly, the protein also interacts with the cell junctions of the multicellular cyanobacterium *Anabaena* suggesting a degree of functional analogy between intercellular communication mechanisms of multicellular prokaryotes and plants. In *Anabaena*, MP induces the formation of MP-associated filaments which traverse the intercellular septa and which may be similar in nature to the fibrous MP-associated material localized to Pd in MP-expressing plants. These observations may suggest the involvement of conserved cytoskeletal elements in the MP-dependent modification and plasticity of intercellular connections in evolutionary divergent species.

Plasmodesmata and Intercellular Communication in Plants

Intercellular communication delivers crucial information for the position-dependent specification of cell fate and, therefore, is an essential biological process during the coordination of development in multicellular organisms. Communication in animals is mediated by receptor:ligand interactions as well as through gap junctions, proteinaceous channels that interconnect contiguous cells and function in transport of small molecules of less that 1 kDa. Moreover, recent observations of 'tunneling nanotubes' suggest that animal cells also have the potential to form cytoplasmic bridges to exchange macromolecules symplasmically.[1] Similar pathways for intercellular communication exist in plants: whereas the apoplasmic pathway mediates communication via receptor:ligand-interactions,[2] the symplasmic pathway allows the direct intercellular exchange of macromolecules through cytoplasmic bridges in the cell wall termed plasmodesmata (Pd).[3-11] The system of Pd is connected to the phloem sieve elements in the vascular veins and stems and, thus, forms a cell-to-cell and long-distance communication network that enables plants to rap-

*Manfred Heinlein—Institut Biologie Moléculaire des Plantes, CNRS UPR 2357, 12, rue du Général Zimmer, F-67084 Strasbourg Cedex, France. Email: manfred.heinlein@ibmp-ulp.u-strasbg.fr

Cell-Cell Channels, edited by Frantisek Baluska, Dieter Volkmann and Peter W. Barlow.
©2006 Landes Bioscience and Springer Science+Business Media.

idly disseminate information and metabolites, thereby coordinating cellular activities at a level above that of the individual cell.[5] The conductivity of Pd is under developmental and physiological control and thus defines cell-to-cell communication within and between 'supracellular domains.[12] For example, when leaves mature and undergo the sink-to-source transition there are conspicuous changes in Pd conductivity and structure.[13] Long-term decreases and increases in Pd conductivity have also been correlated with the local deposition and removal of callose.[14-16] In addition, Pd are intrinsically dynamic and can be gated by non-cell-autonomous proteins (NCAPs).[5,9,17] These rather short-term dynamic changes in the size-exclusion limit (SEL) of Pd may involve Pd-associated cytoskeletal elements.[4,18,19]

The ability of specific NCAPs to gate Pd allows their own trafficking and is involved in the transport of RNA molecules.[5,9] Communication through Pd indeed includes the controlled cell-to-cell and systemic trafficking of a whole range of RNA and protein macromolecules, such as non-cell-autonomous transcription factors, RNA-based silencing signals, and messenger RNAs.[4,9,11,20-24] However, although direct communication via RNA and protein trafficking through Pd has likely evolved to accelerate efficient cell-to-cell and systemic signaling, it also represents an Achilles heel, since RNA viruses use this feature to spread infection. Yet, for the researcher, RNA viruses are excellent keys to the molecular mechanisms that govern intercellular macromolecular trafficking and intercellular communication through Pd.[4,25]

The Movement Protein of TMV Interacts with the Plant Cytoskeleton and Plasmodesmata to Facilitate Intercellular Spread of the Virus

The spread of most, if not all, viruses depends on specialized virus-encoded NCAPs, termed Movement Proteins (MP).[4,5,26] Pioneering studies performed with the MP of *Tobacco mosaic virus* (TMV)[27,28] have demonstrated that this protein targets Pd (Fig. 1A) and modifies their SEL.[29] Furthermore, upon transient expression or microinjection, the protein spreads from cell to cell.[30] Since plants themselves encode endogenous NCAPs that are functionally analogous to viral MPs,[5,31] the function of MP may directly reflect mechanisms of macromolecular Pd transport in normal plants that are exploited by viruses for the movement of their genomes.

The mechanism by which TMV and the plant host coordinate the trafficking of the viral RNA genome (vRNA) is not well understood. Although the targeting of Pd by MP may be sufficient for its own spread, it is in itself not sufficient to drive vRNA movement. Several MP mutants have been described which accumulate in Pd but are dysfunctional with respect to vRNA movement.[32-34] Moreover, sink tissues with Pd characterized by large SELs allow movement of macromolecules of sizes up to 50 kD but still require MP for intercellular transport of vRNA.[13] Apparently, neither the presence of MP in Pd nor an increased SEL of the channel is sufficient for virus movement, indicating that the transport of vRNA depends on additional functions of MP. Indeed, since the size of vRNA amounts to several megadaltons, the MP probably interacts with an active translocation mechanism for transport of the vRNA through the viscous cytoplasm.[35,36]

In vitro studies have revealed that the MP binds to single-stranded nucleic acids which results in the formation of an unfolded, elongated ribonucleoprotein complex (vRNP) with apparent dimensions compatible with translocation through dilated Pd.[37-39] Although in vivo evidence for the formation of an MP:vRNA complex is still lacking, it seems likely that the MP binds to vRNA in vivo, thus forming the core of the infectious particle that spreads between cells. This hypothesis is supported by the fact that the coat protein (CP) of TMV is dispensable for the cell-to-cell movement of infection,[40,41] thus indicating that vRNA is indeed transported in a nonencapsidated form. Recent observations suggest that vRNA moves in the form of larger, membrane-associated replication complexes.[42] Earlier studies reported the association of MP with cytoskeletal elements[43-46] suggesting a role of microtubules, actin filaments, and associated motor proteins in the transport of infection. Further in vivo studies showed that vRNA transport, but not the targeting of MP to Pd, is tightly correlated with the ability of MP to bind microtubules[33,34,47-49] suggesting that MP utilizes microtubules for vRNA transport rather than for its own targeting to Pd. Other subsequent experiments involved treatments of infected *N. benthamiana* and *N. tabacum* plants with cytoskeletal inhibitors

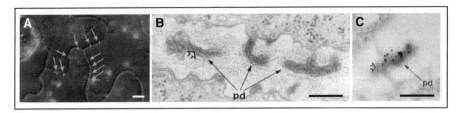

Figure 1. The MP of TMV targets Pd and leads to the deposition of fibrous material within the Pd cavity. A) MP fused to GFP (MP:GFP) localizes to Pd (arrows) in TMV-MP:GFP-infected *Nicotiana benthamiana* epidermal cells. Bar is 10 μm. Reprinted with permission from Heinlein M et al, Plant Cell 10:1107-1120, ©1998 American Society of Plant Biologists.[44] B) Deposition of fibrous material within Pd of MP-expressing *Nicotiana tabacum* plants. Bar is 0.25 μm. C) Presence of MP within Pd of MP-expressing *Nicotiana tabacum* plants, as shown by immuno-labeling. Bar is 0.25 μm. B,C) Reprinted with permission from Moore PJ et al, Protoplasma 170:115-127,[59] ©1992 Springer Verlag.

and indicated that vRNA movement is actin- rather than microtubule-dependent.[42,50] This latter finding supports a role of the actin cytoskeleton during infection.[44,46] Since actin has been localized to Pd,[19] and since the association of MP with Pd seems not to require prior association of MP with microtubules,[32-34] it appears possible that actin filaments are involved in the targeting or anchorage of the MP to Pd.

Microtubules, although involved, may not be essential for movement in *Nicotiana* species. As an alternative or additional function, it has been proposed that microtubules may support virus infection by targeting the protein for degradation,[45] a hypothesis that gained further support by observations indicating that the level and localization of MP is changed in the presence of proteasome inhibitors.[50,51] However, recent studies in our laboratory clearly argue against this possibility by showing that although poly-ubiquitinylated MP accumulates during infection and is detected in total cell extracts, microtubule-associated MP is not ubiquitinylated (manuscript in preparation). We recently localized MP to mobile, microtubule-associated granules in cells at the leading front of infection (manuscript in preparation). However, whether these MP-associated particle movements are indeed related to viral cell-to-cell transport remains to be shown. There is accumulating evidence that microtubules play a role during infection by viruses belonging to diverse families.[52-55] Certainly, further *in planta* studies are required to elucidate the roles of the cytoskeleton during virus replication and the cell-to-cell spread of infection in plants.

Results of studies in vitro (manuscript in preparation), and of studies in which the MP of TMV was expressed in transfected mammalian cells,[34] indicate that the MP intrinsically associates with microtubules and does not require additional factors for this activity. The direct interaction of MP with the microtubule may be mediated by a domain with similarity to the M-loop of α, β, and γ-tubulin, as has been suggested by single amino acid exchange mutations, which rendered the protein temperature-sensitive with respect to in vivo microtubule association and function in vRNA movement.[34] Although further tests are needed to verify that this domain is directly involved in microtubule association, one can speculate that this domain allows the MP to emulate the binding forces of M-loop/N-loop contacts between adjacent microtubule protofilaments and thus to directly manipulate microtubule assembly/disassembly dynamics. Because of the presence of this homology with tubulin, it has been speculated that MP may coassemble with the microtubule and incorporate into the microtubule lattice. However, in vivo experiments employing fluorescence recovery after photobleaching indicate that MP behaves like a MAP and binds to the surface of microtubules rather than moving laterally within the dynamic polymer. Moreover, in vitro, the MP is able to directly bind to preformed and dynamically stabilized microtubules (manuscript in preparation). Thus, the requirement of microtubule assembly dynamics for complex formation appears unlikely.

The potential ability of MP to modify microtubule dynamics is supported by the observation that in transfected mammalian cells, the transient expression of MP has profound effects on the

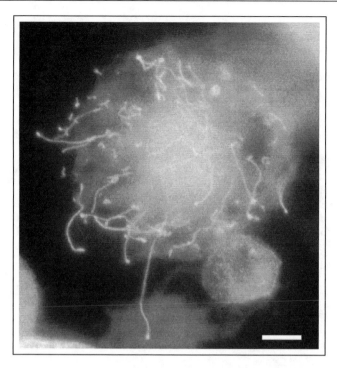

Figure 2. Tobacco BY-2 protoplasts form MP:GFP-containing protrusions upon infection with TMV-MP:GFP. Bar is 5 μm. Reprinted with permission from Heinlein M, Cell Mol Life Sci 59:58-82, ©2002 Birkhäuser Verlag.[25]

structure of the microtubule array and on centrosomal microtubule-nucleation activity.[34] The observed changes correlate with an apparent lack of γ-tubulin from the centrosome, suggesting that the MP might be able to interact with microtubule-organizing complexes to manipulate the cytoskeleton in yet unknown ways.[34] In plants, the MP seems not to induce such strong microtubule rearrangements as those observed in mammalian cells. However, such strong effects may indeed not be expected since, unlike in mammalian cells, microtubule-organizing activity in plant cells is dispersed throughout the cytoskeleton.[56,57] Given that microtubule nucleation events are dispersed, the expression of MP may impose only subtle changes in the overall structure of the array. Preliminary observations made in our laboratory suggest the involvement of local microtubule rearrangements in the formation of plasma membrane projections that protrude of the surface of cell wall-lacking protoplasts infected with TMV[25,44] (Fig. 2). Analogous to the movement mechanism of 'tubule-forming' viruses, these projections may reflect the formation of transport tubules required for the viral transport event through Pd. Although TMV infection has not been reported to result in the formation of tubules within Pd, the interaction of MP with Pd nevertheless results in the deposition of fibrous material within the Pd cavity[58-60] (Fig. 1B). The fibers are associated with MP (Fig. 1C), but whether the fibers solely consist of MP, or are associated with an underlying cytoskeletal element remains unknown. However, the correlation between the possible role of microtubules in the formation of MP-containing membrane protrusions in protoplasts and the occurrence of MP-associated fibrous structures in Pd of plant epidermal cells may suggest that vRNA movement depends upon the local reorganization of the cytoskeletal array and the formation of a virus-induced cytoskeletal transport structure within Pd. The mechanism of cell-to-cell spread of TMV in plants might then be similar to that which facilitates the spread of *Human T-cell leukemia virus* (HTLV-1) or of *Human immunodeficiency virus* (HIV) between lymphocytes, a process which involves the virus-induced reorientation of the microtubule-organizing center (MTOC) and the polarization of the cytoskeleton to induce unique cell-cell adhesion domains (virological synapse) through which the spread of viral material occurs.[61,62]

The Movement Protein of TMV Targets and Modifies Cell-Cell Junctions in *Anabaena*

A potential role of cytoskeletal elements in the manipulation of intercellular communication by MP is also indicated by surprising results obtained upon expression of MP in the cyanobacterium *Anabaena* sp. strain PCC 7120. This is a multicellular prokaryotic organism, in which the patterned differentiation of heterocysts as well as normal growth of the bacterial filament depends on intercellular communication and a set of specific genes.[63-67] One of these genes is *patS*, which encodes a diffusible peptide.[68] The expression of *patS* is localized primarily in proheterocysts and heterocysts, and since its overexpression inhibits heterocyst differentiation, it has been proposed that the processed PatS peptide, originating from differentiating proheterocysts, diffuses between cells, thus creating a gradient of PatS signalling.[68] Although PatS is has been reported to inhibit the DNA binding activity of the transcriptional activator HetR,[69] the mechanism of intercellular communication by PatS has not been further addressed. It has been proposed that PatS may diffuse through the periplasmic space and be taken up by neighbouring cells.[68] However, the peptide may also diffuse directly between physically connected cells. Morphological evidence for direct intercellular connections between cells of filamentous multicellular cyanobacteria has been discovered (Fig. 3) and described as intercellular connections,[70] porelike structures,[71,72] pores,[73] intercalary perforations[74] (Fig. 3), plasmodesmata,[75,76] and microplasmodesmata (mPd,[77,78]).

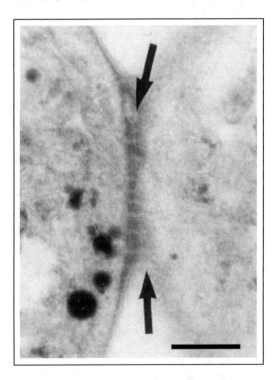

Figure 3. Transmission electron microscopy micrograph of *Anabaena variabilis* showing intercellular perforation (arrows) of ~24 nm in diameter in the cross walls of neighbouring cells. Bar is 0.4 μm. Reprinted with permission from Palinska KA, Krumbein WE, J Phycol 36:139-145, ©2000 Blackwell Publishing.[74]

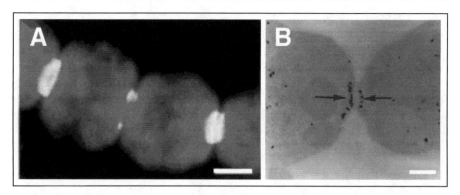

Figure 4. Bacterial filaments of *Anabaena* sp. strain PCC 7120 expressing the MP of TMV at levels, which do not interfere with intercellular communication. A) Immuno-labeled MP (yellow) localizes to the cell junctions. Bar is 1.5 μm. B) Electron microscopical image of a thin section showing the localization of immunogold-labeled MP at the cell junction (arrows). Bar is 0.5 μm. Reprinted with permission from Heinlein M et al, Plant J 14:345-351, ©1998 Blackwell Publishing.[80] A color version of this figure is available online at www.Eurekah.com.

Because of their small outer diameter (10-20 nm) and lack of either a discernable core (porelike structures) or plasma membrane continuity, these intercellular connections are structurally well distinguished from Pd of higher plants. Surprisingly, when MP was expressed in transgenic *Anabaena*, it was found that high level expression of this protein prevented diazotrophic growth and heterocyst differentiation, indicating that the MP perturbs cell-to-cell communication.[79] In a subsequent study, the MP was immunolabelled and found to localize to the cell junctions[80] (Fig. 4). At low concentrations of MP, which did not interfere with heterocyst differentiation, the protein was localized to ring-like structures that flank the intercellular septum at both sides (Fig. 4). In cells expressing the protein at levels that interfere with intercellular communication and heterocyst differentiation, the MP antibody again labelled the ring-like structures, but also labelled a tubular arrangement of filaments that not only traversed the length of the cells, but also extended across the septa connecting adjacent cells (Fig. 5). This observation is exciting because, despite the distant evolutionary relationship between higher plants and multicellular prokaryotes, MP might target both Pd and cyanobacterium cell-cell junctions by a functionally conserved mechanism. This finding indicates that a degree of functional homology exists between cell-cell channels of bacteria and plants. Moreover, the formation of intercellular filaments by MP highlights cytoskeletal elements as being involved in the MP-dependent modification and plasticity of intercellular connections in evolutionary divergent species.

The diameter of the MP-associated filament bundle formed within the cyanobacterial cell junctions was found to be 150 nm and, therefore, much larger than the diameter of mPd (20 nm,[77]) that connect *Anabaena* cells.[63] Conceivable then, the filament constituents of the bundle may traverse the septa by passing individually through a large group of mPd, or the bundle may extend as a whole through much larger septal pores that are either derived from mPd or newly formed in the presence of MP.[80]

The ring-like structures that are localized adjacent to the cell junctions and to which MP also localizes[80] are equally intriguing. This ring-like distribution of the MP may indicate that the protein associates with the *Anabaena* homolog[81] of the essential procaryotic cell division protein FtsZ.[82,83] In *Escherichia coli*, FtsZ is a highly abundant protein[84] that polymerizes in a GTP-dependent manner to form a circumferential ring (the Z-ring) that constricts at the leading edge of the invaginating septum to separate the two daughter cells.[82,85] The molecular structure of FtsZ is congruous with that of eukaryotic tubulin,[86,87] confirming the homology and probable common ancestry of these proteins. Although FtsZ exhibits only low sequence identity to tubulin (10-18% at the amino acid level), its structural homology to tubulin may be sufficient for interactions with MP, either directly

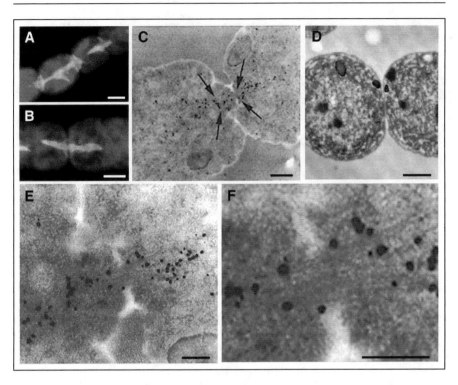

Figure 5. Bacterial filaments of *Anabaena* sp. strain PCC 7120 expressing the MP of TMV at levels that interfere with intercellular communication. A,B) Immuno-labeled MP (yellow) localizes to filaments that span the cells (A) and traverse the septum between cells (B). Bar is 1.5 μm. C) Thin section of a cell junction showing filaments that expand from cell to cell (arrows). The intercellular filaments are immunogold-labeled with antibody against MP. Bar is 0.5 μm. D) The intercellular filaments and MP-specific immunogold labeling are absent in wild-type *Anabaena* cells. Bar is 1 μm. E) Electron micrograph enlargement of the septal junction between contiguous cells. The strand that links both cells is ~ 150 nm in diameter and characterized by a filamentous substructure. The immunogold labeling of the MP can be seen decorating individual filaments of the strand. Stereoelectron micrographs[80] show that the filaments form a tubular arrangement within the bundle. Bar is 0.2 μm. F) Enlargements of the central area of (E) to highlight the filamentous substructure of the bundle traversing the septum. Bar is 0.2 μm. Reprinted with permission from Heinlein M et al, Plant J 14:345-351, ©1998 Blackwell Publishing.[80] A color version of this figure is available online at www.Eurekah.com.

or via FtsZ-associated proteins.[88] With respect to the nature of the MP-induced filaments described above it appears interesting that when overexpressed, the FtsZ protein forms long tubules in E. coli cells.[89] Thus, it may be possible that overexpression of MP in *Anabaena* causes sequestration or displacement of FtsZ-associated factors and subsequent aggregation of overabundant FtsZ and associated MP into filaments.[80] Another conceivable scenario for the formation of the filaments is that MP may mimic bacterial MAPs such as FtsA and ZipA. These proteins bind to FtsZ and support the formation and stabilization of the Z ring.[90,91] Moreover, both proteins carry C-terminal sequences that tether the ring to the membrane. The ability of these proteins to bind to both FtsZ and membrane is indeed reminiscent of MP, which binds to tubulin but also carries hydrophobic transmembrane domains that mediate its interaction with membranes.[92,93] Thus, the MP may function as a modifying cytoskeleton-to-membrane linker, just like FtsA.[91] The ability of FtsA to bind FtsZ depends on its membrane association. Interestingly, when FtsA fails to interact with the membrane and, as a consequence, also with FtsZ, it forms rods.[91] Thus, if the MP is a MAP as described above, the MP-associated filaments observed in *Anabaena* cells over-expressing the protein[80] may be independent of FtsZ and formed by excess MP. Clearly, further supporting evidence is required before a statement concerning the interaction of MP with the FtsZ machinery can be made. In fact, the MP-labelled rings that are found adjacent to the septum in *Anabaena* remain at the cell poles following cytokinesis. This appears to argue against the association of MP with the Z-ring, which, based on studies in *Escherichia coli* and *Bacillus subtilis*, disperses upon completion of septation.[82] Another possibility is that MP interacts with zones of membrane-to-wall adhesions that are referred to as periseptal annuli.[94] This alternative hypothesis would be in agreement with the notion that the Pd of plants also represent sites of membrane-to-cell wall adhesion[95,96] and as such may act as general targets for MP localization.[44]

It should be noted that in addition to the FtsZ ring, bacteria also express other cytoskeletal elements[97,98] with which the MP could potentially interact. One element related to the FtsZ ring is the Min system, which has long been known to be involved in defining the bacterial division site.88 Since the MinD component of this system polymerizes into macromolecular filaments or cables in vitro and in vivo[99-101] it may be related to the observed MP-associated filaments. Evidence for the interaction of MP with plant actin may suggest that the MP-associated filaments in *Anabaena* may also be caused by interaction of the protein with prokaryotic actin orthologs. Bacteria indeed are known to encode such orthologs, such as the cell-shape determinant MreB[102-104] or ParM, which is involved in plasmid segregation.[105,106]

Although at the present time the correlation between the association of MP with cytoskeletal elements of plants and bacteria is highly speculative, the involvement of the cytoskeleton in MP-dependent transport structures in plant and bacterial cell junctions is consistent with infection strategies by certain invasive pathogenic bacteria such as *Listeria*, *Shigella*, and *Rickettsia*, which also use cytoskeleton-induced trasport structures to move intra- and intercellularly. *Listeria monocytogenes* is a facultative intracellular parasite that infects animal cells, causing Listeriosis in humans. Spread of *Listeria* occurs by formation of tubular extensions on the cell surface that extend deep into neighbouring cells. The neighbouring cell then takes up the bacteria by process that is sometimes referred to as paracytophagy.[107] The tubular structures are required for infectivity of *Listeria* and are formed by the ability of the bacterium to nucleate actin from the ActA protein on their surface. This provides the force required to deform and extend the plasma membrane (for reviews, see ref. 108). Another example of tubule formation occurs during spread of *Vaccinia* virus. Following the targeting of the viral particles to the plasma membrane by microtubules, the virus is secreted by exocytosis. Once outside the cell, actin polymerization below the plasma membrane leads to protrusions that push the virus into the adjacent cell.[109]

Future Prospects

The identification of MP-interacting factors in both plants and *Anabaena* represents an exciting perspective for future studies. Since the genome of *Anabaena* has been sequenced[110] and annotated,[111] and a genome database for Cyanobacteria has been created (http://www.kazusa.or.jp/cyano/index.html), *Anabaena* has joined the list of model organisms for studies involving functional genomics. However, even without knowing the detailed underlying molecular mechanisms, the ability of MP to enter and to modify cellular junctions in *Anabaena* as well as in plants is already quite remarkable. Cyanobacteria appeared on Earth approximately 2.7 billion years ago and demonstrate one of the earliest forms of intercellular communication and cellular differentiation in evolution. The ability of MP to interact with cell junctions in higher plants and *Anabaena* may be indicative of a shared mechanism of intercellular communication that has been conserved or reinvented during evolution. The ability of MP to interact with cell junctions in both organisms might also shed light on the origin of non-cell-autonomous viruses. Indeed, these observations might suggest that viral MP-like factors and the ability of viruses to spread between cells evolved as soon as the first multicellular organisms appeared.

References

1. Rustom A, Saffrich R, Markovic I et al. Nanotubular highways for intercellular organelle transport. Science 2004; 303:1007-1010.
2. Boller T. Peptide signalling in plant development and self/nonself perception. Curr Opin Cell Biol 2005; 17:116-122.
3. Epel B. Plasmodesmata: Composition, structure and trafficking. Plant Mol Biol 1994; 26:1343-1356.
4. Heinlein M, Epel BL. Macromolecular transport and signaling through plasmodesmata. Int Rev Cytol 2004; 235:93–164.
5. Lucas WJ, Lee JY. Plasmodesmata as a supracellular control network in plants. Nat Rev Mol Cell Biol 2004; 5:712-726.
6. Zambryski P, Crawford K. Plasmodesmata: Gatekeepers for cell-to-cell transport of developmental signals in plants. Annu Rev Cell Dev Biol 2000; 16:393-421.
7. Waigmann E, Zambryski P. Plasmodesmata. Gateways for rapid information transfer. Curr Biol 1994; 4:713-716.
8. van Bel AJE, van Kesteren WJP. Plasmodesmata, structure, function, role in cell communication. Berlin, Heidelberg, New York: Springer Verlag, 1999.
9. Kim JY. Regulation of short-distance transport of RNA and protein. Curr Opin Plant Biol 2005; 8:45-52.
10. Oparka KJ. Getting the message across: How do plant cells exchange macromolecular complexes? Trends Plant Sci 2004; 9:33-41.
11. Haywood V, Kragler F, Lucas WJ. Plasmodesmata: Pathways for protein and ribonucleoprotein signaling. Plant Cell 2002; (Supp):S303-S325.
12. Lucas WJ, van der Schoot C. Plasmodesmata and the supracellular nature of plants. New Phytol 1993; 125:435-476.
13. Oparka KJ, Roberts AG, Boevink P et al. Simple, but not branched, plasmodesmata allow the nonspecific trafficking of proteins in developing tobacco leaves. Cell 1999; 97:743-754.
14. Radford JE, Vesk M, Overall RL. Callose deposition at plasmodesmata. Protoplasma 1998; 201:30-37.
15. Iglesias VA, Meins Jr F. Movement of plant viruses is delayed in a β-1,3-glucanase-deficient mutant showing a reduced plasmodesmatal size exclusion limit and enhanced callose deposition. Plant J 2000; 21:157-166.
16. Bucher GL, Tarina C, Heinlein M et al. Local expression of enzymatically active class 1 beta-1,3-glucanase enhances symptoms of TMV infection in tobacco. Plant J 2001; 28:361-369.
17. Lee J-Y, Yoo B-C, Rojas MR et al. Selective trafficking of non-cell-autonomous proteins mediated by NtNCAPP1. Science 2003; 299:392-396.
18. Ding B, Kwon M-O, Warnberg L. Evidence that actin filaments are involved in controlling the permeability of plasmodesmata in tobacco mesophyll. Plant J 1996; 10:157-164.
19. Aaziz R, Dinant S, Epel BL. Plasmodesmata and plant cytoskeleton. Trends Plant Sci 2001; 6:326-330.
20. Lucas WJ, Yoo B-C, Kragler F. RNA as a long-distance information macromolecule in plants. Nat Rev Mol Cell Biol 2001; 2:849-857.
21. Heinlein M. Plasmodesmata:dynamic regulation and role in macromolecular cell-to-cell signalling. Curr Opin Plant Biol 2002; 5:543-552.
22. Yoo BC, Kragler F, Varkonyi-Gasic E et al. A systemic small RNA signaling system in plants. Plant Cell 2004; 16:1979-2000.
23. Wu X, Weigel D, Wigge PA. Signaling in plants by intercellular RNA and protein movement. Genes Dev 2002; 16:151-158.
24. Tzfira T, Rhee Y, Chen M-H et al. Nucleic acid transport in plant-microbe interactions: the molecules that walk through the walls. Annu Rev Microbiol 2000; 54:187-219.
25. Heinlein M. The spread of Tobacco mosaic virus infection: Insights into the cellular mechanism of RNA transport. Cell Mol Life Sci 2002; 59:58-82.
26. Deom MC, Lapidot M, Beachy RN. Plant virus movement proteins. Cell 1992; 69:221-224.
27. Deom CM, Oliver MJ, Beachy RN. The 30-kilodalton gene product of tobacco mosaic virus potentiates virus movement. Science 1987; 237:384-389.
28. Citovsky V. Tobacco mosaic virus: A pioneer of cell-to-cell movement. Phil Trans R Soc Lond B 1999; 354:637-643.
29. Wolf S, Deom CM, Beachy RN et al. Movement protein of tobacco mosaic virus modifies plasmodesmatal size exclusion limit. Science 1989; 246:377-379.
30. Waigmann E, Lucas W, Citovsky V et al. Direct functional assay for tobacco mosaic virus cell-to-cell movement protein and identification of a domain involved in increasing plasmodesmal permeability. Proc Natl Acad Sci USA 1994; 91:1433-1437.

31. Xoconostle-Cazares B, Xiang Y, Ruiz-Medrano R et al. Plant paralog to viral movement protein that potentiates transport of mRNA into the phloem. Science 1999; 283:94-98.
32. Kahn TW, Lapidot M, Heinlein M et al. Domains of the TMV movement protein involved in subcellular localization. Plant J 1998; 15:15-25.
33. Boyko V, van der Laak J, Ferralli J et al. Cellular targets of functional and dysfunctional mutants of tobacco mosaic virus movement protein fused to GFP. J Virol 2000; 74:11339-11346.
34. Boyko V, Ferralli J, Ashby J et al. Function of microtubules in intercellular transport of plant virus RNA. Nat Cell Biol 2000; 2:826-832.
35. Sodeik B. Mechanisms of viral transport in the cytoplasm. Trends Microbiol 2000; 8:465-472.
36. Verkman AS. Solute and macromolecule diffusion in cellular aqueous compartments. Trends Biochem Sci 2002; 27(1):27-33.
37. Citovsky V, Knorr D, Schuster G et al. The P30 movement protein of tobacco mosaic virus is a single-stranded nucleic acid binding protein. Cell 1990; 60:637-647.
38. Citovsky V, Wong ML, Shaw AL et al. Visualization and characterization of tobacco mosaic virus movement protein binding to single-standed nucleic acids. Plant Cell 1992; 4:397-411.
39. Kiselyova OI, Yaminsky IV, Karger EM et al. Visualization by atomic force microscopy of tobacco mosaic virus movement protein-RNA complexes formed in vitro. J Gen Virol 2001; 82:1503-1508.
40. Takamatsu K, Ishikawa M, Meshi T et al. Expression of bacterial chloramphenicol acetyltransferase gene in tobacco plants mediated by TMV-RNA. EMBO J 1987; 6:307-311.
41. Dawson WO, Bubrick P, Grantham GL. Modifications of the tobacco mosaic virus coat protein gene affecting replication, movement, and symptomatology. Phytopathology 1988; 78:783-789.
42. Kawakami S, Watanabe Y, Beachy RN. Tobacco mosaic virus infection spreads cell to cell as intact replication complexes. Proc Natl Acad Sci USA 2004; 101:6291-6296.
43. Heinlein M, Epel BL, Padgett HS et al. Interaction of tobamovirus movement proteins with the plant cytoskeleton. Science 1995; 270:1983-1985.
44. Heinlein M, Padgett HS, Gens JS et al. Changing patterns of localization of the tobacco mosaic virus movement protein and replicase to the endoplasmic reticulum and microtubules during infection. Plant Cell 1998; 10:1107-1120.
45. Padgett HS, Epel BL, Kahn TW et al. Distribution of tobamovirus movement protein in infected cells and implications for cell-to-cell spread of infection. Plant J 1996; 10:1079-1088.
46. McLean BG, Zupan J, Zambryski PC. Tobacco mosaic virus movement protein associates with the cytoskeleton in tobacco plants. Plant Cell 1995; 7:2101-2114.
47. Boyko V, Ashby JA, Suslova E et al. Intramolecular complementing mutations in Tobacco mosaic virus movement protein confirm a role for microtubule association in viral RNA transport. J Virol 2002; 76:3974-3980.
48. Boyko V, Ferralli J, Heinlein M. Cell-to-cell movement of TMV RNA is temperature-dependent and corresponds to the association of movement protein with microtubules. Plant J 2000; 22:315-325.
49. Kotlizky G, Katz A, van der Laak J et al. A dysfunctional movement protein of Tobacco mosaic virus interferes with targeting of wild type movement protein to microtubules. Mol Plant Microbe Interact 2001; 7:895-904.
50. Gillespie T, Boevink P, Haupt S et al. Functional analysis of a DNA shuffled movement protein reveals that microtubules are dispensable for the cell-to-cell movement of tobacco mosaic virus. Plant Cell 2002; 14:1207-1222.
51. Reichel C, Beachy RN. Degradation of the tobacco mosaic virus movement protein by the 26S proteasome. J Virol 2000; 74:3330-3337.
52. Serazev TV, Nadezhdina ES, Shanina NA et al. Virions and membrane proteins of the potato X virus interact with microtubules and enables tubulin polymerization in vitro. Mol Biol 2003; 37:919-925.
53. Laporte C, Vetter G, Loudes AM et al. Involvement of the secretory pathway and the cytoskeleton in intracellular targeting and tubule assembly of Grapevine fanleaf virus movement protein in tobacco BY-2 cells. Plant Cell 2003; 15:2058-2075.
54. Karasev AV, Kashina AS, Gelfand VI et al. HSP70-related 65 kDa protein of beet yellows closterovirus is a microtubule-binding protein. FEBS Lett 1992; 304:12-14.
55. Alzhanova DV, Napuli AJ, Creamer R et al. Cell-to-cell movement and assembly of a plant closterovirus: roles for the capsid proteins and Hsp70 homolog. EMBO J 2001; 20:6997-7007.
56. Lloyd C, Chan J. Microtubules and the shape of plants to come. Nat Rev Mol Cell Biol 2004; 5:13-22.
57. Cyr RJ, Palevitz BA. Organization of cortical microtubules in plant cells. Curr Opin Cell Biol 1995; 7:65-71.

58. Ding B, Haudenshield JS, Hull RJ et al. Secondary plasmodesmata are specific sites of localization of the tobacco mosaic virus movement protein in transgenic tobacco plants. Plant Cell 1992; 4:915-928.
59. Moore P, Frenczik CA, Deom CM et al. Developmental changes in plasmodesmata in transgenic tobacco expressing the movement protein of tobacco mosaic virus. Protoplasma 1992; 170:115-127.
60. Lapidot M, Gafny R, Ding B et al. A dysfunctional movement protein of tobacco mosaic virus that partially modifies the plasmodesmata and limits spread in transgenic plants. Plant J 1993; 4:959-970.
61. Igakura T, Stinchcombe JC, Goon PK et al. Spread of HTLV-1 between lymphocytes by virus-induced polarization of the cytoskeleton. Science 2003; 299:1713-1716.
62. Piguet V, Sattentau Q. Dangerous liaisons at the virological synapse. J Clin Invest 2004; 114:605-610.
63. Wilcox M, Mitchison GJ, Smith RJ. Pattern formation in the blue-green alga Anabaena. II. Controlled proheterocyst regression. J Cell Sci 1973; 13:637-649.
64. Wolk CP. Heterocyst formation. Annu Rev Genet 1996; 30:59-78.
65. Adams DG, Duggan PS. Tansley Review No.107. Heterocysts and akinete differentiation in cyanobacteria. New Phytol 1999; 144:3-33.
66. Meeks JC, Elhai J. Regulation of cellular differentiation in filamentous cyanobacteria in free-living and plant-associated symbiotic growth states. Microbiol Mol Biol Rev 2002; 66:94-121.
67. Golden JW, Yoon HS. Heterocyst development in Anabaena. Curr Opin Microbiol 2003; 6:557-563.
68. Yoon HS, Golden JW. Heterocyst pattern formation controlled by a diffusible peptide. Science 1998; 282:935-938.
69. Huang X, Dong Y, Zhao J. HetR homodimer is a DNA-binding protein required for heterocyst differentiation, and the DNA-binding activity is inhibited by PatS. Proc Natl Acad Sci USA 2004; 101:4848-4853.
70. Wildon DC, Mercier FV. The ultrastructure of the vegetative cell of blue-green algae. Aust J Biol Sci 1963; 16:585-596.
71. Drawert H, Metzner I. Fluoreszenz und elektronen-mikroskopische Beobachtungen an Cylindrospermum und einigen anderen Cyanophyceen. Ber Deutsch Bot Ges 1956; 69:291-301.
72. Metzner I. Zur Chemie und zum submikroskopischen Aufbau der Zellwände, Scheiden und Gallerten von Cyanophyceen. Arch Mikrobiol 1955; 22:45-77.
73. Guglielmi G, Cohen-Bezire G. Structure et distribution des pores et des perforations de l'enveloppe de peptidoglycane chez quelques Cyanobacteries. Protistologia 1982; 18:151-165.
74. Palinska KA, Krumbein WE. Perforation patterns in the peptidoglycan wall of filamentous cyanobacteria. J Phycol 2000; 36:139-145.
75. Hagedorn H. Elektonenmikroskopische Untersuchungen an Blaualgen. Naturwissenschaften 1960; 47:430.
76. Pankratz HS, Bowen CC. Cytology of blue-green algae. I. The cells of Symploca muscorum. Am J Bot 1963; 50:387-399.
77. Giddings TH, Staehelin LA. Plasma membrane architecture of Anabaena cylindrica: Occurrence of microplasmodesmata and changes associated with heterocyst development and the cell cycle. Eur J Cell Biol 1978; 16:235-249.
78. Lang NJ, Fay P. The heterocysts of blue-green algae. II. Details of ultrastructure. Proc Roy Soc Lond B 1971; 178:193-203.
79. Zahalak M, Beachy RN, Thiel T. Expression of the movement protein of tobacco mosaic virus in the cyanobacterium Anabaena sp. strain PCC 7120. Mol Plant Micr Interact 1995; 8:192-199.
80. Heinlein M, Wood MR, Thiel T et al. Targeting and modification of prokaryotic cell-cell junctions by tobacco mosaic virus cell-to-cell movement protein. Plant J 1998; 14:345-351.
81. Doeherty HM, Adams DG. Cloning and sequence of ftsZ and flanking regions from the cyanobacterium Anabaena PCC 7120. Gene 1995; 163:93-96.
82. Bi E, Lutkenhaus J. FtsZ ring structure associated with division in Escherichia coli. Nature 1991; 354:161-164.
83. Lutkenhaus J. FtsZ ring in bacterial cytokinesis. Mol Microbiol 1993; 9:403-409.
84. Dai K, Lutgenhaus J. The proper ratio of FtsZ to FtsA is required for cell division to occur in Escherichia coli. J Bacteriol 1992; 174:6145-6151.
85. Wang X, Lutkenhaus J. The FtsZ protein of Bacillus subtilis is localized at the division site and has GTPase activity that is dependent upon FtsZ concentration. Mol Microbiol 1993; 9:435-442.
86. Nogales E, Wolf SG, Downing KH. Structure of the alpha beta tubulin dimer by electron crystallography. Nature 1998; 391:199-203.
87. Löwe J, Amos LA. Crystal structure of the bacterial cell-division protein FtsZ. Nature 1998; 391:203-206.

88. Errington J, Daniel RA, Scheffers DJ. Cytokinesis in bacteria. Microbiol Mol Biol Rev 2003; 67:52-65.
89. Ma X, Ehrhardt DW, Margolin W. Colocalization of cell division proteins FtsZ and FtsA to cytoskeletal structures in living Escherichia coli cells by using green fluorescent protein. Proc Natl Acad Sci USA 1996; 93:12998-13003.
90. Margolin W. Bacterial division: the fellowship of the ring. Curr Biol 2003; 13:R16-18.
91. Pichoff S, Lutkenhaus J. Tethering the Z ring to the membrane through a conserved membrane targeting sequence in FtsA. Mol Microbiol 2005; 55:1722-1734.
92. Brill LM, Nunn RS, Kahn TW et al. Recombinant tobacco mosaic virus movement protein is an RNA-binding, α-helical membrane protein. Proc Natl Acad Sci USA 2000; 97:7112-7117.
93. Reichel C, Beachy RN. Tobacco mosaic virus infection induces severe morphological changes of the endoplasmatic reticulum. Proc Natl Acad Sci USA 1998; 95:11169-11174.
94. Rothfield LI, Cook WR. Periseptal annuli: Organelles involved in the bacterial cell division process. Microbiol Sci 1988; 5:182-185.
95. Oparka KJ. Plasmolysis: New insights into an old process. New Phytol 1994; 126:571-591.
96. Oparka KJ, Prior DAM, Crawford JW. Behaviour of plasma membrane, cortical ER and plasmodesmata during plasmolysis of onion epidermal cells. Plant Cell Env 1994; 17:163-171.
97. Carballido-Lopez R, Errington J. A dynamic bacterial cytoskeleton. Trends Cell Biol 2003; 13:577-583.
98. Moller-Jensen J, Lowe J. Increasing complexity of the bacterial cytoskeleton. Curr Opin Cell Biol 2005; 17:75-81.
99. Hu Z, Gogol EP, Lutkenhaus J. Dynamic assembly of MinD on phospholipid vesicles regulated by ATP and MinE. Proc Natl Acad Sci USA 2002; 99:6761-6766.
100. Suefuji K, Valluzzi R, RayChaudhuri D. Dynamic assembly of MinD into filament bundles modulated by ATP, phospholipids, and MinE. Proc Natl Acad Sci USA 2002; 99:16776-16781.
101. Shih YL, Le T, Rothfield L. Division site selection in Escherichia coli involves dynamic redistribution of Min proteins within coiled structures that extend between the two cell poles. Proc Natl Acad Sci USA 2003; 100:7865-7870.
102. Jones LJ, Carballido-Lopez R, Errington J. Control of cell shape in bacteria: Helical, actin-like filaments in Bacillus subtilis. Cell 2001; 104:913-922.
103. van den Ent F, Amos LA, Lowe J. Prokaryotic origin of the actin cytoskeleton. Nature 2001; 413:39-44.
104. van den Ent F, Amos L, Lowe J. Bacterial ancestry of actin and tubulin. Curr Opin Microbiol 2001; 4:634-638.
105. Moller-Jensen J, Borch J, Dam M et al. Bacterial mitosis: ParM of plasmid R1 moves plasmid DNA by an actin-like insertional polymerization mechanism. Mol Cell 2003; 12:1477-1487.
106. Garner EC, Campbell CS, Mullins RD. Dynamic instability in a DNA-segregating prokaryotic actin homolog. Science 2004; 306:1021-1025.
107. Robbins JR, Barth AI, Marquis H et al. Listeria monocytogenes exploits normal host cell processes to spread from cell to cell. J Cell Biol 1999; 146:1333-1350.
108. Cameron LA, Giardini PA, Soo FS et al. Secrets of actin-based motility revealed by a bacterial pathogen. Nat Rev Mol Cell Biol 2000; 1:110-119.
109. Rietdorf J, Ploubidou A, Reckmann I et al. Kinesin-dependent movement on microtubules precedes actin-based motility of vaccinia virus. Nat Cell Biol 2001; 3:992-1000.
110. Kaneko T, Nakamura Y, Wolk CP et al. Complete genomic sequence of the filamentous nitrogen-fixing cyanobacterium Anabaena sp. strain PCC 7120. DNA Res 2001; 8:205-213; 227-253.
111. Ohmori M, Ikeuchi M, Sato N et al. Characterization of genes encoding multi-domain proteins in the genome of the filamentous nitrogen-fixing Cyanobacterium anabaena sp. strain PCC 7120. DNA Res 2001; 8:271-284.

CHAPTER 11

Viral Movement Proteins Induce Tubule Formation in Plant and Insect Cells

Jan W.M. van Lent* and Corinne Schmitt-Keichinger

Abstract

Plant viruses move from cell to cell through plasmodesmata, which are complex gatable pores in the cell wall. As plasmodesmata normally allow the diffusion of only small molecules, virus movement is only achieved by the action of virus-encoded movement proteins that biochemically or structurally modify these pores to enable the passage of 'naked' viral genomes or virus particles. For a large number of different plant viruses, the movement protein forms a transport tubule inside the plasmodesmal pore to transport mature virus particles. In this review we describe the important factors that seem to play a role in this type of transport and provide a speculative model for this movement mechanism.

Introduction

Once a virus has entered the initial cell of a compatible host plant by vector transmission or mechanical wounding, it rapidly replicates and moves to the neighboring uninfected cells. This cell-to-cell movement requires infectious entities to traverse the rigid cell wall, a plant specific barrier that surrounds every cell and constitutes the "skeleton" of the plant. Plant viruses make use of natural channels, named plasmodesmata (Pd), that are present in these cell walls[1] and provide a route for intercellular communication (recent reviews in refs. 2-9). As such, Pd are of fundamental importance to plant development, but are also crucially involved in disease development, i.e., disease resistance and the spread of RNAi signals and virus. On the basis of their ontogeny, Pd are classified in two categories:[6] primary Pd are formed in the growing cell plate during cytokinesis, whereas secondary Pd are formed after cytokinesis either by neo-formation in preexisting cell walls, and thus can bridge cells deriving from different (not clonally related) tissues or by modification of primary Pd, giving rise to structurally complex, branched Pd (for recent reviews on Pd see refs. 2-6,8).

A general picture of the Pd is that of a pore in the cell wall through which both the plasma and endoplasmic reticulum (ER) membranes of two cells are interconnected. Within the Pd, the connected ER forms a thin so-called desmotubule in the central axial region of the pore (Fig. 1A). Several Pd associated compounds have been identified pointing to a complex molecular structure,[10,11] which results in a limited but variable gating capacity. Pd are dynamic in conductivity and structure, properties that vary with the physiology and the developmental stage of the plant tissue.[12] For tobacco mesophyll cells the physical diameter of the transport channels in Pd was estimated to be 2.5 nm,[30] while microinjection experiments showed a molecular gating limit of ± 1000 Da.[13] These estimates show that Pd are not capable of facilitating the free passage of large molecules[12,14] and that the gating capacity of Pd must drastically increase to enable the passage of viral genomes and virions.

*Corresponding Author: Jan W.M. van Lent—Laboratory of Virology, Department of Plant Sciences, Wageningen University, Binnenhaven 11, 6709 PD Wageningen, The Netherlands. Email: jan.vanlent@wur.nl

Cell-Cell Channels, edited by Frantisek Baluska, Dieter Volkmann and Peter W. Barlow.
©2006 Landes Bioscience and Springer Science+Business Media.

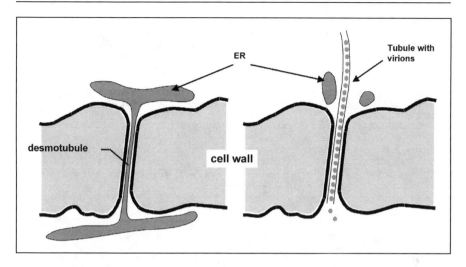

Figure 1. Simplified schematic drawing of a native plasmodesma (A) and a modified plasmodesma (B), lacking the desmotubule and instead containing a viral transport tubule with virions.

Most, if not all, plant viruses encode one or more so-called movement proteins (MPs) that are, amongst other functions, implicated in Pd dilation. These MPs are targeted to the Pd where they modify Pd molecular size exclusion limit (SEL)[13,15-17] and enable passage of viral genomes or even mature virus particles. In spite of this common function there is little or no sequence homology between MPs from different viruses. Based on their primary structure, viral MPs have been divided into several superfamilies, one of which is the '30K' superfamily, related to the 30 kDa *Tobacco mosaic virus* (TMV) MP.[18-21] For viruses with MPs belonging to this superfamily roughly three mechanisms of cell-to-cell movement can be distinguished.[22] In one mechanism, exemplified by TMV, the MPs reversibly adapt Pd to allow transport of a ribonucleoprotein complex.[13,23-26] The viral coat proteins are not required. This "nondestructive" cell-to-cell movement mechanism has been extensively reviewed by Waigmann et al.[11] In the second mechanism, exemplified by *Cowpea mosaic comovirus* (CPMV)[27-30] and *Grapevine fanleaf nepovirus* (GFLV),[31-33] the MPs aggregate into cell wall penetrating nanotubules for transport of mature virions or nucleocapsids. Hence, coat proteins in form of virions are required for this type of viral transport. A third intermediate category can be recognized in viruses that require the coat protein for cell to cell movement, but not in form of a virus particle. The MPs of these viruses are capable of assembling tubules, but tubules are not essentially involved in viral transport. Examples of such viruses are found in the family of the *Bromoviridae*.[34,35] In this chapter, primarily the tubule-guided movement of virions, i.e., those viruses that require an encapsidation competent coat protein, is discussed.

Viruses, Transport Tubules and Plasmodesmata

Whereas viruses that use a TMV-like movement mechanism modify the gating capacity of Pd without noticeable structural alteration (see also Manfred Heinlein's chapter),[11] viruses that employ tubules drastically modify Pd structure to allow assembly and expansion of the transporting tubule within the Pd pore. The presence of virus-induced nanotubules in Pd has been reported for a broad variety of plant viruses from different families and genera and with different types of genomes (RNA, DNA, double stranded or single stranded, plus- or minus-sense). The most conspicuous virus-transporting tubules are those that contain virus particles (Fig. 2A,A'), as have been observed in Pd of plants infected with viruses of the families *Comoviridae* [*Comovirus:* CPMV,[28,36] *Bean pod mottle virus* (BPMV),[37] *Nepovirus: Tomato ringspot virus* (TomRSV),[38] GFLV],[31,39,40] *Bromoviridae* [*Bromovirus: Olive latent virus 2*;[41] *Alfamovirus: Alfalfa mosaic virus* (AMV),[42,43] *Ilarvirus: Tobacco*

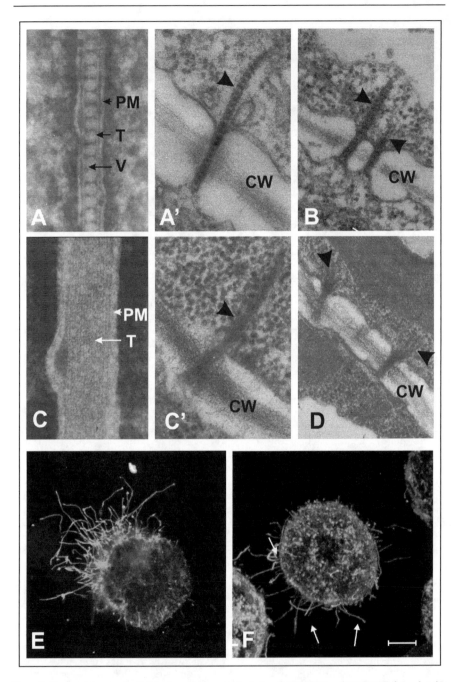

Figure 2. Electron micrographs (A-D) of negatively stained tubules of CPMV (A) and TSWV (C) formed at the protoplast surface and tubules in plasmodesmata of mesophyll cells infected with CPMV (A'), AMV (B), TSWV (C',D). Confocal images (E,F) of tubules protruding from the surface of a protoplast infected with CPMV (E) and of an insect cell expressing the CPMV MP (F). T = tubule, PM = plasma membrane, V = virion, CW = cell wall. Bar in F represents 5 μm.

streak virus (TSV);[44] *Cucumovirus: Tomato aspermi virus* (TAV)],[45] *Sequiviridae* (*Sequivirus*)],[46] *Geminiviridae* [(*Begomovirus*)],[47] and *Caulimoviridae* [*Caulimovirus: Dahlia mosaic virus,*[48,49] *Cauliflower mosaic virus* (CaMV),[50] *Badnavirus*].[51] With *Tomato spotted wilt virus* (TSWV), a plant virus belonging to a family of predominantly animal-infecting viruses (the *Bunyaviridae,* genus *Tospovirus*), tubules have been observed in Pd without a defined particulate content (Fig. 2C,C').[52] Mature TSWV virions are enveloped with a diameter of 80-100 nm and accumulate in the lumen of the endoplasmic reticulum. It is unlikely that TSWV is transported in this form. Studies of early stages of TSWV infection show that few Pd are modified showing a tubular structure (Fig. 2C'), but many Pd contain the MP in form of fibrillar material extending into the cytoplasm of one cell (Fig. 2D; van Lent, unpublished).

A common factor in the structural modification of Pd enabling the assembly of viral transport tubules is the absence of the desmotubule (Figs. 1B, 2A',B). In addition to this, Pd can be dilated to accommodate the passage of larger virions. With small spherical viruses like *Como-* and *Nepoviruses* (particle diameter about 30 nm) no noticeable increase in the diameter of the Pd pore has been reported, but with *Alfalfa mosaic virus* (AMV) the pore of MP-containing Pd appears to be nearly two-fold dilated in cells at the infection front.[42] Also with *Caulimoviruses*[48,50] and TSWV (Fig. 2C,C')[52,53] the Pd diameter seems to become dilated.

Whereas the tubules of *Como-, Nepo-* and *Caulimoviruses* persist in the Pd, steadily growing to lengths of several microns and eventually are embedded in a thick layer of wall deposits,[36,37,40,54-56] with AMV and TSWV the presence of MP and tubules in the Pd, as well as dilation of the Pd pore appears to be transient and is observed only in cells that are in an early stage of infection.[52,57,58] Intercellular movement apparently (and logically) is an early function in the viral infection cycle and with many viruses the structural modification of Pd and the presence of tubules may have been unnoticed due to the early and transient action of the MP. In plant tissues that have a more or less synchronized infection with TSWV and AMV (achieved by differential temperature treatment)[59,60] tubule-like structures are frequently observed in Pd (van Lent, unpublished). Figure 2B,C',D shows such modifications of Pd in cells early (approx. 24 h) after the onset of AMV (Fig. 2B) and TSWV (Fig. 2C',D) infection and the presence of MP in these tubules has been confirmed by immuno-labeling (not shown). Furthermore, at the border of AMV infection, in a layer of approx. 4 cells between infected and noninfected palisade parenchyma cells,[42] nearly 100% of the Pd contain MP (Fig. 3), implicating all Pd transiently in virus movement as no MP is found in Pd outside this infection border.

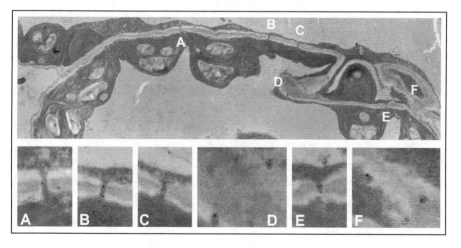

Figure 3. Cell wall between two AMV-infected palisade parenchyma cells within the border of infection. All plasmodesmata contain the viral MP, as indicated by immuno-gold labeling (A-F), and are dilated.

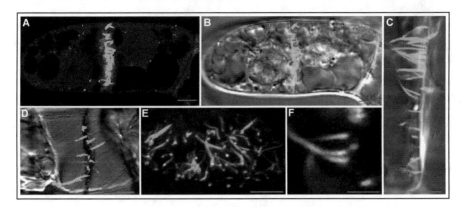

Figure 4. Confocal laser scanning micrographs of BY2 cells expressing GFLV MP fused to GFP (GFP-MP). A-C) tubular structures visible in cross walls. A) fluorescence visualization; B) differential interference contrast image (DIC) superimposed to fluorescence; C) detailed view of oriented tubules in a cross wall (fluorescence and DIC pictures are merged). D) Fluorescent tubules visualized in the membranous Hechtian strands of a plasmolysed cell. (Mannitol (0.45M) was added 10 min. before observation, fluorescence and DIC pictures are merged). E-F) calreticulin immunolabeled foci (red) from which MP tubules emanate (green). E) cross wall in face view, F) cross wall in side view. Images represent single 0.45 μm optical sections except in E representing projection of 23 optical 0.45 μm sections. Bars = 5 μm, except in F (2 μm). Reprinted with permission from Laporte C et al. Plant Cell 15:2058-2075.[62] ©2003 American Society of Plant Biologists. A color version of this figure is available online at www.Eurekah.com.

Apparently, tubule-forming viruses have the capability not only to modify the internal structure of Pd, but also to modulate the cell wall around the Pd and this recalls a so far unresolved question whether these viruses actually use existing (primary or complex secondary) Pd or whether a new (secondary) Pd in the cell wall is induced to assemble their transport tubule.[61] From the reports on transport tubules in Pd (and Figs. 2,3) emerges the picture that preferably unbranched Pd are used for tubule-guided virus transport, as tubules are rarely found in branched Pd. In many cases this could be explained by the fact that movement is an immediate function and viruses rapidly spread to and within sink (young metabolite-importing) tissues that predominantly contain the simple primary Pd.[12] As virion-containing tubules often persist in the Pd they can be observed long after effective cell-to-cell movement has been completed and tissues changed from sink to source (metabolite-exporting).

Alternatively, when these viruses are able to induce the de novo formation of secondary Pd in the cell wall to facilitate the assembly of their transport tubule, such Pd would likely to be simple and not branched. Data obtained with BY2 cells expressing the GFLV MP fused to GFP (GFP-MP) suggest that tubules preferentially target the youngest cross-walls of the cell chain, associating to the cell plate during cytokinesis, and less frequently target preexisting cell walls (Fig. 4A-C).[62] This observation suggests the involvement of primary Pd in tubule-guided movement. However, tubule-guided movement through secondary plasmodesmata, i.e., those Pd that connect tissues originating from different cell lineages, explains how a virus that enters an epidermal cell can move through mesophyll cells to the vasculature for subsequent invasion of the whole plant. However, whether viruses, i.e., their MP can potentiate the formation of Pd remains to be verified.

With TMV,[63] *Cucumber mosaic virus* (CMV)[64,65] and *Apple chlorotic leafspot trichovirus* (ACLSV)[66] MPs have been localized to the central cavity of complex secondary Pd. However, it still remains to be established whether such Pd are active in virus movement.[11,67]

Requirements for the Assembly of the Transport Tubules

In plants, Pd are targeted by viral MPs where they assemble into tubules. However, intact Pd are not required for the assembly of MP into tubules as identical tubules (Fig. 2A,C,E) are readily formed in protoplasts. These isolated plant cells are enzymatically stripped from their cell wall and surrounded only by the plasma membrane. Hence, protoplasts lack structural integer Pd, but the plasma membrane may contain remnants of Pd specific compounds.[68] Many of the aforementioned tubule-forming viruses, CPMV,[29] GFLV,[31] CaMV,[69] TSWV,[58] BMV and AMV[70] also form tubules upon infection of protoplasts or after transfection of these cells with the MP coding genes.[41,58,71-73] These studies have revealed that the MP is the only viral requirement for tubule assembly. In protoplasts, tubules grow from the surface of the cell outwards into the culture medium (Fig. 2E) often reaching lengths of several microns and they are tightly enwrapped by the plasma membrane (Fig. 2A,C). The outward polarity of the plasma membrane engulfed tubules suggests a strong interaction between the tubule-forming MP and the plasma membrane or components herein.

As plant viral MPs fulfil a unique function in plants, i.e., modification of the plant specific Pd to enable virus transport,[74] it is amazing that these MPs are equally capable of inducing tubules with the same polarity in insect (animal) cells (Fig. 2F).[58,75] Recently, similar structures called tunnelling nanotubes (TNT), have been found in animal cells.[76-78] These TNTs permit the selective transport of proteins and endomembrane vesicles between cells and it is suggested that they in some aspects resemble Pd.[78,79] Some factors that are involved in TNT-formation in animal cells may well play a role also in MP tubule-formation. However, with TNTs, nucleating actin is providing the force for deformation and extension of the plasma membrane, while actin is not required for MP-tubule formation (see hereafter).

Among the plant viruses mentioned, a unique position is taken by the Tospoviruses (TSWV). These plant-infecting viruses are also capable of replicating in their insect vector the thrips.[80] Although the NSm movement protein is expressed during infection of the thrips, no aggregation into tubules was ever observed in cells of various thrips tissues.[81] Apparently, this MP is capable of forming tubules in single cultured plant and insect cells, but not in insect cells in the context of a tissue. The ability of MPs to form tubules in both plant and insect cells leads to the conclusion that the host cell requirements for specific targeting of the MPs from the location of synthesis to the plasma membrane, the anchoring at this membrane and the assembly into outward protruding tubules are conserved across the boundaries of the plant and animal kingdom.

In protoplasts, tubule formation seems not to be restricted to viruses that form apparent tubules in planta. Several of the viruses that have a MP belonging to the 30K superfamily form tubule-like protrusions at the surface of protoplasts. The MPs of *Cucumber mosaic virus* (CMV) and *Brome mosaic virus* (BMV), viruses from the *Bromoviridae* family, and of the flexuous *Trichovirus Apple chlorotic leafspot virus* (ACLSV) form tubules in protoplasts, although tubules have never been found in Pd.[70,73,82] Even the MP of TMV, representative of the nondestructive cell-to-cell movement mechanism, forms tubule-like protrusions at the protoplast surface[83,84] (see Fig. 2 in chapter by Heinlein). Either the biological significance of this structural manifestation of the MP has not yet been recognized for these viruses, or it is a more general property of the '30K' MPs, functional in intercellular movement of some viruses (the typical tubule-guided) and not with others.

The MPs That Form Tubules

The MPs of the 30K superfamily share several structural, biochemical and functional properties, despite a very low degree of sequence homology. Only five small regions of moderate sequence conservation were noted.[21] In terms of structural performance experiments with protoplasts suggest that assembly into tubule-like structures maybe a more common function of these MPs. This includes viruses like CMV, BMV and ACLSV, which require the CP for movement and show a discrepancy between plants (no tubules) and protoplasts (tubules), but also TMV MP which forms tubule-like protrusions in protoplasts and does not require the CP for movement. In planta this

MP is found as fibrillar material in secondary Pd cavities,[63] while in bacterial cells fibrils of this MP form tubular arrangements traversing the septum between cells[85] (see Fig. 5 in chapter by Heinlein). With TSWV, high magnification imaging of isolated tubules shows a clear fibrillar substructure (Fig. 2C), which has never been reported for tubules of the *Como-* and *Nepoviruses*. In TSWV-infected plants, Pd that contain the MP often show fibrillar material extending from the Pd-pore (Fig. 2D). Modification of Pd, by lining the interior with a tubular arrangement of MP fibrils may well be a common structural function of the 30K family MPs. Besides these structural similarities, Storms and coworkers showed that the MPs of TSWV and TMV also had similar effects on Pd gating[86] and recently, Lewandowski and Adkins demonstrated that TSWV MP could complement the movement of movement-defective TMV.[87] It was known that TMV-MP could complement the function of defective tubule-forming MPs, as was demonstrated for CPMV,[88] but now the reciprocal complementation has been demonstrated as well.

So besides structural similarities, MPs of the 30K superfamily also share apparent biochemical properties. Evidence is accumulating that these MPs share the capacity to bind nucleic acid in a sequence nonspecific manner. This has been shown for viruses with a CP-independent movement mechanism like TMV,[24] but also for tubule-forming and virion-transporting viruses like CPMV,[89] CaMV,[90] AMV[91,92] and BMV.[93] RNA-binding capacity of MPs is functional in TMV-like movement mechanisms, where the MP forms a movement complex with the viral RNA. It is more difficult to understand this property with viruses that move as virions through tubules. In view of this, it has been proposed that some of these viruses, like CaMV and BMV might use two alternative movement strategies to establish systemic infection: movement as a viral RNA-MP complex or movement as a complete virion by tubule-guided mechanism.[93,94] In other cases, RNA-binding (e.g., for *Como-* and *Nepoviruses*) or tubule-forming capacity (e.g., CMV and TMV) of MPs may be evolutionary remnants.

Another common factor in the '30K' MPs is a very conserved aspartic acid residue, the so-called D-motif, which is contained in a putative rNTP binding site.[21] For TMV, CMV[95] and CPMV[89] it was shown that the MP specifically binds GTP in vitro and for CPMV it was further shown that the D-motif is essential for this binding[89] and that the GTP binding site is required for proper targeting of MP to the cell periphery. On the basis of the observation of Pouwels et al that multimerization of CPMV MP was needed for proper targeting to the plasma membrane,[96] it was speculated that the GTP-binding site could be involved in MP multimerization. Although CPMV is so far the only virus for which a direct link between the GTP binding site and MP targeting to the cell periphery has been established, the highly conserved D-motif suggests that it is a general feature of MPs. This GTP-binding capacity could represent a source of energy, as already suggested by Carrington[1] and by Ghoshroy and coworkers[97] making the assembly of MP tubules a mechanism similar to the formation of microtubules as speculated by Carvalho and coworkers.[89]

Interactions between Tubules and Capsids for Cell-to-Cell Movement

With viruses that move as mature virions through transport tubules, an interaction between MP and coat protein/virion is an obvious requirement. In several cases evidence has been presented for binding of MP and coat protein as well as virus specificity in this process. With most of these viruses, the C-terminal part of the MP is involved in coat protein recognition and binding. For CPMV it was shown that the MP C-terminus is located on the inside of the tubule,[29] in close proximity to the virus particles. Incorporation of virions into the tubule was disturbed with a C-terminal deletion mutant of the MP, giving rise to 'empty' tubules,[98] i.e., tubules without virus particles. Furthermore, Carvalho et al showed specificity of MP binding to CPMV virions, but not to capsids of BMV, TMV or even the related *Comoviruses Cowpea severe mosaic virus* (CPSMV) and *Red clover mottle virus* capsids.[99] Moreover, in blot overlay assays the MP specifically bound to only one of the two CPMV coat proteins. Also GFLV movement is governed by a specific interaction between tubule and virions, as suggested by the results obtained with chimeric constructs between GFLV and the closely related *Arabis mosaic virus*. Virus spread only occurred when the 9 C-terminal residues of the MP were of the same viral origin as the coat protein.[100] Thus, the C-terminus of the

MP protein is believed to be localised towards the inner side of the tubules where it can interact with virions. Similarly, in the case of CaMV, a C-terminal mutant MP was identified that kept its ability to form tubules but was unable to support virus movement[101] suggesting that the 10 last C-terminal amino acids of the MP are involved in interactions with the virus particles. Sanchez-Navarro and coworkers[34] showed that a chimeric BMV MP containing the 44 C-terminal amino-acids of AMV permitted the cell to cell movement of AMV, whereas the wild type BMV MP did not. Furthermore, two alanine scanning mutants in the same C-terminal region of AMV MP, although still competent for AMV cell to cell transport, exhibited a reduced efficiency in that transport as well as an abolition of systemic spread, suggesting that these mutations interfere with MP-coat protein interactions.[102] However, no direct interaction could be detected between the coat protein and the MP in a yeast two-hybrid system.

The data obtained with CPMV, GFLV, CaMV and AMV suggest a high level of specificity of the tubule-guided virion transport for the homologous virus particle. Apparently, although very variable in sequence, the tubule-forming MPs share common functions resulting in a common movement mechanism.

Host Factors Involved in Targeting and Assembly of the Tubules

Data obtained with a collection of CPMV MP mutants fused to GFP and expressed either transiently in protoplasts or upon infection in leaf epidermal cells, provided a sequence of distinguishable events resulting in tubule-guided virus movement. First MP is targeted from the place of synthesis to the plasma membrane. At this membrane the MPs accumulate in punctuate structures in which they further aggregate to assemble tubules that extend into the neighbouring cell. During this assembly process virions are specifically included. In the next cell, the tubule destabilizes thereby releasing the virions for further infection.[96]

As mentioned before, GTP binding and MP multimerization seem to be important viral actors in targeting of MP to the cell periphery. At least for the tubule-forming viruses, MP trafficking should involve cellular mechanisms that are conserved among animals and plants. In recent years increasing evidence has been obtained that implicate the endomembrane system as well as the cytoskeleton as possible routes for intracellular transport of viral MPs (see reviews in refs. 11,26,103).

With tubule-forming viruses like CPMV, the use of cytoskeletal inhibitors like latrunculin B (inhibits the assembly of actin filaments) and oryzalin (inhibits the assembly of microtubules) showed that the targeting of MP to the cell periphery was independent from either cytoskeletal element. Moreover, the formation of tubules was not affected.[104] Treatment with the fungal brefeldin A (BFA), an inhibitor that perturbs the endomembrane system, showed that this system is not required for MP targeting to the plasma membrane, but that it is required for formation of the tubules, possibly by interfering with the targeting of an essential host protein to the plasma membrane or because vesicle transport is needed for tubule formation.[104] Identical results where obtained by Huang and coworkers with the MP of CaMV, another tubule-forming virus.[105] For AMV, it was also shown that MP transport to the cell wall and tubule assembly do not rely on the cytoskeleton,[102] but data on the role of the endomembrane system are not yet available. In the case of the *Nepovirus* GFLV, a role for microtubules in the assembly of tubules is apparent.[62] As tubules were formed preferentially in the youngest cross walls of BY2 cells and MP localised to the growing cell plate during cytokinesis, it was speculated that transport of MP could take place via the route of vesicles along microtubules. Treatment with BFA indeed led to a disturbed distribution of MP, showing that a functional secretory pathway was essential for cross wall targeting and tubule assembly.

Depolymerization of actin filaments or stabilization of microtubules by taxol did not interfere with targeting of MP to the cross walls nor with tubule formation. However, depolymerization of microtubules resulted in tubule formation at improper locations, i.e., the side walls instead of cross walls. Simultaneous distortion of the actin filaments and microtubules led to the formation of aster-like MP structures close to the nucleus. Hence, like CPMV and CaMV, GFLV requires an intact endomembrane system for tubule assembly but unlike CPMV and CaMV, it needs the cytoskeleton for proper targeting of the MP.

Thus the mode of transport of these tubule-forming MPs to the plasma membrane remains an enigma, although it is possible that these viruses exploit both machineries for intracellular transport of their MPs, depending on the relative functionality of these pathways in the infected cell. So far, there is no evidence for such a redundancy in the properties of CaMV, CPMV or GFLV MPs but it is conceivable that viral proteins have evolved to hijack endogenous processes and that they have conserved more than one possibility to be effective in different host cell types (from the first cell entered to the vasculature system) at different physiological stages.

Analysis of the interactions between viral MPs and the host cell proteome will provide further insight in the actors involved in the virus movement process. Several host proteins with affinity for viral MPs have been identified by now. The clearest functional relationship between virus movement and host factors has been reported for TMV, where the MP appears to interact with microtubules and F-actin in infected cells suggesting that these cytoskeleton elements are involved in the intracellular targeting of the MP-RNA complex to plasmodesma.[83,106-112] It has been suggested that a conserved tobamovirus MP sequence with similarity to a tubulin motif is the region that mediates the association of MP with microtubules during cell-to-cell movement.[113] Although the association of MP with microtubules suggests that these play a role in cell-to-cell movement of the viral RNA, movement of TMV can occur independently of microtubules[114] and the microtubules may be involved in targeting the MP for degradation during the latter stages of infection rather than targeting for transport to neighboring cells.[108,110,114,115]

Another host protein, pectin methylesterase (PME) was shown to interact with the MPs of TMV, CaMV and *Turnip vein clearing virus,* and to be localized to the cell wall around Pd.[116,117] PME plays a role in cell wall dynamics (turnover and porosity) and has been implicated in plant responses to pathogen attack.[118] Deletion of the methylesterase binding domain in TMV MP aborts cell-to-cell movement.[117] Furthermore, yeast two-hybrid and Far-Western screens have identified host factors that show affinity to MP, but in most cases the function of these factors in the movement process is not at all understood. Matsushita and coworkers[119] reported a putative transcriptional coactivator (KELP, a protein that modulates host gene expression during pathogenesis) to show affinity for the MP of *Tomato mosaic tobamovirus.*

Also for tubule-forming viruses, host factors with affinity for the MP have been found. In a two-hybrid screening with the MP of TSWV, Soellink and coworkers[120] found interactions with DnaJ-like chaperones. These proteins have functions including protein transport in organelles and the regulation of the chaperone Hsp70,[121] a heat shock protein involved in the translocation of *Closteroviruses.*[122] It was additionally reported that TSWV MP is capable to bind proteins with homologies to myosin and kinesin, suggesting an involvement of molecular chaperones in the attachment of TSWV nucleocapsids to the cytoskeleton for subsequent intracellular trafficking.[123]

Two-hybrid screening with CaMV MP revealed affinity for an *Arabidopsis* protein MPI7. The protein was localized to punctuate spots at the cell periphery, probably representing Pd, and in vivo association between the MP and MPI7 was confirmed by fluorescence resonance energy transfer (FRET).[124] In sequence, MPI7 is related to mammalian Rab acceptor proteins (PRA1), a family of proteins binding Rab GTPases and vSNARE, components implicated in vesicle transport. Thus, although not proven, MP of CaMV might interact, via MPI7, with the vesicle transport machinery to get a hike to the periphery of the cell or to anchor the tubules to the plasma membrane.

As previously mentioned the MP of CPMV binds GTP and this binding is required for MP targeting and tubule formation, however, the MP does not show GTPase activity.[89] This GTP-binding may become significance if we consider the "grab a Rab" model proposed by Oparka for selective transport of MP to the Pd.[125] RabGTPases, that play a role in specificity of vesicle transport, could carry the MP together with a cargo vesicle to the plasma membrane and at the same time, by GTP hydrolysis, it could provide the molecular switch to start MP polymerisation. At the Pd, specific interactions between v-SNARE (soluble N-ethylmaleimide-sensitive factor adaptor protein receptors) and t-SNARE complexes then make the vesicles fuse with the plasma membrane. The vesicles could even transport necessary enzymes for cell wall degradation to enlarge the Pd channel or to form secondary channels for virus transport.

Interaction between MP and Plasma Membrane: Anchoring

In plants transport tubules are located in the Pd pore, replacing the desmotubule, and surrounded by the plasma membrane (Figs. 1B,2A'). In protoplasts tubules protrude from the cell surface, against turgor pressure, and are also engulfed by the plasma membrane (Fig. 2A,C,E). This points to a close interaction between the membrane and tubule. To get some insight in the CPMV MP anchoring and subsequent tubule assembly, Pouwels et al[126] studied the sequence of events from the development of peripheral MP-containing punctuate spots at the plasma membrane until tubule formation, by following the fate of GFP-MP in protoplasts with time laps microscopy and FRAP (fluorescence recovery after photobleaching). These experiments showed that the punctuate spots constitute fixed nucleation sites from which the MPs can aggregate into tubules. The authors further showed that even large membrane-associated proteins are able to diffuse into the membrane surrounding protruding tubules and concluded that the MP does not directly interact with the plasma membrane, but that it probably binds to the membrane via a plasma membrane-intrinsic or peripheral protein. In preliminary studies, using the CPMV/cowpea system Carvalho was able to identify such MP-binding proteins by affinity chromatography of purified plasma membrane fractions using immobilized MP.[127] Several proteins were obtained of which four have been identified up till now as subunits H, D and E of v-ATPase and aquaporin. Both proteins are ubiquitous and conserved membrane proteins and valid candidate MP-binding host proteins, however, the significance of these proteins in MP anchoring or tubule formation remains to be established.

In the case of GFLV, attachment of MP to the cell periphery was shown by its retention in thin membranous Hechtian strands that emanate from Pd during plasmolysis (Fig. 4D).[62] It was speculated that the docking receptor could be calreticulin, a protein found in ER and Pd,[128-130] as MP was targeted to calreticulin-labeled foci in the young cross walls of BY2 cells and tubules appeared to extend from these foci (Fig. 4E,F). Coimmunoprecipitation experiments did not confirm a direct interaction between GFLV MP and calreticulin. Recently, Chen and coworkers were able to show a direct binding between calreticulin and the MP of TMV in vitro and to confirm this interaction also in vivo. Pd targeting was hampered by increased levels of calreticulin, thus showing a functional relationship between the calreticulin, MP and TMV cell-to-cell movement.[131]

A Model for Tubule-Mediated Cell-to-Cell Movement

It is obvious that plant virus movement cannot be grasped in two simplified models like movement of virions through tubules assembled in structurally modified Pd (CPMV) or nondestructive movement of viral genomes complexed with the MP (TMV). The MPs grouped in the 30K family show similarities in many aspects (structural, biochemical, functional) but the viruses do not employ a unique mechanism for cell-to-cell movement. Several viruses clearly move as virions through transport tubules (e.g., CPMV, GFLV, CaMV), while others employ a mechanism best described for TMV. But a number of viruses show characteristics of both extremities, in particular those of the *Bromoviridae* family (AMV, BMV and CMV).

Nevertheless, with the data accumulated so far it is tempting to speculate on a general model for tubule-guided transport of virions.

From the site of virus replication and protein synthesis, which take place in a membranous viral compartment close to the nucleus,[132,133] MP is trafficking to the cell periphery. This process involves the secretory pathway[62,102,104,105,124] and for some viruses also the cytoskeleton (microtubules).[62] Targeted trafficking of the MP depends on multimerization of the protein and GTP-binding,[89] and likely occurs by docking at the membranes of transport vesicles and hiking along the host vesicle transport system. Once such vesicles arrive and fuse at the plasma membrane, the MP migrates to Pd to form nucleation sites in the membrane. From these nucleation sites the MPs further multimerize into tubules inside the Pd pore. Hydrolysis of the MP-bound GTP by host cell GTPases may be the molecular switch for this perpetuate multimerization, and is possibly also required for destabilisation of the tubule in the neighbouring cell to release the virions. Nothing is known about the location where MPs and virions come together in this process. Considering the affinity of the MP for CP and assembled capsids, MP-virion complexes may well be targeted to the plasma membrane in a similar fashion as the MP alone. It is very likely that virions are incorporated during assembly of the tubule, rather than that virions would traverse tubules like cars in a tunnel. When the virion-containing tubule reaches the neighbouring cell, the growing tubule disassembles, delivering the virions in the uninfected cell, where a new infection cycle starts.

Much progress has been made in the past decade in understanding how tubules formed by viral MPs act in cell to cell movement, however many aspects of the tubule-guided movement mechanism remain unresolved. A closer look should be taken at the vesicle transport (secretory pathway) system as possible route for MP and virion delivery at the plasma membrane. Other challenges can be found in the identification of the plasma membrane anchoring place, the analysis of the mechanism of MP multimerization into tubules and the molecular structure of these tubules. No doubt that these questions will soon be addressed and answered.

References

1. Carrington JC, Kasschau KD, Mahajan SK et al. Cell-to-cell and long-distance transport of viruses in plants. Plant Cell 1996; 8:1669-1681.
2. Heinlein M, Epel BL. Macromolecular transport and signaling through plasmodesmata. Int Rev Cytol 2004; 235:93-164.
3. Lucas WJ, Lee JY. Plasmodesmata as a supracellular control network in plants. Nat Rev Mol Cell Biol 2004; 5:712-726.
4. Roberts AG, Oparka KJ. Plasmodesmata and the control of symplastic transport. Plant Cell Environ 2003; 26:103-124.
5. Haywood V, Kragler F, Lucas WJ. Plasmodesmata: Pathways for protein and ribonucleoprotein signaling. Plant Cell 2002; 14:S303-S325.
6. Ehlers K, Kollmann R. Primary and secondary plasmodesmata: Structure, origin, and functioning. Protoplasma 2001; 216:1-30.
7. Blackman LM, Overall RL. Structure and function of plasmodesmata. Aust J Plant Physiol 2001; 28:709-727.
8. Zambryski P, Crawford K. Plasmodesmata: Gatekeepers for cell-to-cell transport of developmental signals in plants. Ann Rev Cell Dev Biol 2000; 16:393-421.
9. Kragler F, Lucas WJ, Monzer J. Plasmodesmata: Dynamics, domains and patterning. Ann Bot 1998; 81:1-10.
10. Escobar NM, Haupt S, Thow G et al. High-throughput viral expression of cDNA-green fluorescent protein fusions reveals novel subcellular addresses and identifies unique proteins that interact with plasmodesmata. Plant Cell 2003; 15:1507-1523.
11. Waigmann E, Ueki S, Trutnyeva K et al. The ins and outs of nondestructive cell-to-cell and systemic movement of plant viruses. Crit Rev Plant Sci 2004; 23:195-250.
12. Oparka KJ, Roberts AG, Boevink P et al. Simple, but not branched, plasmodesmata allow the nonspecific trafficking of proteins in developing tobacco leaves. Cell 1999; 97:743-754.
13. Wolf S, Deom CM, Beachy RN et al. Movement protein of tobacco mosaic virus modifies plasmodesmatal size exclusion limit. Science 1989; 246:377-379.
14. Oparka KJ, Roberts AG. Plasmodesmata: A not so open-and-shut case. Plant Physiol 2001; 125:123-126.
15. Fujiwara T, Giesman-Cookmeyer D, Ding B et al. Cell-to-cell trafficking of macromolecules through plasmodesmata potentiated by the red clover necrotic mosaic virus movement protein. Plant Cell 1993; 5:1783-1794.
16. Poirson A, Turner AP, Giovane C et al. Effect of the alfalfa mosaic virus movement protein expressed in transgenic plants on the permeability of plasmodesmata. J Gen Virol 1993; 74:2459-2461.
17. Vaquero C, Turner AP, Demangeat C et al. The 3a protein from cucumber mosaic virus increases the gating capacity of plasmodesmata in transgenic tobacco plants. J Gen Virol 1994; 75:3193-3197.
18. Melcher U. Similarities between putative transport proteins of plant viruses. J Gen Virol 1990; 71:1009-1018.
19. Koonin EV, Mushegian AR, Ryabov EV et al. Diverse groups of plant RNA and DNA viruses share related movement proteins that may possess chaperone-like activity. J Gen Virol 1991; 72:2895-2904.
20. Mushegian AR, Koonin EV. Cell-to-cell movement of plant viruses. Insights from amino acid sequence comparisons of movement proteins and from analogies with cellular transport systems. Arch Virol 1993; 133:239-257.

76. Rustom A, Saffrich R, Markovic I et al. Nanotubular highways for intercellular organelle transport. Science 2004; 303:1007-1010.
77. Onfelt B, Davis DM. Can membrane nanotubes facilitate communication between immune cells? Biochem Soc Trans 2004; 32:676-678.
78. Baluska F, Hlavacka A, Volkmann D et al. Getting connected: Actin-based cell-to-cell channels in plants and animals. Trends Cell Biol 2004; 14:404-408.
79. Cilia ML, Jackson D. Plasmodesmata form and function. Curr Opin Cell Biol 2004; 16:500-506.
80. Wijkamp I, van Lent J, Kormelink R et al. Multiplication of tomato spotted wilt virus in its insect vector, Frankliniella occidentalis. J Gen Virol 1993; 74:341-349.
81. Storms MMH, Nagata T, Kormelink R et al. Expression of the movement protein of Tomato spotted wilt virus in its insect vector Frankliniella occidentalis. Arch Virol 2002; 147:825-831.
82. Satoh H, Matsuda H, Kawamura T et al. Intracellular distribution, cell-to-cell trafficking and tubule-inducing activity of the 50 kDa movement protein of Apple chlorotic leaf spot virus fused to green fluorescent protein. J Gen Virol 2000; 81:2085-2093.
83. Heinlein M, Padgett HS, Gens JS et al. Changing patterns of localization of the tobacco mosaic virus movement protein and replicase to the endoplasmic reticulum and microtubules during infection. Plant Cell 1998; 10:1107-1120.
84. Mas P, Beachy RN. Distribution of TMV movement protein in single living protoplasts immobilized in agarose. Plant J 1998; 15:835-842.
85. Heinlein M, Wood MR, Thiel T et al. Targeting and modification of prokaryotic cell-cell junctions by tobacco mosaic virus cell-to-cell movement protein. Plant J 1998; 14:345-351.
86. Storms MMH, van der Schoot C, Prins M et al. A comparison of two methods of microinjection for assessing altered plasmodesmal gating in tissues expressing viral movement proteins. Plant J 1998; 13:131-140.
87. Lewandowski DJ, Adkins S. The tubule-forming NSm protein from Tomato spotted wilt virus complements cell-to-cell and long-distance movement of Tobacco mosaic virus hybrids. Virology 2005, (In Press).
88. Taliansky ME, de Jager CP, Wellink J et al. Defective cell-to-cell movement of cowpea mosaic virus mutant N123 is efficiently complemented by sunn-hemp mosaic virus. J Gen Virol 1993; 74:1895-1901.
89. Carvalho CM, Pouwels J, van Lent JWM et al. The movement protein of Cowpea mosaic virus binds GTP and single-stranded nucleic acid in vitro. J Virol 2004; 78:1591-1594.
90. Citovsky V, Knorr D, Zambryski P. Gene I, a potential cell-to-cell movement locus of cauliflower mosaic virus, encodes an RNA-binding protein. Proc Natl Acad Sci USA 1991; 88:2476-2480.
91. Schoumacher F, Erny C, Berna A et al. Nucleic acid-binding properties of the alfalfa mosaic virus movement protein produced in yeast. Virology 1992; 188:896-899.
92. Schoumacher F, Gagey MJ, Maira M et al. Binding of RNA by the alfalfa mosaic virus movement protein is biphasic. FEBS Lett 1992; 308:231-234.
93. Jansen KA, Wolfs CJ, Lohuis H et al. Characterization of the brome mosaic virus movement protein expressed in E. coli. Virology 1998; 242:387-394.
94. Thomas CL, Maule AJ. Identification of the cauliflower mosaic virus movement protein RNA-binding domain. Virology 1995; 206:1145-1149.
95. Li QB, Palukaitis P. Comparison of the nucleic acid- and NTP-binding properties of the movement protein of cucumber mosaic cucumovirus and tobacco mosaic tobamovirus. Virology 1996; 216:71-79.
96. Pouwels J, Kornet N, van Bers N et al. Identification of distinct steps during tubule formation by the movement protein of Cowpea mosaic virus. J Gen Virol 2003; 84:3485-3494.
97. Ghoshroy S, Freedman K, Lartey R et al. Inhibition of plant viral systemic infection by nontoxic concentrations of cadmium. Plant J 1998; 13:591-602.
98. Lekkerkerker A, Wellink J, Yuan P et al. Distinct functional domains in the cowpea mosaic virus movement protein. J Virol 1996; 70:5658-5661.
99. Carvalho CM, Wellink J, Ribeiro SG et al. The C-terminal region of the movement protein of Cowpea mosaic virus is involved in binding to the large but not to the small coat protein. J Gen Virol 2003; 84:2271-2277.
100. Belin C, Schmitt C, Gaire F et al. The nine C-terminal residues of the grapevine fanleaf nepovirus movement protein are critical for systemic virus spread. J Gen Virol 1999; 80:1347-1356.
101. Thomas CL, Maule AJ. Identification of structural domains within the cauliflower mosaic virus movement protein by scanning deletion mutagenesis and epitope tagging. Plant Cell 1995; 7:561-572.

102. Huang M, Jongejan L, Zheng HQ et al. Intracellular localization and movement phenotypes of Alfalfa mosaic virus movement protein mutants. Mol Plant Microb Interact 2001; 14:1063-1074.
103. Boevink P, Oparka KJ. Virus-host interactions during movement processes. Plant Physiol 2005; 138:1815-1821.
104. Pouwels J, Van der Krogt GNM, Van Lent J et al. The cytoskeleton and the secretory pathway are not involved in targeting the cowpea mosaic virus movement protein to the cell periphery. Virology 2002; 297:48-56.
105. Huang Z, Han Y, Howell SH. Formation of surface tubules and fluorescent foci in Arabidopsis thaliana protoplasts expressing a fusion between the green fluorescent protein and the cauliflower mosaic virus movement protein. Virology 2000; 271:58-64.
106. Heinlein M, Epel BL, Padgett HS et al. Interaction of tobamovirus movement proteins with the plant cytoskeleton. Science 1995; 270:1983-1985.
107. McLean BG, Zupan J, Zambryski PC. Tobacco mosaic virus movement protein associates with the cytoskeleton in tobacco cells. Plant Cell 1995; 7:2101-2114.
108. Padgett HS, Epel BL, Kahn TW et al. Distribution of tobamovirus movement protein in infected cells and implications for cell-to-cell spread of infection. Plant J 1996; 10:1079-1088.
109. Lazarowitz SG, Beachy RN. Viral movement proteins as probes for intracellular and intercellular trafficking in plants. Plant Cell 1999; 11:535-548.
110. Mas P, Beachy RN. Replication of tobacco mosaic virus on endoplasmic reticulum and role of the cytoskeleton and virus movement protein in intracellular distribution of viral RNA. J Cell Biol 1999; 147:945-958.
111. Boyko V, Ferralli J, Heinlein M. Cell-to-cell movement of TMV RNA is temperaturedependent and corresponds to the association of movement protein with microtubules. Plant J 2000; 22:315-325.
112. Boyko V, van der LJ, Ferralli J et al. Cellular targets of functional and dysfunctional mutants of tobacco mosaic virus movement protein fused to green fluorescent protein. J Virol 2000; 74:11339-11346.
113. Boyko V, Ferralli J, Ashby J et al. Function of microtubules in intercellular transport of plant virus RNA. Nat Cell Biol 2000; 2:826-832.
114. Gillespie T, Boevink P, Haupt S et al. Functional analysis of a DNA-shuffled movement protein reveals that microtubules are dispensable for the cell-to-cell movement of Tobacco mosaic virus. Plant Cell 2002; 14:1207-1222.
115. Reichel C, Beachy RN. Tobacco mosaic virus infection induces severe morphological changes of the endoplasmic reticulum. Proc Natl Acad Sci USA 1998; 95:11169-11174.
116. Dorokhov YL, Makinen K, Frolova OY et al. A novel function for a ubiquitous plant enzyme pectin methylesterase: The host-cell receptor for the tobacco mosaic virus movement protein. FEBS Lett 1999; 461:223-228.
117. Chen MH, Sheng JS, Hind G et al. Interaction between the tobacco mosaic virus movement protein and host cell pectin methylesterases is required for viral cell-to-cell movement. EMBO J 2000; 19:913-920.
118. Markovic O, Jornvall H. Pectinesterase: The primary structure of the tomato enzyme. Eur J Biochem 1986; 158:455-462.
119. Matsushita Y, Deguchi M, Youda M et al. The tomato mosaic tobamovirus movement protein interacts with a putative transcriptional coactivator KELP. Mol Cells 2001; 12:57-66.
120. Soellick TR, Uhrig JF, Bucher GL et al. The movement protein NSm of tomato spotted wilt tospovirus (TSWV): RNA binding, interaction with the TSWV N protein, and identification of interacting plant proteins. Proc Nat Acad Sci USA 2000; 97:2373-2378.
121. Kelley WL. Molecular chaperones: How J domains turn on Hsp70s. Curr Biol 1999; 9:R305-308.
122. Alzhanova DV, Napuli AJ, Creamer R et al. Cell-to-cell movement and assembly of a plant closterovirus: Roles for the capsid proteins and Hsp70 homolog. EMBO J 2001; 20:6997-7007.
123. von Bargen S, Salchert K, Paape M et al. Interactions between the tomato spotted wilt virus movement protein and plant proteins showing homologies to myosin, kinesin and DnaJ-like chaperones. Plant Physiol Biochem 2001; 39:1083-1093.
124. Huang Z, Andrianov VM, Han Y et al. Identification of Arabidopsis proteins that interact with the cauliflower mosaic virus (CaMV) movement protein. Plant Mol Biol 2001; 47:663-675.
125. Oparka KJ. Getting the message across: How do plant cells exchange macromolecular complexes? Trends Plant Sci 2004; 9:33-41.
126. Pouwels J, van der Velden T, Willemse J et al. Studies on the origin and structure of tubules made by the movement protein of Cowpea mosaic virus. J Gen Virol 2004; 85:3787-3796.

127. Carvalho CM. The Cowpea mosaic virus movement protein: Analysis of its interactions with viral and host proteins. PhD Thesis. The Netherlands: Wageningen University, 2003:88.
128. Baluska F, Cvrckova F, Kendrick Jones J et al. Sink plasmodesmata as gateways for phloem unloading. Myosin VIII and calreticulin as molecular determinants of sink strength? Plant Physiol 2001; 126:39-46.
129. Baluska F, Samaj J, Napier R et al. Maize calreticulin localizes preferentially to plasmodesmata in root apex. Plant J 1999; 19:481-488.
130. Torres E, Gonzalez-Melendi P, Stoger E et al. Native and artificial reticuloplasmins coaccumulate in distinct domains of the endoplasmic reticulum and in post-endoplasmic reticulum compartments. Plant Physiol 2001; 127:1212-1223.
131. Chen MH, Tian GW, Gafni Y et al. Effects of calreticulin on viral cell-to-cell movement. Plant Physiol 2005; 138:1866-1876.
132. Carette JE, Stuiver M, Van Lent J et al. Cowpea mosaic virus infection induces a massive proliferation of endoplasmic reticulum but not Golgi membranes and is dependent on de novo membrane synthesis. J Virol 2000; 74:6556-6563.
133. Ritzenthaler C, Laporte C, Gaire F et al. Grapevine fanleaf virus replication occurs on endoplasmic reticulum-derived membranes. J Virol 2002; 76:8808-8819.

Cell-Cell Movements of Transcription Factors in Plants

Alexis Maizel*

Abstract

In the last few years, the intercellular trafficking of regulatory proteins has emerged as a novel mechanism of cell-to-cell communication in plant development. Here I present a review of the documented cases of transcription factors movement in plants and examine the common themes underlying these different examples.

Introduction

Plant growth differs from that of animals in that most organs originate post-embryonically from meristems, which are groups of undifferentiated cells set aside during embryogenesis. In these structures, the morphogenetic fields that guide organogenesis are established and maintained. Because plant cells differentiate mostly according to positional clues rather than based on lineage,[1] cell-to-cell communication is crucial to initiate and maintain these fields, as well as to coordinate cellular responses within each field. As in animals, intercellular communication can occur through the extra-cellular space, via secreted ligands and interactions with receptors (apoplastic communication).[2] In plants, communication also occurs wholly within the cytosol (symplastic communication) via specialized channels called plasmodesmata (PD), which provide cytoplasmic continuity between plant cells.[3] The structure and the mechanisms of protein translocation through PD is reviewed extensively in a different chapter, thus I will here exclusively focus on a subset of the genes that have been shown to have non cell-autonomous functions: genes encoding transcription factors.[4] Transcription factor orchestrate the development of organisms by regulating the expression of thousand of genes thus the first explanation for their nonautonomous effects is the modulation of expression of downstream diffusible targets that affect neighboring cells. There is however an alternative hypothesis: direct trafficking of the encoded protein is responsible for noncell autonomy. Numerous examples support this latter hypothesis. In this chapter I will first review the current examples of intercellular movement of transcription factors in plants and then outline the common themes to all these examples.

Noncell Autonomous Action of Transcription Factors by Direct Protein Transfer in Plants

The maize KNOTTED-1 (KN1) homeodomain protein was the first plant protein found to traffic cell-to-cell. KN1 can be detected in the epidermal layer (L1) of the maize SAM, even though its RNA is limited to the L2 and L3.[5-7] This activity explains the noncell autonomy of

*Alexis Maizel—Department of Molecular Biology, Max Planck Institute of Developmental Biology, 37-39 Spemannstrass, D-72076 Tübingen, Germany.
Email: maizel@tuebingen.mpg.de

Cell-Cell Channels, edited by Frantisek Baluska, Dieter Volkmann and Peter W. Barlow.
©2006 Landes Bioscience and Springer Science+Business Media.

dominant neomorphic *KN1* alleles in maize leaf development,[8-10] and trafficking may also contribute to the normal function of *KN1* in stem cell maintenance.[7] The modification of PD size exclusion limit by KN1 argues in favor of trafficking through PDs.[5] In addition to its classical DNA binding function, the homeodomain acts as an intercellular trafficking signal able to drive movement of an unrelated protein in cis.[11] In addition to KN1 protein trafficking, noncell autonomy of the KN1 mRNA has also been reported,[5,12] and the KN1 homeodomain is sufficient to mediate the specific intercellular trafficking of its mRNA.[11] RNA binding activity of transcription factors has already been reported.[13] The best-studied example is the homeodomain protein Bicoïd, which binds *caudal* mRNA and represses its translation in the *Drosophila* embryo.[14-17] The homeodomain is sufficient to bind to both RNA and DNA molecules, and a specific position of the homeodomain DNA recognition helix discriminates RNA versus DNA binding.[18] However noncell autonomous activity of Bicoïd has never been reported, which make the observation that KN1 can escort its own messenger more intriguing and raises questions about the molecular mechanism and the signification of this transfer.

Soon after this first example of transcription factor movement, Perbal et al[19] found that the *Antirrhinum* MADS domain protein DEFICIENS (DEF) involved in floral organs identity can move within developing flowers. Using genetic chimera, they showed that the DEF movement was directional within the meristem. DEF can move from the inside L2 layer toward the outermost epidermal L1 layer, but not in the other direction.

A third transcription factor shown to move within floral meristems is LEAFY (LFY) in *Arabidopsis*.[20] LFY is a plant specific transcription factor specifying floral faith to newly formed meristems by repressing stem characteristics and triggering expression of floral organ identity genes.[21,22] The most dramatic results were seen when LFY was expressed from the epidermis-specific ML1 promoter: although the RNA was, as expected, restricted to the L1, LFY protein was found in a gradient that extended into several interior cell layers. The functionality of trafficked LFY protein was shown both by complete rescue of the mutant phenotype, and by the ability of LFY to activate a direct target gene in interior layers. By monitoring the extent of protein movement in the meristem with GFP fusion, Wu et al established that the extent of LFY movement is constrained in a similar fashion to the movement of a 2xGFP, a paradigm for movement by diffusion (non targeted), and correlate with the cytoplasmic abundance of the protein.[23] Suspicion that LFY movement is actually non targeted and occurs in absence of any specific determinant in the protein other than its size, is strengthen by the observation that all fragments of LFY are equally able to move within the meristem. This result suggests that either specific movement determinants in the protein are multiple or that none exist.

Contrarily to LFY, recent work indicate that the movement of the SHORT ROOT (SHR) transcription factor, a member of the GRAS family, is likely to be determined by specific mechanisms. Similarly to the shoot, the root is formed by a meristem comprising the root initials. There are distinct initials that give rise to several radially organized layers in the mature root, including from the center to the periphery, the stele, the endodermis, the cortex, and the epidermis (Fig. 1A). The *SHR* gene is required both for correct specification of the endodermal cell layer and normal patterning of the root.[24] In situ hybridization had revealed that *SHR* RNA is expressed only in the stele,[24] but subsequent studies using both GFP fusions and immunohistochemistry showed that the protein product is found in both the stele and the adjacent endodermis[25] and thus moves one cell away from its source of expression to the cell layer where its function is required (Fig. 1B). Using an SHR-GFP fusion protein expressed from its own promoter, Nakajima et al showed that SHR is both nuclear and cytoplasmic in the stele, but strictly nuclear in the endodermis. SHR movement was shown to be dependent upon its presence in the cytoplasm.[26] Forcing nuclear targeting of SHR by adding a strong nuclear localization signal to SHR-GFP (SHR-NLS-GFP), blocks protein movement.[26] Interestingly, expression of SHR-GFP in epidermal cells where the protein is also efficiently nuclear localized, correlate with absence of movement into adjacent layers.[27] Altogether nuclear targeting seems to be a potential mechanism by which SHR trafficking is limited to the endodermis. SCARECROW, a GRAS family protein that functions downstream of *SHR* and whose expression is specific

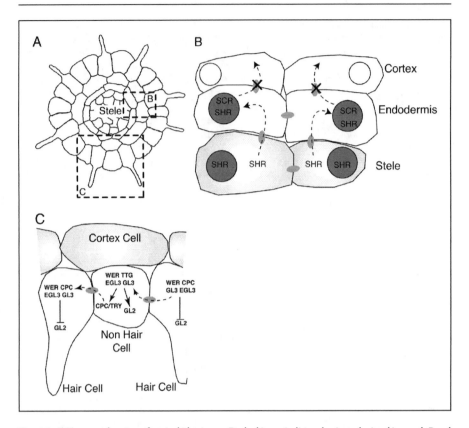

Figure 1. A) Transversal section of an *Arabidopsis* root. Dashed insets indicate the tissue depicted in panels B and C. B) Specification of root endodermis identity by intercellular transfer of SHR from the stele to the endodermis. SHR is expressed in the stele where it localizes to the cytoplasm (light gray shading) and nucleus (dark gray circles) and accumulates in the nucleus of endodermis cells (dark gray circles) where it promotes SCR expression. SHR transfer is blocked between endodermis and cortex cells. Plasmodesmata are grey ovals. C) Patterning of *Arabidopsis* root epidermis by intercellular transfer of transcription factors. The network of transcription factors involved in nonhair and hair cell faith specification is indicated. Full arrow between genes indicates activation or repression, while dashed ones indicate intercellular transfer of protein through plasmodesmata.

of the endodermis, can control SCR cellular retention in this cell layer. When *SHR* is expressed in the root epidermis in a *scr* mutant background, the SHR protein does not efficiently nuclear localize and is able to move into the adjacent cell layer. These data provide direct evidence that SHR must localize to the cytoplasm in order to move and indicate a role for SCR in limiting SHR movement, potentially by regulating the subcellular localization of SHR. The missense allele *shr5* of *SHR* is entirely cytoplasmic and still does not move indicating that, contrarily to LFY, SHR movement is not occurring by diffusion of the cytoplasmic pool of the protein but require interaction with specific cell components.

Another beautiful example of how transcriptional regulator movement orchestrate a correct patterning is the development of root hairs in *Arabidopsis*. Root hair development results from a lateral inhibition mechanism where epidermal cells located at covering the junction between two underlying cortical cells acquire root hair identity while the surrounding epidermal cells do not develop root hairs. This patterning process involves several transcriptional regulators: a homeodomain protein, GLABRA2 (GL2); a WD-repeat protein, TRANSPARENT TESTA

GLABRA (TTG); an R2R3 MYB-type transcription factor, WEREWOLF (WER); two closely related basic helix-loop-helix proteins, GLABRA3 (GL3) and ENHANCER OF GLABRA3 (EGL3); and three small MYB proteins, CAPRICE (CPC), TRIPTYCHON (TRY) and EN-HANCER OF TRIPTYCHON AND CAPRICE (ETC1) (for review see ref. 28). These factors specify epidermal cell fate through bi-directional signaling mechanism between adjacent epidermal cells (Fig. 1C). Development of hair cells is the ground state for root and acquisition of an hairless faith is promoted by expression of *GL2* which expression is positively regulated by the TTG/WER/GL3/EGL3 complex. This positive regulation can be antagonized by the CPC/TRY/ ETC1 proteins that are thought to displace WER from the TTG/WER/GL3/EGL3 complex and thus inhibit GL2 expression. Interestingly the *CPC/TRY* inhibitors genes are only expressed in non root hair cells but act noncell autonomously in the root hair cells. Expression of a CPC-GFP fusion from its native promoter elegantly established that the protein traffic from nonhair cells to hair cells where it accumulates in the nucleus.[29] Even though not formally established, it is postulated that TRY is also moving the same direction. Thus lateral inhibition, mediated by CPC/TRY, spread from the nonhair cell to its neighbor and repress hairless faith. Interestingly the hairless faith promoting factors GL3 and EGL3 are expressed preferentially in hair cell but acts noncell autonomously in hairless cell.[30] Here also a GL3-GFP fusion has established that GL3 traffic from hair cells to hairless cells. This mutual exchange of transcriptional activator and inhibitor between neighboring cells with opposite faith provide a nice example of patterning by bidirectional cross talk.

Common Mechanistic Trends in Plant Transcription Factors Movement?

Beyond the diversity of transcription factor movement just exposed, can general trends be identified? The first question that arises is whether or not movement is due to specific signal in the protein (targeted movement) or the sole reflection of diffusion of protein from the producing cells as a consequence of PD size exclusion limit (non targeted)? The work of Wu et al[23] strongly favors that LEAFY movement is non targeted. This result indicate that endogenous proteins movement, in a similar way to GFP, can be due to passive diffusion of the free cytoplasmic pool of protein. Whether this conclusion holds true for all transcription factors require further study, but indicates that contrarily to what could have been originally thought, movement of transcription factor could be by default and that what's matter is retention. Limitation of transcription factor spreading is an important issue. Anarchical diffusion of these regulators could have disastrous effects on development. Two levels of regulation can be envisioned. The first is cellular retention by association with cofactors or addressing to organelles.[31] In the case of LFY the more the cytoplasmic pool of LFY, the more extensive is the movement. Along this line, the MADS box AP1 transcription factor which cannot move is essentially nuclear and is known to be part of multi-proteins complexes with other MADS box proteins.[32] The second level can be the regulation of PD size exclusion limit. Degree of PD aperture in the embryo and meristems defines multiple symplasmic fields with a highly dynamic geometry.[33-35] Although the physiological mechanisms of PD aperture regulation are totally unknown, they could represent an efficient way to constraint transcription factors in specific territories. These mechanisms could also account for the interspecific differences in trafficking observed between *Arabidopsis* and *Antirrhinum*. Although the MADS box proteins DEF and GLO can move in *Antirrhinum*, the DEF homologue APETALA3 (AP3) in *Arabidopsis* does not.[19,36] This different of comportment is likely to reflect differences of PD aperture in *Antirrhinum* and *Arabidopsis* as cell autonomy of both AP3 and DEF is dependant of the host.[37]

The work of the Benfey and Jackson labs on intercellular transfer of SCR and KN1 respectively indicates that the movement of these proteins is most likely due to specific mechanisms and does not reflect diffusion. The KN1 homeodomain (the DNA binding domain) is necessary and sufficient for intercellular trafficking[11] and the M6 mutation, which maps to the N-terminal part of this domain, can block KN1 transfer.[5] Interestingly this mutation affects a putative nuclear localization signal (NLS) although the effect of this mutation on KN1 subcellular localization has not been studied in plant cell, it drives the protein to the cytoplasm in animal cells (A. Joliot, personal communication), suggesting a role for nuclear targeting in addressing KN1 to the PD pathway. Although no formal demonstration of an association between transcription factors with targeted movement and PD has been established, the modification of PD size exclusion limit by KN1 argues in favor of trafficking through PD.[5]

Nuclear targeting is a key component of SHR movement mechanism. SHR is both nuclear and cytoplasmic in the stele but strictly nuclear in the endodermis where it activates its target *SCR*. SHR movement can be restrained by forcing the protein to the nucleus[26] and SHR nuclear targeting is dependant on SCR activity.[27] Thus a tentative model to constraint SHR movement in the endodermis could involve the activation by SHR of a cofactor in a *SCR* dependent manner that would drive SHR to the nucleus of endodermis cells and block SHR movement. A plausible candidate for this cofactor could be SCR itself. SCR is an exclusively nuclear protein and protein-protein interactions between GRAS protein members have been already reported.[38] This parsimonious model is formally equivalent to a variant where the cofactor is a gene expressed in response to both SCR and SHR. Cytoplasmic components are also required to trigger SHR movement in the stele as indicated by the *shr5* allele, which although entirely cytoplasmic is movement-deficient. This allele correspond to a T → I amino-acid substitution targeting a putative phosphorylation site. The nature of this cofactor needs to be precised.

Beyond the targeted or non targeted nature of transcription factors movement, it is clear that many regulations take place at the level of the PD itself and that these structures, similarly to the nuclear pore complexes should not be consider as simple holes between cells.[39] KN1 and DEF show polarity in their transport, the proteins can traffic from L2 to L1 but not from L1 to L2.[7] Directionality in trafficking through PD is a perfect example of such complex regulation occurring at the boundary between tissues.

Unicity of PD pathway can also be questioned. The identification in tobacco by Lee et al of the Non Cell Autonomous Pathway Protein (NtNCAPP1) as a player of intercellular protein trafficking[40] indicate that more than one pathway of trafficking through PD may exist. Over-expression of a truncated form of NtNCAPP1 in tobacco (acting as dominant negative form) blocks the intercellular trafficking of Tobacco Mosaic Virus movement proteins (TMV-MP) and non cell autonomous protein CmPP16 but did not block transfer of the Cucumber Mosaic virus movement protein (CMV-MP) and KN1. Over-expression of NtNCAPP1 dominant negative form phenocopies some aspect of LFY over-expression and expansion of tobacco LFY homolog protein in the meristem is observed. Thus NtNCAPP1 dominant negative form expression can block the transfer of some protein with targeted movement (TMV MP and CmPP16), has no effect of other (KN1) and favor the transfer of a protein with non targeted movement (LFY). These pleiotropic effects of targeted and nontargeted protein movement strongly suggest that more than one pathway exist.

Biological Significance of Transcription Factor Movement in Plants?

Independently of the mechanism of intercellular transfer, one can speculate about the biological significance of transcription factor movement in plants. In the meristem, there are no striking differences in RNA and protein patterns of DEF and LFY,[41] which make the biological significance of their movement puzzling. However it has been speculated that LFY movement presents a redundant mechanism to ensure that all three tissue layers in the meristem coordinately adopt floral fate.

Direct trafficking of transcription factors is a parsimonious way to achieve specification of cell faith. During root epidermis development, the bi-directional signaling from hairless cell to root-hair cell (via CPC and possibly also TRY/ETC1) and in the other direction via GL3 (and possibly EGL3), is required for appropriate accumulation of GL3/EGL3 in the nonhair cell position and thus hairless faith acquisition. This 'back and forth' signaling between cells is conceptually similar to bidirectional signaling identified in other patterning models such as larval vulval cell specification in *C. elegans*.[42] However it does not involve any receptor-mediated signal transduction pathway and directly influence gene expression. Interestingly this situation has been formulated and theorized by Prochiantz and Joliot[43] in the case of homeodomain transcription factors that can traffic from cell to cell in animals.

References

1. van den Berg C, Willemsen V, Hage W et al. Cell fate in the Arabidopsis root meristem determined by directional signalling. Nature 1995; 378:62-65.
2. Fletcher JC, Brand U, Running MP et al. Signaling of cell fate decisions by CLAVATA3 in Arabidopsis shoot meristems. Science 1999; 283:1911-1914.
3. Heinlein M, Epel BL. Macromolecular transport and signaling through plasmodesmata. Int Rev Cytol 2004; 235:93-164.
4. Lucas WJ, Lee J-Y. Plasmodesmata as a supracellular control network in plants. Nat Rev Mol Cell Biol 2004; 5:712.
5. Lucas WJ, Bouche-Pillon S, Jackson DP et al. Selective trafficking of KNOTTED1 homeodomain protein and its mRNA through plasmodesmata. Science 1995; 270:1980-1983.
6. Kim JY, Yuan Z, Cilia M et al. Intercellular trafficking of a KNOTTED1 green fluorescent protein fusion in the leaf and shoot meristem of Arabidopsis. Proc Natl Acad Sci USA 2002; 99:4103-4108.
7. Kim JY, Yuan Z, Jackson D. Developmental regulation and significance of KNOX protein trafficking in Arabidopsis. Development 2003; 130:4351-4362.
8. Vollbrecht E, Veit B, Sinha N et al. The developmental gene Knotted-1 is a member of a maize homeobox gene family. Nature 1991; 350:241-243.
9. Jackson D, Hake S. Morphogenesis on the move: Cell-to-cell trafficking of plant regulatory proteins. Curr Opin Genet Dev 1997; 7:495-500.
10. Jackson D, Veit B, Hake S. Expression of maizel KNOTTED1 related homeobox genes in the shoot apical meristem predicts patterns of morphogenesis in the vegetative shoot. Development 1994; 120:405-413.
11. Kim JY, Rim Y, Wang J et al. A novel cell-to-cell trafficking assay indicates that the KNOX homeodomain is necessary and sufficient for intercellular protein and mRNA trafficking. Genes Dev 2005; 19:788-793.
12. Kim M, Canio W, Kessler S et al. Developmental changes due to long-distance movement of a homeobox fusion transcript in tomato. Science 2001; 293:287-289.
13. Cassiday LA, Maher LJ. Having it both ways: Transcription factors that bind DNA and RNA. Nucl Acids Res 2002; 30:4118-4126.
14. Dubnau J, Struhl G. RNA recognition and translational regulation by a homeodomain protein. Nature 1996; 379:694-699.
15. Rivera-Pomar R, Niessing D, Schmiddt-ott U et al. RNA binding and translational regulation by Bicoid. Nature 1996; 379:746-748.
16. Chan SK, Struhl G. Sequence-specific RNA binding by bicoid. Nature 1997; 388:634.
17. Niessing D, Driever W, Sprenger F et al. Homeodomain position 54 specifies transcriptional versus translational control by Bicoid. Mol Cell 2000; 5:395-401.
18. Niessing D, Dostatni N, H Jc et al. Sequence interval within the PEST motif of Bicoid is important for translational repression of caudal mRNA in the anterior region of the Drosophila embryo. EMBO J 1999; 18:1966-1973.
19. Perbal MC, Haughn G, Saedler H et al. Noncell-autonomous function of the Antirrhinum floral homeotic proteins DEFICIENS and GLOBOSA is exerted by their polar cell-to-cell trafficking. Development 1996; 122:3433-3441.
20. Sessions A, Yanofsky MF, Weigel D. Cell-cell signaling and movement by the floral transcription factors LEAFY and APETALA1. Science 2000; 289:779-782.
21. Weigel D, Alvarez J, Smyth DR et al. LEAFY controls floral meristem identity in Arabidopsis. Cell 1992; 69:843-859.
22. Lohmann JU, Weigel D. Building beauty: The genetic control of floral patterning. Dev Cell 2002; 2:135-142.
23. Wu X, Dinneny JR, Crawford KM et al. Modes of intercellular transcription factor movement in the Arabidopsis apex. Development 2003; 130:3735-3745.
24. Helariutta Y, Fukaki H, Wysocka-Diller J et al. The SHORT-ROOT gene controls radial patterning of the Arabidopsis root through radial signaling. Cell 2000; 101:555-567.
25. Nakajima K, Sena G, Nawy T et al. Intercellular movement of the putative transcription factor SHR in root patterning. Nature 2001; 413:307-311.
26. Gallagher KL, Paquette AJ, Nakajima K et al. Mechanisms regulating SHORT-ROOT intercellular movement. Curr Biol 2004; 14:1847.
27. Sena G, Jung JW, Benfey PN. A broad competence to respond to SHORT ROOT revealed by tissue-specific ectopic expression. Development 2004; 131:2817-2826.

28. Pesch M, Hülskamp M. Creating a two-dimensional pattern de novo during Arabidopsis trichome and root hair initiation. Curr Opin Gen Dev 2004; 14:422-427.
29. Wada T, Kurata T, Tominaga R et al. Role of a positive regulator of root hair development, CAPRICE, in Arabidopsis root epidermal cell differentiation. Development 2002; 129:5409-5419.
30. Esch JJ, Chen M, Sanders M et al. A contradictory GLABRA3 allele helps define gene interactions controlling trichome development in Arabidopsis. Development 2003; 130:5885-5894.
31. Crawford KM, Zambryski PC. Subcellular localization determines the availability of nontargeted proteins to plasmodesmatal transport. Curr Biol 2000; 10:1032-1040.
32. Honma T, Goto K. Complexes of MADS-box proteins are sufficient to convert leaves into floral organs. Nature 2001; 409:525.
33. Gisel A, Barella S, Hempel FD et al. Temporal and spatial regulation of symplastic trafficking during development in Arabidopsis thaliana apices. Development 1999; 126:1879-1889.
34. Kim I, Cho E, Crawford K et al. Cell-to-cell movement of GFP during embryogenesis and early seedling development in Arabidopsis. Proc Natl Acad Sci USA 2005; 102:2227-2231.
35. Rinne PL, van der Schoot C. Symplasmic fields in the tunica of the shoot apical meristem coordinate morphogenetic events. Development 1998; 125:1477-1485.
36. Jenik PD, Irish VF. The Arabidopsis floral homeotic gene APETALA3 differentially regulates intercellular signaling required for petal and stamen development. Development 2001; 128:13-23.
37. Efremova N, Perbal M-C, Yephremov A et al. Epidermal control of floral organ identity by class B homeotic genes in Antirrhinum and Arabidopsis. Development 2001; 128:2661-2671.
38. Itoh H, Ueguchi-Tanaka M, Sato Y et al. The gibberellin signaling pathway is regulated by the appearance and disappearance of SLENDER RICE1 in nuclei. Plant Cell 2002; 14:57-70.
39. Lee J-Y, Yoo B-C, Lucas WJ. Parallels between nuclear-pore and plasmodesmal trafficking of information molecules. Planta 2000; 210:177.
40. Lee JY, Yoo BC, Rojas MR et al. Selective trafficking of noncell-autonomous proteins mediated by NtNCAPP1. Science 2003; 299:392-396.
41. Parcy F, Nilsson O, Busch MA et al. A genetic framework for floral patterning. Nature 1998; 395:561-566.
42. Yoo AS, Bais C, Greenwald I. Crosstalk between the EGFR and LIN-12/Notch pathways in C. elegans vulval development. Science 2004; 303:663-666.
43. Prochiantz A, Joliot A. Can transcription factors function as cell-cell signalling molecules? Nat Rev Mol Cell Biol 2003; 4:814-819.

SECTION VI
Animal Cells

CHAPTER 13

Gap Junctions:
Cell-Cell Channels in Animals

Fabio Mammano*

Abstract

Gap junctions provide one of the most common forms of intercellular communication. The structures underlying these communicating cell junctions[1] were soon resolved in membrane associated particles forming aggregates of six subunits.[2] They are composed of membrane proteins that form a channel that is permeable to ions and small molecules, connecting the cytoplasm of acdjacent cells. Two unrelated protein families are involved in this function; **connexins**, which are found only in chordates, and **pannexins**, which are ubiquitous and present in both chordate and invertebrate genomes.[3] In this chapter, structural and functional issues of gap junction channels are reviewed. Several types of pathologies associated to channel dysfunction, with an emphasis on deafness, are also examined.

Connexins

Structure of Connexin Channels

Connexin-based gap junction channels are composed of two hemichannels, or connexons, each provided by one of two neighbouring cells. Two connexons join in the gap between the cells to form a gap junction channel that connects the cytoplasms of the two cells. A connexon is composed of six transmembrane proteins, called connexins (Cx), which form a multigene family. Connexons may be **homomeric** (composed of six identical connexin subunits) or **heteromeric** (composed of more than one species of connexins). The channel may be **homotypic** (if connexons are identical) or **heterotypic** (if the two connexons are different).[4]

At least 20 distinct human Cx isoforms have been cloned.[5] Connexins are classified according to their molecular mass or grouped into α, β and γ subtypes, based on sequence similarities.[6,7] They have highly conserved sequences and may have originated from a common ancestor. In the human genome, the majority of β connexin genes map to two gene clusters at either 1p34-p35 or 13q11-q12.[8] Connexins have four α helical transmembrane domains (M1 to M4), intracellular N- and C-termini, two extracellular loops (E1 and E2), and a cytoplasmic loop (CL).[9] E1 and E2 mediate the docking of the two hemichannels.[7] The CL and the C-terminus constitute the least homologous regions across the connexin family, suggesting that many of the functional differences between connexins reside there.[10]

Electron cryomicroscopy of two-dimensional crystals[11] permitted to derive a three-dimensional density map at 5.7 angstroms in-plane and 19.8 angstroms vertical resolution, and to identify the positions and tilt angles for the 24 alpha helices within each hemichannel. The four hydrophobic segments in connexin sequences were assigned to the alpha helices in the map based on biochemical

*Fabio Mammano—Venetian Institute of Molecular Medicine, University of Padova, via G. Orus 2, 35129 Padua, Italy. Email: fabio.mammano@unipd.it

Cell-Cell Channels, edited by Frantisek Baluska, Dieter Volkmann and Peter W. Barlow.
©2006 Landes Bioscience and Springer Science+Business Media.

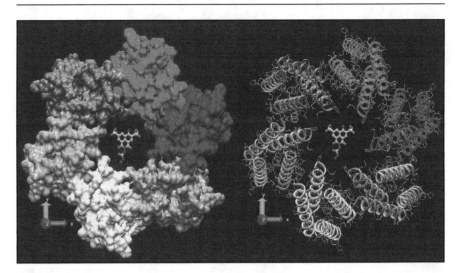

Figure 1. Atomic model of the transmembrane segment of a Cx26 connexon. A molecule of Lucifer Yellow, a fluorescent tracer commonly employed to assay cell-cell communication, is shown in the middle of the pore.

and phylogenetic data. The final model specifies the coordinates of C-alpha atoms in the transmembrane domain.[12] Figure 1 presents a model of the transmembrane segment of a single Cx26 connexon derived by the coordinates of the Cα model deposited in the Protein Data Bank (accession code 1TXH). The six connexin molecules are currently thought to be arranged so that M3 is the major pore-lining helix, together with M1.[13,14] However, due to the low resolution of the density map, the estimate of the helices orientation around their principal axis is affected by a large error, possibly larger than 90°.

Oligomerization of connexins into hemichannels occurs in the ER–Golgi. Vesicles transport hemichannels to the plasma membrane shearing the general secretory pathway with other membrane proteins.[15,16] Although each connexin exhibits a distinct tissue distribution, many cell types express more than one connexin isoform. The variations in the structural composition of gap junction channels allow for a greater versatility of their physiological properties, however not all connexins participate in the formation of hetero-connexons.[17] Indeed, it has been suggested that an ''assembly'' signal allows connexin subunits to recognize each other, thus preventing the interaction of incompatible connexins.

Permeability

Although gap junctions have been traditionally described as nonselective pores, various studies have revealed a high degree of selectivity among the different connexins.[4] The unitary conductance (i.e., the conductance of a single gap junction channel) varies widely (25 pS for Cx45, 350 pS for Cx37), yet, the sequence of monovalent cation selectivity is the same (K>Na>Li>TEA) for different connexins.[18] Gap junction channels are permeable to soluble molecules such as metabolites or second messengers (cAMP, Ca^{2+} and IP_3).[19-22]

A classical method for evaluating connexin permeability is the intercellular transfer of membrane impermeant fluorescent molecules (e.g., Lucifer Yellow, see Fig. 1) delivered intracellularly to a single cell.[23] By employing a wide array of molecular tracers, it has been demonstrated that the selective transfer of molecules through gap junction channels is not dictated by size alone but is affected also by other parameters such as charge or rigidity.[18,24] Recently, this type of analysis has been extended to signalling molecules (Fig. 2).[19,25,26]

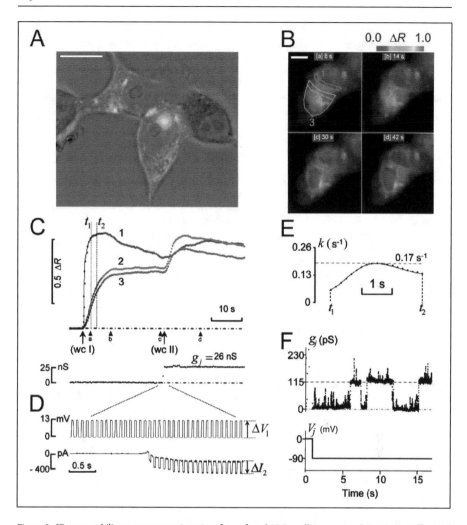

Figure 2. IP$_3$ permeability measurement in pairs of transfected HeLa cells expressing hCx26. A) Differential interference contrast (DIC) image merged with the fluorescence image of the same field illuminated at 500 nm showing a small cluster of HeLa cells, two of which are positive for transfection with hCx26 tagged with EYFP; scale bar, 15 μm. B) False color images sampling the time course of fura-2 fluorescence ratio changes, ΔR, at the times shown following intracellular delivery of IP$_3$ first to cell 1 (at time t=5s, frames [a-c]) and, later, also to cell 2 (frame [d]); three color-coded ROIs used in the computations are superimposed on frame [a]; scale bar, 10 μm. C) Top: traces from the corresponding ROIs in (B), located on the two sides of the gap junction (traces n. 1 and 2) and over the entire cell 2 (trace n.3); vertical arrows point at the onset of IP$_3$ delivery to cell 1 (wc I) and cell 2 (wc II); arrowheads, labeled a-d, mark acquisition times of the corresponding frames in (B); thin lines through data points are least square fits with suitable functions f_1, f_2, f_3. Bottom: junctional conductance (g_j) monitored during image acquisition. D) Voltage steps applied to cell 1 (ΔV_1, top trace) to elicit junctional current responses (ΔI_2, bottom trace) during delivery of IP$_3$ to cell 2. E) Time course of transfer rate $k = df_3/dt/(f_1-f_2)$, during time interval (t_1, t_2) in (B); gray dashed line indicates stationary value used to estimate permeability coefficient, i.e., the mean total permeability due to a single gap junction channel. F) Conductance changes (g_j, top trace) due to gating of a single channel formed by hCx26 following application of a transjunctional potential difference V_j = 90 mV (bottom trace); estimated unitary conductance, γ = 115 pS (gray dashed line).

The rate of permeation of metabolites through gap junctions has been shown to be conditioned by connexin composition. For instance, Cx32 is permeable to both cAMP and cGMP, whereas heteromeric connexons composed of Cx32 and Cx26 lose permeability to cAMP, but not to cGMP.[27] IP$_3$ has been shown, in transfected HeLa cells loaded with fura-2 to monitor [Ca^{2+}]$_i$, to permeate homotypic Cx32 gap junctions approximately four times more efficiently than Cx26 and 2.5 times more efficiently than Cx43 gap junctions.[28] Indeed, it has been suggested that the most physiologically consequential difference between different connexin channels is their differing abilities to permit intercellular passage of molecules (such as second messengers) considerably larger than current carrying ions.[29]

Voltage Gating

Gap junction channels exhibit **voltage gating**, a property shared by several ion channels, whereby conductance is sensitive to voltage. Voltage gating depends primarily on transjunctional voltage V_j, i.e., the potential difference between the cytoplasm of the two adjacent cells (Fig. 3).[30] Over the last two decades, the view has changed from one with gap junction channel having a single transjunctional voltage-sensitive (V_j-sensitive) gating mechanism to one with each hemichannel of a formed channel, as well as unapposed hemichannels, containing two, molecularly distinct gating mechanisms.[30] In few connexin channels, changes in membrane potential (V_m) may activate an additional gate (V_m gate).[31]

The voltage sensitivity exhibited by gap junctions has the potential to uncouple communicating cells.[32] In homotypic channels, the intercellular conductance, g_j is an even function of V_j (i.e., it is symmetric about $V_j = 0$): the same **reduction** in g_j, is obtained either by hyperpolarizing one cell of a connected pair, or by depolarizing the other cell (by the same amount). In some cases, g_j does not decline to zero with increasing V_j, but reaches a plateau or **residual conductance** that varies depending on the Cx isoform, indicating that a wide range of voltage gating behaviours exists.[30,33] Some channels exhibit fast V_j gating transitions (~1 ms) to the residual state and slow V_j gate transitions (~10 ms) to the fully closed state, which are mediated by molecularly distinct processes (see next section). It has recently been proposed that the passage of large molecules between cells, such as fluorescent tracer molecules and cAMP, may be controlled by transjunctional voltage differences that activate the connexin voltage gate while having little effect on the electrical coupling arising from the passage of small electrolytes.[26]

Analysis of connexin sequences failed to reveal a region similar to ion-channels' S4, thereby implying that the molecular mechanisms which mediate voltage-induced closure of gap junction channels must be different.[9] A proline in the second transmembrane domain, which is conserved among all members of the connexin family, may play a central role in a conformational change that links the voltage sensor and the voltage gate of intercellular channels.[34] However, published reports suggest that both sensorial and gating elements of the fast gating mechanism are formed by transmembrane and cytoplasmic components of connexins among which the N terminus is most essential and which determines gating polarity.[30] Recent data are consistent with a gating model in which the voltage sensor is positioned in the N-terminus and its inward movement initiates V_j-gating.[35]

Chemical Gating

Transjunctional/transmembrane voltages are neither the sole nor the most important parameters capable of influencing the state of gap junction channels. Permeability can be altered also by specific changes in cytosolic ion composition, as well as by post-translational modifications. These alternative mechanisms are termed slow or 'loop' gating. A distinguishing feature of the slow gate is that the gating transitions consist of a series of transient substates en route to opening and closing.[30] The slow gating mechanism is also sensitive to V_j, but there is evidence that this gate may mediate gating by transmembrane voltage, V_m, intracellular Ca^{2+} and pH, chemical uncouplers (Fig. 4) and gap junction channel opening during de novo channel formation.[30] At the single channel level, the chemical/slow gate closes the channels slowly and completely, whereas the fast V_j gate closes the channels rapidly and incompletely.[36] Chemical agents that reduce coupling usually do not leave a residual conductance and their effect is readily reversible.

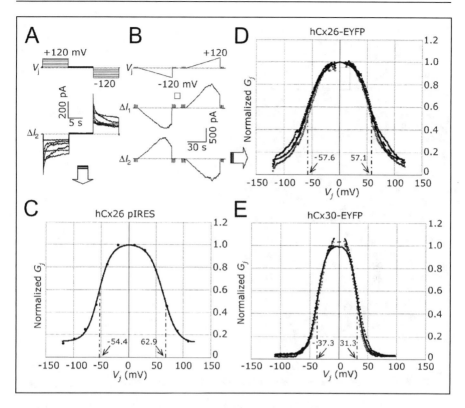

Figure 3. Voltage dependence of gap junction conductance. Human connexins 26 and 30 were expressed either through the bicistronic pIRES-EGFP expression vector, or as EYFP-tagged chimera in isolated HeLa cell pairs. A) top: voltage commands applied to one of two neighbouring cells (conventionally, Cell 1), each one separately patch-clamped with a different amplifier; bottom: junctional currents recorded from the adjacent cell (Cell 2), which was kept at the common prestimulus holding potential (-20 mV). Cells in A were transiently transfected with hCx26 cDNA hosted in a bicistronic vector that carried also the cDNA of EGFP. Dotted lines indicate prestimulus values of voltage and current. B) Top: voltage ramps applied to Cell 1 in a culture stably transfected with mCx26; bottom: whole cell currents recorded simultaneously from Cell 1 and the adjacent Cell 2, which was kept at the common prestimulus holding potential (-20 mV). C) Normalised conductance G_j (circles) vs. transjunctional potential V_j (abscissa) from steady-state data in A (currents measured 10 ms before the end of each voltage step). D) Normalised conductance (ordinates) vs. transjunctional voltage (abscissa) measured from ramp responses in cell pairs transiently transfected with the fusion product hCx26-EYFP: mean (squares), minima (open circles) and maxima (closed circles) of n=25 pairs. E, same as D for hCx30-EYFP (n=11). (See also ref. 102).

Chemical gating of gap junction channels is a complex phenomenon that may involve intra- and intermolecular interactions among connexin domains and a cytosolic molecule (possibly calmodulin) that may function as channel plug.[37] Some evidence suggests that low pH_i affects gating via an increase in $[Ca^{2+}]_i$; in turn, Ca^{2+} is likely to induce gating by activation of CaM, which may act directly as a gating particle. The effective concentrations of both Ca^{2+} and H^+ vary depending on cell type, type of connexin expressed and procedure employed to increase their cytosolic concentrations; however, pH_i as high as 7.2 and $[Ca^{2+}]_i$ as low as 150 nM or lower have been reported to be effective in some cells. At least three molecular models of channel gating have been proposed, but all of them are mostly based on circumstantial evidence.[31]

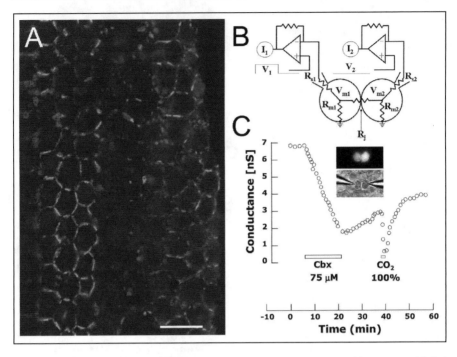

Figure 4. Example of chemical gating. A) Confocal section through the organ of Corti of the guinea pig, labelled with an antibody against connexin 26 (Cx26); scale bar, 30 μm. B) Schematics of dual patch clamp recordings. C) effect of gap junction inhibitors, carbenoxolone (Cbx) and CO_2 on junctional conductance. Inset: fluorescence (top) and bright field image (bottom) of pair of rat cochlear Hensen's cell loaded in situ with Oregon green 488 BAPTA-1 through the patch-clamp pipettes.

Regulation of Cell-to-Cell Communication

The number and distribution of gap junction channels is generally relatively stable under physiological conditions. However, the flux of connexins into and out of gap junctions is frequently reported to be highly dynamic.[16] New channels are added to the outside of the junctional plaque while older channels are internalized from the centre to be degraded.[16,38] The regulation of gap junction trafficking, assembly/disassembly and degradation is likely to be critical in the control of intercellular communication and phosphorylation has been implicated in the regulation of the connexin "lifecycle" at several stages.[39] The C-terminal region of connexins appears to be the primary region that becomes phosphorylated on serine or threonine residues.[39,40] Cx26 is the only connexin that has been reported not to be phosphorylated.[41] This may be due to the fact that it is the shortest connexin and only has a few C-terminal tail amino acids (a.a.) that could interact with cytoplasmic signalling elements. Clearly, connexin phosphorylation is not required for the formation of all gap junction channels, as recombinant homomeric-homotypic channels formed by wild-type Cx26 are functional.[42]

Intercellular Calcium Waves

Connexin channels are permeable to some of the more important secondary messengers involved in cell signalling, such as cAMP,[20] and IP_3,[21] the first and the principal inositol phosphate that is formed from inositol lipid hydrolysis due to G protein-linked receptor stimulation of phospholipase C.[43] IP_3 molecules diffuse rapidly throughout the cell,[44] interact with specific receptors (IP_3R) present in the endoplasmic reticulum and Ca^{2+} is liberated, raising its concentration ($[Ca^{2+}]_i$) in the cytosol.[45,46] In some cells, these $[Ca^{2+}]_i$ signals are targeted to control processes in limited

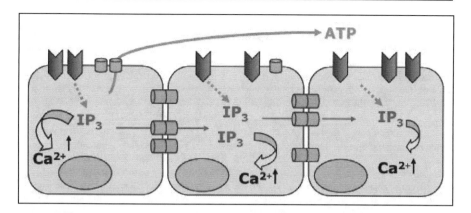

Figure 5. Mechanisms underlying the propagation of intercellular Ca^{2+} waves. Waves can be mediated by the intracellular transfer of signalling molecules (red arrows), such as IP_3, trough gap junction channels (in green). The extracellular pathway involves the diffusion of different molecules, such as ATP, through the extracellular space, activating P2 receptors (in blue). A color version of this figure is available online at www.Eurekah.com.

cytoplasmic domains, but in other systems long-range signalling involves intercellular Ca^{2+} waves.[47] Two pathways have been implicated so far in intercellular Ca^{2+} signalling (Fig. 5): (i) the extracellular diffusion of molecules[48,49] such as adenosine nucleotides, glutamate and other neurotransmitters which, by interacting with receptors on nearby cells alter their intracellular calcium levels[50] and (ii) the diffusion of cytoplasmic messenger molecules through gap junction.[51]

Models of Ca^{2+} waves that involve the production of IP_3 are based on the diffusion of IP_3 from a stimulated cell through gap junction channels into neighboring cells where it elicits Ca^{2+} release from intracellular stores.[52,53] However, more complex schemes have been proposed to account for the long-range periodic Ca^{2+} waves in the liver, where a messenger diffusing through gap junction channels appears to be regenerated in each participating cell.[54]

Intercellular propagation of Ca^{2+} waves has been described in a wide variety of cell types and is considered a mechanism by which cell activity is coordinated.[55] In the central nervous system, such waves occur among and between neurons and glial cells both under normal and pathological conditions.[56] In summary, direct propagation of intercellular Ca^{2+} waves from the cytosol of one cell to an adjacent cell requires the presence of gap junction channels, which allow signaling molecules, such as IP_3,[21,57] to be transferred across coupled cells. The alternative, but not mutually exclusive, pathway for communication of the Ca^{2+} signals involves the diffusion of signaling molecules, such as ATP, through the extracellular space, activating P2 receptors in neighboring cells that may or may not be in contact.[48,49]

Connexin-Related Pathologies

Recent evidence has shown that mutations within several connexins give rise to various hereditary diseases in humans such as Charcot–Marie–Tooth disease and erythrokeratodermia variabilis. In addition to these disorders, connexin mutations are also involved in deafness, cataracts, and numerous skin diseases. The association of connexin mutations with an increasing number of human pathologies, provides an unequivocal demonstration that gap junctional communication is crucial for diverse physiological processes.

Skin Diseases

In human skin, intercellular communication is mediated by all known β-connexin genes. Autosomal dominant mutations in connexins genes have been linked with several skin disorders which involve an increased thickness of the skin outer layers. This would indicate a critical role for connexins in maintaining the balance between proliferation and differentiation of the epidermis.[58,59] *GJB3* was the first connexin gene reported to cause the autosomal dominant skin disorder erythrokeratodermia variabilis (EKV) and several mutations have been identified so far in its protein product, Cx31: G12R, G12D, C86S, R42P and F137L.[60] Dominant mutations in Cx26 have also been described in syndromic deafness associated with skin disease. These disorders include keratitis-ichthyosis-deafness syndrome (KID), palmoplantar keratoderma with deafness (PPK), Vohwinkel syndrome (VS) and hystrix-like ichthyosis-deafness syndrome (HID).[61,62] Dominant mutations in *GJB2* affect a.a residues positioned in the highly conserved first extracellular domain (G59A and D66H), or at the boundary of this domain with the first (DelE42) or second (R75W) transmembrane domain. The exact role of Cx26 in skin disease, however, can not be attributed to loss of Cx26 channel activity alone due to the fact that many Cx26 mutations result in nonsyndromic deafness (see below). This would indicate that mutant forms of Cx26 associated with skin disorders must impact additional genes in a manner that upsets tissue homeostasis and results in disease.

Different missense mutations in Cx30, G11R, V37E and A88V, affecting conserved a.a. residues positioned in the N-terminal end and in the second transmembrane domains, have been identified as cause of ectodermal dysplasia (HED)[63] or Clouston syndrome. These mutations impair trafficking of the protein to the cell membrane and lead to cytoplasmic accumulation when transiently expressed in cultured keratinocytes, thus resulting in a complete loss of gap junction function.[64]

Peripheral Neuropathy and Congenital Cataract

The first disease associated to mutations in connexin genes was a form of Charcot-Marie-Tooth disease (CMT). Dominant and recessive X-linked mutations in the Cx32 sequence result in progressive degeneration of peripheral nerve and are characterized by distal muscle weakness and atrophy.[65] Cx32 is present in Schmidt–Lanterman incisures and nodes of Ranvier of myelinating Schwann cells,[66] providing continuity between the Schwann cell body and the cytoplasmic collar of the myelin sheath adjacent to the axon. It has been suggested that the S26L mutation of Cx32, implicated in X-linked CMT, alters the permeability of the gap junction to cyclic nucleotides.[67] At least another connexin, possibly Cx31, participates in forming gap junctions in these cells.[68] Cx31 was reported to be expressed in mouse auditory and sciatic nerves in a pattern similar to that of mouse Cx32.[69]

In the ocular lens, gap junctional communication is a key component of homeostatic mechanisms preventing cataract formation.[70] Altered biochemical coupling, resulting from inappropriate mixing of Cx46 and Cx50, perturbed cellular homeostasis but not lens growth, suggesting that unique biochemical modes of gap junctional communication influence lens clarity and lens growth, and that biochemical coupling is modulated by the connexin composition of the gap junction channels.[71,72]

Deafness

In developed countries deafness has an important genetic origin and at least 60% of the cases are inherited. In particular, the early onset forms of hearing impairment in developed countries are almost exclusively genetic in origin and not due to undiagnosed infections. Thirty-seven genes responsible for isolated hearing impairment in humans are known to date. Mutations in one of them, underlying the DFNB1 form of deafness, have been found to be responsible for about half of all cases of human deafness in countries surrounding the Mediterranean Sea. DFNB1 is thus almost as frequent as cystic fibrosis. DFNB1 can be caused by mutations in the *GJB2* gene (121011), which encodes the gap junction protein Cx26. However the complex DFNB1 locus (13q11-q12) has been shown to contain another gap junction gene *GJB6* (604418, which maps to 13q12), and encodes Cx30.[73] Specifically, DFNB1 can also be due to a deletion of 342 Kb involving *GJB6* (Ballana E, Ventayol M, Rabionet R, Gasparini P, Estivill X. Connexins and deafness Homepage. World wide web URL: http://www.crg.es/deafness) and which has been related to a dominant type of deafness in an Italian family.[74,75]

In the cochlea, gap junctions are found in two networks of cells: the **epithelial network** of supporting cells and the **connective tissue network** of fibrocytes and cells of the stria vascularis.[76,77] Cx26 is the most prominent connexin expressed in the inner ear, although it is also expressed in various other organs, including the proximal tubules of the kidney, the liver and the rat placenta.[8]

Cx26 colocalizes with Cx30 between supporting cells, in the spiral limbus, the spiral ligament and the stria vascularis, where they may form heteromeric connexons.[78,79] Cx30 is found also in adult mouse brain and skin,[80] and its a.a. sequence shares 77% identity with that of Cx26.[81]

More than 50 distinct recessive mutations of *GJB2* have been described, including nonsense, missense, splicing, frame-shift mutations and in-frame deletions.[82] A large French dominant family affected by prelingual deafness showed linkage to chromosome 13q12 (DFNA3), the same region to which DFNB1 had been mapped.[83] An interaction between *GJB2* mutations and a mitochondrial mutation appears to be the cause of hearing impairment in some heterozygote patients.[84]

In addition to Cx26 and Cx30, several other connexins are known to be expressed in the inner ear. Cx31 (*GJB3*, 1p32-p36) localize to the inner ear connective tissue. It is expressed at P12 and reaches the adult pattern at P60.[85] Moreover, measurable mRNA levels have been detected in the stria vascularis for Cx43, Cx37, Cx30.2 and Cx46,[86] although is not clear if these connexins are expressed in the cochlea. Cx43 is expressed in the connective tissues only during development. However, from P8, Cx43 is almost exclusively expressed in the bone of the otic capsule.[87]

Not surprisingly, other connexin genes are also involved in deafness: *GJB1* (Cx32), which is also responsible for X-linked Charcot-Marie-Tooth disease type I;[88] *GJB3* (Cx31), involved in deafness[89] or a skin disease[60] (erythrokeratodermia variabilis) depending on the location of the mutation; *GJA1* (Cx43), involved in recessive deafness.[90]

Targeted Ablation of Inner Ear Connexins

Gene targeting of connexins in mice has provided new insights into connexin function and the significance of connexin diversity.[65] Complete removal of the Cx26 gene results in neonatal lethality, thereby preventing analysis of its function in hearing.[91] This limitation was overcome through the generation of a cochlear specific knockout of Cx26 using the Cre–*loxP* system,[92] successfully deleting Cx26 in the epithelial gap junction network without effecting Cx26 expression in other organs. Animals with this deletion are a model of recessive deafness and displayed normal patterns of cochlear development, but showed an increase in postnatal cell death within the cochlea along with significant hearing loss. The initiation of cell death was found to occur in supporting cells proximal to the inner hair cells, and coincided with the onset of audition. It was hypothesized that loss of Cx26 prevented recycling of K^+ after sound stimulation, and that elevated K^+ in the extracellular perilymph inhibited uptake of the neurotransmitter glutamate, which accumulated and resulted in cell death.[92] While the evidence concerning Cx26 is compelling, this hypothesis becomes more complex when considering the activity of Cx30 in the inner ear. Lautermann et al[93] have shown that Cx26 and Cx30 normally colocalize within the cochlea, and tissue-specific deletion of the Cx26 gene did not alter expression patterns of Cx30.[92] This observation complicates the role of gap junctional dysfunction in the onset of deafness due to the fact that Cx30 passes K^+ in an efficient manner[94] but could not prevent hearing loss in the absence of Cx26. Thus, the presence of a single type of connexin with similar ionic selectivity was unable to rescue the phenotype observed in these mice.

Similar to the tissue-specific loss of Cx26, deletion of Cx30 in mice resulted in hearing loss, but did not alter the development of the inner ear.[95] Although development was normal, Cx30 knockouts lacked the endocochlear potential, but maintained the $[K^+]$ of the endolymph. In addition, Cx30 knockouts also presented increased apoptosis within the cochlear sensory epithelium. Deletion of Cx30 did not alter the cochlear expression of Cx26, again raising the question of why the continued expression of Cx26 was not able to compensate for the loss of Cx30.[95]

One possible explanation for these phenotypes is that gap junctions may have other roles in addition to recycling K^+. This idea is supported by data from functional studies showing that, while the K^+ conductances of Cx26 and Cx30 are similar, the channels display significant differences regarding the permeability to fluorescent dyes that are similar to cyclic nucleotides and second messengers in both size and charge. These findings indicate that specific loss of either Cx26 or Cx30 within cochlear epithelial cells would not simply reduce the intercellular passage of K^+, but would also significantly alter the availability of larger solutes that could be exchanged between the coupled cells.[59]

Connexin Permeability Defects and Genetic Deafness

Connexin permeability defects may produce a deafness phenotype by interfering with potassium spatial buffering by cochlear supporting cells,[96] which most probably requires a coordinated activity. The identification in DFNB1 patients of a recessive Cx26 mutation, V84L, that did not appreciably affect basic channel properties, has recently provided some insight into the molecular mechanisms that underlie the disease. Using state-of-the-art techniques we demonstrated that V84L channels exhibit an impaired permeability to IP_3. The apparent permeability coefficient of V84L for IP_3 is only 8% of that of wild-type Cx26. Thus mutated channels must have a structural modification that reduces the passage of IP_3, without compromising that of other ions. Our report[25] is the first direct demonstration that a connexin mutation results selectively in a defective transfer of a second-messenger molecule. Modification of gap-junction permeability has been suggested, on the basis of indirect evidence, to be at the basis of some pathological connexin mutation, e.g., the S26L mutation of Cx32,[67] but no direct proof that this may be the case had yet been offered. In the organ of Corti, gap junction blockade impairs the spreading of Ca^{2+} waves (Fig. 6) and the formation of a functional

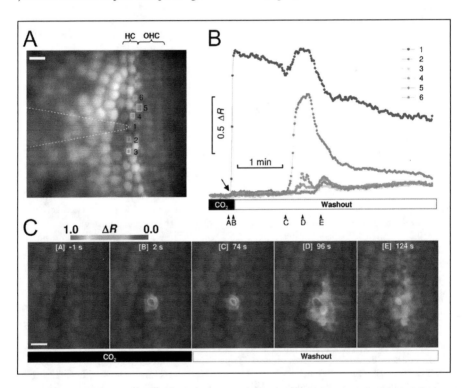

Figure 6. IP_3 injection elicits intercellular Ca^{2+} waves in supporting cells of the organ of Corti. A) Cell 1 was contacted by a patch pipette (outlined) loaded with 500 μM IP_3. B) Fura-2 ratio changes, ΔR, from the corresponding regions of interest in (A); intracellular delivery of IP_3 started 30 s after the onset of the recording, while the culture was exposed to CO_2 (solid black bar) to block gap junction channels. During washout (empty bar) the 1.5 ml chamber was superfused with normal extracellular saline, flowing at 2 ml/min. C) Selected frames from the sequence in (B) captured at the times shown, measured from the onset of IP_3 delivery. Letters in square brackets in (C) and below the arrowheads in (B) are matched. Suramin was omitted on purpose to allow paracrine propagation of the stimulus. Note spread of the Ca^{2+} rise to cells at distances >60 μm from the patch-clamped cell 2-3 min after washout. Scale bars, 20 μm. Reproduced with permission from Figure S1, Supplementary Information, Nat Cell Biol 2005; 7(1):63-9. A color version of this figure is available online at www.Eurekah.com.

syncytium. Wave propagation necessitates also a regenerative mechanism mediated by P2Y receptors[97] and this may represent a fundamental mechanism by which cochlear supporting cells coordinate their responses to sound.

Pannexins

Database search has led to the identification of a family of proteins, the pannexins, which share some structural features with the gap junction forming proteins of invertebrates and vertebrates.[98] The pannexin genes *PANX1*, *PANX2* and *PANX3*, encoding putative gap junction proteins homologous to invertebrate innexins, constitute a new family of mammalian proteins. In the brain, the pannexins may represent a novel class of electrical synapses.

Phylogenetic analysis revealed that pannexins are highly conserved in worms, molluscs, insects and mammals, pointing to their important function. Both innexins and pannexins are predicted to have four transmembrane regions, two extracellular loops, one intracellular loop and intracellular N and C termini. Both the human and mouse genomes contain three pannexin-encoding genes. Mammalian pannexins *PANX1* and *PANX3* are closely related, with *PANX2* more distant. The human and mouse *PANX1* mRNAs are ubiquitously, although disproportionately, expressed in normal tissues. Human *PANX2* is a brain-specific gene; its mouse orthologue, *Panx2*, is also expressed in certain cell types in developing brain. In silico evaluation of *Panx3* expression predicts gene expression in osteoblasts and synovial fibroblasts.[99]

On expression in Xenopus oocytes, pannexin 1 (Px1), but not Px2 forms functional hemichannels. Coinjection of both pannexin RNAs results in hemichannels with functional properties that are different from those formed by Px1 only. In paired oocytes, Px1, alone and in combination with Px2, induces the formation of intercellular channels. The functional characteristics of homomeric Px1 versus heteromeric Px1/Px2 channels and the different expression patterns of Px1 and Px2 in the brain indicate that pannexins form cell type-specific gap junctions with distinct properties that may subserve different functions.[98] There is evidence for the interaction of Px1 with Px2 and that the pharmacological sensitivity of heteromeric Px1/Px2 is similar to that of homomeric Px1 channels. In contrast to most connexins, both Px1 and Px1/Px2 hemichannels were not gated by external Ca^{2+}. In addition, they exhibited a remarkable sensitivity to blockade by carbenoxolone (with an IC_{50} of approximately 5 μM), whereas flufenamic acid exerted only a modest inhibitory effect, thus indicating that gap junction blockers are able to selectively modulate pannexin and connexin channels.[100] In the light of these findings, it is now necessary to consider pannexins as an alternative to connexins in vertebrate intercellular communication.[3]

Intercellular calcium wave propagation initiated by mechanical stress is a phenomenon found in nearly all cell types. However, the conduit for ATP has remained elusive and both a vesicular and a channel mediated release have been considered. Px1 channels are of large conductance and have been shown to be (a) permeant for ATP, and (b) mechanosensitive, suggesting that pannexins are candidates for the release of ATP to the extracellular space upon mechanical stress.[101]

Acknowledgements

Supported by grants from the Telethon Foundation, (Project n. GGP02043), Ministero dell'Università e Ricerca Scientifica (MIUR, FIRB n. RBAU01Z2Z8, PRIN-COFIN 2002067312_002) to F.M., Centro di Eccellenza (coordinator, Tullio Pozzan) and the Italian Health Ministry. I thank all present and past graduate students and post-doctoral fellows, directly or indirectly involved in work on the connexins, and in particular: Martina Beltramello, Valeria Piazza, Catalin D. Ciubotaru, Mario M. Bortolozzi, Victor H. Hernandez, Andrea Lelli, Sergio Pantano and Stefano Bastianello (Venetian Institute of Molecular Medicine, Padova, Italy). I also thank: Feliksas F. Bukauskas, Vytas Verselis, Michael V. L. Bennett (Department of Neuroscience, Albert Einstein College of Medicine, Bronx, New York), Paola D'Andrea (University of Trieste, Italy) and, last but not least, Roberto Bruzzone (Institut Pasteur, Paris, France) for helpful discussions.

References

1. Bennett MV. Physiology of electrotonic junctions. Ann NY Acad Sci 1966; 137:509-539.
2. Revel JP, Karnovsky MJ. Hexagonal array of subunits in intercellular junctions of the mouse heart and liver. J Cell Biol 1967; 33:C7-C12.
3. Panchin YV. Evolution of gap junction proteins—the pannexin alternative. J Exp Biol 2005; 208:1415-1419.
4. Harris AL. Emerging issues of connexin channels: Biophysics fills the gap. Q Rev Biophys 2001; 34:325-472.
5. Willecke K, Eiberger J, Degen J et al. Structural and functional diversity of connexin genes in the mouse and human genome. Biol Chem 2002; 383:725-737.
6. Beyer EC, Paul DL, Goodenough DA. Connexin family of gap junction proteins. J Membr Biol 1990; 116:187-194.
7. Kumar NM, Gilula NB. The gap junction communication channel. Cell 1996; 84:381-388.
8. Beyer EC, Willecke K. Gap junction genes and their regulation. In: Bittar EE, Hel, eds. Gap Junctions. Vol 30. Stamford, Connecticut: Jai Press Inc., 2000:1-29.
9. Bruzzone R, White TW, Paul DL. Connections with connexins: The molecular basis of direct intercellular signaling. Eur J Biochem 1996; 238:1-27.
10. Martin PE, Evans WH. Incorporation of connexins into plasma membranes and gap junctions: Cardiovasc. Res 2004; 62:378-387.
11. Unger VM, Kumar NM, Gilula NB et al. Three-dimensional structure of a recombinant gap junction membrane channel. Science 1999; 283:1176-1180.
12. Yeager M, Unger VM. Culturing of mammalian cells expressing recombinant connexins and two-dimensional crystallization of the isolated gap junctions. Methods Mol Biol 2001; 154:77-89.
13. Skerrett IM, Aronowitz J, Shin JH et al. Identification of amino acid residues lining the pore of a gap junction channel. J Cell Biol 2002; 159:349-360.
14. Fleishman SJ, Unger VM, Yeager M et al. A c-alpha model for the transmembrane alpha helices of gap junction intercellular channels. Mol Cell 2004; 15:879-888.
15. George CH, Kendall JM, Evans WH. Intracellular trafficking pathways in the assembly of connexins into gap junctions. J Biol Chem 1999; 274:8678-8685.
16. Segretain D, Falk MM. Regulation of connexin biosynthesis, assembly, gap junction formation, and removal. Biochim Biophys Acta 2004; 1662:3-21.
17. Falk MM, Buehler LK, Kumar NM et al. Cell-free synthesis and assembly of connexins into functional gap junction membrane channels. EMBO J 1997; 16:2703-2716.
18. Goldberg GS, Valiunas V, Brink PR. Selective permeability of gap junction channels. Biochim Biophys Acta 2004; 1662:96-101.
19. Bedner P, Niessen H, Odermatt B et al. A method to determine the relative camp permeability of connexin channels. Exp Cell Res 2003; 291:25-35.
20. Lawrence TS, Beers WH, Gilula NB. Transmission of hormonal stimulation by cell-to-cell communication. Nature 1978; 272:501-506.
21. Saez JC, Connor JA, Spray DC et al. Hepatocyte gap junctions are permeable to the second messenger, inositol 1,4,5-trisphosphate, and to calcium ions. Proc Natl Acad Sci USA 1989; 86:2708-2712.
22. Sanderson MJ. Intercellular calcium waves mediated by inositol trisphosphate. Ciba Found Symp 1995; 188:175-189, (189-194).
23. Weber PA, Chang HC, Spaeth KE et al. The permeability of gap junction channels to probes of different size is dependent on connexin composition and permeant-pore affinities. Biophys J 2004; 87:958-973.
24. Elfgang C, Eckert R, Lichtenberg-Frate H et al. Specific permeability and selective formation of gap junction channels in connexin-transfected hela cells. J Cell Biol 1995; 129:805-817.
25. Beltramello M, Piazza V, Bukauskas FF et al. Impaired permeability to ins(1,4,5)p(3) in a mutant connexin underlies recessive hereditary deafness. Nat Cell Biol 2005; 7:63-69.
26. Qu Y, Dahl G. Function of the voltage gate of gap junction channels: Selective exclusion of molecules. Proc Natl Acad Sci USA 2002; 99:697-702.
27. Niessen H, Willecke K. Strongly decreased gap junctional permeability to inositol 1,4, 5-trisphosphate in connexin32 deficient hepatocytes. Febs Lett 2000; 466:112-114.
28. Niessen H, Harz H, Bedner P et al. Selective permeability of different connexin channels to the second messenger inositol 1,4,5-trisphosphate. J Cell Sci 2000; 113:1365-1372.
29. Valiunas V, Beyer EC, Brink PR. Cardiac gap junction channels show quantitative differences in selectivity. Circ Res 2002; 91(2):104-111.
30. Bukauskas FF, Verselis VK. Gap junction channel gating. Biochim Biophys Acta 2004; 1662:42-60.

31. Peracchia C. Chemical gating of gap junction channels; roles of calcium, ph and calmodulin. Biochim Biophys Acta 2004; 1662:61-80.
32. Verselis V, White RL, Spray DC et al. Gap junctional conductance and permeability are linearly related. Science 1986; 234:461-464.
33. Bennett MV. Gap junctions as electrical synapses. J Neurocytol 1997; 26:349-366.
34. Suchyna TM, Xu LX, Gao F et al. Identification of a proline residue as a transduction element involved in voltage gating of gap junctions. Nature 1993; 365:847-849.
35. Oh S, Rivkin S, Tang Q et al. Determinants of gating polarity of a connexin 32 hemichannel. Biophys J 2004; 87:912-928.
36. Veenstra RD, Dehaan RL. Measurement of single channel currents from cardiac gap junctions. Science 1986; 233:972-974.
37. Peracchia C, Wang XG, Peracchia LL. Chemical gating of gap junction channels. Methods 2000; 20:188-195.
38. Sosinsky GE, Gaietta GM, Hand G et al. Tetracysteine genetic tags complexed with biarsenical ligands as a tool for investigating gap junction structure and dynamics. Cell Commun Adhes 2003; 10:181-186.
39. Lampe PD, Lau AF. The effects of connexin phosphorylation on gap junctional communication. Int J Biochem Cell Biol 2004; 36:1171-1186.
40. Warn-Cramer BJ, Lau AF. Regulation of gap junctions by tyrosine protein kinases. Biochim Biophys Acta 2004; 1662:81-95.
41. Traub O, Look J, Dermietzel R et al. Comparative characterization of the 21-kD and 26-kD gap junction proteins in murine liver and cultured hepatocytes. J Cell Biol 1989; 108:1039-1051.
42. Barrio LC, Suchyna T, Bargiello T et al. Gap junctions formed by connexins 26 and 32 alone and in combination are differently affected by applied voltage. Proc Natl Acad Sci USA 1991; 88:8410-8414.
43. Berridge MJ, Dawson RM, Downes CP et al. Changes in the levels of inositol phosphates after agonist-dependent hydrolysis of membrane phosphoinositides. Biochem J 1983; 212:473-482.
44. Allbritton NL, Meyer T, Stryer L. Range of messenger action of calcium ion and inositol 1,4,5-trisphosphate. Science 1992; 258:1812-1815.
45. Streb H, Irvine RF, Berridge MJ et al. Release of Ca^{2+} from a nonmitochondrial intracellular store in pancreatic acinar cells by inositol-1,4,5-trisphosphate. Nature 1983; 306:67-69.
46. Irvine RF. 20 Years of ins(1,4,5)p3, and 40 years before. Nat Rev Mol Cell Biol 2003; 4:586-590.
47. Cornell-Bell AH, Finkbeiner SM, Cooper MS et al. Glutamate induces calcium waves in cultured astrocytes: Long-range glial signaling. Science 1990; 247:470-473.
48. Osipchuk Y, Cahalan M. Cell-to-cell spread of calcium signals mediated by ATP receptors in mast cells. Nature 1992; 359:241-244.
49. Hassinger TD, Guthrie PB, Atkinson PB et al. An extracellular signaling component in propagation of astrocytic calcium waves. Proc Natl Acad Sci USA 1996; 93:13268-13273.
50. Schuster S, Marhl M, Hofer T. Modelling of simple and complex calcium oscillations. From single-cell responses to intercellular signalling. Eur J Biochem 2002; 269:1333-1355.
51. Charles AC, Naus CC, Zhu D et al. Intercellular calcium signaling via gap junctions in glioma cells. J Cell Biol 1992; 118:195-201.
52. Sneyd J, Charles AC, Sanderson MJ. A model for the propagation of intercellular calcium waves. Am J Physiol 1994; 266:C293-302.
53. Sneyd J, Wetton BT, Charles AC et al. Intercellular calcium waves mediated by diffusion of inositol trisphosphate: A two-dimensional model. Am J Physiol 1995; 268:C1537-1545.
54. Robb-Gaspers LD, Thomas AP. Coordination of Ca^{2+} signaling by intercellular propagation of Ca^{2+} waves in the intact liver. J Biol Chem 1995; 270:8102-8107.
55. Sanderson MJ, Charles AC, Boitano S et al. Mechanisms and function of intercellular calcium signaling. Mol Cell Endocrinol 1994; 98:173-187.
56. Carmignoto G. Reciprocal communication systems between astrocytes and neurones. Prog Neurobiol 2000; 62:561-581.
57. Churchill GC, Louis CF. Roles of Ca^{2+}, inositol trisphosphate and cyclic adp-ribose in mediating intercellular Ca^{2+} signaling in sheep lens cells. J Cell Sci 1998; 111:1217-1225.
58. Brissette JL, Kumar NM, Gilula NB et al. Switch in gap junction protein expression is associated with selective changes in junctional permeability during keratinocyte differentiation. Proc Natl Acad Sci USA 1994; 91:6453-6457.
59. Gerido DA, White TW. Connexin disorders of the ear, skin, and lens. Biochim Biophys Acta 2004; 1662:159-170.
60. Richard G, Smith LE, Bailey RA et al. Mutations in the human connexin gene GJB3 cause erythrokeratodermia variabilis. Nat Genet 1998; 20:366-369.

61. Kelsell DP, Di WL, Houseman MJ. Connexin mutations in skin disease and hearing loss. Am J Hum Genet 2001; 68:559-568.
62. Richard G. Connexin disorders of the skin. Adv Dermatol 2001; 17:243-277.
63. Lamartine J, Munhoz Essenfelder G, Kibar Z et al. Mutations in GJB6 cause hidrotic ectodermal dysplasia. Nat Genet 2000; 26:142-144.
64. Common JE, Becker D, Di WL et al. Functional studies of human skin disease- and deafness-associated connexin 30 mutations. Biochem Biophys Res Commun 2002; 298:651-656.
65. White TW, Paul DL. Genetic diseases and gene knockouts reveal diverse connexin functions. Annu Rev Physiol 1999; 61:283-310.
66. Bergoffen J, Scherer SS, Wang S et al. Connexin mutations in x-linked charcot-marie-tooth disease. Science 1993; 262:2039-2042.
67. Oh S, Ri Y, Bennett MV et al. Changes in permeability caused by connexin 32 mutations underlie x-linked charcot-marie-tooth disease. Neuron 1997; 19:927-938.
68. Balice-Gordon RJ, Bone LJ, Scherer SS. Functional gap junctions in the schwann cell myelin sheath. J Cell Biol 1998; 142:1095-1104.
69. Lopez-Bigas N, Olive M, Rabionet R et al. Connexin 31 (GJB3) is expressed in the peripheral and auditory nerves and causes neuropathy and hearing impairment. Hum Mol Genet 2001; 10:947-952.
70. White TW, Goodenough DA, Paul DL. Targeted ablation of connexin50 in mice results in microphthalmia and zonular pulverulent cataracts. J Cell Biol 1998; 143:815-825.
71. White TW, Bruzzone R. Gap Junctions: Fates Worse Than Death? Curr Biol 2000; 10:R685-688.
72. Martinez-Wittinghan FJ, Sellitto C, Li L et al. Dominant cataracts result from incongruous mixing of wild-type lens connexins. J Cell Biol 2003; 161:969-978.
73. Petit C, Levilliers J, Hardelin JP. Molecular genetics of hearing loss. Annu Rev Genet 2001; 35:589-646.
74. Xia JH, Liu CY, Tang BS et al. Mutations in the gene encoding gap junction protein beta-3 associated with autosomal dominant hearing impairment. Nat Genet 1998; 20:370-373.
75. Grifa A, Wagner CA, D'ambrosio L et al. Mutations in GJB6 cause nonsyndromic autosomal dominant deafness at DFNA3 locus. Nat Genet 1999; 23:16-18.
76. Kikuchi T, Adams JC, Paul DL et al. Gap junction systems in the rat vestibular labyrinth: Immunohistochemical and ultrastructural analysis. Acta Otolaryngol 1994; 114:520-528.
77. Kikuchi T, Kimura RS, Paul DL et al. Gap junctions in the rat cochlea: Immunohistochemical and ultrastructural analysis. Anat Embryol (Berl) 1995; 191:101-118.
78. Forge A, Marziano NK, Casalotti SO et al. The inner ear contains heteromeric channels composed of Cx26 and Cx30 and deafness-related mutations in Cx26 have a dominant negative effect on Cx30. Cell Commun Adhes 2003; 10:341-346.
79. Rabionet R, Gasparini P, Estivill X. Molecular genetics of hearing impairment due to mutations in gap junction genes encoding beta connexins. Hum Mutat 2000; 16:190-202.
80. Dahl E, Manthey D, Chen Y et al. Molecular cloning and functional expression of mouse connexin-30,a gap junction gene highly expressed in adult brain and skin. J Biol Chem 1996; 271:17903-17910.
81. Kelley PM, Abe S, Askew JW et al. Human connexin 30 (GJB6), a candidate gene for nonsyndromic hearing loss: molecular cloning, tissue-specific expression, and assignment to chromosome 13q12. Genomics 1999; 62:172-176.
82. Richard G. Connexin gene pathology. Clin Exp Dermatol 2003; 28:397-409.
83. Chaib H, Lina-Granade G, Guilford P et al. A gene responsible for a dominant form of neurosensory nonsyndromic deafness maps to the NSRD1 recessive deafness gene interval. Hum Mol Genet 1994; 3:2219-2222.
84. Abe S, Kelley PM, Kimberling WJ et al. Connexin 26 gene (GJB2) mutation modulates the severity of hearing loss associated with the 1555a—>G mitochondrial mutation. Am J Med Genet 2001; 103:334-338.
85. Forge A, Becker D, Casalotti S et al. Gap junctions in the inner ear: Comparison of distribution patterns in different vertebrates and assessement of connexin composition in mammals. J Comp Neurol 2003; 467:207-231.
86. Buniello A, Montanaro D, Volinia S et al. An expression atlas of connexin genes in the mouse. Genomics 2004; 83:812-820.
87. Cohen-Salmon M, Maxeiner S, Kruger O et al. Expression of the connexin43- and connexin45-encoding genes in the developing and mature mouse inner ear. Cell Tissue Res 2004; 316:15-22.
88. Stojkovic T, Latour P, Vandenberghe A et al. Sensorineural Deafness In X-Linked Charcot-Marie-Tooth Disease With Connexin 32 Mutation (R142q). Neurology 1999; 52:1010-1014.
89. Liu XZ, Xia XJ, Xu LR et al. Mutations in connexin31 underlie recessive as well as dominant nonsyndromic hearing loss. Hum Mol Genet 2000; 9:63-67.

90. Liu XZ, Xia XJ, Adams J et al. Mutations in GJA1 (Connexin 43) are associated with nonsyndromic autosomal recessive deafness. Hum Mol Genet 2001; 10:2945-2951.
91. Gabriel HD, Jung D, Butzler C et al. Transplacental uptake of glucose is decreased in embryonic lethal connexin26-deficient mice. J Cell Biol 1998; 140:1453-1461.
92. Cohen-Salmon M, Ott T, Michel V et al. Targeted ablation of connexin26 in the inner ear epithelial gap junction network causes hearing impairment and cell death. Curr Biol 2002; 12:1106-1111.
93. Lautermann J, Ten Cate WJ, Altenhoff P et al. Expression of the gap-junction connexins 26 and 30 in the rat cochlea. Cell Tissue Res 1998; 294:415-420.
94. Valiunas V, Manthey D, Vogel R et al. Biophysical properties of mouse connexin30 gap junction channels studied in transfected human hela cells. J Physiol 1999; 519:631-644.
95. Teubner B, Michel V, Pesch J et al. Connexin30 (Gjb6)-deficiency causes severe hearing impairment and lack of endocochlear potential. Hum Mol Genet 2003; 12:13-21.
96. Jentsch TJ. Neuronal KCNQ potassium channels: Physiology and role in disease. Nat Rev Neurosci 2000; 1:21-30.
97. Gale JE, Piazza V, Ciubotaru CD et al. A mechanism for sensing noise damage in the inner ear. Curr Biol 2004; 14:526-529.
98. Bruzzone R, Hormuzdi SG, Barbe MT et al. Pannexins, a family of gap junction proteins expressed in brain. Proc Natl Acad Sci USA 2003; 100:13644-13649.
99. Baranova A, Ivanov D, Petrash N et al. The mammalian pannexin family is homologous to the invertebrate innexin gap junction proteins. Genomics 2004; 83:706-716.
100. Bruzzone R, Barbe MT, Jakob NJ et al. Pharmacological properties of homomeric and heteromeric pannexin hemichannels expressed in xenopus oocytes. J Neurochem 2005; 92:1033-1043.
101. Bao L, Locovei S, Dahl G. Pannexin membrane channels are mechanosensitive conduits for ATP. FEBS Lett 2004; 572:65-68.
102. Beltramello M, Bicego M, Piazza V et al. Permeability and gating properties of human connexins 26 and 30 expressed in HeLa cells. Biochem Biophys Res Commun 2003; 305:1024-1033.

CHAPTER 14

Tunneling Nanotubes:
Membranous Channels between Animal Cells

Hans-Hermann Gerdes* and Amin Rustom

Abstract

Intercellular communication is a major requirement for the development and maintainance of multicellular organisms. Diverse mechanisms for the exchange of signals between cells during evolution have been established. These mechanisms include intercellular membrane channels between plant cells, called plasmodesmata, and proteinaceous channels of animal cells, called gap junctions. Recently, highly sensitive nanotubular structures have been described which are formed de novo between animal cells resulting in the formation of complex cellular networks. These membrane channels mediate membrane continuity between connected cells and are referred to as tunneling nanotubes (TNTs). They have been shown to facilitate the intercellular transfer of organelles as well as, on a limited scale, of membrane components and cytoplasmic molecules. It has been proposed that TNTs represent a novel and general biological principle of cell interaction based on membrane continuity and the intercellular exchange of organelles. It is increasingly apparent that TNTs and TNT-related structures fulfill important functions in the physiological processes of multicellular organisms.

Introduction

Since Schleiden and Schwann formulated the cell theory in the years 1838-39, animals have been regarded as an assembly of individual, membrane-bounded entities representing the building blocks of all higher organisms. Exceptions to this rule only became apparent during specific developmental stages (this book) or under certain conditions such as viral infection when animal cells are caused to form syncytia. The theory of individual building blocks contrasts with the view of multicellular organisms in the plant kingdom which, since the discovery of plasmodesmata more than 100 years ago, are regarded as an intercellular continuum where cells are interconnected via membranous channels. These differing concepts now seem to converge in an exciting new concept. The motivation for this rethinking was the discovery that animal cells also establish thin membrane channels which, like the structurally related plasmodesmata in plants, mediate membrane continuity between connected cells.[1] With respect to their peculiar architecture, these structures were termed tunneling nanotubes (TNTs). This discovery has initiated reconsiderations of previous concepts of cellular interactions. The primary proposal that TNTs represent a general phenomenon of cell-to-cell communication in the animal kingdom was soon supported by subsequent studies showing comparable membrane channels in variegated cellular systems (Table 1), including primary cultures such

*Corresponding Author: Hans-Hermann Gerdes—Department of Biomedicine, Section for Biochemistry and Molecular Biology, University of Bergen, Jonas Lies vei 91, Bergen 5009, Norway, and Interdisciplinary Center of Neuroscience (IZN), Institute of Neurobiology, University of Heidelberg, INF 364, Heidelberg 69120, Germany.
Email: hans-hermann.gerdes@biomed.uib.no

Cell-Cell Channels, edited by Frantisek Baluska, Dieter Volkmann and Peter W. Barlow.
©2006 Landes Bioscience and Springer Science+Business Media.

Table 1. Cell cultures shown to form TNTs

Permanent Cell-Lines	Primary Cultures
- rat pheochromocytoma PC12 cells[1]	- between adult human endothelial progenitor cells and neonatal rat cardiomyocytes[2]
- human embryonic kidney (HEK) cells[1]	- transformed human B cell line 721.221[3]
- normal rat kidney (NRK) cells[1]	- human macrophages prepared from peripheral blood[3]
- between PC12 and HEK cells*	- murin macrophage J774 cells[3]
- between HEK and NRK cells*	- human peripheral blood NK cells + 721.221[3]
- between PC12 and NRK cells*	- primary cultures of medulla*
- THP-1 monocytes[4]	- primary cultures of astrocytes[11]
	- dendritic cells cultured from peripheral blood monocytes[4]

* S. Gurke, A. Rustom, H.H. Gerdes unpublished data.

as cardiac myocytes.[2] Additionally, these studies point to important physiological functions of these novel connections in cell-to-cell communication. In the following, we summarize the current knowledge of TNT and TNT-related membrane channels by focusing on three of their major aspects: structure, formation and function.

Structure

TNTs were first characterized between cultured rat pheochromocytoma PC12 cells (Fig. 1A).[1] Here they represent thin membranous channels mediating continuity between the plasma membranes of connected cells (Fig. 1).[1] The diameter of these channels ranges between 50 and 200 nm and their length reaches up to several cell diameters,[1] in rare cases more than 300 μm (A. Rustom, H.H. Gerdes, unpublished data). TNT-related structures described for other cultured cell lines possess similar dimensions. For macrophages, Epstein Barr Virus (EBV) transformed human B cells or peripheral blood natural killer (NK) cells, an average length of 30 μm and in some cases above 140 μm has been reported.[3] In the case of human endothelial progenitor cells (EPCs) and neonatal rat cardiac myocytes, their diameter ranges from 50 to 800 nm and their length from 5 to 120 μm.[2] The length of TNTs between THP-1 monocytes was observed to be up to 100 μm.[4] For PC12 cells, TNTs are stretched between interconnected cells attached at their nearest distance.[1] Thus, in contrast to other cellular protrusions, TNTs are not associated with the substrate, but hover freely in the cell culture medium. This peculiar morphology of TNTs is less evident the flatter and the more irregular-shaped the respective cells are (S. Gurke, A. Rustom, H.H. Gerdes, unpublished data). In retrospect, we believe that the apparent spheroidal shape of PC12 cells significantly contributed to the discovery of TNTs by emphasizing their unique characteristics among the great variety of different cellular protrusions. It is of note that these characteristics share striking similarities with artificial phospholipid nanotubes generated between giant unilamellar liposomes mediating membrane continuity.[5] Whereas in most cases TNTs and TNT-related structures appeared as straight lines, the occurrence of branched connections could only be observed in rare cases.[1,2]

Another structural property of TNTs in PC12 cell cultures is the prominent F-actin bundle tightly wrapped by the surrounding plasma membrane.[1] Microtubules have not been found in TNTs connecting PC12 cells.[1] In the case of immune cells, some membrane tubes appear to contain F-actin while others appear to have both microtubules and F-actin (D. Davis, unpublished data). The presence of microtubules was also observed in membrane channels between cultured prostate cancer cells.[6]

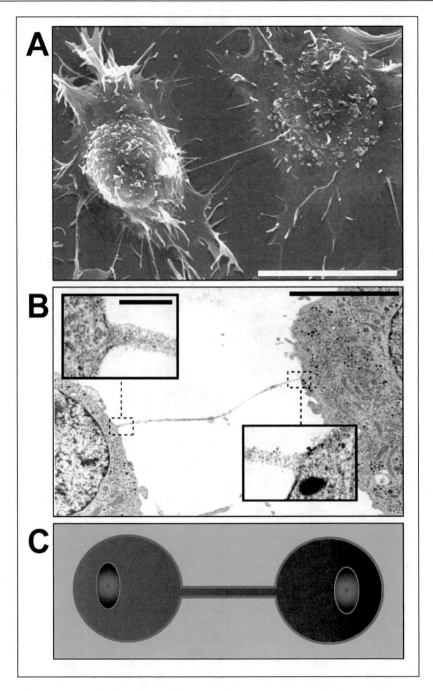

Figure 1. Ultrastructure and schematic representation of tunneling nanotubes (TNTs). A) Scanning electron microscopic (SEM) or B) transmission electron microscopic (TEM) image of a TNT connecting two cultured PC12 cells. The inserts in (B) depict higher resolution images of both TNT bases. C) Schematic representation of a TNT. Bars: 20 μm (A); 10 μm (B); 200 nm (insets).

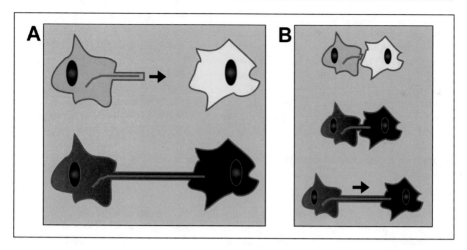

Figure 2. Model of TNT formation. A) One cell forms an actin-driven protrusion directed towards the target cell (top). Fusion of the cell protrusion with the membrane of the target cell results in TNT formation (bottom). B) TNTs may form between adjacent cells which subsequently diverge.

A notable property of TNTs between PC12 cells is their extreme sensitivity to mechanical stress, chemical fixation and prolonged light exposure.[1] The latter leads to visible vibrations and subsequent rupture of the thin membrane tubes.[1] Likewise, TNT-related structures between other cell types, e.g., cardiac myocytes, endothelial progenitor cells and THP-1 monocytes, have been shown to display a similar sensitivity.[2,4] This pronounced sensitivity may be one reason why theses structures were not recognized earlier and could hamper further structural and functional analyses.

Formation

One of the most exciting characteristics of TNTs in PC12 cell cultures is that they can be formed de novo between dislodged cells.[1] This fact contrasts with the generation of other known cell-to-cell channels of animal cells such as the formation of ring canals[7] during *Drosophila* gametogenesis as well as the generation of primary plasmodesmata of plant cells which derive from incomplete cytokinesis.[8]

The de novo formation of TNTs and TNT-related structures seems to be initiated by the outgrowth of filopodia-like cell protrusions (Fig. 2A, top). In the case of macrophages, the velocity of this outgrowth was determined to be approximately 0.2 μm per second,[3,6] which is in the velocity range of actin-driven processes.[9] Interestingly, in the case of PC12 cells, the growing filopodia-like structures seem to be directed towards the target cell.[1] This may point to a chemo-tactical guidance of the protrusions or the involvement of other signals. In support of this view, a directed outgrowth of long and thin cellular protrusions referred to as cytonemes towards a basic fibroblast growth factor (bFGF) gradient has been observed in in vitro cultures of drosophila cells.[10] The presence of F-actin in TNTs suggests that the directed outgrowth is mediated by actin polymerization. This notion is further corroborated by the absence of TNTs between PC12 cells upon treatment with latrunculin B, an actin-depolymerising drug.[1]

After a filopodia-like structure encounters the target cell, both membranes fuse to form a stretched membrane connection (Fig. 2A, bottom). In the case of PC12 cells, this process was accompanied by the degeneration of the remaining protrusions, possibly reflecting the existence of a feedback mechanism.[1] The complete process of TNT formation between PC12 cells is accomplished within a few minutes.[1] Interestingly, TNT formation is not only a single event restricted to pairs of cells, but frequently leads to the formation of local networks between groups of cells.[1,4] Theoretical calculations based on transfer experiments using fluorescent membrane dyes suggest that most cultured

PC12 cells are connected via TNTs (A. Rustom, H.H. Gerdes, unpublished data). In the case of PC12 cells, network formation starts directly after plating of individual cells and culminates after two hours of plating.[1] However, this time schedule presumably differs with respect to the growth conditions and/or cell lines which are under investigation. This is exemplified for cardiac myocytes and EPCs, where TNT-related structures first became visible after a few hours, and whose formation culminated after 24 hours and declined after 48 hours.[2] Because TNTs and TNT-related structures were frequently found between diverging cells, it is possible that they also exist between associated cells. This may indicate that a preexisting membrane channel between cells which are in close contact is elongated as they start to diverge from one another (Fig. 2C). It can be envisioned that the formation of TNTs is a regulated process. This view is supported by the recent finding that H_2O_2, suggested to trigger the phosphorylation of the p38 mitogen-activated protein kinase (MAPK), promoted the formation of TNT-like structures in primary astrocyte cultures of rat.[11]

Function(s)

One of the most important questions concerning intercellular membrane channels concerns the functions they play in cell-to-cell communication. In this context, the intercellular transfer of cellular components, as has been shown for plasmodesmata and gap junctions, is one apparent function. In contrast to plasmodesmata and gap junctions, which mediate the transfer of molecules between adjacent cells, TNTs and TNT-related structures were found also to accomplish long-range communication between dislodged cells. This function was first suggested by video-microscopic studies of neuroendocrine PC12 cells. Differential interference contrast (DIC) microscopy showed bulges moving uni-directionally along TNTs at a speed of approximately 25 nm/s.[1] This process showed striking similarities to the demonstrated transfer of lipid containers between liposomes connected via phospholipid bilayer nanotubes.[5] The assumption that the moving bulges of PC12 cells indeed represent membrane containers was subsequently verified by transfer experiments employing fluorescent markers for different cellular organelles. These experiments, performed with two differentially labeled PC12 cell populations, showed that the moving containers represent small organelles belonging to the endosomal/lysosomal system (Fig. 3).[1] This, together with the finding that F-actin is present inside TNTs, suggests that the movement of organelles along TNTs is facilitated by an actin/myosin-dependent mechanism (Fig. 3). Such a view is further corroborated by the presence of the actin-specific motor protein myosin Va in TNTs, partly colocalising with the respective organelles.[1] Alternatively, the transfer of organelles could be driven by the polymerization of actin itself, which could be regarded as a process of intercellular "treadmilling". According to this model, organelles are transferred by their association with the constantly sliding F-actin rod. Irrespective of the existent mechanism, the uni-directional transport observed implies that the individual actin fibers involved inside the tubes have the same polarity.

Apart from the transfer of small organelles, subsequent studies showed that TNT-related structures also facilitate the intercellular transfer of larger organelles such as mitochondria, from cardiac myocytes to endothelial progenitor cells.[2] Like the uni-directional transport of endosome-related structures between PC12 cells, this transport seems to occur in one direction only.[2] It is tempting to speculate that the cell which initiates the formation of a membrane tube, is also the donor cell for the delivery of organelles. In this respect, organelle transfer along membrane tubes could also be an explanation for the thus far unexplained transfer of melanosomes from melanocytes to keratinocytes[12] as well as for the intercellular transfer of prion proteins incorporated into vesicular structures of neuronal cells.[13] More generally, it is conceivable that any molecules packaged into or associated with endosomal structures can be transferred over long distances between cells. This opens an alternative route for e.g., the selective distribution of morphogens during developmental processes. Intercellular transport of morphogens through tubular connections via membrane containers, in addition to passive diffusion through extracellular space, seems plausible because morphogens such as *sonic hedgehog* are often closely membrane-associated.[14-16]

The occurrence of an intercellular transfer of organelles raises the question as to what kind of information is transferred and how this information is integrated in the target cells. First, it is notable that the early endosomal system in particular is one of the major reloading points for a

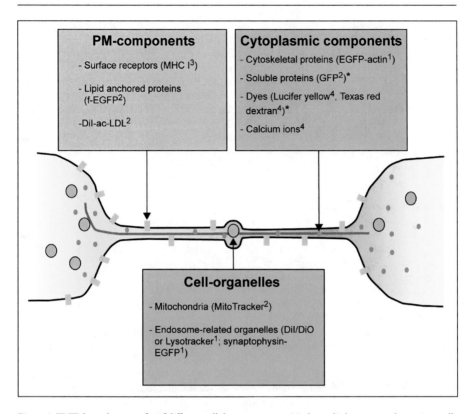

Figure 3. TNT-dependent transfer of different cellular components. Markers which were tested experimentally appear in brackets. The asterisk indicates the transfer of a marker which appears to be cell-type dependent.

variety of signaling complexes functioning at the cell surface. Second, transfer studies using PC12 cells revealed that after being transferred along TNTs, endosome-related structures are able to fuse with their counterparts in the respective target cells.[1] These facts provoke the intriguing model that the transfer of organelles and their subsequent fusion with membranes of the target cell conveys information between connected cells.

Initial studies of PC12 cells showed that in addition to organelles, plasma membrane components such as lipid-anchored proteins are also transported along TNTs to the connected target cells (Fig. 3).[1] This lateral transfer in the plane of the cell surface is consistent with membrane continuity between the connected cells. In the case of PC12 cells, membrane transfer is quite slow as has been shown by the intercellular transport of the lipid-anchored membrane marker f-EGFP.[1] The slow transfer of membrane components from thin tubular projections to spherical cell surfaces has also been predicted by mathematical models of the process.[17] Nevertheless, in the case of cell signaling molecules, the lateral transfer of only a few molecules can fulfill important roles in intercellular communication. In this respect, the detection of MHC I complexes in TNTs between immune cells could point to a crucial role in the proper functioning of the immune system, in particular of the immunological synapse.[3,18]

The existence of membrane continuity mediated by membrane channels should result in the occurrence of a cytoplasmic bridge that, in principle, facilitates the transfer of cytoplasmic material between interconnected cells. For this reason, it was surprising to find that TNTs between PC12 cells seem to impede the intercellular flow of even small molecules such as the 400 dalton dye molecule calcein.[1] This suggests that the passive diffusion of small cytoplasmic molecules through

TNTs is significantly slowed. However, the transfer of actin as the structural backbone of TNTs could be detected,[1] a finding which is consistent with the proposed intercellular "treadmilling" process. It is possible that the F-actin rod tightly wrapped by the surrounding membrane functions as a plug for passive diffusion. However, Lucifer yellow was transferred TNT-dependent between THP-1 cells[4] and GFP seems to be transferred between TNT-connected cardiac myocytes and EPCs.[2] Furthermore, calcium fluxes through relatively large diameter TNTs were observed between dendritic cells and THP-1 monocytes.[4] This may indicate that TNTs or TNT-related structures possess different permeabilities for soluble molecules which in turn may reflect the existence of a gating mechanism or structural differences between the respective intercellular structures of different cell types. In agreement with this view, different permeabilities were also found for structurally related plasmodesmata, e.g., during different stages of plant development.[8] In addition, gating mechanism(s) have been proposed.[8]

Implications and Outlook

The discovery of TNTs has revealed a novel type of intercellular communication in animal organisms. In particular, the ability to establish selective long-range transfer of various cellular components such as organelles, membrane constitutents and soluble molecules between dislodged cells points to important physiological functions which remain to be investigated. In this respect, the striking similarities between TNTs and plasmodesmata strongly suggests that knowledge of the functions of the latter can be instructive for understanding the role of the former and vice versa. In light of the finding that plasmodesmata play fundamental roles in pattern-formation during plant development, e.g., by RNA transfer,[8] it can be anticipated that TNTs and TNT-related structures also orchestrate essential parts of developmental and differentiation processes in animal organisms, as has recently been shown for the differentiation of endothelial progenitor cells.[2] Apart from these processes, TNTs could also play a crucial role in the maintenance of animal organisms and especially in defence mechanisms. In support of this notion, two recent studies present evidence for a nanotube-dependent long distance communication between various cell of the immune system.[3,4] If TNTs and related structures indeed play key roles in cell-to-cell signaling, it can be expected that alterations in their structure and function may lead to severe pathological defects in the respective organisms. This may be relevant for e.g., the development and growth of cancer as one of the most severe modern disorders. In this respect, it has been reported recently that the P-glycoprotein, which mediates multidrug resistance, is transferred between different tumor cells in vitro and in vivo.[19] Although the underlying mechanism for this intercellular transfer is still unknown,[19] the data presented is consistent with a TNT-dependent transfer mechanism.

Another conceivable scenario is that TNTs are misused as backdoors for the intercellular spread of pathogens like viruses, bacteria and prions. In a recent study showing the intercellular transfer of the protease resistant prion protein (PrP-res) between mouse septum neurons and primary cultures of adult hamster cortical neurons, it has been speculated that such a misuse of TNTs may indeed occur.[13] Here, the analogy to the plant system, where the transfer of pathogens through plasmodesmata has been characterized in much detail,[8] is once again striking.

Aside from all the speculation about the physiological role of TNTs, the main challenge for the future will be to demonstrate their existence in tissue. If they indeed exist in cell assemblies of multicellular organisms, the question emerges as to how they are configured. It is almost certain that, due to space limitations, TNTs in tissue will not display the stretched configuration shown in dissociated cell cultures but may have a wound configuration following the clefts between cells. Innovative approaches have to come into operation to trace TNTs in their natural settings.

Acknowledgements

We thank S. Dimmeler and D. Davis for sharing unpublished data, S. Leichtle for critical comments on the manuscript, and H. Bading for space and infrastructure at the Department of Neurobiology Heidelberg.

References

1. Rustom A, Saffrich R, Markovic I et al. Nanotubular Highways for Intercellular Organelle Transport. Science 2004; 303:1007-1010.
2. Koyanagi M, Brandes RP, Haendeler J et al. Cell-to-cell connection of endothelial progenitor cells with cardiac myocytes by nanotubes: a novel mechanism for cell fate changes? Circ Res 2005; 96(10):1039-1041.
3. Onfelt B, Nedvetzki S, Yanagi K et al. Cutting edge: Membrane nanotubes connect immune cells. J Immunol 2004; 173(3):1511-1513.
4. Watkins SC, Salter RD. Functional Connectivity between Immune Cells Mediated by Tunneling Nanotubules. Immunity 2005; 23(3):309-318.
5. Karlsson A, Karlsson R, Karlsson M et al. Networks of nanotubes and containers. Nature 2001; 409(6817):150-152.
6. Vidulescu C, Clejan S, O'Connor K C. Vesicle traffic through intercellular bridges in DU 145 human prostate cancer cells. J Cell Mol Med 2004; 8(3):388-396.
7. Burgos MH, Fawcett DW. Studies on the fine structure of the mammalian testis. I. Differentiation of the spermatids in the cat (Felis domestica). J Biophys Biochem Cytol 1955; 1(4):287-300.
8. Oparka K. Plasmodesmata; 2005.
9. Cameron LA, Footer MJ, van Oudenaarden A et al. Motility of ActA protein-coated microspheres driven by actin polymerization. Proc Natl Acad Sci USA 1999; 96(9):4908-4913.
10. Ramirez-Weber FA, Kornberg TB. Cytonemes: cellular processes that project to the principal signaling center in Drosophila imaginal discs. Cell 1999;97(5):599-607.
11. Zhu D, Tan KS, Zhang X et al. Hydrogen peroxide alters membrane and cytoskeleton properties and increases intercellular connections in astrocytes. J Cell Sci 2005; 118(Pt 16):3695-703.
12. Scott G, Leopardi S, Printup S et al. Filopodia are conduits for melanosome transfer to keratinocytes. J Cell Sci 2002; 115(Pt 7):1441-1451.
13. Magalhaes AC, Baron GS, Lee KS et al. Uptake and neuritic transport of scrapie prion protein coincident with infection of neuronal cells. J Neurosci 2005; 25(21):5207-5216.
14. Incardona JP, Lee JH, Robertson CP et al. Receptor-mediated endocytosis of soluble and membrane-tethered Sonic hedgehog by Patched-1. Proc Natl Acad Sci USA 2000; 97(22):12044-12049.
15. Pepinsky RB, Zeng C, Wen D et al. Identification of a palmitic acid-modified form of human Sonic hedgehog. J Biol Chem 1998; 273(22):14037-14045.
16. Porter JA, Young KE, Beachy PA. Cholesterol modification of hedgehog signaling proteins in animal development. Science 1996; 274(5285):255-259.
17. Berk DA, Clark A Jr, Hochmuth RM. Analysis of lateral diffusion from a spherical cell surface to a tubular projection. Biophys J 1992; 61(1):1-8.
18. Davis DM, Dustin ML. What is the importance of the immunological synapse? Trends Immunol Jun 2004; 25(6):323-327.
19. Levchenko A, Mehta BM, Niu X et al. Intercellular transfer of P-glycoprotein mediates acquired multidrug resistance in tumor cells. Proc Natl Acad Sci USA 2005; 102(6):1933-1938.

CHAPTER 15

Cytoplasmic Bridges as Cell-Cell Channels of Germ Cells

Sami Ventelä*

Abstract

Transient intercellular bridges are seen between a wide variety of cells before the completion of cytokinesis.[1] However, these are distinct from stable intercellular bridges that remain persistent after incomplete cytokinesis.[2] The diameter of the cytoplasmic bridges is rather big, 1-10 μm, compared with the very tiny gap junctions which allow passage of only small molecules or peptides (< 1-2 kDa). Among somatic cells there are a number of examples of intercellular bridges, for instance in muscle cells and neurons. The best studied entity at both functional and molecular level is cytoplasmic bridges connecting germ cells. Many conserved features exist in cytoplasmic bridge formation and function during germ cell development: the diameters of the bridges increase during gametogenesis and is 1-10 μm in Drosophila oogenesis and 1-3 μm in mammalian spermatogenesis depending on the developmental stage of the gametes. The transportation mechanisms, e.g., the importance of cytoskeleton during transportation, are quite similar in both sexes from insects to mammals. Obviously the function of cytoplasmic bridges is to facilitate the sharing of cytoplasmic constituents between neighbouring cells.[3] This is probably most energy-efficient way and allows germ cell differentiation to be directed by the products of both parental chromosomes. In this article special features and recent investigations of cytoplasmic bridges as cell-cell channels during gametogenesis are reviewed.

Cytoplasmic Bridges during Gametogenesis

Stable intercellular bridges exist both in female and male gametogenesis, in which the incomplete cytokinesis gives a possibility for sharing of nutrients and relatively large cytoplasmic granules and organelles. During oogenesis female germ cells are developed in clusters of interconnected cells called cysts. Cyst formation has recently been demonstrated to be very conserved, with small varieties from species to species. Cyst formation are demonstrated to occur for example in *Drosophila*,[4] *C. elegans*,[5] mice[6] and *Hydra*.[7] During oogenesis each daughter cell, called cystocyte, is joined by cytoplasmic bridges allowing the transport of organelles and specific mRNAs into one of the cystocytes. During oogenesis, syncytial development gives rise to only one mature oocyte while the other cells retract and die after contributing their cytoplasmic contents through cytoplasmic bridges to the oocyte.[8] Most closely the cytoplasmic bridges have been studied in oogenesis of *drosophila melanogaster* in which only one cystocyte developes to an oocyte and all others become nurse cells.[9] These nurse cells then undergo a programmed breakdown, often at a specific time in germ cell development.

A characteristic feature of spermatogenesis is that the dividing germ cells fail to complete cell division resulting in formation of stable cytoplasmic bridges that interconnect a large number of

*Sami Ventelä—Departments of Otorhinolaryngology, Head and Neck Surgery, Physiology and Pediatrics, University of Turku FIN-20520 Turku, Finland. Email: satuve@utu.fi

Cell-Cell Channels, edited by Frantisek Baluska, Dieter Volkmann and Peter W. Barlow.

cells.[10,11] Although all the spermatids (step 1-19 in rats) contain only haploid genome; each spermatid will finally develop into fully maturing spermatozoa. It is obvious that the spermatids need an efficient intercellular trafficking system where the gene products of haploid cells are shared between the neighbouring cells. Braun et al[12] showed with transgenic mouse strain that chimaeric gene product expressed only by postmeiotic cells, are evenly distributed between genotypically haploid spermatids. Recent findings have demonstrated that many genes, such as TRA54,[13] are expressed only in haploid cells. This makes the role of cytoplasmic bridges in sharing of gene products an even more intriguing subject to study.

Development and Molecular Structures of Cytoplasmic Bridges

At molecular level Drosophila oogenesesis is well characterized, although the early phases of cytoplasmic bridge development are not well understood.[2] Early studies using electron microscopy revealed a highly vesiculated structure called fusome extending through cytoplasmic bridges, possibly functioning to stabilize the cleavage furrow until the cytoplasmic bridge is established.[14,15] In these stabilized cleavage furrows the formation of cytoplasmic bridges starts from the outer rim containing F-actin, anillin, a mucin-like glycoprotein and a protein with high content of phosphorylated tyrosine residues.[16] Then the inner rim is formed by the deposition of actin, the Hu-li tai shao protein or Hts-RC,[17] the filamin encoded by the cheerio gene,[18,19] the Tec29 and Src64 tyrosine kinases[20,21] and the Kelch protein.[22,23] Concurrently as the size of cystocytes increase the cytoplasmic bridges also grow in diameter from 0.5 μm up to 10 μm.[24]

Similarly as in *Drosophila* oogenesis the presence of an electron-dense outer layer of cytoplasmic bridges is demonstrated to occur in rat spermatogenesis supposing to give rigidity to the bridge.[11] Diameter of cytoplasmic bridges in premature spermatogonial stem cells are from 1 to 1.3 μm in rat.[25] Likewise in *Drosophila* the diameter of the cytoplasmic bridges increases as cells progress during spermatogenesis: at step 1 spermatids the diameter of the bridge is approximately 1.8 μm and most developed spermatids, step 19, have a bridge diameter of 3.0 μm. In mammals only few proteins have been demonstrated in the walls of the cytoplasmic bridges: cytoskeletal proteins actin, sak 57 and a heat shock factor 2.[26-28] Interestingly, a mutation in specific gene in *Drosophila*, called cheerio,[29] prevents the expansion of cytoplasmic bridges during oogenesis, resulting in an obvious malfunction of cytoplasmic material sharing. Moreover, the decrease of the size of oocyte and female sterility occur in cheerio mutants demonstrating the great importance of the cytoplasmic bridges during gametogenesis. According to the function of the found proteins it can be suggested that regulatory mechanisms of the transport, and possibly packaging (chaperone) functions for transported materials, occur in the cytoplasmic bridges.

The Cytoplasmic Bridges in Action

The diameters of the cytoplasmic bridges in germ cells are relatively large in all species allowing transportation of rather big vesicles and organelles. Kinetic analyses with time-laps video enhanced phase contrast microscope or scanning confocal microscopy have revealed that small vesicle and granule transportation occurs via cytoplasmic bridges between nurse cells and oocyte in *Drosophila*[30,31] and between haploid spermatids in the rat.[32] Throughout oogenesis the nurse cells provide cytoplasmic components such as mRNAs, proteins and organelles to the steadily growing oocyte via cytoplasmic bridges.[33] The process of this transfer can be divided into two main phases. During the first phase the cytoplasmic constituents are transported very selectively from nurse cells to the oocyte, whereas during the second nonselective phase the nurse cell cytoplasm stream rapidly (dumping) into the oocyte leading to apoptotic cell death of nurse cells.

During rat spermiogenesis there is very active cytoplasmic material transportation via cytoplasmic bridges between haploid spermatids.[32] The cytoplasm of living early spermatids contained a multitude of granules with diameters of approximately 0.5 μm. They move along defined paths in a nonrandom fashion in varying speeds and have frequent contacts with each other and with larger organelles, such as the Golgi complex and the male germ cell specific organelle chromatoid body (Fig. 1). During the migration through the cytoplasmic bridge the average velocity of the granules decreased and was approximately 47.7% lower than that in the cyto-

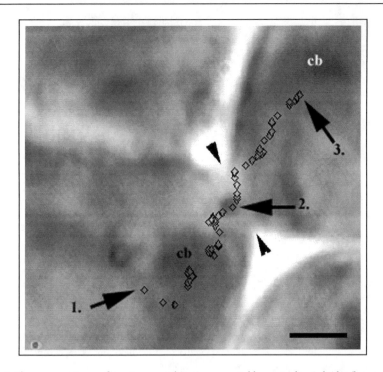

Figure 1. Phase contrast image of step 2 spermatid ex vivo connected by a cytoplasmic bridge (between black arrowheads), showing a unidirectional movement of a granule through the cytoplasmic bridge. Representative positions of the granule are shown by arrows: the starting point of analysis (1), the granule localization inside the cytoplasmic bridge (2), the end point of the analysis (3). cb = chromatoid body, Bar = 2 μm.

plasm. Similarly as in cytoplasmic bridges connecting the ovary and nurse cells in *Drosophila,*[30] only few (ca 30%) of the granules circulating in the immediate vicinity of the bridge actually were seen to pass through. This suggests that selection and regulation of material takes place in the cytoplasmic bridges, which is probably carried out by the proteins attached to the bridge.

Content of the Vesicles

An essential question is what gene products are packed in the small vesicles passing through the cytoplasmic bridges. The oocyte is transcriptionally inactive during most of *Drosophila* oogenesis, suggesting that oocyte specific mRNAs as well as the factors required for RNA localization are synthesized in the nurse cells and then transported into the oocyte.[33-35] Recently it has been demonstrated with video microscope that at least products of exu gene[36] and mitochondria[37] are passing through the cytoplasmic bridges between the nurse cells and oocyte in *Drosophila*. The exu gene is recuired during oogenesis for bcd mRNA localization to the anterior pole of oocyte which is needed in axis specification. Selective transportation mechanisms providing mitchondria from nurse cells to the oocyte is intriguing because oocytes provide all the mitochondria of the zygote at the time of fertilization.

Immunohistochemical analyses have revealed that haploid cell specific monoclonal antibody TRA54 was localized in the granules and transported through the cytoplasmic bridges into neighbouring spermatid in the rat (Fig. 2). Translation of the antigenic epitope (sugar moiety) of TRA54 starts at early spermiogenesis and its first localization was at the Golgi complex. Then TRA54 is packed to small granules and transported to the acrosome system of a neighbouring spermatid. The number of TRA54 positive granules inside the cytoplasm decreased while the spermatids

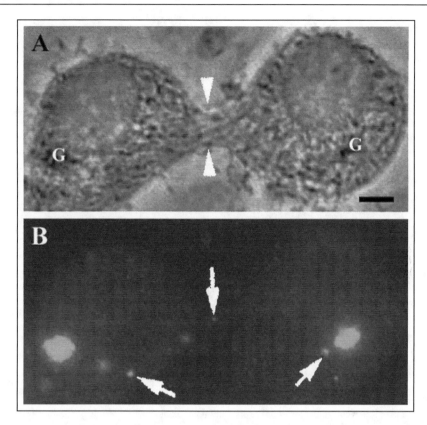

Figure 2. TRA54 localization in early spermatids. Squash preparations from stage I were fixed and immunostained with TRA54. Phase contrast (A), immunofuorescence of TRA54 (B), cytoplasmic bridges between arrowheads (A). The Golgi complex (G) was highly TRA54 positive. Small granules inside the cytoplasm and cytoplasmic bridge were TRA54 positive (white arrows, B). Bar = 2 μm.

developed. Finally TRA54 was located in the middle part of the membranes surrounding the acrosomic system at the head of spermatid. It might be needed in recognition or fertilization process existing between sperm and oocyte. The observations of selective passage of cytoplasmic material together with morphological alterations of cytoplasmic bridges during gametogenesis suggest that these channels are rather active organelles connecting the germ cells.

Organelle Transportation

Another evolutionarily conserved feature at haploid phase of germ cell differentiation is a large perinuclear cytoplasmic organelle. During spermatogenesis this organelle is called chromatoid body. In *Drosophila* oocytes, an analogous organelle is called sponge body[38] or yolk nucleus in humans fetal oocytes.[39] Recent investigations have suggested similar functions for these organelles in both sexes. During *Drosophila* oogenesis sponge body is demonstrated to locate inside the nurse cells, inside the oocytes and inside the cytoplasmic bridge between nurse cells and oocytes, suggesting that this subcellular structure is transported between the cells.[38] Sponge body contains RNA and is therefore probably needed in assembly and transport of mRNA during *Drosophila* oogenesis.

Male germ cell specific organelle chromatoid body appears in pachytene spermatocytes at stage VIII of the rat and is biggest in size at the early spermiogenesis. Figueroa and Burzio[40] have shown that isolated chromatoid bodies contained a complex population of RNAs; mRNA, 5.8 and 5 S ribosomal RNA, but no tRNA. Previously it has also been demonstrated that chromosomal protein histone H4,[41] germ cell-specific DNA and RNA binding protein p48/52,[42] hnRNPs and ribosomal proteins,[43] mouse VASA-homologue[44,45] and actin[46] are localized in the chromatoid body. When living early spermatids are studied ex vivo under the phase contrast microscope this organelle is wide moving, phase contrast positive and contains two main components of the rapid movement either parallel or perpendicular in relation to the nuclear envelope.[47,48] Moreover during parallel movement chromatoid body has pinocytosis-like transient engulfments towards nuclear pale chromatin areas.[48] It has been proposed that parallel movements over the haploid nucleus are needed for collection of gene products from various parts of the haploid nucleus from different chromosome territories.[49] Recent demonstration that chromatoid body moves back and forth through the cytoplasmic bridge between neighbouring spermatids with transient contacts with nuclei of both spermatids gives explanation to perpendicular movement of chromatoid body (Fig. 3). The thresholded image series (Fig. 3, A2-E2) shows the changing lobulated morphology of the chromatoid body and the close transient interactions between the chromatoid body and both nuclei. The velocity of the movement of the chromatoid body inside the bridge was rather constant, in average 0.20 μm/s. Electron microscopic analyses of snap-frozen specimens have revealed a very close relationship between nuclear pore complex with a large contact area between chromatoid body and the nuclear envelope supporting assumption that intranuclear material is exchanged between the nucleus and chromatoid body.[48,50] Interestingly there exist many molecular and functional similarities between chromatoid body and sponge body. Obviously these perinuclear organelles of both sexes have an important role during gametogenesis, since targeted disruption of VASA or its homologue genes results in sterility and serious defects in germ cell development in both sexes.[44,51]

Mechanisms Needed for Cytoplasmic Material Transportation between the Germ Cells

Inhibition studies in *Drosophila* have demonstrated that polarized microtubule cytoskeleton reveal steps in the oocyte differentiation pathway and plays a critical role in oocyte development.[52-55] Moreover, immunofluorescence analyses and microtubule depolymerization studies suggested that minus end of the microtubules locate at the oocyte and plus ends extend through cytoplasmic bridges into the surrounding nurse cells providing a scaffold for the transport of cytoplasmic material from nurse cells to the oocyte.[31,56] The first phase of cytoplasmic material transportation at the beginning of oogenesis is microtubule dependent and sensitive to the microtubule inhibitors. During the second rapid phase of cytoplasm transport at the end of oogenesis remaining nurse cell contents are pushed into the oocyte requiring a cortical actomyosin network to drive nurse cell contraction. Moreover cytoplasmic bridge stability and growth are dependent on the accumulation of an inner rim consisting of loosely bundled actin filaments.[24] Finally, a dense network of hexagonally packed parallel actin bundles is required in the cytoplasm of nurse cells to restrain their large polyploid nuclei from being pushed into the cytoplasmic bridge, which would otherwise block further cytoplasm transport. Also many genes are demonstrated to be crucial during cytoplasmic transportation between nurse cells and oocyte,[8,34] in cytoplasmic bridge formation[55] and in organization of the microtubule network.[57,58]

A number of similarities has been demonstrated in cytoplasmic material exchange of *Drosophila* oogenesis and rat spermiogenesis.[33] The most obvious difference between these two models is the number of end results: while oogenesis directs to develop one mature oocyte, the goal of spermatogenesis is to achieve millions of mature spermatozoa. This explains why cytoplasmic material transportation is unidirectional during oogenesis, from nurse cells to oocyte, and bi-directional during spermatogenesis. Same way as in Drosophila oogenesis cytoskeleton plays a major role during cytoplasmic transportation between spermatids.[33] Microtubule inhibitors, nocodazole and vincristine, turned the nonrandom cytoplasmic granule movement into random brownian motion demonstrating the importance of the microtubules during spermatogenesis. Moreover microtubule inhibitors disintegrate the chromatoid body and induce a clear separation of chromatoid material and ribosome-like granules. Likewise microtubule inhibitors prevented the movement of the chromatoid body.[32] Immunohistochemical

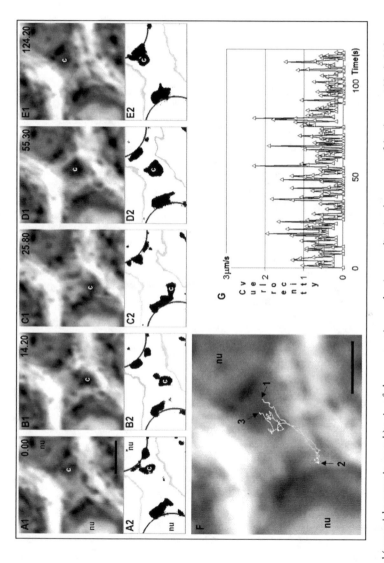

Figure 3. Time-scaled (upper right corner in seconds) series of phase contrast images showing two-directional movement of the chromatoid body (c) between two step 1 spermatids (A1-E1) through the cytoplasmic bridge. The same series is thresholded in A2-E2 to show in detail the chromatoid body and associated granules at the nuclear (nu) envelope. Moreover, in threshold image series the edges of the cells (gray) and nucleus (black) are presented. The movement path of the centroid of the chromatoid body is shown in panel F. Arrow 1 shows the starting point of the movement and arrow 2 the point where chromatoid body was close to the nucleus of the lower cell. Arrow 3 indicates the end point of chromatoid body movement during the 124.2 s recording period. The velocity of chromatoid body at various time points is shown in panel G. Bars = 2μm. Reprinted from: Mol Biol Cell 2003; 14:2768-2780; published online before print as 10.1091/mbc.E02-10-0647; with permission of The American Society for Cell Biology.[32]

analyses revealed after incubation with nocodazole that two types of spermatids developed: one showed haploid specific TRA54 immunostaining in the Golgi complex while the other lacked TRA54 labelling, demonstrating the block of the transportation of TRA54 between the spermatids. Microfilament inhibitor cytochalasin D did not affect to morphology or the movement of the chromatoid body even though previously it has been demonstrated that actin is a component of the chromatoid body.[46] This suggests that actin inside the chromatoid body is most probably involved in mRNA binding to the cytoskeletal framework demonstrated to exist in many type of cells (for reviews see refs. 59,60) and not in the normal movement of chromatoid body. Based on these facts the microtubules are involved in the normal integrity of the chromatoid body, in its movements, and in the normal Golgi complex derived granule traffic in the cytoplasm of spermatids and between neighbouring cells. Moreover, it is known that the sorting of proteins according to their final destination is performed at the Golgi complex and that the Golgi complex plays a main role in vesicle and granule trafficking (for reviews see refs. 61,62). During spermatogenesis the role of Golgi complex is crucial, because of its role in formation of the acrosome system, which contains hydrolytic enzymes involved in fertilization and penetration through zona pellucida.[63,64] When the Golgi complex was disturbed by brefeldin A, the formation of the acrosome system was inhibited.[65] Also formation and transportation of TRA54 positive granules were inhibited by brefeldin A.[32] These findings suggest that the Golgi complex plays a key role in small granule transportation during spermatogenesis.

Concluding Remarks

Digital techniques for quantification of organelle movements in living cells are useful for clarification of the possible function of certain organelles, such as the chromatoid body and Golgi complex in living cells. Recent studies have opened new insights to the cytoplasmic bridges: Hamer et al[66] showed that apoptotic signals are shared between interconnected spermatogonia, and it has also been demonstrated that certain cancer cells can assemble cytoplasmic bridges and use those in cytoplasmic material sharing.[67] In the future, we will have more detailed information about the function of cytoplasmic bridges in living cells by using new quantitative video analyses combined with toxicological compounds affecting on cytoplasmic bridge formation or selective material exchange between the cells.

Acknowledgement

I am grateful to professors Jorma Toppari and Martti Parvinen for encouragement and critical reading of the manuscript. In addition, Jorma Toppari kindly arranged financial support from The Department of Pediatrics, University Hospital of Turku. This project received a long-time support from Turku Graduate School of Biomedical Sciences (TuBS).

References

1. Sanders SL, Field CM. Cell division. Septins in common? Curr Biol 1994; 4:907-910.
2. Robinson DN, Cooley L. Stable intercellular bridges in development: The cytoskeleton lining the tunnel. Trends Cell Biol 1996; 6:474-479.
3. Erickson RP. Haploid gene expression versus meiotic drive: The relevance of intercellular bridges during spermatogenesis. Nat New Biol 1973; 243:210-212.
4. de Cuevas M, Lilly MA, Spradling AC. Germline cyst formation in Drosophila. Annu Rev Genet 1997; 31:405-428.
5. Gumienny TL, Lambie E, Hartwieg E et al. Genetic control of programmed cell death in the Caenorhabditis elegans hermaphrodite germline. Development 1999; 126:1011-1022.
6. Pepling ME, Spradling AC. Female mouse germ cells form synchronously dividing cysts. Development 1998; 125:3323-3328.
7. Alexandrova O, Schade M, Bottger A et al. Oogenesis in Hydra: Nurse cells transfer cytoplasm directly to the growing oocyte. Dev Biol 2005; 281:91-101.
8. Cooley L, Verheyen E, Ayers K. Chickadee encodes a profilin required for intercellular cytoplasm transport during Drosophila oogenesis. Cell 1992; 69:173-184.
9. Telfer W. Development and physiology of the oocyte-nurse cell syncytium. Adv Insect Physiol 1975; 11:223-319.
10. Burgos MH, Fawcett DW. Studies on the fine structure of the mammalian testis. J Biophys Biochem Cytol 1955; 1:287-300.

11. Fawcett DW, Ito S, Slautterback DL. The occurrence of intercellular bridges in groups of cells exhibiting synchronous differentiation. J Biophys Biochem Cytol 1959; 5:453-460.
12. Braun RE, Behringer RR, Peschon JJ et al. Genetically haploid spermatids are phenotypically diploid. Nature 1989; 337:373-376.
13. Pereira LA, Tanaka H, Nagata Y et al. Characterization and expression of a stage specific antigen by monoclonal antibody TRA 54 in testicular germ cells. Int J Androl 1998; 21:34-40.
14. Koch EA, King RC. The origin and early differentiation of the egg chamber of Drosophila melanogaster. J Morphol 1966; 119:283-303.
15. King RC, Storto PD. The role of the otu gene in Drosophila oogenesis. Bioessays 1988; 8:18-24.
16. Kramerov AA, Mikhaleva EA, Rozovsky Ya M et al. Insect mucin-type glycoprotein: Immunodetection of the O-glycosylated epitope in Drosophila melanogaster cells and tissues. Insect Biochem Mol Biol 1997; 27:513-521.
17. Yue L, Spradling AC. hu-li tai shao, a gene required for ring canal formation during Drosophila oogenesis, encodes a homolog of adducin. Genes Dev 1992; 6:2443-2454.
18. Sokol NS, Cooley L. Drosophila filamin encoded by the cheerio locus is a component of ovarian ring canals. Curr Biol 1999; 9:1221-1230.
19. Li MG, Serr M, Edwards K et al. Filamin is required for ring canal assembly and actin organization during Drosophila oogenesis. J Cell Biol 1999; 146:1061-1074.
20. Dodson GS, Guarnieri DJ, Simon MA. Src64 is required for ovarian ring canal morphogenesis during Drosophila oogenesis. Development 1998; 125:2883-2892.
21. Roulier EM, Panzer S, Beckendorf SK. The Tec29 tyrosine kinase is required during Drosophila embryogenesis and interacts with Src64 in ring canal development. Mol Cell 1998; 1:819-829.
22. Xue F, Cooley L. Kelch encodes a component of intercellular bridges in Drosophila egg chambers. Cell 1993; 72:681-793.
23. Robinson DN, Cooley L. Examination of the function of two kelch proteins generated by stop codon suppression. Development 1997; 124:1405-1417.
24. Tilney LG, Tilney MS, Guild GM. Formation of actin filament bundles in the ring canals of developing Drosophila follicles. J Cell Biol 1996; 133:61-74.
25. Weber J, Russel L. A study of intercellular bridges during spermatogenesis in the rat. Am J Anat 1987; 180:1-24.
26. Russell LD, Vogl AW, Weber JE. Actin localization in male germ cell intercellular bridges in the rat and ground squirrel and disruption of bridges by cytochalasin D. Am J Anat 1987; 180:25-40.
27. Tres LL, Rivkin E, Kierszenbaum AL. Sak 57, an intermediate filament keratin present in intercellular bridges of rat primary spermatocytes. Mol Reprod Dev 1996; 45:93-105.
28. Alastalo TP, Lönnström M, Leppä S et al. Stage-specific expression and cellular localization of the heat shock factor 2 isoforms in the rat seminiferous epithelium. Exp Cell Res 1998; 240:16-27.
29. Robinson DN, Smith-Leiker TA, Sokol NS et al. Formation of the Drosophila ovarian ring canal inner rim depends on cheerio. Genetics 1997; 145:1063-1072.
30. Bohrmann J, Biber K. Cytoskeleton-dependent transport of cytoplasmic particles in previtellogenic to mid-vitellogenic ovarian follicles of drosophila: Time-lapse analysis using video-enhanced contrast microscopy. J Cell Sci 1994; 107:849-858.
31. Theurkauf WE, Hazelrigg TI. In vivo analyses of cytoplasmic transport and cytoskeletal organization during Drosophila oogenesis: Characterization of a multi-step anterior localization pathway. Development 1998; 125:3655-3666.
32. Ventelä S, Toppari J, Parvinen M. Intercellular organelle traffic through cytoplasmic bridges in early spermatids of the rat: Mechanisms of haploid gene product sharing. Mol Biol Cell 2003; 14:2768-2780.
33. Mahajan-Miklos S, Cooley L. Intercellular cytoplasm transport during Drosophila oogenesis. Dev Biol 1994; 165:336-351.
34. Cooley L, Theurkauf WE. Cytoskeletal functions during Drosophila oogenesis. Science 1994; 266:590-596.
35. Cheung HK, Serano TL, Cohen RS. Evidence for a highly selective RNA transport system and its role in establishing the dorsoventral axis of the Drosophila egg. Development 1992; 114:653-661.
36. Theurkauf WE, Hazelrigg TI. In vivo analyses of cytoplasmic transport and cytoskeletal organization during Drosophila oogenesis: Characterization of a multi-step anterior localization pathway. Development 1998; 125:3655-3666.
37. Cox RT, Spradling AC. A Balbiani body and the fusome mediate mitochondrial inheritance during Drosophila oogenesis. Development 2003; 130:1579-1590.
38. Wilsch-Bräuninger M, Schwarz H, Nusslein-Volhard C. A sponge-like structure involved in the association and transport of maternal products during Drosophila oogenesis. J Cell Biol 1997; 139:817-829.
39. Hertig AT, Adams EC. Studies on the human oocyte and its follicle. I. Ultrastructural and histochemical observations on the primordial follicle stage. J Cell Biol 1967; 34:647-675.

40. Figueroa J, Burzio LO. Polysome-like structures in the chromatoid body of rat spermatids. Cell Tissue Res 1998; 291:575-579.
41. Werner G, Werner K. Immunocytochemical localization of histone H4 in the chromatoid body of rat spermatids. J Submicrosc Cytol Pathol 1995; 27:325-330.
42. Oko R, Korley R, Murray MT et al. Germ cell-specific DNA and RNA binding proteins p48/52 are expressed at specific stages of male germ cell development and are present in the chromatoid body. Mol Reprod Dev 1996; 44:1-13.
43. Biggiogera M, Fakan S, Leser G et al. Immunoelectron microscopical visualization of ribonucleoproteins in the chromatoid body of mouse spermatids. Mol Reprod Dev 1990; 26:150-158.
44. Tanaka SS, Toyooka Y, Akasu R et al. The mouse homolog of Drosophila Vasa is required for the development of male germ cells. Genes Dev 2000; 14:841-853.
45. Toyooka YN, Tsunekawa Y, Takahashi Y et al. Expression and intracellular localization of mouse Vasa-homologue protein during germ cell development. Mech Dev 2000; 93:139-149.
46. Walt H, Armbruster BL. Actin and RNA are components of the chromatoid bodies in spermatids of the rat. Cell Tissue Res 1984; 236:487-490.
47. Parvinen M, Parvinen LM. Active movements of the chromatoid body: A possible transport mechanism for haploid gene products. J Cell Biol 1979; 80:621-628.
48. Parvinen M, Salo J, Toivonen M et al. Computer analysis of living cells: Movements of the chromatoid body in early spermatids compared with ultrastructure in snap-frozen preparations. Histochem Cell Biol 1997; 108:77-81.
49. Cremer T, Kurz A, Zirbel R et al. Role of chromosome territories in the functional compartmentalization of the cell nucleus. Cold Spring Harb Symp Quant Biol 1993; 58:777-792.
50. Parvinen M. The chromatoid body in spermatogenesis. Int J Androl 2005; 28:189-201.
51. Styhler S, Nakakamura A, Swan A et al. Vasa is required for GURKEN accumulation in the oocyte, and is involved in oocyte differentiation and germline cyst development. Development 1998; 125:1569-1578.
52. Theurkauf WE, Alberts BM, Jan YN et al. A central role for microtubules in the differentiation of Drosophila oocytes. Development 1993; 118:1169-1180.
53. Pokrywka NJ, Stephenson EC. Microtubules are a general component of mRNA localization systems in Drosophila oocytes. Dev Biol 1995; 167:363-370.
54. Mathe E, Inoue YH, Palframan W et al. Orbit/Mast, the CLASP orthologue of Drosophila, is required for asymmetric stem cell and cystocyte divisions and development of the polarised microtubule network that interconnects oocyte and nurse cells during oogenesis. Development 2003; 130:901-915.
55. Theurkauf WE, Smiley S, Wong ML et al. Reorganization of the cytoskeleton during Drosophila oogenesis: Implications for axis specification and intercellular transport. Development 1992; 115:923-936.
56. Robinson DN, Cant K, Cooley L. Morphogenesis of Drosophila ovarian ring canals. Development 1994; 120:2015-2025.
57. Theurkauf WE. Premature microtubule-dependent cytoplasmic streaming in cappuccino and spire mutant oocytes. Science 1994; 265:2093-2096.
58. Huynh JR, St Johnston D. The role of BicD, Egl, Orb and the microtubules in the restriction of meiosis to the Drosophila oocyte. Development 2000; 127:785-794.
59. Ornelles DA, Fey EG, Penman S. Cytochalasin releases mRNA from the cytoskeletal framework and inhibits protein synthesis. Mol Cell Biol 1986; 6:1650-1662.
60. Jansen RP. RNA-cytoskeletal associations. FASEB J 1999; 13:455-466.
61. Lippincott-Schwartz J, Roberts TH, Hirschberg K. Secretory protein trafficking and organelle dynamics in living cells. Annu Rev Cell Dev Biol 2000; 16:557-589.
62. Moreno RD, Ramalho-Santos J, Sutovsky P et al. Vesicular traffic and golgi apparatus dynamics during mammalian spermatogenesis: Implications for acrosome architecture. Biol Reprod 2000; 63:89-98.
63. Hermo L, Rambourg A, Clermont Y. Three-dimensional architecture of the cortical region of the Golgi apparatus in rat spermatids. Am J Anat 1980; 157:357-373.
64. Moreno RD, Ramalho-Santos J, Chan EK et al. The Golgi apparatus segregates from the lysosomal/acrosomal vesicle during rhesus spermiogenesis: Structural alterations. Dev Biol 2000; 219:334-349.
65. Ventelä S, Mulari M, Okabe M et al. Regulation of acrosome formation in mice expressing green fluorescent protein as a marker. Tissue Cell 2000; 32:501-507.
66. Hamer G, Roepers-Gajadien HL, Gademan IS et al. Intercellular bridges and apoptosis in clones of male germ cells. Int J Androl 2003; 26:348-353.
67. Vidulescu C, Clejan S, O'Connor KC. Vesicle traffic through intercellular bridges in DU 145 human prostate cancer cells. J Cell Mol Med 2004; 8:388-396.

CHAPTER 16

Fusome as a Cell-Cell Communication Channel of *Drosophila* Ovarian Cyst

Jean-René Huynh*

Abstract

In most animal species, female and male gametes are produced within clusters of germ cells which share a common cytoplasm through cell-cell channels. In *Drosophila* ovaries, these cells synchronise their divisions and specialise one cell of the cluster as the future egg. Both processes are organised by a germline-specific organelle of communication called the fusome. Until recently, the fusome has remained largely mysterious despite a hundred years of research on its composition, formation and functions. Novel results have now suggested several molecular mechanisms to explain how the fusome synchronises the divisions by controlling cell-cycle regulators and how it determines and polarises the future egg by organising the microtubule cytoskeleton. Importantly, a structure similar to the fusome has been identified during *Xenopus* oogenesis, suggesting that it is widely conserved from invertebrates to vertebrates, and that it thus serves an essential function.

Introduction

In most animal species, male and female gametes start their differentiation as groups of germ cells with synchronous phases of development, indicating extensive communication between these cells.[1] Accordingly, it was found that these cells (called "cystocytes") share a common cytoplasm through cytoplasmic bridges, thus forming a syncytium referred to as a cyst. This specific form of cell-cell channel originates from a series of incomplete cell divisions of a cyst-founder cell called a cystoblast. The bridges are derived from persistent mid-bodies in the arrested cleavage furrows.[2,3] Male gametes retain these intercellular bridges during most of their differentiation. It has been suggested that they may serve not only to synchronise their development, but also to share gene products after the completion of meiosis, as half of the sperms will inherit one X chromosome and the other half a Y chromosome.[4,5] Synchronisation and efficient sharing might thus alleviate the differences between advantaged and disadvantaged sperm. In females, intercellular bridges can be limited to the synchronous phase of divisions, as in mammalian ovaries, or they can form stable structures called ring canals as shown in *Drosophila*.[1] In this fruitfly, communication between the germ cells has an additional function as it allows the specialization of one cell of the cluster as the oocyte (the future egg), while the remaining germ cells of the cyst become nurse cells, which provide the oocyte with nutrients and cytoplasmic components through the ring canals.

A second and equally important feature of *Drosophila* cyst development is the presence of a large cytoplasmic structure called the fusome, which links all the cells of the cluster through the ring canals.[6] Although the behaviour of the fusome has been best studied in insects,[7] a similar structure has been recently described in *Xenopus laevis* during the formation of female germline cysts.[8] The fusome seems thus widely conserved from invertebrates to vertebrates, which suggests that it serves

*Jean-Rene Huynh—Institut Jacques-Monod, CNRS, Universités Paris 6 et 7, 2, place Jussieu, F-75251 Paris Cedex 05, France. Email: huynh@ijm.jussieu.fr

Cell-Cell Channels, edited by Frantisek Baluska, Dieter Volkmann and Peter W. Barlow.
©2006 Landes Bioscience and Springer Science+Business Media.

an important function. The fusome is mainly made of densely packed vesicles filling up the ring canals. It was thus first described as a "plug" within the ring canals and thought to block any communication between cells of the cyst.[6] Surprisingly, it was found in *Drosophila* ovaries that in the absence of the fusome, cells of the same cluster divide asynchronously and fail to specify an oocyte, despite the presence of ring canals.[9,10] This demonstrates a key role of the fusome as a channel of communication between the cells for the synchronisation and differentiation of the germline cyst. In this review, I will mainly focus on the fusome of the *Drosophila* ovaries, where it has been possible to study its formation and functions at the genetic, cellular and molecular levels. Other aspects of *Drosophila* oogenesis have been summarized elsewhere, including ring canals formation (this issue and[11,12]), germline stem cells regulation[13] and differentiation of the cyst.[14,15]

The *Drosophila* ovary is composed of 16-20 ovarioles, each of which contains a chain of progressively more and more mature egg chambers.[16] Germline cysts are produced throughout the life of the adult fly, by the divisions of germline stem cells localised in a specialised region called the germarium situated at the anterior tip of each ovariole (Fig. 1). The germarium has been divided into 4 regions according to the developmental stage of the cyst. Oogenesis begins in region 1, when a germline stem cell (GSC) divides asymmetrically to produce a posterior cystoblast, and a new GSC, which remains attached to the neighbouring somatic cells at the anterior.[13,17] The cystoblast then undergoes precisely four rounds of mitosis with incomplete cytokinesis to form a cyst of 16 germline cells, which are interconnected by ring canals. Once the 16 cell cyst has formed, it enters the region 2a of the germarium. At this stage, all the cells of one cyst look the same, but by the time it reaches region 2b, one cell will have differentiated as an oocyte. This differentiation can be followed with several types of marker summarized in Figure 1. In region 2b, the cyst changes shape and becomes a one cell-thick disc that spans the whole width of the germarium. Oocyte-specific factors are now concentrated into the oocyte and a microtubule organising centre (MTOC) is clearly seen in this cell. At the same time, somatic follicle cells start to migrate and surround the cyst. As the cyst moves down to region 3 (also called stage 1), it rounds up to form a sphere with the oocyte always lying at the posterior pole. The cyst then leaves the germarium into the vitellarium.

Formation of the Fusome

The fusome was first described in 1886 by Platner as Verbindungsbrücken ("bridging connections") in the spermatocytes of several insects using iron hematoxylin staining and light microscopy.[18] In 1901, Giardina recorded the origin and formation of the fusome in the female germline cyst of the diving beetle *Dytiscus*.[2] His observations and conclusions remain still valid and turn out to be general principles of fusome formation. Between 1960 and 1970, King, Mahowald and colleagues described a region of the cytoplasm of the *Drosophila* cysts as rich in vesicles and fibrils using electron microscopy.[19-21] This structure was then recognized as the fusome by Telfer in 1975.[6] More recently Allan Spradling's lab, taking advantage of *Drosophila* new genetic tools, described the fusome at the molecular and cellular levels using fluorescent immunostaining and confocal microscopy.[9,10,22-25] Finally, using transgenic flies expressing GFP-tagged markers, Mary Lilly's group was able to observe the fusome using live confocal microscopy and photobleaching techniques.[26] After more than a century of observations and experiments, a highly dynamic picture of the formation of fusome has emerged.

A Complex Choreography

The fusome sensu stricto appears during the first division of the cystoblast. It arises from a similar structure present in the germline stem cells (GSC) called a spectrosome, which is also made of vesicles and whose molecular composition differs only slightly form that of the fusome.[24] The first appearance of the spectrosome itself has been reported in the germ cells of the gastrulating embryo.[27] In the adult, the spectrosome is stably anchored at the anterior side of the GSC, in contact with the adherens junctions between the GSC and the overlying cap cells (Figs. 1,2).[28] During the GSC division, one pole of the mitotic spindle is anchored by the spectrosome thus orientating the division along the antero-posterior axis of the germarium (Fig. 2).[27,29] The cystoblast

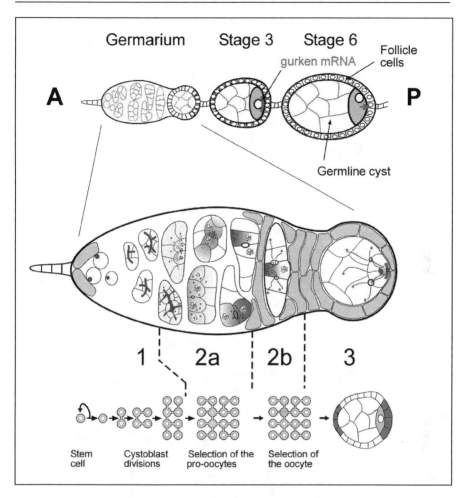

Figures. 1 Drosophila early oogenesis. Each ovariole is made of a chain of more and more mature egg chambers towards the posterior (P). An egg chamber comprises 16 germline cells surrounded by a monolayer of follicle cells. The egg chambers are produced in a specialised structure, called the germarium, at the anterior (A) of the ovariole. The germarium is divided into 4 morphological regions along the antero-posterior axis. The germline stem cells reside at the anterior tip of the germarium in contact with the somatic cap cells (grey cells on the left) and divide to produce cytoblasts, which divide four more times in region 1 to produce 16 cell germline cysts that are connected by ring canals. The stem cells and cystoblasts contain a spectrosome (red circles), which develops into a branched structure called the fusome, which orients each division of the cyst. In early region 2a, the synaptonemal complex (SC, red lines) forms along the chromosomes of the 2 cells with 4 ring canals (the pro-oocytes) as they enter meiosis. The SC then appears transiently in the 2 cells with 3 ring canals, before becoming restricted to the pro-oocytes in late region 2a. By region 2b, the oocyte has been selected, and is the only cell to remain in meiosis. In region 2a, cytoplasmic proteins, mRNAs and mitochondria (green), and the centrosomes (blue circles) progressively accumulate at the anterior of the oocyte. In region 2b, the minus-ends of the microtubules are focused in the oocyte, and the plus-ends extend through the ring canals into the nurse cells. The follicle cells (grey) also start to migrate and surround the germline cells. As the cyst moves down to region 3, the oocyte adheres strongly to the posterior follicle cells and repolarises along its antero-posterior axis, with the MT minus-ends and specific cytoplasmic components now localised at the posterior cortex. Modified from reference 15. A color version of this figure is available online at www.Eurekah.com.

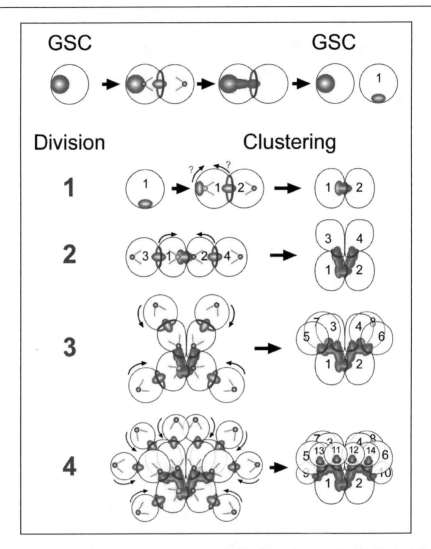

Figures 2. Formation of the fusome. The spectrosome (red) of the GSC anchors one pole of the mitotic spindle and orients the division along the anterior-posterior axis. A new fusome plug forms into the transient ring canal. The spectrosome elongates to fuse with the plug. The cytoplasmic bridge is then severed and one third of the spectrosome/fusome is inherited by the cystoblast while two-third remains in the GSC, marking this division as clearly asymmetric. The spectrosome/fusome (red) of the cystoblast (1) interacts with one of the centrosomes (green and blue spheres) to anchor one pole of the mitotic spindle (green lines), during the first incomplete division. A fusome plug (red) forms in the arrested furrow or ring canal (blue). The spectrosome (or "original" fusome) and the fusome plug come together to fuse. The direction of these movements is not known (?). The same mechanism is repeated for division 2, 3 and 4: (1) One pole of each mitotic spindle is anchored by the fusome. (2) A new fusome plug forms into each ring canal. (3) the ring canals move centripetally for the fusome plugs to fuse with the central fusome (black arrows). This behaviour has several crucial consequences: (1) cystocyte (1) has more fusome than the other cystocytes; (2) the same centrosome (green sphere) could be inherited by the cystocyte (1) from division 1 through division 4; and (3) the fusome always marks the anterior of cystocyte (1), after the clustering of the ring canals. Modified from reference 15. A color version of this figure is available online at www.Eurekah.com.

is then produced toward the posterior and the renewed GSC stays at the anterior of the germarium, in a specialized cellular environment called a "niche".[13] At the end of telophase, a transient ring canal forms between the GSC and the cystoblast. New fusome material accumulates in this ring canal and the spectrosome elongates from the anterior side of the GSC to fuse with the new fusome.[22,29] The cytoplasmic bridge is then severed and one third of the spectrosome/fusome is inherited by the cystoblast while two-third remains in the GSC, marking this division as clearly asymmetric.[27,29] The spectrosome relocalises to the anterior side of the GSC, while the fusome takes a spherical shape at one end of the cystoblast.

The first mitosis of the cystoblast is very similar to the GSC division (Fig. 2). One pole of the mitotic spindle is anchored by the spherical fusome, a new fusome "plug" forms into the arrested furrow at the other end of the cell and comes to fuse with the "original" fusome. However, in contrast to the GSC division, cytokinesis is incomplete and both cells remain linked by a stable ring canal. Furthermore, although it is not known how the plug and the "original" fusome move to fuse together, the "original" fusome does not seem to elongate from one side of the cystoblast, as in the GSC.[22] This division is also asymmetric as one cell contains the "original" fusome plus half of the plug, whereas the other cell only retains the other half of the fusome plug. At the next division, the two mitotic spindles again orient with one pole close to the fusome and new fusome plugs form in the two ring canals at opposite ends.[22,30-32] The fusome plugs then move with their ring canal to fuse with the central fusome, which thus remains in the two previous cells. This asymmetric behaviour of the fusome is then repeated in the next two divisions (Figs. 2,3). The oldest cell, therefore, retains the original fusome and accumulates four halves of fusome plugs. Thus, this cell has more fusome than all the other cells and can be identified throughout the divisions. The current model suggests that this cell will become the oocyte (see below). Once the 16 cells-cyst is formed, the fusome starts to break down and disappears. The behaviour of the *Drosophila* fusome reveals general morphogenetic principles of how germline cysts form and has important consequences on the polarisation of the female cyst.[6,14]

Firstly, a consequence of the synchronous divisions is that the number of cells per cyst is a power of two: 2^n, with n being the number of divisions. In *Drosophila* male and female n equals 4 (2^4=16), but n can range from 1 to at least 8 (256 cells) in the Strepsipteran *Elenchinus japonicus*.[14] The 2^n rule seems also to apply to vertebrates as *Xenopus* cysts go through four synchronous divisions and contain 16 cells.[8]

A second general principle is that germline cysts do not form linear clusters of cells but are organised along a "maximally branched" pattern of interconnections. It is a consequence of the anchoring of one pole of each mitotic spindle by the fusome.[22,30-32] This orientation of the divisions ensures that one cell inherits all the previous interconnections (ring canals), while a new one is formed at the opposite end of the cell and thus branched off the central fusome. This orientation also leads to an invariant pattern of interconnections between the cells, with the two central or oldest cells having n ring canals, their daughter cells n-1, etc. This pattern is important for the polarisation of the cyst as the oocyte always arises from one of the two cells with the greatest number (n) of ring canals.[7]

At each interphase, the fusome plugs and the associated ring canals move centripetally to fuse with the central fusome. This clustering of the ring canals at the centre of the cyst with the cells protruding outward creates a rosette-like structure (Fig. 2). This specific geometry is a third characteristic of the germline cysts.[6,7] One function of the rosette shape could be to minimize the distances between cells inside the cyst and thus make cell-cell communication more efficient.[14] This shape is later on stabilised by the formation of adherens junctions around the ring canals, as seen by the localisation of Armadillo, E-Cadherin and Bazooka and confirmed by EM sections (ref. 33 and A.P. Mahowald, personal communication).

Finally, the behaviour of the fusome reveals several levels of polarisation within the female cyst. As already mentioned, the invariant pattern of divisions restricts the oocyte fate to the two cells with the greatest number of ring canals. In addition, the centripetal movement of the ring canals along the cell membrane to fuse with the central fusome uncovers a polarity within each cell of the cyst

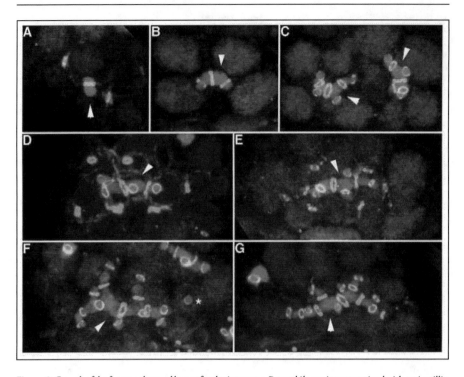

Figures 3. Growth of the fusome observed by confocal microscopy. *Drosophila* ovaries were stained with anti-anillin (ring canal, green) and anti-Hts (fusome, red) antibodies. Anterior is at the top in each panel. A) A 4-cell cyst in early-mid interphase. Two fusome plugs have formed, one in each new ring canal. B) A 4-cell cyst in late interphase. The plugs have fused with the original fusome, forming one continuous structure that spans the three ring canals. C) A pair of 8-cell cysts in late interphase. In each cyst, the four new plugs, in the four smallest ring canals, have already fused. D) A 16-cell cyst in early interphase. Fusomal material is beginning to accumulate in the new ring canals, but has not yet formed distinct plugs. Only seven of the eight new ring canals are visible here. E,F) 16-cell cysts in mid interphase. The plugs have enlarged and are beginning to fuse, though not all are fusing at the same rate. In E, only 14 ring canals are visible, the other is out of focus. In F, only one of the plugs (asterisk) has not yet fuse with the rest of the fusome. G) A 16-cell cyst in late interphase. All fusome plugs have fused; the fusome spans all 15 ring canals. One of the two cells with the most ring canals always contains more fusomal material than its sisters (arrowhead). Thus, throughout cyst formation, the fusome is asymmetrically distributed within the cyst. (Images courtesy of Allan Spradling, Department of Embryology, Carnegie Institution of Washington and Development, the Company of Biologists, UK, and reprinted with permission from: de Cuevas M, Spradling AC. Development 1998; 125:2781-2789.[22])

(cystocytes). It is not known what drives the movement of the fusome plugs, but several evidences suggest that it could be microtubules based. Mutations in MT-binding proteins, Orbit/Mast and *abnormal spindle* (*asp*), or in an MT minus-end motor, dynein, and its associated regulator, *Lis1*, affect the formation of the fusome, which appears less branched.[31,34-37] Consequently, mutant cysts are produced with less than 16 cells. It is not yet clear, however, if these mutations affect the movement and formation of the plugs, or the anchoring of the mitotic spindles to the fusome, or both. A direct role of the microtubules in fusome biogenesis has not yet been tested, using depolymerising drugs. The polarity of the plugs movement may be given by a difference between the two centrosomes. The centrosome anchored to the fusome may have the ability to attract the plug, in contrast to the opposite centrosome. Whatever the mechanism, it shows that the plugs are able to read a polarity within each cell.

Furthermore, it is reasonable to assume that as early as the two-cell stage, the central fusome occupies a fix position, while it is the plugs which move toward it. The fusome therefore marks one side of each cell, which is at the center of the cyst (the "central" surface of each cell). In particular, in region 2b, the fusome remnant will always be located at the anterior of the oocyte, where all the cytoplasmic components accumulate to form a Balbiani body.[38] Therefore, as early as the two-cell stage, the "central" localisation of the fusome marks the future anterior side of the oocyte. An anterior-posterior axis in the oocyte is thus inherited from the polarised cyst divisions. It is tempting to try to go one step earlier and to relate this polarity to the cystoblast division or even to the GSC division. However, there is no fix referential in the cystoblast, and for the moment it is not possible to tell if the "original" fusome is always at the posterior of the cystoblast, or if after the first cyst division, it is the plug, or the "original" fusome, or both, which move to fuse.[22,29] Live imaging of these processes should help to solve these questions.

Fusome Composition

The Fusome Is Made of ER-Derived Vesicles

In 1901, Giardina described the fusome as a fibrillar material accumulated inside the arrested cleavage furrow and named it *residue fusoriale* to indicate that it was a remnant of the mitotic spindle.[2] However, light microscopy could not allow the nature of the fibrillar material to be determined. A real breakthrough came when King, Mahowald and colleagues used electronic microscopy (EM) to describe the *Drosophila* germarium.[19-21] They found an unusual organelle made of vesicles linking all the cells of the cyst through their ring canals (Fig. 4). This organelle excluded other components of the cytoplasm such as mitochondria and microtubules. This structure was then recognised by Telfer in 1975 as the fusome described earlier by Giardina and Hirschler.[2,39] The next step was then to identify the nature of these vesicles and it remained a mystery until very recently. Mary Lilly and colleagues made transgenic flies expressing GFP-tagged proteins targeted to specific cellular subcompartments such as the Golgi and the endoplasmic reticulum (ER).[26] They found that only the proteins targeted to the ER (Lys-GFP-KDEL, a lumenal ER protein and Sec61α-GFP an ER membrane protein) colocalised with the fusome, indicating that the fusome tubules are most similar to ER vesicles (Fig. 5). In an elegant experiment, they further showed that the fusome is continuous across all the cells of the cyst and is not made of independent vesicles packed together. They used Fluorescence Recovery after Photobleaching (FRAP) of Lys-GFP-KDEL and Sec61α-GFP to calculate their diffusion coefficient (Deff) within the fusome. They reasoned that if the fusome is made of discrete vesicles, both proteins should diffuse with the same speed, because their movements would be limited by the diffusion of the whole vesicle. In contrast, if the fusome is made of one continuous network of interconnected tubules, the lumenal protein Lys-GFP-KDEL should diffuse faster than Sec61α-GFP embedded in the more viscous ER membrane. They found the later case to be true and concluded that the fusome is continuous across the cyst. Moreover, they could also deplete the fluorescence of both proteins inside the fusome by repeatedly photobleaching the cytoplasmic ER of one cell of the cyst. This technique called FLIP (Fluorescence Loss in Photobleaching) demonstrates that the fusome is also interconnected with the cytoplasmic ER of each cell, which means that all the cells within a single cyst could share a common ER. These results are of primary importance as they could explain how the fusome synchronizes the development of the cells within the cyst by acting as a direct channel of communication. Interestingly, they further found that the continuity of the fusome is lost when the development of the cyst becomes asynchronous in region 2 of the germarium.[26]

However, we are still far from understanding how these ER-derived vesicles can form a plug inside a ring canal and then fuse to make a fusome. The final step of normal cytokinesis requires the insertion of new membranes at the cleavage furrow to separate the newly formed cells. Vesicles are transported to a compact structure rich in microtubules called the midbody, where they fuse with the existing plasma membrane.[40] Interestingly, the fusome plugs are also formed and associated with the microtubules at the midbody during early interphase after each cyst division.[3] It is thus possible that the plug is made of vesicles which are normally transported to the midbody but which, for an

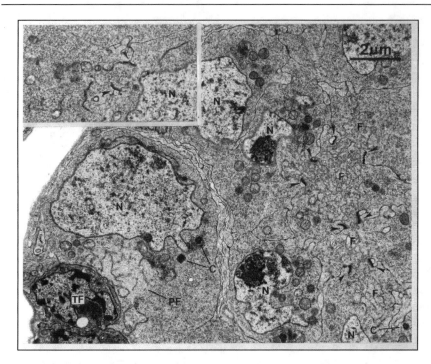

Figures 4. Electronic microscopy of the fusome and spectrosome in wild type and mutant germarium. Thin sections of the anterior region of the germarium from wild type and *hts* mutant (inset) females shown at the same magnification. Anterior is to the left and posterior to the right. In the wild type germarium, a germline stem cell (GSC) with a large nucleus (N) lies closed to the somatic terminal filament cells (TF) at the anterior. The spectrosome (PF) is the region rich in vesicles between the GSC nucleus and the TF. A wild type cyst contains a large fusome (F) that extends through four ring canals (labeled by four pair of arrows) and associated with a centriole (C). The fusome can be seen to contain small membranous vesicles and to exclude most ribosomes and mitochondria. In *hts* mutant female, these vesicles are absent (inset). The ring canal pictured appears to lack fusomal material entirely. (Image courtesy of Allan Spradling, Department of Embryology, Carnegie Institution of Washington and Development, the Company of Biologists, UK, and reprinted with permission from: Lin H, Yue L, Spradling A, Development 1994; 120:947-956.)[24]

unknown reason, do not fuse with the plasma membrane. This could be due to the intrinsic nature of these vesicles or because they are coated with specific proteins.

Molecular Composition of the Fusome

The first hint about the molecular composition of the fusome came from the genetic study of the *hts* (hu-li-tai shao) gene in *Drosophila*.[10,24] *hts* mutant flies produce germline cysts with a number of cells not following the 2^n rule and which usually lack an oocyte and are only made of nurse cells. EM analysis of *hts* mutant germarium revealed a complete lack of fusome vesicles, further indicating that Hts was a good candidate as a fusome component (Fig. 4). The molecular cloning of Hts identified it as the *Drosophila* homologue of adducin, an actin and spectrin-binding protein constituting part of the membrane cytoskeleton.[10] The membrane cytoskeleton is made of α and β-spectrin dimers linked together into a meshwork by junctional complexes composed of adducin, actin, protein 4.1 and tropomyosin. This lattice is attached to the plasma membrane by ankyrin and glycophorin. In erythrocytes, where it has been best studied, the sprectrin-based cytoskeleton was shown to modulate the mechanical properties of the plasma membrane.[41] In addition, these membrane skeletal proteins are also associated with internal organelles such as the

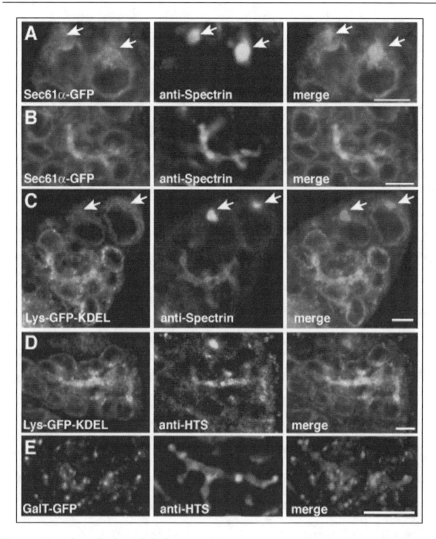

Figures 5. ER markers colocalise with the fusome. A-D) Germaria expressing the ER markers Sec61α-GFP (A,B) or Lys-GFP-KDEL (C,D) were fixed and labeled with the fusome markers anti-spectrin (A-C) or anti-Hts (D,E). Regions of colocalisation in the merged images (yellow) appear within the ER accumulations between cystocytes, but not within the cystocyte cytoplasm to reveal the ER origin of fusome membranes. In A and C, large spectrin accumulations (spectrosomes) in stem cells are indicated by white arrows. These structure do contain ER-labeling, though the structure are not substantially enriched with ER accumulations. E) A cyst from a GalTase-GFP (Golgi marker) ovary stained with anti-Hts. No significant colocalization between these markers was observed. Bars, 5 µm. (Image courtesy of Mary Lilly, Cell Biology and Metabolism branch, NIH. Reprinted from: Snapp EL, Iida T, Frescas D et al, Mol Biol Cell 2004, 15:4512-4521, with permission from The American Society for Cell Biology.[26]) A color version of this figure is available online at www.Eurekah.com.

Golgi complex and the ER.[42] Accordingly, α and β-spectrin were found to be part of the spectrosome and fusome, while ankyrin was mostly found in the spectrosome, suggesting that the molecular nature of the spectrosome/fusome is developmentally regulated (Fig. 3).[9,24] In contrast, *Drosophila* homologues of protein 4.1 such as Coracle, Moesin or Merlin were not seen on the fusome.[9,43]

F-actin is not detected reliably on the fusome in *Drosophila* female ovaries, whereas it is a major component of the male fusome.[5] Strong fixation protocols seem to be required to detect F-actin on the female fusome, suggesting that F-actin is unstable.[44] Furthermore, it has recently been shown that mutations in *Scr64*, *Tec29* and *kelch*, which are involved in crosslinking actin filaments, affect the branching and maintenance of the fusome.[44] Scr64 and Tec29 are tyrosine kinases which can phosphorylate Kelch and inhibit its actin-bundling ability. It has thus been proposed that the branching defects in *Scr64*, *Tec29* and *kelch* mutants could be due to a failure to loosen actin filaments crosslinking around the fusome plugs to allow their fusion with the central fusome. Whatever the mechanism, these results at least suggest that the actin cytoskeleton could play an important role in the formation of the fusome.

In *hts* mutant most of the vesicles are gone, while Hts itself is only faintly detected on fusome remnants of α-*spectrin* mutant cysts.[9,24] Although it suggests a direct role of the membrane skeletal proteins in the biogenesis of the fusome, their precise function remains unknown. They could be involved in the formation of the fusome by coating some ER-vesicles to give them a specific identity and/or affinity. This would create a separate subdomain of the ER allowing the fusion of these vesicles and the formation of a distinct continuous structure. Alternatively, they could be required for the maintenance of the fusome by surrounding it with a rigid cytoskeleton, prohibiting its fusion with the plasma membrane and the dispersal of the vesicles.

Bam and its associated protein TER94 constitute a second class of components of the fusome. Bam is a novel protein, which localises both to the fusome of all germ cells and to the cytoplasm of dividing cysts.[45,46] *bam* mutant cysts are mostly stopped at the one-cell cystoblast stage and contain a greatly reduced fusome, leading to the suggestion that Bam could be involved in the recruitment of vesicles to the fusome. Mutations in the *benign gonial cell neoplasm* (*bgcn*) gene induce an almost identical phenotype and interestingly, *bgcn* cysts also fail to localise Bam protein on the fusome, while Hts and α-spectrin are normally localised.[47] These results suggest that Bam localisation on the fusome is required for its function.

TER94 was found to directly interact with Bam in a two-hybrid screen and also to localise to the fusome.[48] TER94 is a fly orthologue of the vertebrate TER proteins, required for vesicles fusion during the biogenesis of the ER and Golgi. This conserved function of TER94 supports the view that the fusome is formed by the fusion of ER-derived vesicles. However, TER94 function during the formation of the fusome could not be assessed as the existing alleles are cell lethal. Furthermore, as the *bam* mutant cysts do not progress beyond the one-cell stage, it remains unclear whether they contain a spectrosome or a fusome. One major difference between the fusome and the spectrosome is that the spectrosome is less dense in vesicles, which is similar to what has been described for the "reduced fusome" of *bam* mutant cysts.[24,45] It is thus not clear if Bam is directly required for the formation of the fusome, or if the defects seen in *bam* mutant cysts are due to an early arrest of the cystoblast development.

A third class of fusome components is made of microtubules-associated proteins including the plus-end binding proteins, Orbit/Mast and CLIP-190, the MT minus-end motor, dynein, and the *Drosophila* homologue of Spectraplakin, shot. Orbit/Mast shows a cell-cycle dependent localisation on the fusome.[36] During mitosis, it localises on the mitotic spindles and then on the spindle remnants at anaphase. During interphase, it is incorporated into the newly formed ring canals, the fusome plugs and then into the fusome. Dynein has been reported to localise on the fusome only during mitosis, however a clear colocalisation with known fusome components has not been demonstrated yet.[31] Furthermore, as shown previously (see II.1), mutations in the dynein and Orbit/Mast genes lead to the formation of fragmented and unbranched fusomes indicating that both proteins and probably the microtubule cytoskeleton are required for the formation of the fusome.

Finally, Shot and several other proteins have been reported to associate at least transiently with the fusome such as Cyclin A, P-Cyclin E, Cul1, 19S-S1 and Par-1, LKB1 (Par-4), Par-5.[49-54] However, these proteins seem to be required for the differentiation of the germline cyst rather than the formation of the fusome per se. Their function will thus be discussed in this context below.

Fusome Functions in the Formation and Differentiation of the Germline Cyst

In the absence of a fusome such as in *hts* and *α-spectrin* mutant cysts, the formation and differentiation of the germline cyst is greatly affected.[9,10] These mutant cysts contain a variable number of cells, which are usually less than 16 and not a factor of 2^n. In addition, they frequently lack an oocyte and are only made of nurse cells. These defects arise despite the presence of ring canals, indicating that the fusome is likely to be the main channel of cell-cell communication for the control of the cyst divisions and the differentiation of the oocyte during normal development. I will review recent exciting results which shed new light on the role of the fusome in both of these processes.

The Fusome and the Regulation of the Cell-Cycle

Drosophila germline cysts are produced by four rounds of synchronous divisions with incomplete cytokinesis. This requires at least two levels of regulation: one mechanism synchronising the cell cycle of each cystocyte and a second mechanism counting the number of division to precisely four. Direct and reverse genetic approaches have been used to find proteins involved in both mechanisms. The main cell-cycle regulators, Cyclins, were obvious candidates.

In *Drosophila*, CyclinA (CycA) associates with the cyclin-dependent kinase, cdk1, to promote the transition from G2 phase to mitosis. Interestingly, CycA was shown to localise on the fusome precisely during late G2 and early prophase, while it is degraded after metaphase (Fig. 6).[55] The

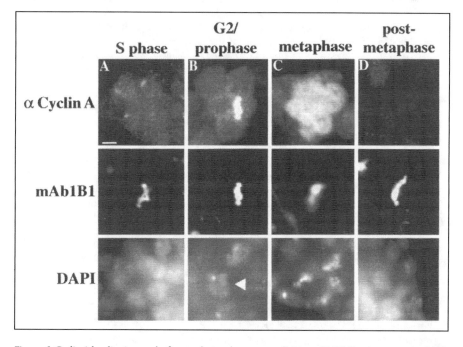

Figures 6. Cyclin A localisation on the fusome during the cystocyte divisions. (A-D) Ovaries were stained with antibodies raised against Drosophila CycA and the monoclonal antibody mAb1B1 which marks the fusome. DNA is labeled with DAPI. A single cyst in the indicated cell cycle stage is shown in each part. The DNA begins to condense at the G2/prophase transition (arrowhead). CycA precisely localises on the fusome at the G2/prophase transition and starts to be degraded at metaphase. For all images, not all cells within the dividing cyst are in the plane of focus. Bars, 5 μm. (Image courtesy of Mary Lilly, Cell Biology and Metabolism branch, NIH and the Academic Press, and reprinted with permission from: Lilly MA et al. Dev Biol 218:53-63, ©2000 Elsevier.[55])

association of CycA with the whole fusome in late G2 phase could provide a mechanism to synchronise the entry of all cystocytes into mitosis. However, it remains unclear how the fusome could spatially equalize the activation of the CycA-cdk1 in all cells. One possibility is that phosphatases and kinases required for the activation of CycA-cdk1 could also be localised on the fusome. The activation of CycA-cdk1 at one point of the fusome could then feedback to the kinases and phosphatases and spread to all the cells by using in some way the fusome as a physical support. As mentioned earlier, in late G2 the fusome is continuous across the cyst and connect all the cells.[26] The propagation of the active state of CycA-cdk1 would then synchronise the entry into mitosis.

The degradation of the cyclins appears equally important for the counting mechanism. Mutations in the *encore* gene induce one extra-division of the cyst and the formation of egg chambers containing 32 cells and one oocyte with five ring canals.[56] Levels of CycA and Cyclin E, a G1 to S phase promoting cyclin, were shown to be abnormally high and to persist longer in region 2 of *encore* mutant germarium than in wild type, suggesting that an excess of activated cyclins could cause the extra division.[53] Accordingly, the direct overexpression of CycA and CycE with a heat inducible promoter also induces the formation of 32-cells cysts, albeit at a low frequency.[53,55,57] In addition, mutations affecting the degradation machinery also lead to high levels of CycA and CycE and greatly enhance the frequency on the extra round of mitosis in *encore* mutant ovaries. The intriguing twist of the story is that several members of the ubiquitin-dependent degradation pathway such as Cul1 and the proteasome 19S subunit S1, as well as phosphorylated forms of CyclinE (P-cyclinE) marked for degradation, localise on the fusome.[53] It was further shown that Encore physically interacts with the proteasome and CyclinE and was required for their localisation to the fusome, although Encore biochemical function remains unknown. Thus, the model proposes that Encore recruits the proteasome to the fusome to degrade simultaneously CyclinE in all cells, which would end the four rounds of mitosis.[53] However, these results and the model do not explain why the cyst stops dividing after precisely four cycles. This question is central for developmental biologists, because many other processes such as cellularization of the *Drosophila* embryo and differentiation of the peripheral nervous system are also triggered after a precise number of divisions. How cells count the number of mitosis is a fascinating mystery and the divisions of the *Drosophila* cyst provides a powerful model to address it.

The Fusome and the Determination of the Oocyte

In *Drosophila* ovaries, as soon as the four divisions are complete and a 16-cells cyst is formed, the development of each cell of the cyst becomes asynchronous and an oocyte starts to differentiate. The determination of the oocyte has puzzled biologists for more than a century, because it arises from a syncytium of 16 sister cells that share the same cytoplasm. Several steps are required for one cell to become the female gamete (Fig. 1). Firstly, the oocyte needs to be selected from 16 sister cells in region 1 of the germarium. Secondly, the oocyte has to differentiate by accumulating oocyte-specific proteins and mRNAs in region 2. Finally, the oocyte needs to be correctly polarised to maintain its identity in region 3. Although the fusome is only present in region 1 of the germarium, recent data suggest that it may well be required at several steps for the correct determination of the oocyte (Fig. 7).

The Fusome and the Selection of the Oocyte

Two main models have been proposed to explain how the oocyte is selected. One model is based on the symmetrical behaviour of the two pro-oocytes until mid-late region 2a, and proposes that there is a competition between the two pro-oocytes to become the oocyte.[58,59] The "winning" cell would become the oocyte, while the "losing" cell would revert to the nurse cell fate. However, the factor that could control this decision has remained elusive. A second model suggests that the choice of the oocyte is biased by the establishment of some asymmetry as early as the first cystoblast division, which is maintained until the overt differentiation of the oocyte.[30,60] As described earlier, the formation of the fusome provides the strongest evidence in support of the second model.[22,29-31,61] Indeed, the asymmetric inheritance of the "original fusome" during the cyst divisions could play the role of such an asymmetric cue. This model is further supported by analogy with the diving beetle

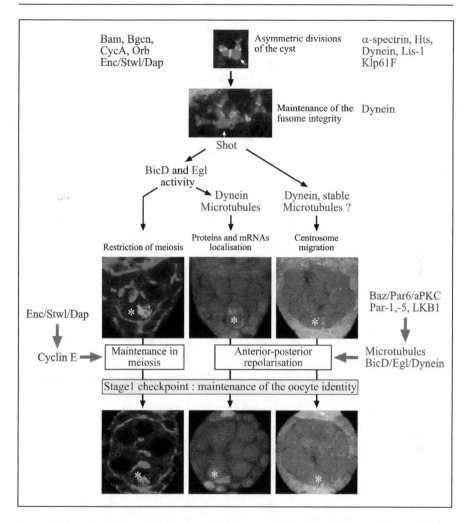

Figures 7. The early steps in the determination and polarisation of the *Drosophila* oocyte. The early differentiation of the oocyte is a multistep process. Genes involved at each step are indicated on the figure. The lists are not exhaustive. Actors of the cytoplasmic differentiation are in red, whereas regulators of the cell cycle are in blue. The top panel shows a 4-cell cyst (α-spectrin, in red, marks the fusome and anilin, in green, marks the ring canals). One cell has more fusome than the other cells (arrow). The panel below shows a 16-cell cyst after the last division with one cell having more fusome (arrow). There are three different pathways to restrict the oocyte identity to one cell (asterik). The left panel shows the actin in red and the synaptonemal complex in green. In the middle panels, a nuclear GFP is in green, and Orb (an oocyte-specific cytoplasmic protein) is in red. On the right panels, the γ-tubulin marks the centrosomes in red, the α-spectrin marks the fusome in blue, and a nuclear GFP is in green. Orb and the centrosomes are clearly seen migrating from the anterior of the oocyte to the posterior, revealing the repolarisation of the oocyte. (Reprinted with permission from: Huynh JR et al. Curr Biol 14:R438-449, ©2004 Elsevier.[15]) A color version of this figure is available online at www.Eurekah.com.

Dytiscus, in which oogenesis is very similar to *Drosophila*.[2,6] A *Dytiscus* cyst is formed of 15 nurse cells and one oocyte, resulting from four incomplete and synchronous divisions of a cystoblast. However, unlike *Drosophila*, the oocyte can be distinguished as early as the 2-cell stage because it contains a large ring of highly amplified rDNA. Moreover, the cell that inherits that ring of rDNA, also inherits the fusome. This also suggests that the early selection of the oocyte could be a general feature among insects.[3,22] Unfortunately, in *Drosophila* most of the fusome has already degenerated by the time oocyte-specific proteins such BicD or Orb accumulate in a single cell in late region 2a. However, the preferential accumulation of the centrosomes and *osk*, and *orb* mRNAs in one cell can be detected earlier in region 2a, and this is always the cell with the most fusome.[3,38] This is particularly obvious in *egl* and *BicD* mutants, in which the fusome perdures longer, and where the centrosomes clearly accumulate in the cell with the largest piece of fusome remnant.[34] These data strongly suggest that the "original" fusome marks the future oocyte, in support of the second model. It does not rule out the possibility that both pro-oocytes can become the oocyte, but shows that if there is a competition, it is strongly biased.

What is the link between the asymmetric inheritance of the fusome and the selection of the oocyte? The simplest model is that an oocyte determinant is asymmetrically distributed at each division with the "original" fusome into the future oocyte. It has been proposed that one of the cystoblast centrioles could stay in contact with the fusome during each division, and because of the semi-conservative replication of the centrosome, could be inherited by the oocyte (Green centrosome in Fig. 2).[60] Consequently, oocyte determinants could cosegregate with this centriole. Such a mechanism has been shown to mediate the segregation of *dpp* and *eve* mRNAs into specific cells during the asymmetric divisions of the early *Ilyanassa obsoleta* embryo.[62] Alternatively, the oocyte could inherit more of some protein or activity associated with the fusome, and this early bias could initiate a feedback loop that induces the transport of oocyte determinants towards this cell.

The Fusome and the Differentiation of the Oocyte

Even though strong evidence suggest that the oocyte is selected early in region 1, its identity only becomes obvious two days later, in late region 2a. The cyst differentiation in region 2a is gradual and two-fold. In the cytoplasm, the oocyte accumulates specific components and organelles, and in the nucleus, it enters meiosis and arrests in prophase I. Recent results have uncovered a complex picture in which multiple processes function in parallel to restrict different types of oocyte-specific components to one cell (for a detailed information see our recent review).[15] However, all of these probably depend on the initial polarity of the fusome, which may act in three distinct ways for the differentiation of the oocyte.

Firstly, the fusome organises a polarised network of dynamic microtubules (MTs) that direct the localisation of oocyte-specific proteins and mRNAs into one cell. Indeed, one of the main features of cyst differentiation is the formation of a microtubule array that is polarised towards the oocyte and extends through the ring canals into the other cells of the cyst.[63] Since the MT form along the fusome, this polarisation has been suggested to be a direct read out of the fusome polarity.[3] MTs are essential for the determination of the oocyte, since treatments with the MT depolymerising drug, colchicine, result in the failure to localise the oocyte-specific proteins or mRNAs asymmetrically within these cysts and in the formation of egg chambers with 16 nurse cells and no oocyte.[64] Furthermore, mutations in the adaptor proteins *Bicaudal-D* (*Bic-D*), *egalitarian* (*egl*) and the dynein light chain produce a very similar phenotype, suggesting that the transport of these proteins and mRNAs is MT-dependent and uses the dynein-dependent transport of cargoes that are linked to the motor through BicD, Egl and the dynein light chain.[31,65-67] Secondly, the fusome also nucleates stable MTs that are associated with the *Drosophila* spectraplakin, Shot. Indeed, colchicine disrupts the localisation of proteins and mRNAs to the oocyte, but not the migration of the centrioles or the restriction of meiosis to one cell.[34,68] Since MT-destabilising drugs like colchicine only affect dynamic MTs, one possibility is that the centrioles migrate along stable MTs that are not affected by these treatments. In support of this view, it has recently been found that antibodies against acetylated tubulin, a marker for stable MTs, label a population of MTs associated with the fusome.[54] Furthermore, mutations in the *Drosophila* spectroplakin, Shot disrupt these stable microtubules without affecting the fusome itself, and this blocks the migration of the centrioles, as well as mRNAs and proteins, into one cell. Shot is a component of the fusome, and contains a GAS domain, which has been shown to bind and stabilise MTs. This suggests a model in which Shot assembles stable MTs on the fusome, along which the centrioles migrate. The situation is less clear for the restriction of meiosis, as it is only partially disrupted in *shot* germline clones. This raises the possibility that a third MT-independent pathway reads the fusome polarity directly to control the entry into meiosis.[34,68]

In summary, the current model suggests that although the fusome starts to degenerate in early region 2a, it acts as a matrix to organise the restriction of oocyte-specific proteins, centrioles and meiosis to a single cell by at least three distinct pathways (Fig. 7). However, what molecular mechanisms regulate the different pathways and how these pathways interpret the fusome polarity remains unknown.

The Fusome and the Early Polarisation of the Oocyte

When a germline cyst reaches region 2b, the oocyte specific proteins and mRNAs, as well as the centrosomes and mitochondria have been transported along the fusome into the presumptive oocyte. These components remain associated with the fusome remnants and therefore accumulate at the anterior of the oocyte to form a Balbiani body.[38] When the oocyte moves through region 3, all of the components of the Balbiani body disassociate and move around the oocyte nucleus to form a tight crescent at the posterior cortex. This movement is the first sign of anterior-posterior polarity in the oocyte, and is a crucial step in the maintenance of its identity.[51,69]

The importance of this polarisation step was first demonstrated by the analysis of mutations affecting members of a conserved group of genes called the *par* genes, which specifically disrupt this process (reviewed in ref. 70). The *Drosophila par* genes, which include *par-1*, *par-3* (*bazooka*), *LKB1* (*par-4*), *par-5* (*14-3-3*), *par-6* and *aPKC*, were first discovered in *C. elegans* where they are required for the polarisation of the one-cell zygote. They were later shown to be widely required from *C. elegans* to *Drosophila* and Human for the polarisation of many cell types.[71] In *Drosophila*, egg chambers mutant for any of the *par* genes contains 16 nurse cells and lack an oocyte. The oocyte appears to be selected normally, as the centrosomes, meiotic markers, and Orb accumulate in one cell in region 2b/3.[51] These components do not translocate to the posterior of the oocyte in region 3, however, and the oocyte de-differentiates as a nurse cell, by exiting meiosis and becoming polyploid. Intriguingly, Par-1, LKB1 (Par-4) and Par-5 localise on the fusome, while Par-3 and aPKC form rings of adherens junctions around the ring canal.[33,49,51,52] However, none of these genes seem to be required for the formation of the fusome itself, nor during the cyst divisions. What could be the link between the fusome and the early polarisation of the oocyte? One possibility is that the *par* genes are required with the fusome to polarize each cell of the cysts during its divisions in region 1. This polarisation of the oocyte would then organise its cytoskeleton required for the translocation of the Balbiani body to the posterior pole in region 3. As already mentioned, the fusome marks the "central" surface of each cell and could thus serve to asymmetrically localised the PAR proteins and polarise each cell of the cyst, including the future oocyte. Alternatively, as the Balbiani body forms around the fusome remnants, the localisation of the PAR proteins on the fusome and its remnants could be required directly for the MT-dependent transport of the Balbiani vesicles to the posterior of the oocyte. Although mainly unexplored, the links between the fusome, the *par* genes and the early polarisation of the oocyte are promising.

Conclusions and Perspectives

Although I have focused on *Drosophila* oogenesis in this review, the role of the fusome as a channel of communication is likely to be widely conserved. Indeed, groups of germ cells dividing synchronously have been reported in females and males of most if not all animal species. Given the primary role of the fusome in synchronising these cells, it suggests that a fusome could exist in those same species at least during the divisions. In support of this view, a structure similar to the *Drosophila* fusome at the morphological and molecular levels has been recently found in *Xenopus*.[8] Indeed, *Xenopus* germline cysts are formed by four rounds of incomplete mitosis and contain a spectrin-rich structure branching across the cyst (Fig. 8). However, in *Xenopus* all the germ cells become oocyte.

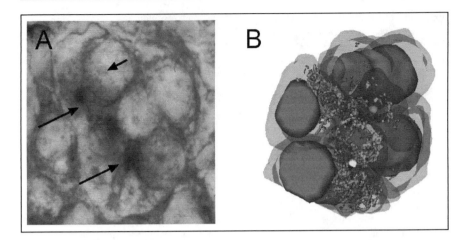

Figures 8. Fusome of *Xenopus laevis* female ovaries. A) *Xenopus* 8-cell cyst stained with an antibody raised against spectrin. Long arrows point to spectrin in the cytoplasm where the primary mitochondrial clouds (PMC) and the fusome localise. Short arrow indicates the nucleus. B) Three-dimensional reconstruction of an 8-cell cyst. Cytoplasm is grey, nuclei are red, mitochondria of PMC are green, centrioles blue and ring canals are yellow. PMC, ring canals and centrioles face each other and are located centripetally in the "rosette" conformation. This reconstruction was made from 38 serial ultrathin sections. (Images courtesy of Malgorzata Kloc and Laurence Etkin, Department of Molecular Genetics, University of Texas, and reprinted with permission from: Kloc M et al. Dev Biol 266:43-61, ©2004 Elsevier.[8]) A color version of this figure is available online at www.Eurekah.com.

This suggests that the role of the fusome in the selection and differentiation of the oocyte may be a secondary function restricted to species which differentiate accessory germ cells such as the nurse cells in *Drosophila*.

We are only starting to understand the origin, formation and functions of the fusome and many questions remain unanswered. However, rapid progresses have been made recently thanks to the use of powerful genetic tools available in *Drosophila*. Indeed, several genes required for the formation of the fusome and proteins associated with the fusome have been found in large scale genetic screens and reverse genetic approaches. In addition, it is now possible to study the fusome in vivo using confocal live imaging and transgenic flies expressing specific fluorescently tagged proteins.[26] Furthermore, the recent success in culturing GSCs in a cell medium opens exciting new possibilities to study the differentiation of the germline cyst in vitro.[72] The early steps of *Drosophila* oogenesis thus provide a useful paradigm to comprehend the role of the fusome and therefore to better understand the formation of functional gametes crucial to the sexual reproduction.

Acknowledgements

I thank Jean-Antoine Lepesant for critical reading of the manuscript. I also thank Laurence Etkin, Malgorzata Kloc, Mary Lilly and Allan Spradling for the images illustrating this review. I am supported by the CNRS and a grant from Association pour la Recherche sur le Cancer (A.R.C.).

References

1. Pepling ME, de Cuevas M, Spradling AC. Germline cysts: A conserved phase of germ cell development. Trends Cell Biol 1999; 9:257-262.
2. Giardina A. Origine dell'oocite e delle cellule nutrici nel Dytiscus. Int Mschr Anat Physiol 1901; 18:417-484.
3. Grieder N, De Cuevas M, Spradling A. The fusome organizes the microtubules network during oocyte differentiation in Drosophila. Development 2000; 127:4253-4264.
4. Braun RE, Behringer RR, Peschon JJ et al. Genetically haploid spermatids are phenotypically diploid. Nature 1989; 337:373-376.

5. Hime G, Brill J, Fuller M. Assembly of ring canals in the male germ line from structural components of the contractile ring. J Cell Sci 1996; 109:2779-2788.
6. Telfer W. Development and physiology of the oocyte-nurse cell syncytium. Adv Insect Physiol 1975; 11:223-319.
7. Buning J. The Insect Ovary: Ultrastructure, previtellogenic growth and evolution. New-York: Chapman and Hall, 1994.
8. Kloc M, Bilinski S, Dougherty MT et al. Formation, architecture and polarity of female germline cyst in Xenopus. Dev Biol 2004; 266:43-61.
9. de Cuevas M, Lee JL, Spradling AC. α-spectrin is required for germline cell division and differentiation in the Drosophila ovary. Development 1996; 124:3959-3968.
10. Yue L, Spradling A. hu-li tai shao, a gene required for ring canal formation during Drosophila oogenesis, encodes a homolog of adducin. Genes Dev 1992; 6:2443-2454.
11. Cooley L. Drosophila ring canal growth requires Src and Tec kinases. Cell 1998; 93:913-915.
12. Hudson AM, Cooley L. Understanding the function of actin-binding proteins through genetic analysis of Drosophila oogenesis. Annu Rev Genet 2002; 36:455-488.
13. Spradling A, Drummond-Barbosa D, Kai T. Stem cells find their niche. Nature 2001; 414:98-104.
14. de Cuevas M, Lilly MA, Spradling AC. Germline cyst formation in Drosophila. Annu Rev Genet 1997; 31:405-428.
15. Huynh JR, St Johnston D. The origin of asymmetry: Early polarisation of the Drosophila germline cyst and oocyte. Curr Biol 2004; 14:R438-449.
16. Spradling A. Developmental genetics of oogenesis. In: Bate M, Martinez-Arias A, eds. The Development of Drosophila melanogaster. New-York: Cold Spring Harbor Laboratory Press, 1993:1-70.
17. Gonzalez-Reyes A. Stem cells, niches and cadherins: A view from Drosophila. J Cell Sci 2003; 116:949-954.
18. Platner G. Die Karyokinese bie den Lepidopteran als Grundlage fur eine Theorie der Zellteilung. Int Mschr Anat Physiol 1886; 3:341-398.
19. Koch E, King R. The origin and early differentiation of the egg chamber of Drosophila. J Morph 1966; 119:283-304.
20. Mahowald AP. The formation of ring canals by cell furrows in Drosophila. Z Zellforsch Mikrosk Anat 1971; 118:162-167.
21. Mahowald AP. Ultrastructural observations on oogenesis in Drosophila. J Morphol 1972; 137:29-48.
22. de Cuevas M, Spradling AC. Morphogenesis of the Drosophila fusome and its implications for oocyte specification. Development 1998; 125:2781-2789.
23. Lin H, Spradling A. Fusome asymmetry and oocyte determination in Drosophila. Dev Genet 1995; 16:6-12.
24. Lin H, Yue L, Spradling A. The Drosophila fusome, a germline-specific organelle, contains membrane skeletal proteins and functions in cyst formation. Development 1994; 120:947-956.
25. Zaccai M, Lipshitz H. Role of Adducin-like (hu-li tai shao) mRNA and protein localization in regulating cytoskeletal structure and function during Drosophila oogenesis and early embryogenesis. Dev Genet 1996; 19:249-257.
26. Snapp EL, Iida T, Frescas D et al. The fusome mediates intercellular endoplasmic reticulum connectivity in Drosophila ovarian cysts. Mol Biol Cell 2004; 15:4512-4521.
27. Lin H, Spradling A. A novel group of pumilio mutations affects the asymmetric division of germline stem cells in the Drosophila ovary. Development 1997; 124:2463-2476.
28. Xie TAS. A niche maintaining germline stem cells in the Drosophila ovary. Science 2000; 290:328-330.
29. Deng W, Lin H. Spectrosomes and fusomes anchor mitotic spindles during asymmetric germ cell divisions and facilitate the formation of a polarized microtubule array for oocyte specification in Drosophila. Dev Biol 1997; 189:79-94.
30. Lin H, Spradling AC. Fusome asymmetry and oocyte determination in Drosophila. Dev Gen 1995; 16:6-12.
31. McGrail M, Hays TS. The microtubule motor cytoplasmic dynein is required for spindle orientation during germline cell divisions and oocyte differentiation in Drosophila. Development 1997; 124:2409-2419.
32. Storto PD, King RC. The role of polyfusomes in generating branched chains of cystocytes during Drosophila oogenesis. Dev Genet 1989; 10:70-86.
33. Huynh JR, Petronczki M, Knoblich JA et al. Bazooka and PAR-6 are required with PAR-1 for the maintenance of oocyte fate in Drosophila. Curr Biol 2001; 11:901-906.
34. Bolivar J, Huynh JR, Lopez-Schier H et al. Centrosome migration into the Drosophila oocyte is independent of BicD and egl, and of the organisation of the microtubule cytoskeleton. Development 2001; 128:1889-1897.

35. Liu Z, Xie T, Steward R. Lis1, the Drosophila homolog of a human lissencephaly disease gene, is required for germline cell division and oocyte differentiation. Development 1999; 126:4477-4488.
36. Mathe E, Inoue YH, Palframan W et al. Orbit/Mast, the CLASP orthologue of Drosophila, is required for asymmetric stem cell and cystocyte divisions and development of the polarised microtubule network that interconnects oocyte and nurse cells during oogenesis. Development 2003; 130:901-915.
37. Riparbelli MG, Massarelli C, Robbins LG et al. The abnormal spindle protein is required for germ cell mitosis and oocyte differentiation during Drosophila oogenesis. Exp Cell Res 2004; 298:96-106.
38. Cox RT, Spradling AC. A Balbiani body and the fusome mediate mitochondrial inheritance during Drosophila oogenesis. Development 2003; 130:1579-1590.
39. Hirschler J. Gesetzmassigkeiten in den Ei-Nahrzellenverbanden. Zool Jb Abt Allg Zool Physiol 1945; 61:141-236.
40. Straight AF, Field CM. Microtubules, membranes and cytokinesis. Curr Biol 2000; 10:R760-770.
41. Bennett V, Baines AJ. Spectrin and ankyrin-based pathways: Metazoan inventions for integrating cells into tissues. Physiol Rev 2001; 81:1353-1392.
42. Devarajan P, Stabach PR, Mann AS et al. Identification of a small cytoplasmic ankyrin (AnkG119) in the kidney and muscle that binds beta I sigma spectrin and associates with the Golgi apparatus. J Cell Biol 1996; 133:819-830.
43. McCartney BM, Fehon RG. Distinct cellular and subcellular patterns of expression imply distinct functions for the Drosophila homologues of moesin and the neurofibromatosis 2 tumor suppressor, merlin. J Cell Biol 1996; 133:843-852.
44. Djagaeva I, Doronkin S, Beckendorf SK. Src64 is involved in fusome development and karyosome formation during Drosophila oogenesis. Dev Biol 2005.
45. McKearin D, Ohlstein B. A role for the Drosophila bag-of-marbles protein in the differentiation of cystoblasts from germline stem cells. Development 1995; 121:2937-2947.
46. McKearin D, Spradling A. bag-of-marbles: A Drosophila gene required to initiate both male and female gametogenesis. Genes Dev 1990; 4:2242-2251.
47. Lavoie CA, Ohlstein B, McKearin DM. Localization and function of Bam protein require the benign gonial cell neoplasm gene product. Dev Biol 1999; 212:405-413.
48. Leon A, McKearin D. Identification of TER94, an AAA ATPase protein, as a Bam-dependent component of the Drosophila fusome. Mol Biol Cell 1999; 10:3825-3834.
49. Benton R, Palacios IM, St Johnston D. Drosophila 14-3-3/PAR-5 is an essential mediator of PAR-1 function in axis formation. Dev Cell 2002; 3:659-671.
50. Cox DN, Lu B, Sun T et al. Drosophila par-1 is required for oocyte differentiation and microtubule organization. Curr Biol 2001; 11:75-87.
51. Huynh JR, Shulman JM, Benton R et al. PAR-1 is required for the maintenance of oocyte fate in Drosophila. Development 2001; 128:1201-1209.
52. Martin SG, St Johnston D. A role for Drosophila LKB1 in anterior-posterior axis formation and epithelial polarity. Nature 2003; 421:379-384.
53. Ohlmeyer JT, Schupbach T. Encore facilitates SCF-Ubiquitin-proteasome-dependent proteolysis during Drosophila oogenesis. Development 2003; 130:6339-6349.
54. Roper K, Brown NH. A spectraplakin is enriched on the fusome and organizes microtubules during oocyte specification in Drosophila. Curr Biol 2004; 14:99-110.
55. Lilly MA, de Cuevas M, Spradling AC. Cyclin A associates with the fusome during germline cyst formation in the Drosophila ovary. Dev Biol 2000; 218:53-63.
56. Hawkins N, Thorpe J, Schupbach T. Encore, a gene required for the regulation of germ line mitosis and oocyte differentiation during Drosophila oogenesis. Development 1996; 122:281-290.
57. Lilly M, Spradling A. The Drosophila endocycle is controlled by Cyclin E and lacks a checkpoint ensuring S-phase completion. Genes Dev 1996; 10:2514-2526.
58. Carpenter A. Electron microscopy of meiosis in Drosophila melanogaster females. I Structure, arrengement, and temporal change of the synaptonemal complex in wild-type. Chromosoma 1975; 51:157-182.
59. Carpenter A. Egalitarian and the choice of cell fates in Drosophila melanogaster oogenesis. Ciba Found Symp 1994; 182:223-246, (246-254).
60. Theurkauf WE. Microtubules and cytoplasm organisation during Drosophila oogenesis. Dev Biol 1994; 165:352-360.
61. Storto P, King R. The role of polyfusomes in generating branched chains of cystocytes during Drosophila oogenesis. Dev Genet 1989; 10:70-86.
62. Lambert JD, Nagy LM. Asymmetric inheritance of centrosomally localized mRNAs during embryonic cleavages. Nature 2002; 420:682-686.

63. Theurkauf W, Alberts M, Jan Y et al. A central role for microtubules in the differentiation of Drosophila oocytes. Development 1993; 118:1169-1180.
64. Koch E, Spitzer R. Multiple effects of colchicine on oogenesis in Drosophila; induced sterility and switch of potencial oocyte to nurse-cell developmental pathway. Cell Tissue Res 1983; 228:21-32.
65. Mach JM, Lehmann R. An Egalitarian-BicaudalD complex is essential for oocyte specification and axis determination in Drosophila. Genes Dev 1997; 11:423-435.
66. Navarro C, Puthalakath H, Adams JM et al. Egalitarian binds dynein light chain to establish oocyte polarity and maintain oocyte fate. Nat Cell Biol 2004; 6:427-435.
67. Ran B, Bopp R, Suter B. Null alleles reveal novel requirements for Bic-D during Drosophila oogenesis and zygotic development. Development 1994; 120:1233-1242.
68. Huynh JR, St Johnston D. The role of BicD, Egl, Orb and the microtubules in the restriction of meiosis to the Drosophila oocyte. Development 2000; 127:2785-2794.
69. Paré C, Suter B. Subcellular localization of Bic-D:GFP is linked to an asymmetric oocyte nucleus. J Cell Sci 2000; 113:2119-2127.
70. Pellettieri J, Seydoux G. Anterior-posterior polarity in C. elegans and Drosophila -PARallels and differences. Science 2002; 298:1946-1950.
71. Ohno S. Intercellular junctions and cellular polarity: The PAR-aPKC complex, a conserved core cassette playing fundamental roles in cell polarity. Curr Opin Cell Biol 2001; 13:641-648.
72. Kai T, Williams D, Spradling AC. The expression profile of purified Drosophila germline stem cells. Dev Biol 2005.

CHAPTER 17

Cytonemes as Cell-Cell Channels in Human Blood Cells

Svetlana Ivanovna Galkina,* Anatoly Georgievich Bogdanov,
Georgy Natanovich Davidovich and Galina Fedorovna Sud'ina

Abstract

Human blood cells similar to embryonic and nerve cells can project thin and very long extensions having the same diameter along the entire length called cytonemes. Cytonemes were shown to connect blood cells over a distance of several cell diameters and transport membrane proteins, lipids and ions from one of connected cells to another one, thus executing long range intercellular communications. Formation and breakage of cytonemes upon neutrophil rolling along the vessel walls regulate rolling velocity and control neutrophil adhesion to the endothelium. Direct interaction of neutrophils with platelets over a distance seems to be of great importance in thrombosis. Cytonemes of B cells, peripheral NK cells, monocytes and dendritic cells can play a critical role in long range cellular signalling upon antigen presentation and formation of immune response.

Introduction

Blood cells secret and accept signalling molecules to communicate with other cells over a distance. Recent investigations demonstrate that cells can directly interact with cells over a distance of several cell diameters using very long and very thin filopodia. Such filopodia were observed firstly in a variety of embryonic cells.[1-7] Along with thin filopodia, a number of synonyms are used to call very thin intercellular connectives, such as membrane tethers, tubulovesicular extensions or thread-like projections, tunnelling nanotubes or nanotubular highways. Based on a connectives thickness, authors unite sometimes together under the same term very thin taper and thread-like filopodia, which differ in origin, behaviour and functions.

The term cytoneme was firstly used to describe "slightly beaded" thread-like filopodia with a diameter of approximately 0.2 μm projected by *Drosophila* wing imaginal disc cells.[6] Cytonemes in the present work can be determined as tubular or tubulovesicular cellular filopodia with uniform diameter along the entire length. Diameter of cytonemes can vary from 0.035 to 0.4 μm and more, but always has the same size at their basal, middle and distal parts. The main peculiarities of cytonemes are their length (several cell diameters), high rate of development (1 μm/min-40 μm/s), flexibility and motility (for cytonemes with free tips).

*Corresponding Author: Svetlana Ivanovna Galkina—A.N. Belozcrsky Institute of Physico-Chemical Biology of the M. V. Lomonosov Moscow State University, 119992, Moscow, Vorobievi gory, Bldg. A, Moscow, 119992, Russia. Email: galkina@genebee.msu.ru

Cell-Cell Channels, edited by Frantisek Baluska, Dieter Volkmann and Peter W. Barlow.
©2006 Landes Bioscience and Springer Science+Business Media.

Cytonemes of Embryonic Cells

Cytonemes of blood cells resemble in size, behaviour and functions thin dynamic filopodia of sea urchin primary mesenchyme cells (PMCs). These filopodia seem to serve as cellular probing and adhesive organelles and play a role in signaling and patterning at gastrulation.[1,5] During gastrulation of the sea urchin embryo PMCs migrate from the vegetal pole to a site below the equator of the embryo where they form a ring-like structure and begin producing the larval skeleton. As these cells migrate, they extend and retract dynamic thin filopodia which interact with the basal lamina which lines the blastocoel and with underlying ectoderm. For normal skeleton development PMCs have been shown to obtain extensive positional information from ectoderm. Thin, dynamic and rapidly elongating filopodia of PMCs seem to probe the inner surface of the outer epithelial cells and establish cell-cell contacts with them. Bulges, sometimes moving along filopodia to the PMCs cell bodies, could represent transport of signaling substances from filopodia tips.[5] Cells of ectoderma also have a capability to develop similar filopodia.[5]

Isolated PMCs are also capable of growing thin, elongated, active filopodia upon adhesion to extracellular matrix or fibronectin in the presence of deposits of extracellular material prepared from mesenchyme blastula.[2-5] These in vitro filopodia closely resemble the filopodia seen in vivo and exhibit several apparent functions: as sensory organelles, as anchoring appendages and as intercellular connectives. Filopodia can extend at a rate 1 μm/min from cells migrating in vitro, or as rapidly as 10-25 μm/min in vivo and can reach 30-80 μm in length. The filopodia diameter varies from 0.2 to 0.4 μm.[1-5]

Similar cellular extensions were observed in insect embryonic cells. To develop wing imaginal disc cells in *Drosophila* need information from signaling center associated with the anterioir/posterior (A/P) and dorsal/ventral (D/V) compartment borders. This information can be delivered by cytonemes—long, polarized cellular extensions that were found to extend from outlining disc epithelial cells to the signaling A/P and D/V centres.[6,7] Decapentaplegic (Dpp) signaling was shown to determine appearance and orientation of cytonemes. The average length of A/P and D/V oriented cytonemes was 20 μm and 9 μm respectively. Dpp receptor, Thick-veins (Tkv), revealed as Tkv-GFP was found in the plasma membrane of expressing cells and in punctae that move along cytonemes in both anterograde and retrograde directions at a rate 5-7 μm/s.[7]

Cytonemes in cultured wing disc cells can be induced in vitro in response to tissue fragments containing the anterior/posterior boundary. The rate of cytonemes development can overcome 15 μm/min and cytonemes can reach 700 μm in length. Their diameter does not exceed 0.2 μm.[6]

During *Drosophila* oogenesis, a small cluster of border cells migrate over 120 μm along the central surface of the nurse cells to oocyte.[8] Posterior germ cells extend cell processes similar to cytonemes that might help them to communicate with the border cells. The germ cell processes are considerably larger in diameter (1-2 μm), probably reflecting the proportionately larger size of germ cells compared with disc epithelial cells.

Cytonemes of Human Blood Cells

Human leukocytes can produce very long and thin membrane structures resembling cytonemes of embryonic cells: ultralong membrane tethers that were found to attach neutrophils (polymorphonuclear leukocytes) to platelets under physiological flow;[9-11] tubulovesicular extensions, connecting neutrophils to other neutrophils, erythrocytes and objects for phagocytosis;[12,13] thread-like projections of activated B cells;[14] membrane nanotubes connecting peripheral blood NK cells, macrophages and cells of EBV-transformed human B cell line;[15] tunnelling nanotubules in myeloid lineage dendritic cells and monocytes.[16]

Ultralong membrane tethers developed from neutrophil cell bodies were shown to connect human neutrophils to spread platelets or immobilizes P-selectin under shear stress.[9-11] Adhesive receptors of neutrophils belong to integrin and selectin families. Circulating leukocytes have a round cell shape and roll along the vessel walls, temporarily adhering to endothelial cells by selectins. L-selectin and PSGL-1 (P-selectin glycoprotein ligand-1, common for L-, E- and P-selectins) on neutrophils, E- and P-selectin on the endothelium mediate rolling of neutrophils. L-selectin and PSGL-1 are shown to be clustered on the tips of neutrophil microvillus.[17-19]

Formation of long membrane tethers in human neutrophils was observed by high speed, high resolution videomicroscopy of flowing neutrophils interacting with spread platelets.[9] Thin membrane tethers with the average length 6 μm were pulled from the neutrophil bodies at the average rate of 6-40 μm/s after capture by spread platelets as the wall shear rate was 100-250 s^{-1}.[9] Antibodies against P-selectin or PSGL-1, but not anti-CD18 (common β subunit of β$_2$ integrin) antibodies, blocked tether formation. The average tether lifetime was consistent with P-selectin/PSGL-1 bond dynamics under shear stress. It was supposed that formation of thin membrane tethers could occur as a result of binding of PSGL-1 located on the neutrophil microvillus tips to platelet P-selectin and following microvillus elongation under shear stress.[9] Such long range contact interactions of neutrophils to platelets could play a critical role in thrombosis.

When neutrophils were perfused over P-selectin surfaces at a wall shear rate 150 s^{-1}, they rolled over the surface and formed thin membrane tethers similar to those observed upon interactions with platelets.[9-11] Tethers reached 40 μm and more in length with the average length 8.9 μm. The characteristic jerking motion of the neutrophil over P-selectin coexisted with tether growth, whereas tether breakage caused an acute jump in the rolling velocity. Neutrophil seem to stabilize rolling velocity by rapidly adjusting tether number in response to changes in wall shear stress.[10,11] Tether number was rapidly increased as wall shear stress rose and decreased as wall shear stress declined.

Similar mechanism can stabilize neutrophil rolling velocity in blood vessels. Tether formation in circulation could occur due to binding of L-selectin and PGPL-1 clustered on the neutrophil microvillus tips to E- and P-selectin on endothelial cells, lining vessel walls, and following microvillus elongation under shear stress.

Scanning electron microscopy revealed cytonemes formation in human neutrophils upon adhesion to fibronectin-coated substrata, when cell spreading was blocked. Long microvillus-like highly dynamic tubulovesicular extensions developed from the cell bodies of neutrophils, plated to fibronectin-coated substrata in Na$^+$-free extracellular medium or in the presence of drugs, capable of blocking neutrophil spreading, like the rather unspecific phospholipase A$_2$ inhibitor 4-bromophenacyl bromide (BPB), presumed inhibitors of vacuolar-type ATPases N-ethylmaleimide (NEM) and 7-chloro-4-nitrobenz-2-oxa-1,3-diazole (NBD-Cl) and cytochalasin D.[11] Those extensions, attached cells to substrata in β$_1$-, β$_2$-integrin-independent, but L-selectin-dependent manner and connected neutrophils to each other.[12,13] Tubulovesicular extensions can reach 10-80 μm in length during 20 minutes and displayed uniform diameter along the entire length. A variety of cytonemes connecting neutrophils are represented on Figure 1.

In contrast to selectin, β$_2$-integrin Mac-1 (CD11b/CD18) that mediates firm adhesion of neutrophils is mainly located on the membrane of the cell bodies.[17] Integrin-mediated adhesion of neutrophils to the vessel walls occurs upon metabolic disorders such as diabetes or ischemia and following reperfusion and leads to the capillary closure and endothelium injury by attached neutrophils. Leukocyte-endothelium interactions are tightly regulated by a number of mediators, including nitric oxide (NO) constitutively produced by both, neutrophils and endothelial cells. NO is shown to reduce β$_2$-integrin-dependent leukocyte adhesion to endothelium, thus protecting vessel walls from the injury induced by adherent leukocytes. At the same time NO did not affect selectin-dependent rolling of leukocytes.[20,21]

How NO affect leukocyte adhesion remains to be elucidated. NO was shown to induce formation of tubulovesicular extensions in human neutrophils.[13] Development of multiple very long and highly dynamic tubulovesicular extensions, executing cell adhesive interactions over a distance through L-selectin, could reduce involvement of β$_2$-integrin located on the cell bodies in cell-cell interactions. NO could alter integrin-dependent adhesion of neutrophils to endothelial cells for selectin dependent rolling along endothelium due to development of cytonemes.

Phagocytosis of micro organisms is the main neutrophil function. Leukocytes are also known to scavenge old erythrocytes from the blood stream by phagocytosis. Tubulovesicular extensions demonstrated a property to stick and coil around serum opsonized zymosan particles and erythrocytes (Fig. 2). Development of extensions increases the contact area of the cells thus stimulating finding and catching of objects for phagocytosis.[13]

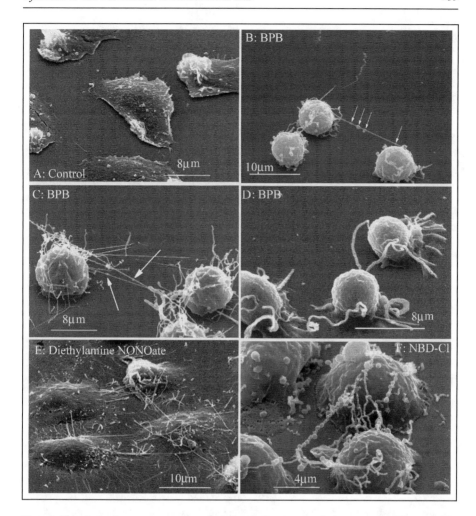

Figure 1. This picture demonstrates a variety of tubulovesicular extensions in human neutrophils revealed by scanning electron microscopy. Flexible extensions with unattached tips (B,C,E), flexible (B-F) and strait (B,C,E) extensions connecting neutrophils are presented. Flexible extensions consist of tubular and/or vesicular interconnected fragments. Strait extensions have a tubular form and contain often a number of bulges along their length (B,C arrows). Diameter of strait (tubular) extensions varied from 125 to 200 nm, diameter of flexible extensions (consisted of tubular and vesicular fragments of the same diameter) varied from 180 to 290 nm. The diameter of extensions did not depend on stimulating agent, since it can vary even in the same conditions (C,D). Neutrophils were isolated from freshly drawn donor blood on a bilayer gradient of Ficoll-Paque.[12] Cells were plated to fibronectin-coated substrata during 20 min at 37°C in: Hank's –HEPES solution (A, control); 10 µM BPB (B,C,D); 1 mM donor of NO diethylamine NONOate (E); 100 µM NBD-Cl (F).

In lymphocytes engagement of B-cell antigen receptor (BCR) was shown to induce formation of cytoneme-like extensions.[14] Cytonemes induced by IgM (surrogate for antigen) were observed in primary splenic lymphocyte and Bal 17 cells. Cytonemes had a thickness of 0.2-0.4 µm and lengths reaching up 10 cell diameters (= 80 µm) during 30 min. The induction of cytonemes on B cells suggests that that they may participate in long-distance communication between the antigen-stimulated B cells and other immune cells in the lymphoid organs such as

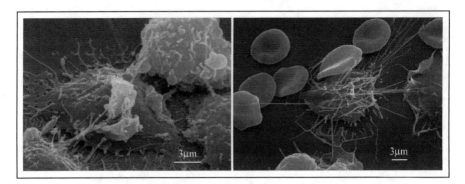

Figure 2. The picture demonstrates involvement of cytonemes in catching and holding of objects for phagocytosis by human neutrophils revealed by scanning electron microscopy. Neutrophils were plated to fibronectin-coated cover-slips during 15 min incubation at 37°C in the presence of 1 mM donor of NO diethylamine NONOate. Serum-opsonized symosan particles (right panel) or erythrocytes (left panel) were added for 5 min at 37°C.[13]

follicular dendritic cells and T cells. The time course of appearance of long, stable cytonemes is consistent with a possible role in presentation of antigen taken up via the BCR to helper T cells.[14]

Membrane nanotubes were found also to connect human peripheral blood NK cells, macrophages and EBV-transformed B cells.[15] Nanotubes seem to be pulled from the cell bodies during disassembly of the immunological synapse, as cells move apart. Nanotubes grew with a speed 0.2 μm/s, lasted over 15 min and reached 140 μm in length. Nanotubes were capable of transporting membrane GPI-anchored proteins, along the surface of the tube, from one of connected cells onto the surface of another cell. GFP-tagged (green fluorescent protein-tagged) cell surface class I MHC proteins expressed in one of the connected cells was found on the nanotubes membrane.[15] The data demonstrate a role of nanotubules in cell signalling communication over distance upon formation of immune response.

A fluorescently labelled antibody against class I MHC were used to visualize tunnelling nanotubules in myeloid-lineage dendritic cells and monocytes by confocal microscopy.[16] A network of tunnelling nanotubules (TNT) with the average diameter 35 nm were shown to transmit calcium fluxes stimulated by chemical or mechanical stimuli between interconnected cells. Microinjected fluid phase marker Lucifer yellow was demonstrated to be transported between cells through tunnelling nanotubules.[16] Nanotubules are suggested to execute long range communication of immune cells, including antigen presentation.

Formation of cytonemes-like long and thin filopodia can be induced in a variety of eukaryotic cells due to transfection of B144/LST1 gene.[22] The structures are dynamically rearranging and sometimes connect one cell with another over a distance of 300 μm. B144/LST1 is a gene encoded in human major histocompatibility complex. It is highly expressed in vitro and in vivo in dendritic cells of the immune system. Dendritic cells are professional antigen-presenting cells and have the problem of finding the occasional T cell whose receptor structure is present to recognize the antigen the dendritic cell is presenting. The occurrence of dynamic long cellular extensions offers a possible means for increasing the efficiency of this process.

Formation and Properties of Cytonemes Connecting Blood Cells

Human neutrophils or sea urchin PMCs upon adhesion to fibronectin-coated substrata were shown to develop long dynamic filopodia with unattached tips. These filopodia can probe the environment and establish contacts with the neighbouring cells.[5,12] Appearance of multiple dynamic filopodia with unattached tips and their capability to stick to the neighbouring neutrophils, erythrocytes or zymosan particles supports this suggestion (Figs. 1,2).

Electron microscopy studies revealed two types of cytonemes, connecting neutrophils—strait tubular and flexible tubulovesicular extensions (Fig. 1).[12] Strait extensions could be derived from flexible tubulovesicular extensions due to applied tension after attachment of the tips. But strait filopodia or membrane tether can be pulled from the cell bodies after binding of selectin receptors, as it described for neutrophil-spread platelets interactions under shear stress.[9] Similar membrane tether can be pulled from neutrophil cell bodies by micropipette manipulation, including suction of latex beads coated with antibodies to proteins on the neutrophil membrane surface.[23,24]

Cytonemes contained a number of bulges along their length (Fig. 1; see also refs. 5-7,9,12,13,22). A constant retrograde and/or anterograde motion of these bulges, observed by optic or confocal microscopy, is supposed to represent cytoplasm or lipid flow representing cell-cell communication. The rate of bulges movement coincides with the rate of cytoneme growth. As a size of cytonemes is near the limit of resolution for optical microscopy, it is impossible to distinguish bulged movement along cytonemes from cytonemes (containing bulges) movement between connected cells. Movement of cytonemes, consisting of tubular and vesicular fragments, between cells could look like bulges movement along cytonemes.

Tunnelling nanotubes of myeloid lineage were found to transport calcium ions and small molecules between cells.[16] As it was shown for GPI-anchored green fluorescent protein, cell surface proteins also can be transported along the surface of the tube, from one of connected cells onto the surface of another cell.[15] Cytonemes movement as a unit can execute transport of signalling membrane proteins and lipids, as well as cytoplasm ions and solutes, between interconnected cells.

Cytonemes were shown to grow unhindered in the presence of the microtubule-destabilizing agents nocodazole or colchicines and found to be tubuline-negative.[2,12,22] There is no agreement on a role of actin filaments in cytonemes. As was revealed by rodamine-phalloidin staining thin filopodia of PMCs and long cell filopodia of cells transfected with B144/LST1 contained actin.[5,6,22] To examine the involvement of actin cytoskeleton in the formation and maintenance of B cell cytonemes, an actin-GFP fusion protein was expressed in Bal 17 cells. BCR cross-linking resulted in the induction of cytonemes with punctuate staining for actin-GFP along the length of extension.[13] All these data are often used to confirm that cytonemes are actin-driven protrusions.

Cytonemes are easily destroyed during fixation. Additional treatment with detergent is required for rodamine-phalloidin staining. In experiments with neutrophils treatment with a very low concentration of detergent eliminated all cytonemes from the cell bodies. It indicated the absence of filamentous actin resistible to detergents, in cytonemes. Moreover, disruption of actin filaments with cytochalasin D induced formation of cytonemes-like dynamic filopodia in neutrophils.[12] Treatment with latrunculin A was shown to relieve pulling of membrane tethers from the neutrophil bodies by micropipette suction of latex beads coated with antibodies to proteins on the neutrophil membrane surface.[24] .Those results indicate that actin filaments do not drive, but rather hinder development of cytonemes or membrane tethers.

To summarize all data on actin cytoskeleton together we can suppose, that cytonemes contain monomeric G-actin unable to drive cytonemes formation, but not filamentous F-actin. G-actin is diffusely distributed in cytoplasm and punctuate staining at that can reflect vesicular structure of cytonemes.

Cytoneme formation seems not depend on cytoskeleton. Model experiments, demonstrating formation of lipid nanotubes resembling cytonemes in size and rate of development in protein-free system can confirm this suggestion. Very long tubular membrane tethers with diameter 200-400 nm in length were observed to connect two daughter liposomes, which were formed upon mechanical fission of multilammelular liposomes (5-20 μm in diameter) of different lipid composition. Mechanical pushing of one of the daughter liposome at a rate of about 5-15 μm/s leaded to elongation of tubes to several hundreds of micrometers.[25]

Materials (particles, small liposomes) may be transported between two nanotube-connected daughter liposomes by creating a difference in the surface tension of the membrane, for example by microinjection of buffer solution in one liposome. The similar mechanism of stimulation of intracellular transport through tunneling nanotubules was described for the cells of myeloid leneage.[16]

Mechanical modulation of membrane tension triggered dendritic cells and monocytes to flux calcium through tunneling nanotubules to other cells at distance of hundreds microns away.[17]

One can suppose that electrolytes and water transport, regulating membrane tension, could provide a driving force for cytonemes formation. Na+-free medium is the most effective inductors of cytonemes in neutrophils.[12] Na+-free medium could inhibit multiple electrogenic cotransporters and antiporters, using Na+ gradient to drive the uphill transport of solute and water into cells, thus inducing extrusion of cytonemes.

Origin and Degradation of Cytonemes

The intracellular origin of the membrane in cytonemes and its traffic pathways remain to be elucidated. Neutrophils or PMCs have diameter 7-8 μm but can extend a number of cytonemes reaching sometimes 80-700 μm. A rate of cytoneme growth can reach 40 μm/s.[9] To built numerous and long membrane extracellular structures, additional membrane have to be quickly delivered from intracellular pools and inserted into the plasma membrane. Early endosomes, endoplasmic reticulum, lysosomes and secretory granules can supply plasmalemma with additional membrane for spreading, phagocytosis or surface wound repairing.[26-30] Tubulovesicular structures of endosomes, lysosomes and trans Golgi network in some cases can fuse into apparently continuous interconnected tubular structures in the cells.[31] Similar integral reorganization of the cellular membrane pool seems to occur upon cytoneme extensions.

As PMCs migrate, they extend and retract thin filopodia which appear to interact with the basal lamina and underlying ectoderm.[2-4] Retraction of cytonemes in blood cells was not observed. In neutrophils formation of tubulovesicular extensions was shown to be an alternative for neutrophil spreading.[12] Neutrophils, having developed numerous extensions, became normally spread in few minutes after Na+ ions addition or in 30 min after removing of the chemical inhibitors of spreading by washing.

Tubulovesicular extension of neutrophils can be observed in 15-40 min after neutrophil adhesion to fibronectin in the Na+-free medium or in the presence of inhibitors of spreading.[12] Further incubation leaded to degradation of extensions by shedding (Fig. 3, left panel), swelling and lysis.[12,13] Inspection of the P-selectin-coated surface after neutrophil rolling on the surface under shear stress also demonstrated the existence of shed tethers overcoming 100 μm in length with a beaded appearance.[9] Heparitinase I was shown to accelerate degradation of extensions, thus leading to appearance of specific holes on the neutrophil cell bodies (Fig. 3F, right panel).[13] Such holes on the neutrophil cell surface resemble in size and appearance porosome—a structure universally present in secretory cells, from the exocrine pancreas to the neurons, where membrane-bound secretory vesicles transiently dock and fuse to expel vesicular contents.[32]

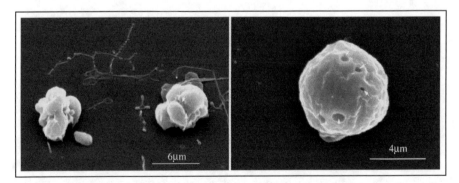

Figure 3. Scanning electron microscopy images of neutrophils, plated to fibronectin in the presence of 100 μM NEM (right panel) or in the presence of 100 μM NEM plus 0.25 units/ml heparitinase (left panel).[13]

Neutrophil tubulovesicular extensions could represent protrusions of exocytotic membrane trafficking, which fuse with the plasma membrane of neutrophils upon spreading, supporting it with additional membrane and adhesive receptors. Inhibition of fusion of tubulovesicular extensions with plasma membrane can lead to extrusion of exocytotic carriers from the cell bodies. The presence of the holes on cell bodies after shedding and disruption of cytonemes leads to suggestion that cytonemes can be protruded through these structures. Those holes in the cell bodies can represent the common opening for "compound exocytosis" of tubular and vesicular exocytotic carriers described for neutrophils.[33]

Conclusions

Formation of cytonemes seems to be a wide spread phenomenon, observed in embryonic, nerve and blood cells. But studies of cytonemes are very complicated. Cytonemes are too small to be observed by optic microscopy. They are easily destroyed upon fixation for electron microscopy. Immunostaining of cytonemes has troubles with fixation and detergent treatments. Due to these difficulties recent investigations demonstrate mainly existence and possible functions of cytonemes in cells of different types. Mechanisms of cytonemes formation and intracellular sources of membranes for building of cytonemes remain to be elucidated. A manner of intercellular signalling between cells mediated by cytonemes also needs further studies.

Acknowledgements

This work was supported by Grants of Russian Foundation of Basic Research 03-04-49270 and 04-04-48495.

References

1. Gustafson T, Wolpert L. Cellular movement and contact in sea urchin morphogenesis. Biol Rev Camb Philos Soc 1967; 42:442-498.
2. Karp GC, Solursh M. Dynamic activity of the filopodia of sea urchin embryonic cells and their role in directed migration of the primary mesenchyme in vitro. Dev Biol 1985; 112:276-283.
3. Solursh M, Lane MC. Extracellular matrix triggers a directed cell migratory response in sea urchin primary mesenchyme cells. Dev Biol 1988; 130:397-401.
4. Malinda KM, Fisher GW, Ettensohn CA. Four-dimensional microscopic analysis of the filopodial behaviour of primary mesenchyme cells during gastrulation in the sea urchin embryo. Dev Biol 1995; 172:552-566.
5. Miller J, Fraser SE, McClay D. Dynamics of thin filopodia during sea urchin gastrulation. Development 1995; 121:2501-2511.
6. Ramirez-Weber FA, Kornberg TB. Cytonemes: cellular processes that project to the principle signaling centre in Drosophila imaginal discs. Cell 1999; 97:599-607.
7. Hsiung F, Ramirez-Weber FA, Iwaki DD et al. Dependence of Drosophila wing imaginal disc cytonemes on Decapentaplegic. Nature 2005; 437:560-563.
8. Goode S. Germ cell cytonemes? Trends Cell Biol 2000; 10:89-90.
9. Schmidtke DW, Diamond SL. Direct observation of membrane tethers formed during neutrophil attachment to platelets or P-selectin under physiological flow. J Cell Biol 2000; 149:719-729.
10. Park EY, Smith MJ, Stropp ES et al. Comparison of PSGL-1 microbead and neutrophil rolling: Microvillus elongation stabilizes P-selectin bond clusters. Biophys J 2002; 82:1835-1847.
11. Ramachandran V, Williams M, Yago T et al. Dynamic alterations of membrane tether stabilize leukocyte rolling on P-selectin. Proc Natl Acad Sci USA 2004; 101:13519-13524.
12. Galkina SI, Sudina GF, Ullrich V. Inhibition of neutrophil spreading during adhesion to fibronectin reveals formation of long tubulovesicular cell extensions (cytonemes). Exp Cell Res 2001; 66:222-228.
13. Galkina SI, Molotkovsky JG, Ullrich V et al. Scanning electron microscopy study of neutrophil membrane tubulovesicular extensions (cytonemes) and their role in anchoring, aggregation and phagocytosis. The effect of nitric oxide. Exp Cell Res 2005; 304:620-629.
14. Gupta N, DeFranco AL. Visualizing lipid raft dynamics and early signaling events during antigen receptor-mediated B-lymphocyte activation. Mol Biol Cell 2003; 14:432-444.
15. Onfelt B, Nedvetzki S, Yanagi K et al. Cutting edge: Membrane nanotubes connect immune cells. J Immunol 2004; 173:1511-1513.
16. Watkins SC, Salter RD. Functional connectivity between immune cells mediated by tunnelling nanotubules. Immunity 2005; 23:309-318.

17. Erlandsen SL, Hasslen SR, Nelson RD. Detection and spatial distribution of the beta 2 integrin (Mac-1) and L-selectin (LECAM-1) adherence receptors on human neutrophils by high-resolution field emission SEM. J Histochem Cytochem 1993; 41:327-333.
18. von Andrian UH, Hasslen SR, Nelson RD et al. A central role for microvillous receptor presentation in leukocyte adhesion under flow. Cell 1995; 82:989-999.
19. Moore KL, Patel KD, Bruehl RE et al. P-selectin glycoprotein ligand-1 mediates rolling of human neutrophils on P-selectin. J Cell Biol 1995; 128:661-667.
20. Moncada S, Palmer RM, Higgs EA. Nitric oxide: Physiology, pathophysiology, and pharmacology. Pharmacol Rev 1991; 43:109-142.
21. Kubes P, Kurose I, Granger DN. NO donors prevent integrin-induced leukocyte adhesion but not P-selectin-dependent rolling in postischemic venules. Am J Physiol 1994; 267:H931-H937.
22. Raghunathan A, Sivakamasundari R, Wolenski J et al. Functional analysis of B144/LST1: A gene in the tumor necrosis factor cluster that induces formation of long filopodia in eukaryotic cells. Exp Cell Res 2001; 268:230-244.
23. Shao JY, Ting-Beall HP, Hochmuth RM. Static and dynamic lengths of neutrophil microvilli. Proc Natl Acad Sci USA 1998; 95:6797-6802.
24. Marcus WD, Hochmuth RM. Experimental studies of membrane tethers formed from human neutrophils. Ann Biomed Eng 2002; 30:1273-1280.
25. Karlsson A, Karlsson R, Karlsson M et al. Networks of nanotubes and containers. Nature 2001; 409:150-152.
26. Buys SS, Keogh EA, Kaplan J. Fusion of intracellular membrane pools with cell surfaces of macrophages stimulated by phorbol esters and calcium ionophores. Cell 1984; 38:569-576.
27. Gagnon E, Duclos S, Rondeau C et al. Endoplasmic reticulum-mediated phagocytosis is a mechanism of entry into macrophages. Cell 2002; 110:119-131.
28. Miyake K, McNeil PL. Vesicle accumulation and exocytosis at sites of plasma membrane disruption. J Cell Biol 1995; 131:1737-1745.
29. Reddy A, Caler EV, Andrews NW. Plasma membrane repair is mediated by Ca^{2+}-regulated exocytosis of lysosomes. Cell 2001; 106:157-169.
30. Borregaard N, Kjeldsen L, Lollike K et al. Granules and vesicles of human neutrophils. The role of endomembranes as source of plasma membrane proteins. Eur J Haematol 1993; 51:318-322.
31. Lippincott-Schwartz J, Yuan L, Tipper C et al. Brefeldin A's effects on endosomes, lysosomes, and the TGN suggest a general mechanism for regulating organelle structure and membrane traffic. Cell 1991; 67:601-616.
32. Jena BP. Discovery of the porosome: revealing the molecular mechanism of secretion and membrane fusion in cells. J Cell Mol Med 2004; 8:1-21.
33. Lollike K, Lindau M, Calafat J et al. Compound exocytosis of granules in human neutrophils. J Leukoc Biol 2002; 71:973-980.

CHAPTER 18

Paracellular Pores in Endothelial Barriers

Luca Manzi and Gianfranco Bazzoni*

Abstract

The endothelium is an efficient barrier located at the boundary between vascular and perivascular compartments. However, it is also regarded to as a permeable filter that contains aqueous pores and allows selective passage of solutes between these compartments. Two distinct routes of permeability have been identified, namely the transcellular pathway (which crosses the apical and basolateral membranes of individual cells) and the paracellular pathway (which passes through the intercellular tight junctions between contacting cells). Here, we focus on the molecular architecture of the tight junctions, as an initial attempt to outline the molecular determinants of paracellular permeability.

Endothelial Permeability

The endothelial lining of the vessel wall is located at the interface between circulating blood and perivascular tissues. Because of this location, the endothelium serves two distinct functions. On one side, it separates tissues from blood, thus maintaining tissue homeostasis and preventing thrombosis. On the other side, it allows communications between the vascular and perivascular compartments, thus playing a central role in vascular physiology. Hence, the endothelium acts as a barrier but, at the same time, behaves as a permeable filter. Permeability, which consists in the transport of solutes across the endothelial layer, follows two distinct pathways. The paracellular route passes through intercellular spaces between contacting cells and is mediated by the tight junctions (TJ). At variance, the transcellular route crosses apical and basal cell membranes and is mediated by vesicles, channels, carriers, and pumps. We focus here on paracellular permeability and its molecular basis. For a detailed analysis of transcellular permeability in the endothelium, the reader is referred to excellent reviews references 1-3.

A comparison of the two pathways is discussed in other reviews references 4-5. Here, we will just briefly mention that, in addition to the cellular and molecular components, the two pathways also differ with respect to the physical properties. Briefly, transport through the paracellular pathway is exclusively passive (while the transcellular pathway is either passive or active), is characterized by higher conductance and lower selectivity, and is not rectified (with similar conductance and selectivity in either apical-to-basal or basal-to-apical directions). Nonetheless, also the paracellular pathway displays well-defined values of electrical conductance, charge selectivity and size selectivity.[6]

Tight Junctions and Paracellular Permeability

In endothelial and epithelial cells, the junctional complex mediates adhesion and communication between contacting cells. Specifically, the junctional complex is composed of TJ, adherens junctions, desmosomes, and gap junctions. TJ, which are the most apical component of the junc-

*Corresponding Author: Gianfranco Bazzoni—Department of Biochemistry and Molecular Pharmacology, Istituto di Ricerche Farmacologiche Mario Negri, Via Eritrea 62, I-20157 Milano, Italy. Email: bazzoni@marionegri.it

Cell-Cell Channels, edited by Frantisek Baluska, Dieter Volkmann and Peter W. Barlow.

tional complex, have been analyzed by electron microscopy more than forty years ago and have been also named *zonulae occludentes*. In thin sections, TJ were initially described as fusions between the plasma membranes of adjacent cells. Nowadays, it is clear that the membranes are not fused, but just in close contact to each other. The appearance of TJ as fusions between adjacent membranes, right at the boundary between apical and basolateral domains, suggest that TJ serve the major function of tightly occluding the lateral intercellular space, thus restricting both diffusion of solutes across the intercellular spaces (barrier function) and movement of membrane molecules between the apical and basolateral domains of the plasma membrane (fence function). Still by electron microscopy, but in freeze-fracture preparations, TJ look like anastomosis of fibrillar strands within the plasma membrane. The strands are composed of particles that are preferentially localized to the internal leaflet. Complementary grooves are localized in correspondence of the strands, mostly on the external leaflet.[7] It is commonly thought that the strands are the morphological and functional units of the TJ.

Molecular Architecture of the TJ

The junctional strands are primarily composed of proteins.[8] As the molecular architecture of the TJ has been reviewed in detail,[9-14] here we will just make a brief mention to the best-characterized TJ molecules, which include both transmembrane and intracellular proteins. It should be noted that these two subclasses of proteins contribute not only to the establishment of the TJ structure and to the control of paracellular permeability, but also to cell polarity,[14] as well as to signal transduction and gene expression.[10]

Transmembrane Proteins of the TJ

Among the transmembrane proteins of the TJ, both occludin and the more than twenty members of the claudin family contain two extracellular loops, four membrane-spanning domains and two cytoplasmic termini.[15,16] However, claudins are smaller than and display no sequence similarity with occludin. Indirect evidence suggests that occludin and claudins might contribute to cell-cell adhesion. For instance, upon expression in fibroblasts (which are devoid of intercellular junctions), exogenous occludin,[17] claudin-1 and claudin-2[18] localize to the adhesive points of cell-cell contact. Notably, both claudin-1 and -2 reconstitute junctional strands, even though claudin-1-based strands are continuous, whereas claudin-2-based strands are fragmented. The finding suggests that claudin-2 forms more leaky TJ than claudin-1, as it will be discussed below. It is also worth mentioning that claudin-5 expression is restricted to endothelial cells[19] and that the expression levels of endothelial claudin-5 and occludin are noticeably high in brain vessels,[19,20] which are characterized by very low levels of paracellular permeability.

Another transmembrane component of the TJ is the glycoprotein Junctional Adhesion Molecule-A (JAM-A), which is composed of an extracellular region, a transmembrane segment, and a cytoplasmic tail.[12,21] The extracellular segment of JAM-A comprises two immunoglobulin-like folds and mediates homophilic adhesion.[22] Similarly to occludin and claudins, upon transfection in Chinese Hamster Ovary cells, JAM-A localizes to the adhesive sites of cell-cell contact and reduces paracellular permeability. In addition, JAM-A contributes to the migration of leukocytes between adjacent endothelial cells, which occurs during inflammation.[21] More recently, several groups have identified molecules that are homologous to JAM-A and that localize to the endothelial junctions, such as JAM-B,[23,24] JAM-C,[25,26] and JAM-4,[27] as well as Endothelial Cell-Selective Adhesion Molecule[28,29] and Coxsackie- and adenovirus receptor.[30]

Intracellular Molecules of the TJ

Among the intracellular TJ molecules, the Zonula Occludens (ZO) proteins were named after their localization to the *zonulae occludentes* of epithelial and endothelial cells. They comprise ZO-1,[31] ZO-2[32] and ZO-3.[33] In addition to the ZO proteins, endothelial TJ comprise several other cytoplasmic molecules, such as AF-6/Afadin,[34] PAR-3/ASIP,[35] cingulin[36] and 7H6.[37]

In the context of this discussion, the primary importance of defining the molecular architecture of the TJ relates to the need of identifying which molecular mechanisms control paracellular permeability. It is known that, in endothelial cells, several permeability-inducing agents cause Myosin Light Chain Kinase-dependent phosphorylation of a nonmuscle isoform of myosin light chain, contraction of the perijunctional actomyosin ring, partial opening of the TJ, and ultimately increased permeability.[38,39] Predictably, centripetal contraction of the actomyosin ring may exert pulling forces on the transmembrane proteins of the TJ. So far, however, it remains unclear which intracellular molecules actually transmit these forces from the contractile cytoskeleton to the adhesive proteins at the cell surface. It is noteworthy that several (if not all) transmembrane proteins of the TJ are indirectly linked to actin-associated proteins. For instance, occludin,[40] claudins,[41] and JAM-A[42,43] bind ZO-1. In turn, ZO-1 associates with F-actin,[44] the actin-associated proteins α-catenin[45] and cortactin,[46] and the myosin-associated protein cingulin.[47] In addition, JAM-A binds AF-6/Afadin[43] and CASK/Lin2,[48] which in turn associate with actin. Interestingly, inflammatory cytokines, besides enhancing permeability, reduce the solubility of some TJ molecules in nonionic detergents,[49,50] which is a likely consequence of reduced association with filamentous actin.

The Role of Claudins in Paracellular Permeability

As the extracellular domains of the transmembrane TJ proteins abut the lateral intercellular space, it can be assumed that these domains come in close contact with the hydrophilic solutes that are transported along the paracellular pathway. In particular, evidence indicates that claudins (which are important components of the TJ strands) play a major role in paracellular permeability. Thus, it is reasonable to expect that claudin transfection should invariably reduce paracellular permeability. However, contrary to the expectation, transfection of exogenous claudins not always decreases paracellular permeability. For instance, expression of claudin-2 in the high-resistance kidney epithelial cells MDCK-I and MDCK-C7 enhances paracellular permeability.[51,52] A plausible explanation for these apparently counterintuitive results comes from the "pores in the wall" hypothesis. According to the hypothesis, TJ strands are indeed impermeable barriers (the "walls"). Nonetheless, a series of permeable "pores" perforate the walls.[53] Thus, in transfection experiments, the different ability of each individual claudin to predominantly act as either "wall" or "pore" determines the observed decrease and increase in permeability, respectively.

Actually, if claudin-2 acts as a "pore", some claudins do behaves as "walls". As a consequence, transfection of these claudins (mostly in the low-resistance MDCK-II cells) reduces permeability. For instance, expression of claudin-1,[54] claudin-4,[55] claudin-8[56] and claudin-15[57] was found to reduce paracellular permeability. As a word of caution, it should be mentioned that the recipient cell lines already express a background of endogenous claudins. Thus, these transfection studies do not define the absolute permeability of the exogenous claudins, but rather relative changes compared to the background.

Besides quantitatively regulating permeability, some claudins also regulate charge selectivity. As reported by Colegio et al, mutations that replace three negative to positive residues in the first extracellular loop of claudin-15 increase permeability to chlorine anions. At variance, a mutation that replaces a positive to a negative charge in the first extracellular loop of claudin-4 increases permeability to sodium.[57] Thus, claudin-15 and claudin-4 likely act as selective electrostatic barriers for anions and cations, respectively. Finally, some claudins also regulate size selectivity. The in vivo analysis of claudin-5 provides an interesting example of this notion. Nitta et al reported that deletion of *claudin-5* in mice is not associated with altered development of brain vessels. In these mice, however, the blood brain barrier looses the selective ability to block the passage of small molecular weight tracers.[19]

Even though individual claudins may act as either walls or pores, interpretation of results from the transfection studies requires theoretical models that keep into account not only the exogenous claudins, but also the background of endogenous claudins. As described by Yu,[4] there are at least two possible ways, whereby exogenous claudins in theory can change permeability with respect to the background. In the *series model*, novel junctional strands (which are formed by exogenous claudins) behave as additional barriers in series with the preexisting strands (which are formed by endogenous claudins), thus decreasing paracellular permeability. For instance, claudin-1 transfection increases both strand number and electrical resistance.[54] Along the same line, it has long been observed that electrical resistance (at least in epithelia) increases with the number of strands.[58]

At variance, in the *parallel replacement model*, exogenous claudins do not form novel strands, but rather replace endogenous claudins in the preexisting strands. Although the model predicts that claudin transfection leaves the number of strands unchanged, it cannot predict the net effect on paracellular permeability. Actually, the effect is dependent upon the balance between intrinsic conductances of endogenous and exogenous claudins (that are displaced from and inserted in the preexisting junctional strands, respectively). The reduced levels of endogenous claudin-2 (together with the unchanged number of junctional strands) that follow transfection of the exogenous claudin-8 provide a likely example of parallel replacement.[56] In this model system, transfected claudin-8 reduces sodium permeability, probably because exogenous claudin-8 monomers replace the endogenous and "leaky" claudin-2 monomers from the existing strands, and not because they form novel strands (composed of "tight" claudin-8 monomers).

Open Issues

Over the past two decades several studies have described functional properties and molecular basis of endothelial permeability. However, there are still several important questions that await further analysis. In particular, it will be essential to understand how the complex networks of molecular interactions among the numerous TJ proteins control endothelial permeability. In addition, defining the real nature of the aqueous pores within the junctional strands will undoubtedly necessitate structural studies at high resolution, even though the purification of transmembrane molecules and molecular assemblies for these studies is a daunting challenge for structural biologists.

Acknowledgements

The generous support by AICR (Association International for Cancer Research, St. Andrews, United Kingdom; Grant 04-095) and MIUR (Ministry of Education, University and Research, Rome, Italy; Grants RBNE01T8C8_004 and RBAU01E5F5) is gratefully acknowledged.

References

1. Mann GE, Yudilevich DL, Sobrevia L. Regulation of amino acid and glucose transporters in endothelial and smooth muscle cells. Physiol Rev 2003; 83:183-252.
2. Nilius B, Droogmans G. Ion channels and their functional role in vascular endothelium. Physiol Rev 2001; 81:1415-1459.
3. Predescu D, Vogel SM, Malik AB. Functional and morphological studies of protein transcytosis in continuous endothelia. Am J Physiol Lung Cell Mol Physiol 2004; 287:L895-901.
4. Yu AS. Claudins and epithelial paracellular transport: The end of the beginning. Curr Opin Nephrol Hypertens 2003; 12:503-509.
5. Van Itallie CM, Anderson JM. The molecular physiology of tight junction pores. Physiology (Bethesda) 2004; 19:331-338.
6. Powell DW. Barrier function of epithelia. Am J Physiol 1981; 241:G275-288.
7. Farquhar MG, Palade GE. Junctional complexes in various epithelia. J Cell Biol 1963; 17:375-412.
8. Tsukita S, Furuse M, Itoh M. Multifunctional strands in tight junctions. Nat Rev Mol Cell Biol 2001; 2:285-293.
9. D'Atri F, Citi S. Molecular complexity of vertebrate tight junctions. Mol Membr Biol 2002; 19:103-112.
10. Matter K, Balda MS. Signalling to and from tight junctions. Nat Rev Mol Cell Biol 2003; 4:225-236.
11. Schneeberger EE, Lynch RD. The tight junction: A multifunctional complex. Am J Physiol Cell Physiol 2004; 286:C1213-1228.
12. Bazzoni G. The JAM family of junctional adhesion molecules. Curr Opin Cell Biol 2003; 15:525-530.
13. Bazzoni G, Dejana E. Endothelial cell-to-cell junctions: Molecular organization and role in vascular homeostasis. Physiol Rev 2004; 84:869-901.
14. Cereijido M, Contreras RG, Shoshani L. Cell adhesion, polarity, and epithelia in the dawn of metazoans. Physiol Rev 2004; 84:1229-1262.

15. Furuse M, Hirase T, Itoh M et al. Occludin: A novel integral membrane protein localizing at tight junctions. J Cell Biol 1993; 123:1777-1788.
16. Morita K, Furuse M, Fujimoto K et al. Claudin multigene family encoding four-transmembrane domain protein components of tight junction strands. Proc Natl Acad Sci USA 1999; 96:511-516.
17. Van Itallie CM, Anderson JM. Occludin confers adhesiveness when expressed in fibroblasts. J Cell Sci 1997; 110:1113-1121.
18. Furuse M, Sasaki H, Fujimoto K et al. A single gene product, claudin-1 or -2, reconstitutes tight junction strands and recruits occludin in fibroblasts. J Cell Biol 1998; 143:391-401.
19. Nitta T, Hata M, Gotoh S et al. Size-selective loosening of the blood-brain barrier in claudin-5-deficient mice. J Cell Biol 2003; 161:653-660.
20. Hirase T, Staddon JM, Saitou M et al. Occludin as a possible determinant of tight junction permeability in endothelial cells. J Cell Sci 1997; 110:1603-1613.
21. Martin-Padura I, Lostaglio S, Schneemann M et al. Junctional adhesion molecule, a novel member of the immunoglobulin superfamily that distributes at intercellular junctions and modulates monocyte transmigration. J Cell Biol 1998; 142:117-127.
22. Bazzoni G, Martinez-Estrada OM, Mueller F et al. Homophilic interaction of junctional adhesion molecule. J Biol Chem 2000; 275:30970-30976.
23. Palmeri D, van Zante A, Huang CC et al. Vascular endothelial junction-associated molecule, a novel member of the immunoglobulin superfamily, is localized to intercellular boundaries of endothelial cells. J Biol Chem 2000; 275:19139-19145.
24. Cunningham SA, Arrate MP, Rodriguez JM et al. A novel protein with homology to the junctional adhesion molecule. Characterization of leukocyte interactions. J Biol Chem 2000; 275:34750-34756.
25. Arrate MP, Rodriguez JM, Tran TM et al. Cloning of human junctional adhesion molecule 3 (JAM3) and its identification as the JAM2 counter-receptor. J Biol Chem 2001; 276:45826-45832.
26. Aurrand-Lions M, Duncan L, Ballestrem C et al. JAM-2, a novel immunoglobulin superfamily molecule, expressed by endothelial and lymphatic cells. J Biol Chem 2001; 276:2733-2741.
27. Hirabayashi S, Tajima M, Yao I et al. JAM4, a junctional cell adhesion molecule interacting with a tight junction protein, MAGI-1. Mol Cell Biol 2003; 23:4267-4282.
28. Hirata Ki, Ishida T, Penta K et al. Cloning of an immunoglobulin family adhesion molecule selectively expressed by endothelial cells. J Biol Chem 2001; 276:16223-16231.
29. Nasdala I, Wolburg-Buchholz K, Wolburg H et al. A transmembrane tight junction protein selectively expressed on endothelial cells and platelets. J Biol Chem 2002; 277:16294-16303.
30. Cohen CJ, Shieh JT, Pickles RJ et al. The coxsackievirus and adenovirus receptor is a transmembrane component of the tight junction. Proc Natl Acad Sci USA 2001; 98:15191-15196.
31. Stevenson BR, Siliciano JD, Mooseker MS et al. Identification of ZO-1: a high molecular weight polypeptide associated with the tight junction (zonula occludens) in a variety of epithelia. J Cell Biol 1986; 103:755-766.
32. Jesaitis LA, Goodenough DA. Molecular characterization and tissue distribution of ZO-2, a tight junction protein homologous to ZO-1 and the Drosophila discs-large tumor suppressor protein. J Cell Biol 1994; 124:949-961.
33. Haskins J, Gu L, Wittchen ES et al. ZO-3, a novel member of the MAGUK protein family found at the tight junction, interacts with ZO-1 and occludin. J Cell Biol 1998; 141:199-208.
34. Yamamoto T, Harada N, Kano K et al. The Ras target AF-6 interacts with ZO-1 and serves as a peripheral component of tight junctions in epithelial cells. J Cell Biol 1997; 139:785-795.
35. Izumi Y, Hirose T, Tamai Y et al. An atypical PKC directly associates and colocalizes at the epithelial tight junction with ASIP, a mammalian homologue of Caenorhabditis elegans polarity protein PAR-3. J Cell Biol 1998; 143:95-106.
36. Citi S, Sabanay H, Jakes R et al. Cingulin, a new peripheral component of tight junctions. Nature 1988; 333:272-276.
37. Satoh H, Zhong Y, Isomura H et al. Localization of 7H6 tight junction-associated antigen along the cell border of vascular endothelial cells correlates with paracellular barrier function against ions, large molecules, and cancer cells. Exp Cell Res 1996; 222:269-274.
38. Hecht G, Pestic L, Nikcevic G et al. Expression of the catalytic domain of myosin light chain kinase increases paracellular permeability. Am J Physiol 1996; 271:C1678-1684.
39. Turner JR, Rill BK, Carlson SL et al. Physiological regulation of epithelial tight junctions is associated with myosin light-chain phosphorylation. Am J Physiol 1997; 273:C1378-1385.
40. Furuse M, Itoh M, Hirase T et al. Direct association of occludin with ZO-1 and its possible involvement in the localization of occludin at tight junctions. J Cell Biol 1994; 127:1617-1626.
41. Itoh M, Furuse M, Morita K et al. Direct binding of three tight junction-associated MAGUKs, ZO-1, ZO-2, and ZO-3, with the COOH termini of claudins. J Cell Biol 1999; 147:1351-1363.

42. Bazzoni G, Martinez-Estrada OM, Orsenigo F et al. Interaction of junctional adhesion molecule with the tight junction components ZO-1, cingulin, and occludin. J Biol Chem 2000; 275:20520-20526.
43. Ebnet K, Schulz CU, Meyer ZU et al. Junctional adhesion molecule interacts with the PDZ domain-containing proteins AF-6 and ZO-1. J Biol Chem 2000; 275:27979-27988.
44. Fanning AS, Jameson BJ, Jesaitis LA et al. The tight junction protein ZO-1 establishes a link between the transmembrane protein occludin and the actin cytoskeleton. J Biol Chem 1998; 273:29745-29753.
45. Itoh M, Nagafuchi A, Moroi S et al. Involvement of ZO-1 in cadherin-based cell adhesion through its direct binding to alpha catenin and actin filaments. J Cell Biol 1997; 138:181-192.
46. Katsube T, Takahisa M, Ueda R et al. Cortactin associates with the cell-cell junction protein ZO-1 in both Drosophila and mouse. J Biol Chem 1998; 273:29672-29677.
47. Cordenonsi M, D'Atri F, Hammar E et al. Cingulin contains globular and coiled-coil domains and interacts with ZO-1, ZO-2, ZO-3, and myosin. J Cell Biol 1999; 147:1569-1582.
48. Martinez-Estrada OM, Villa A, Breviario F et al. Association of junctional adhesion molecule with calcium/calmodulin-dependent serine protein kinase (CASK/LIN-2) in human epithelial caco2 cells. J Biol Chem 2001; 276:9291-9296.
49. Bruewer M, Luegering A, Kucharzik T et al. Proinflammatory cytokines disrupt epithelial barrier function by apoptosis-independent mechanisms. J Immunol 2003; 171:6164-6172.
50. Martinez-Estrada OM, Manzi L, Tonetti P et al. Opposite effects of tumor necrosis factor and soluble fibronectin on junctional adhesion molecule-A in endothelial cells. Am J Physiol Lung Cell Mol Physiol 2005; 288:L1081-1088.
51. Furuse M, Furuse K, Sasaki H et al. Conversion of zonulae occludentes from tight to leaky strand type by introducing claudin-2 into Madin-Darby canine kidney I cells. J Cell Biol 2001; 153:263-272.
52. Amasheh S, Meiri N, Gitter AH et al. Claudin-2 expression induces cation-selective channels in tight junctions of epithelial cells. J Cell Sci 2002; 115:4969-4976.
53. Tsukita S, Furuse M. Pores in the wall: Claudins constitute tight junction strands containing aqueous pores. J Cell Biol 2000; 149:13-16.
54. McCarthy KM, Francis SA, McCormack JM et al. Inducible expression of claudin-1-myc but not occludin-VSV-G results in aberrant tight junction strand formation in MDCK cells. J Cell Sci 2000; 113:3387-3398.
55. Van Itallie C, Rahner C, Anderson JM. Regulated expression of claudin-4 decreases paracellular conductance through a selective decrease in sodium permeability. J Clin Invest 2001; 107:1319-1327.
56. Yu AS, Enck AH, Lencer WI et al. Claudin-8 expression in Madin-Darby canine kidney cells augments the paracellular barrier to cation permeation. J Biol Chem 2003; 278:17350-17359.
57. Colegio OR, Van Itallie CM, McCrea HJ et al. Claudins create charge-selective channels in the paracellular pathway between epithelial cells. Am J Physiol Cell Physiol 2002; 283:C142-147.
58. Claude P, Goodenough DA. Fracture faces of zonulae occludentes from "tight" and "leaky" epithelia. J Cell Biol 1973; 58:390-400.

CHAPTER 19

Channels across Endothelial Cells

Radu V. Stan*

Abstract

The evolution of multicellular organisms entailed the formation of biological compartments separated by epithelial cellular barriers. As a consequence, strategies have evolved to move/exchange materials between these compartments for nutrition, wastes removal and information exchange, while still maintaining their identity. These exchanges occur either across barrier cells (transcellular) or in between adjacent cells (paracellular). Among the strategies employed for transcellular exchange are the formation of physical patent channels/pores that cut through the cells and the formation of "functional channels" by shuttling vesicles during transcytosis. In this chapter, I will summarize the knowledge of the transcellular exchange systems using the vascular endothelium as a barrier model. We will focus on the components, structure and function of the endothelial organelles such as fenestrae, vesiculo-vacuolar organelles and transendothelial channels as well as the vesicular carriers involved in transcytosis (i.e., plasmalemmal vesicles or caveolae), which form either physical or functional channels across endothelial cells.

Introduction

A critical step in the evolution of multicellular organisms (metazoans and higher) was the establishment of different biological compartments to carry out specific functions. This entailed the development of epithelial cellular barriers between different environments as well as the development of strategies to selectively move molecules between compartments (i.e., for delivery of nutrients, elimination of wastes as well as ensuring molecular communication between distant cells and tissues) while maintaining their distinct composition.

The movement of material is generally accomplished via routes situated either across (i.e., transcellular) or in between (i.e., paracellular) the cells forming the barrier. While the **paracellular transport** occurs via the intercellular junctions, there are at least three **transcellular** routes described so far: (a) water and small molecules are moved via **membrane transporters** that have a different distribution/orientation on opposite fronts of a barrier cell; (b) **transcytosis**, defined as the transport of macromolecular cargo from one front of a polarized cell to the other within membrane-bounded carrier(s), and (c) **pores** or **channels** which are patent openings through the barrier cells with or without selective permeability for different molecules. Together, these processes (Fig. 1A) contribute to the success of multicellular organisms (reviewed in refs. 1,2).

The transcellular exchange systems have achieved a high level of complexity and sophistication in mammals. Among different types of epithelia, the vascular endothelium has the most diversified repertoire of specific structures that might carry out these exchanges. During decades of research on the topic, several endothelial subcellular structures that might participate in this function have been identified, such as caveolae, transendothelial channels (TEC), vesiculo-vacuolar organelles (VVO),

*Radu V. Stan—Dartmouth Medical School, Angiogenesis Research Center, Department of Pathology, HB 7600, Borwell 502W, 1 Medical Center Drive, Hanover, New Hampshire 92093-0651, U.S.A. Email: Radu.V.Stan@Dartmouth.edu

Cell-Cell Channels, edited by Frantisek Baluska, Dieter Volkmann and Peter W. Barlow.
©2006 Landes Bioscience and Springer Science+Business Media.

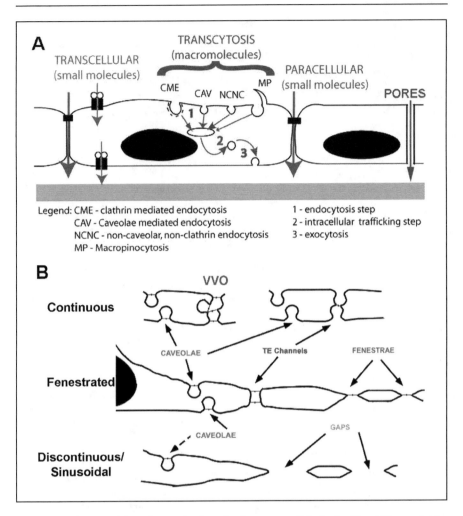

Figure 1. A) Schematic of the pathways taken by molecules across a cellular barrier. Material is moved either between the cells via the paracellular pathway or across the cells via transcytosis, pores or membrane transporters. B) Schematic of the endothelial phenotypes and their content of structures that were involved in transport across endothelium; TEC—Transendothelial channels, VVO—vesiculo-vacuolar organelles.

fenestrae and intercellular junctions. We will visit the endothelial structures involved in mediating the exchanges of water and solutes between the blood plasma and the interstitial fluid / tissues across endothelial cells.

Vascular Permeability—Pathways of Transendothelial Exchange

Vascular endothelium is a highly differentiated cellular monolayer with the organization of a simple squamous epithelium that lines the entire cardiovascular system. It constitutes a quasi-ubiquitous presence in organs and tissues throughout the body forming one of the largest vertebrate "organs" (3). Endothelia have been classically defined into three main structural types, depending on their content of specific structures: the continuous, fenestrated and discontinuous endothelium (Fig. 1B, reviewed in ref. 1). The **continuous** endothelium forms an uninterrupted barrier between the blood and tissues. It features a large population of caveolae or plasmalemmal vesicles,[4,5] extremely few TEC[6] and virtually no fenestrae. In addition to relatively much fewer caveolae (Fig. 3), the **fenestrated** endothelium features pores such as fenestrae (Fig. 4), transendothelial channels (Fig. 4) and endothelial pockets.[3,7-11] **Discontinuous** or **sinusoidal** endothelium lines the sinusoids in the liver and bone marrow. It has extremely few caveolae and sinusoidal gaps and is characterized by the presence of pores called fenestrae or gaps, which are much larger and heterogeneous in size than the fenestrae of the fenestrated endothelium (reviewed in ref. 12).

Endothelium performs important functions in the physiology of vertebrates.[1,13-16] Its primary function is the control of transendothelial exchanges of water and solutes (both small and large molecules) between the blood plasma and the underlying tissues, also known as microvascular permeability. By this endothelium has a critical role in the delivery of nutrients to tissues and clearing of wastes, as well as in the functional integration at organ and organism level allowing the information exchange between tissues situated at distance one from another (i.e, hormones etc). The dominant character of transendothelial transport is clearly expressed in the average thickness of endothelium of less than 0.3 µm as well as at the expression of specific structures that mediate this function.

Although endothelium is a highly dynamic tissue, it is generally recognized that it mediates a **"basal" microvascular permeability**, defined as the permeability of microvessels to fluid and hydrophilic solutes that occurs in normal tissues and an **increased microvascular permeability**, which is the large increase in permeability to fluid and plasma proteins that occurs in acutely or chronically inflamed tissues or in tumors. The basal permeability is a continuous process that occurs with small physiological variations by mechanisms that are largely unknown. It is also not known whether the increased permeability is an exaggeration in rate of the basal mechanisms or it occurs via completely different mechanisms.

Endothelium differs from other epithelia on its much higher (2-3 orders of magnitude) permeability to water and solutes (both small and especially large molecules). This higher permeability was found by physiologists to have both a diffusive and convective component. The best fitting model (known as the "pore theory") (Figs. 1A, 2A) was that the exchange takes place through two types of patent (open), water-filled, cylindrical or slit-like channels: small pores with diameter of <11 nm and ~15.1 units/μm^2 and large pores with diameter of ~50 nm and ~0.2-0.02 units/μm^2.[17-19]

The structural aspects of the fluid and solute exchanges across endothelium have been approached by cell biologists mainly by electron microscopy alone or combined with in vivo tracer studies with a variety of probes of variable chemical and physical properties (Fig. 2B, reviewed in refs. 20,21). The above-mentioned tracer studies are in agreement with the pore theory in the case of the small pores being represented by the paracellular pathway. Also there is agreement in the case of fenestrated and discontinuous endothelia where structural equivalents of large pores have been found such as TEC and fenestrae. No porelike structural equivalents were found for the large pores by electron microscopy in continuous endothelia (where most of the data supporting the "pore theory" had been obtained) with the exception of VVOs in the venular endothelium.[22] This established a long-standing controversy (schematized in Fig. 2A) regarding the structural basis of capillary permeability for macromolecules between the capillary physiologists and the cell biologists/morphologists interested in the topic. In the case of the continuous endothelium, the tracer studies have instead documented the alternative concept/process of transcytosis the equivalent of a "functional pore" across epithelia (see below).[23-25]

Functional Pores across Endothelium—Transcytosis

At its simplest, transcytosis is the transport of macromolecular cargo from one side of a polarized cell to the other (e.g., apical to basolateral) within membrane-bounded carriers (i.e., vesicles). It is a landmark function of all epithelia where it is performed in several distinct steps (Fig. 1A) such as: (1) **endocytosis**, or uptake of cargo at one front of the cells forming barrier, (2) **intracellular trafficking** of cargo in membrane bound vesicles, and, (3) **exocytosis** or fusion of these vesicles with the opposite front of the barrier cells and discharge of the cargo. The endocytotic step was shown to occur by either caveolae-mediated endocytosis (CAV-ME), clathrin-mediated endocytosis (CME), macropinocytosis (MP) or noncaveolae, nonclathrin pathways (NCNC).[2,20,26]

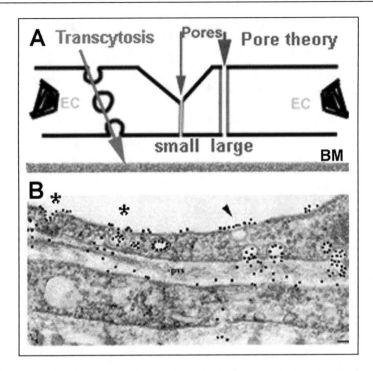

Figure 2. A) Schematic of the controversy of transendothelial exchange between the "pore theory" and transcytosis via plasmalemmal vesicles. B) EM micrograph demonstrating the involvement of vesicles in the size range of caveolae in the transport of tracers. Courtesy of D. Predescu (U. Illinois Chicago).

All these pathways have been shown to operate in transcytosis in the endothelium as well. Clathrin-mediated endocytosis was shown to be important in endothelium of the testis for receptor-mediated transport of hormones such as LH/CG and FSH, as a mechanism for specific delivery to target organs.[27-30] Macropinocytosis-like pathways operate via the megalin (gp330)/ LDL-receptor binding protein (LRP) system involved in transport across the brain endothelium of beta-amyloid binding Apolipoprotein J,[31-36] Receptor Associated Protein (RAP) as well as transport of IgG and nutrients from mother to fetus in mammalian placenta.[37-41] In has been shown that the syncitiotrophoblast (STB) layer has drastically reduced levels of caveolin 1 and caveolae[39,42] and that the transport of IgG across the STB layer occurs via the Fcγ Receptor IIb2 in by a NCNC pathway.[43] The same NCNC pathway seems to operate in the fetal endothelium where the IgG and FcγRIIb2.[41,43,44] This system might also play a role in the lung[45] and brain[46] where Fc Receptors were demonstrated on the endothelium. A very interesting finding was that one of the IgG receptors FcRn binds albumin at an acidic pH, that the lifespan of albumin is shortened in FcRn-deficient mice, and that the plasma albumin concentration of FcRn-deficient mice (i.e., either the α-chain or β2-microglobulin of the FcRn heterodimer) is less than half that of wild-type mice.[40] The FcRn receptor functions in an acidic prelysosomal internal compartment where molecules destined to transcytosis might be sorted. Its function is to protect albumin and IgG from degradation in lysosomes diverting them to the opposite front of the cells. However, a direct demonstration of the trafficking of albumin and other relevant plasma proteins in endothelium is still lacking.

The above transcytotic pathways exist for regulated, receptor-mediated transport across cells in all polarized epithelia,[2,47,48] a broader tissue category to which the vascular endothelium belongs.[49] However, this does not explain the higher (2-3 orders of magnitude) permeability of endothelium to plasma proteins that occurs in an adsorptive manner (low affinity) and does not seem to make use of

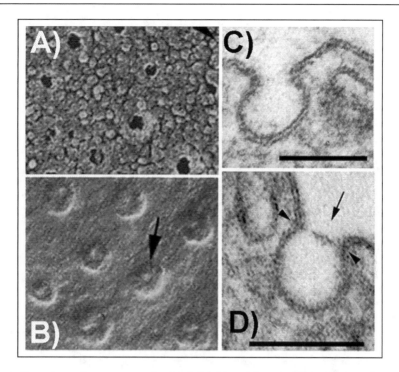

Figure 3. Transmission electron micrographs of endothelial caveolae demonstrating the presence (B,D) and absence (A,C) of stomatal diaphragms by either deep-etch of rapidly frozen samples (A-B) routine plastic embedding (C-D). A and B reproduced from reference 7, with permission.

receptor mediated systems, unless these pathways could be modulated as to account for such an increase. Caveolae-mediated transcytosis has been proposed to fulfill this gap and to be major players in basal vascular permeability. This put them at the center of the controversy outlined above and thereby the subject of intense study. Data pro or against caveolar were obtained in attempts to answer whether; (1) vesicular transport (transcytosis), although admittedly occurring, has a meaningful contribution to vascular permeability, and (2) if yes, whether caveolae are implicated in this transport (for more detailed reviews see refs. 1,17,19-21,26).

Caveolae (or plasmalemmal vesicles) (Fig. 3) were first described in the endothelium of continuous type by Palade[4] but they occur in all types of endothelia as well as most mammalian cell types (for reviews see refs. 1,20,26). They are morphologically defined as spherical invaginations of plasma membrane of regular shape and size (~70 nm outer diameter),[5,50,51] which can occur single or in grape-like clusters attached to either front of the endothelium. Only in select endothelia of continuous type (i.e., in lung, tongue, kidney vasa recta) and in all fenestrated and sinusoidal endothelia, caveolae are provided with a Stomatal Diaphragm (SD).[52,53] The SD is a thin (~5-7 nm) protein barrier of variable diameter (<40 nm), which occurs in their stomata (hence the name) or in between vesicles that are part of a cluster. The function of the SD is not known (for reviews see refs. 1,20,26).

Among the most important arguments for caveolae as transcytotic carriers are:
a. Endothelial caveolae are numerous, commensurate with the massive exchanges of water and solutes that occur continuously. By EM, rarely seen transendothelial pores in the continuous endothelium are at a much lower surface density than postulated by the pore theory.[3,54-60] Morphologically defined caveolae were shown to mediate the transport of different tracers across endothelium in situ.[50,51,61-74] Increased numbers of caveolae correlate with increased transcytosis of tracers in experimental diabetes.[75] Most importantly,

caveolae can be targeted for transport of cargo in tissue specific manner; a monoclonal antibody against a lung endothelial caveolar antigen was taken up and transported across the endothelial barrier via caveolae only in the lung thus establishing proof of principle for tissue targeted delivery of drugs via transcytosis.[72,76]

b. The transport of tracers is inhibited by N-ethyl maleimide (NEM), known to interfere with vesicular docking and fusion to target membranes[65,67] and filipin, a cholesterol-binding drug known to disrupt caveolae.[67] Caveolae from lung vasculature contain molecules involved in docking and fusion such as VAMP-2,[72,77,78] NSF and SNAP 25.[74,79]

c. Caveolae contain molecules involved in vesicle fission such as dynamin[80-83] or intersectin.[84] Importantly, endothelial caveolae pinch off from plasma membranes in cell-free assays and their detachment requires dynamin2 and GTP hydrolysis in Ecs.[85,86]

d. Caveolae are capable of endocytosis in a dynamin2, glycolipid, tyrosine kinase and actin-dependent manner. Once internalized caveolae can either: fuse with an intracellular compartment called a "caveosome" or fuse with Rab5 containing early endosomes as shown in other cell types[87-98] or in endothelium.[99-107] The finding that the internalization of vesicles that participate in transcytosis can be modulated obviates an advantage over a system of rigid, permanent pores insensitive to such modulations.

e. Evidence of a transcellular route across endothelium was obtained by live microscopy in situ in the rat cremaster[108] who showed different kinetics for the transport of a small pore probe as compared with a vesicular transport probe (a stearyl dye that fluoresces only in membrane bound vesicles).

Taken together, these data show that the transcellular transport via uncoated vesicles is a fact, that the integrity of lipid rafts is necessary for the transport of proteins across endothelium, that caveolae contain the molecules necessary for either membrane fission or membrane docking and fusion and that they are internalized in an inducible, regulated manner. All of these lines of evidence are consistent with the participation of caveolae in transcytosis.

There are arguments against the importance of transcytosis in vascular permeability as well as against caveolae-mediated transcytosis.

a. Three-dimensional reconstructions by serial sectioning of fixed tissues showed that most endothelial vesicles would either be fused to the plasma membrane or form branching structures whose lumen communicate with one cell front or another but not with both at the same time. Few free vesicles were encountered, which was taken as evidence against vesicular transport. However, only occasional TEC were recorded as well, as already pointed out. A counterargument used by the proponents of transcytosis was that the chemical fixation is a slow process and it involves mainly proteins. During its progress, many morphological changes could occur that could lead to result the formation of the racemose structures as a manifestation of the "lowest free energy" in the system.[23] The volume density of caveolae in chemically fixed samples seems to be significantly higher (~70%) as compared with rapidly frozen specimens,[109,110] which was used by the opponents of vesicular transport as an argument against its importance.[52,109,110] Also, in continuous endothelium the frequency of caveolae varies widely depending on the organ. And this variation does not correlate with permeability, suggesting that caveolae have also other functions. Indeed caveolae have been also involved in eNOS signaling,[111] lipoprotein metabolism[112-114] and mechanotransduction.[115,116]

b. Capillary permeability increases with the increase in the vascular blood pressure, which has been explained by the existence of patent channels. An alternative explanation could be that transcytosis is pressure sensitive. There is some evidence that indicates that the number of endothelial caveolae and their loading of tracers are increased and the thickness of endothelium is decreased in high blood pressure.[117-123] The transport rate is insensitive to temperature downshifts and to chemical fixation by aldehydes suggesting that it is not active and it does not depend on proteins, respectively.[18,124-126] No morphological evidence as to the integrity of the endothelium was presented. Recent papers have suggested that NEM and filipin have actually an increasing effect on permeability to macromolecules in an ex-vivo lung preparation as well as in a peritoneal dialysis model.[17,127,128] The perfusion times used were rather lengthy and no morphological documentation of the capillary endothelium is presented to assess its integrity in such preparations (discussed in refs. 21,66). Some indication for this comes from the measurements of albumin permeability in isolated rat hindquarters,[124] which were several fold increased as compared with measurements in intact animals.[129]

c. Cav1 is essential for the formation of caveolae yet its overexpression seems to have an inhibitory effect on the caveolar internalization and cargo uptake into other cell types than ECs.[130,131] In the same vein treatment of mice with Cav1 scaffolding domain fused to the transducing domain of antennapedia lowered the increased permeability in inflammation as well as that in tumors. VEGF increased permeability is also inhibited by the Cav1 scaffolding domain.[132,133] It is not clear however, whether the basal and the increased permeability of the endothelium occur through the same mechanisms.

d. A serious challenge to the importance of transcytosis via caveolae in basal permeability comes from data obtained in caveolin1 -/- mice. These mice lack morphologically distinct caveolae in all nonmuscle cells, endothelium included[134-136] thereby showing that caveolin and caveolae are not essential either for the survival of these mice or for the normal development of their vasculature. In absence of caveolae the albumin concentration in the cerebrospinal fluid is the same as in wild type. The albumin uptake into cells (i.e., fibroblasts) or tissues (i.e., aortic rings) was found decreased in these mice.[137] The baseline permeability to albumin of large vessels (i.e., aorta) was found as decreased whereas the permeability of the capillaries was increased.[138] In keeping with this, albumin concentration in serum was found slightly but significantly decreased. The raise in basal capillary permeability of CAV1-/- mice was reduced by L-nitro arginine methyl ester (L-NAME), a nonspecific inhibitor of endothelial nitric oxide synthase (eNOS).[138] This study, tied in with data showing that the negative regulation of eNOS by caveolin 1 decreases the increased permeability in tumors and inflammation.[132,133,139-142] This suggests that in absence of caveolae, the mechanisms of increased permeability are turned on, which could compensate the loss of caveolae and to mediate an increase in basal permeability. Lastly, the failure of L-NAME to decrease the permeability in wild-type mice suggests that NO-independent mechanisms concur to the basal, physiological permeability. In other words, the increased permeability in inflammation or tumors occurs by a different mechanism and is not just a modulation (i.e., increase) of the mechanism of basal permeability.

Taken together, the consensus is that, depending on their physical properties molecules cross the continuous endothelium as follows: water and small molecules (e.g., ions, glucose) use a paracellular pathway as well as a transcellular pathway via transporters, whereas the pathway employed by macromolecules is transcellular. A role for membrane-bound vesicles in the transcellular pathway is not disputed. It is still debated whether the exchanges occur through transient TECs formed by vesicular fusion or by shuttling vesicles during transcytosis.[19] The role of caveolae in transendothelial exchanges is still unresolved but in the light of data from CAV1-/- mice it is unclear whether the control of permeability occurs via caveolae participating in transcytosis, via caveolin regulation of eNOS or both. In summary, the data imply several possibilities with respect to caveolar involvement in basal permeability: (i) Caveolae play the major role in transcytosis and in CAV1-/- mice compensatory mechanisms are switched on in their absence. (ii) Exchange of macromolecules across endothelium occurs via multiple partially redundant vesicular pathways, caveolae being one of them. (iii) Transcytosis of macromolecules occurs via a pathway other than caveolae that is sensitive to lipid rafts disassembly (i.e., macropinocytosis or related novel NCNC pathway).

Pores across Endothelium

As mentioned above, physical pores across endothelium have been found. These are the fenestrae, gaps, TEC and VVOs. Due to space constraints, I will discuss only the morphology and functional data, referring the reader to more comprehensive reviews.

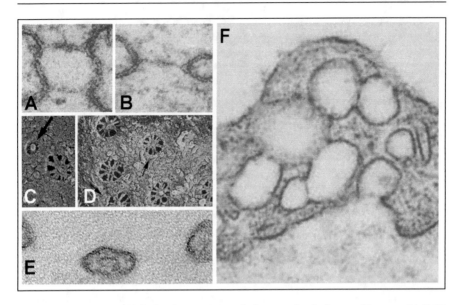

Figure 4. Pores across endothelial cells. Electron micrographs demonstrating the features of fenestrae with (B-D) and without (E) fenestral diaphragms, transendothelial channels (A, C) and vesiculo-vacuolar organelles (F). C and D reproduced from reference 7 with permission.

Endothelial Fenestrae

Endothelial fenestrae (Fig. 4B-E), the landmarks of the endothelia of fenestrated type, are circular windows resembling "boat portholes" that cut through the cell body. They are arranged in ordered linear arrays within large planar clusters called "sieve plates". Individual fenestral pores have a remarkably constant diameter (~62-68 nm).[3,8,9] Their circumference is delineated by the fenestral rim that occurs where the luminal aspect of plasma membrane continues with the abluminal one under a sharp angle. As in the case of the SDs, electron dense material has been demonstrated in the cytoplasm within the sharp angle formed by the plasma membrane. The circumference of most fenestral pores is apparently round, but as demonstrated by Maul, 20-30% of them have an octagonal symmetry, which can be photographically enhanced by rotating the image several times around its center.[143] The octagonal symmetry has been clearly established by deep-etch of rapidly frozen specimens[53] and high resolution scanning EM.[144]

The fenestrae are provided with a Fenestral Diaphragm (FD)[7] in all cases (i.e., in kidney peritubullar capillaries, all endocrine and exocrine glands, intestinal villi) except in capillaries of kidney glomerulus (Fig. 4E).[145-148] Although of a slightly larger diameter, the FDs are morphologically similar to the SDs by TEM. They appear as thin protein barriers anchored in the fenestral rim and provided with a central knob or density, as demonstrated in parallel and perpendicular sections. Moreover, in en face views obtained in oblique or grazing sections, the central density seems to be connected to the fenestral rim by thin fibrils.[149] Bearer and Orci have exquisitely demonstrated the intimate organization of the FD.[53] In rapidly frozen deep-etch specimens, they have shown the FD to consist of radial fibrils, starting at the rim and interweaving in a central mesh, the equivalent of the central density/knob seen by TEM. Unfortunately, the SDs of caveolae and TEC could not be resolved in the same detail due to their smaller size and experimental conditions (i.e., angle of shadowing).[53] It was readily apparent that the caveolar diaphragms resembled those of the transendothelial channels both featuring a central particle.[53] Close examination also reveals hints to the same radial fibrils pattern as for FDs. PV1 protein was shown to be a key component of the FDs, as in the case of the SDs.[150-153]

Finally, perfusion fixation of vascular beds provided with fenestrated endothelia where the fixatives have been dissolved in fluorocarbons (oxygen carriers used as blood plasma expanders) combined with tannic acid staining, revealed the presence of large (up to 400 nm) tufts or fascinae fenestrae on the luminal side of FDs and their absence from the luminal side of fenestrae.[154] These tufts were absent from the caveolar SDs. The fibers forming these tufts have been interpreted as the morphologic equivalent of the heparan sulfate proteoglycans (HSPGs) shown to reside on the luminal surface of fenestrae by other means (see below).

Fenestrated capillaries occur in organs that are involved in reabsorbtion of water and small solutes (e.g., kidney and intestine) or hormones (e.g., endocrine glands) from the interstitium into the blood. Most of the data in the literature argues that the fenestrated capillaries are highly permeable to water and small hydrophilic solutes but their permeability to macromolecules is basically the same as that of the continuous endothelium.[155] Although, the relative contributions to permeability of caveolae and their SDs, TEC, fenestrae and VVOs is not clear, these permeability characteristics have been attributed to fenestrae, which are considered the equivalents of the functional pores of the pore theory of capillary permeability in the fenestrated endothelium (reviewed in ref. 19). TEC were considered just double diaphragmed fenestrae and VVOs did not exist as a concept yet. The low albumin penetrance was explained by the contribution of the proteoglycans tufts as well as the basement membranes to the overall permeability. It has been calculated that the permeability to water and small solutes could be explained with just a fraction of the fenestrae being open, therefore the function of the proteoglycans would be that of a permeability-reducing plug.[156] Tracer experiments show that all fenestrae are permeable to horseradish peroxidase (∅ 4 nm)[8] whereas a few permit the exit of ferritin (∅ 11 nm),[8] dextrans (∅ 12.5-22.5 nm) and glycogen (∅ 22 and 30 nm).[157] As shown by Bearer and Orci,[53] the pores in between the fibrils of fenestrae have dimensions of 5-6 nm, therefore the passage of the larger tracers must be accommodated by fenestrae with larger pores. As in the case of the fenestrae, all these tracers gained access to few caveolae past their SDs.[157,158] An interpretation of these data would be that the diaphragms are dynamic structures and the degree of their permeability could be modulated by yet unknown factors. Another possibility that cannot be discarded at present would be the artifact creation in perfusion experiments or during processing in experiments where the tracers were injected in circulation.

In a mouse model the fenestrae and TEC were shown to disappear from the pancreatic islet vessels upon targeted genetic ablation of VEGF-A secretion from the pancreatic islet cells. The endothelium of islet capillaries became continuous with increased number of vesicles resembling caveolae. Functional assays of absorption of hormones secreted by the islet cells (e.g., insulin) showed that the effect of lack of fenestrae was a delayed absorption of these hormones. This constitutes strong evidence in favor of fenestrae as involved in the rapid exchange of water and small solutes such as peptide hormones.[159]

The functions of VVOs have been extensively reviewed recently.[22] They have been involved in increased permeability related to inflammation and tumors, permeability to macromolecules as well as particulates. Interestingly, their SDs seem to play a restrictive role in the permeability of particulates.[160] The results obtained in cells in culture[161] bolster the claims for VVOs as an organelle involved in the permeability induced by cytokines. However, it is not clear how their restricted localization to the venular endothelium would contribute to the overall permeability in a particular vascular bed, accepting the hypothesis that the bulk of transendothelial exchanges would occur at the level of the capillaries.

Transendothelial Channels

Transendothelial channels (TEC) (Figs. 4A,C) are patent pores spanning the endothelial cell (EC) body from one front to the other. TEC are rarely found in continuous endothelia where they seem to be formed by the fusion of either one caveola/plasmalemmal vesicle with both luminal and abluminal aspects of the plasmalemma or by chains of usually two to four caveolae.[5,6,162] In the fenestrated endothelia, TEC occur in the attenuated part of the endothelial cell and are provided with two SD (one luminal and one abluminal).[3,9,162,163] As in the case of caveolar SDs, the fine structure of the SDs of TEC could not be clearly resolved by the deep-etch procedure of Bearer and Orci due to technical limitations.[53] However, close inspections of the micrographs reveal the same pattern of radial fibrils and a central density. PV1 protein is also present in these diaphragms at both fronts opf the endothelial cell.

Vesiculo-Vacuolar Organelles (VVO)

Vesiculo-vacuolar organelles (VVOs) are morphologically defined as chains of interconnected vesicles of variable size that form intricate transendothelial channels spanning the cytoplasm of the endothelial cells from one front of EC to the other.[164] They are usually provided with SDs at the connection points between vesicles and vacuoles as well as at the level of their stoma or communication with the extracellular space.[160] The SDs of the VVOs closely-resemble those of caveolae by TEM. A recent review discusses the biology of VVOs in great detail.[22]

VVOs have been first described in the tumor vasculature[165] and subsequently in the normal endothelium of the post-capillary venules.[164-166] The definition of VVOs is purely morphological therefore there is a difficulty in discerning VVOs from clusters of caveolae or other vesicles that occur in most endothelia[3,5] in absence of specific biochemical markers. This raised doubts as to their status as a bona fide, novel organelle,[163] but this status might be bolstered by recent EM data showing that in caveolin 1 -/- mice (lacking caveolae in all nonmuscle cell types, endothelia included), the venules were still provided with structures resembling VVOs (Fig. 4F).[26,134,135] However, the relationship between these structures and VVOs remains to be clarified following biochemical evaluation.

Endothelial Pockets

These are infrequent structures that by EM resemble a pocket or a large vacuole formed by cellular processes that contain fenestrae with the usual structure[10] (reviewed in ref. 1). The information on these is scarce and, so far, they seem to occur in very low numbers only in the fenestrated endothelia. No functional data is available on these structures.

Conclusions

Undoubtedly our understanding of the mechanisms of the exchanges across cellular barriers has greatly advanced in the last years. What is clear is that there are many pathways by which material can be moved from one side of the barrier to the other. Which pathways are important for each individual molecule is a matter of further inquiry. As for the mechanisms of transendothelial exchange, the controversy is still there, as we still do not have clear information on the molecular players that allow its increased and continuous protein leakage

Acknowledgements

I would like to thank G.E. Palade, C. Carriere, B. Jacobson and L. Ghitescu for advice and long and fruitful collaborations. I would like to apologize for relying on reviews and not being able to cite all the primary sources of the data due to constraints of space.

References

1. Stan RV. Endothelial diaphragms: Nanogates of vascular permeability. Anat Rec 2005; In press.
2. Tuma PL, Hubbard AL. Transcytosis: Crossing cellular barriers. Physiol Rev 2003; 83:871-932.
3. Simionescu M, Simionescu N, Palade GE. Morphometric data on the endothelium of blood capillaries. J Cell Biol 1974; 60:128-152.
4. Palade GE. Fine structure of blood capillaries. J Applied Physics 1953; 24:1424.
5. Palade GE, Bruns RR. Structural modulations of plasmalemmal vesicles. J Cell Biol 1968; 37:633-649.
6. Simionescu N, Siminoescu M, Palade GE. Permeability of muscle capillaries to small heme-peptides. Evidence for the existence of patent transendothelial channels. J Cell Biol 1975; 64:586-607.
7. Gautier A, Bernhard W, Oberling C. Sur l'existence d'un appareil lacunaire pericapillaire du glomerule de Malpighi, revele par la microscopie electronique. Comptes rendus de la seances de la Societe de Biologie 1950; 144:1605-1607.
8. Clementi F, Palade GE. Intestinal capillaries. I. Permeability to peroxidase and ferritin. J Cell Biol 1969; 41:33-58.

9. Clementi F, Palade GE. Intestinal capillaries. II. Structural effects ofEDTA and histamine. J Cell Biol 1969; 42:706-714.
10. Milici AJ, Peters KR, Palade GE. The endothelial pocket. A new structure in fenestrated endothelia. Cell Tissue Res 1986; 244:493-499.
11. Milici AJ, Furie MB, Carley WW. The formation of fenestrations and channels by capillary endothelium in vitro. Proc Natl Acad Sci USA 1985; 82:6181-6185.
12. Braet F, Wisse E. Structural and functional aspects of liver sinusoidal endothelial cell fenestrae: A review. Comp Hepatol 2002; 1:1.
13. Aird WC. Spatial and temporal dynamics of the endothelium. J Thromb Haemost 2005; 3:1392-1406.
14. Stevens T, Rosenberg R, Aird W et al. NHLBI workshop report: Endothelial cell phenotypes in heart, lung, and blood diseases. Am J Physiol Cell Physiol 2001; 281:C1422-1433.
15. Ghitescu L, Robert M. Diversity in unity: The biochemical composition of the endothelial cell surface varies between the vascular beds. Microsc Res Tech 2002; 57:381-389.
16. Cines DB, Pollak ES, Buck CA et al. Endothelial cells in physiology and in the pathophysiology of vascular disorders. Blood 1998; 91:3527-3561.
17. Rippe B, Rosengren BI, Carlsson O et al. Transendothelial transport: The vesicle controversy. J Vasc Res 2002; 39:375-390.
18. Rippe B, Haraldsson B. Transport of macromolecules across microvascular walls: The two-pore theory. Physiol Rev 1994; 74:163-219.
19. Michel CC, Curry FE. Microvascular permeability. Physiol Rev 1999; 79:703-761.
20. Stan RV. Structure and function of endothelial caveolae. Microsc Res Tech 2002; 57:350-364.
21. Predescu D, Vogel SM, Malik AB. Functional and morphological studies of protein transcytosis in continuous endothelia. Am J Physiol Lung Cell Mol Physiol 2004; 287:L895-901.
22. Feng D, Nagy JA, Dvorak HF et al. Ultrastructural studies define soluble macromolecular, particulate, and cellular transendothelial cell pathways in venules, lymphatic vessels, and tumor-associated microvessels in man and animals. Microsc Res Tech 2002; 57:289-326.
23. Palade GE. Role of plasmalemmal vesicles. In: Crystal RG, West JB, eds. The Lung: Scientific Foundations. New York: Raven Press Ltd, 1991:359-367.
24. Simionescu N, ed. Transcytosis and Traffic of Membranes in Endothelial Cells. Berlin: Springer-Verlag, 1981.
25. Simionescu N. Cellular aspects of transcapillary exchange. Physiol Rev 1983; 63:1536-1579.
26. Stan RV. Structure of caveolae. Biochim Biophys Acta 2005.
27. Misrahi M, Beau I, Ghinea N et al. The LH/CG and FSH receptors: Different molecular forms and intracellular traffic. Mol Cell Endocrinol 1996; 125:161-167.
28. Ghinea N, Milgrom E. A new function for the LH/CG receptor: Transcytosis of hormone across the endothelial barrier in target organs. Semin Reprod Med 2001; 19:97-101.
29. Vu Hai MT, Lescop P, Loosfelt H et al. Receptor-mediated transcytosis of follicle-stimulating hormone through the rat testicular microvasculature. Biol Cell 2004; 96:133-144.
30. Ghinea N, Mai TV, Groyer-Picard MT et al. How protein hormones reach their target cells. Receptor-mediated transcytosis of hCG through endothelial cells. J Cell Biol 1994; 125:87-97.
31. Zlokovic BV, Martel CL, Mackic JB et al. Brain uptake of circulating apolipoproteins J and E complexed to Alzheimer's amyloid beta. Biochem Biophys Res Commun 1994; 205:1431-1437.
32. Zlokovic BV. Cerebrovascular transport of Alzheimer's amyloid beta and apolipoproteins J and E: Possible anti-amyloidogenic role of the blood-brain barrier. Life Sci 1996; 59:1483-1497.
33. Zlokovic BV, Martel CL, Matsubara E et al. Glycoprotein 330/megalin: Probable role in receptor-mediated transport of apolipoprotein J alone and in a complex with Alzheimer disease amyloid beta at the blood-brain and blood-cerebrospinal fluid barriers. Proc Natl Acad Sci USA 1996; 93:4229-4234.
34. Chun JT, Wang L, Pasinetti GM et al. Glycoprotein 330/megalin (LRP-2) has low prevalence as mRNA and protein in brain microvessels and choroid plexus. Exp Neurol 1999; 157:194-201.
35. Calero M, Rostagno A, Matsubara E et al. Apolipoprotein J (clusterin) and Alzheimer's disease. Microsc Res Tech 2000; 50:305-315.
36. Shayo M, McLay RN, Kastin AJ et al. The putative blood-brain barrier transporter for the beta-amyloid binding protein apolipoprotein j is saturated at physiological concentrations. Life Sci 1997; 60:PL115-118.
37. Fuchs R, Ellinger I. Endocytic and transcytotic processes in villous syncytiotrophoblast: Role in nutrient transport to the human fetus. Traffic 2004; 5:725-738.
38. Van de Perre P. Transfer of antibody via mother's milk. Vaccine 2003; 21:3374-3376.
39. Linton EA, Rodriguez-Linares B, Rashid-Doubell F et al. Caveolae and caveolin-1 in human term villous trophoblast. Placenta 2003; 24:745-757.

40. Chaudhury C, Mehnaz S, Robinson JM et al. The major histocompatibility complex-related Fc receptor for IgG (FcRn) binds albumin and prolongs its lifespan. J Exp Med 2003; 197:315-322.
41. Takizawa T, Anderson CL, Robinson JM. A novel Fc{gamma}R-defined, IgG-containing organelle in placental endothelium. J Immunol 2005; 175:2331-2339.
42. Lyden TW, Anderson CL, Robinson JM. The endothelium but not the syncytiotrophoblast of human placenta expresses caveolae. Placenta 2002; 23:640-652.
43. Lyden TW, Robinson JM, Tridandapani S et al. The Fc receptor for IgG expressed in the villus endothelium of human placenta is Fc gamma RIIb2. J Immunol 2001; 166:3882-3889.
44. Gafencu A, Heltianu C, Burlacu A et al. Investigation of IgG receptors expressed on the surface of human placental endothelial cells. Placenta 2003; 24:664-676.
45. Kim KJ, Malik AB. Protein transport across the lung epithelial barrier. Am J Physiol Lung Cell Mol Physiol 2003; 284:L247-259.
46. Schlachetzki F, Zhu C, Pardridge WM. Expression of the neonatal Fc receptor (FcRn) at the blood-brain barrier. J Neurochem 2002; 81:203-206.
47. Simons K, Wandinger-Ness A. Polarized sorting in epithelia. Cell 1990; 62:207-210.
48. Caplan MJ. Membrane polarity in epithelial cells: Protein sorting and establishment of polarized domains. Am J Physiol 1997; 272:F425-429.
49. Muller WA, Gimbrone Jr MA. Plasmalemmal proteins of cultured vascular endothelial cells exhibit apical-basal polarity: Analysis by surface-selective iodination. J Cell Biol 1986; 103:2389-2402.
50. Bruns RR, Palade GE. Studies on blood capillaries. I. General organization of blood capillaries in muscle. J Cell Biol 1968; 37:244-276.
51. Bruns RR, Palade GE. Studies on blood capillaries. II. Transport of ferritin molecules across the wall of muscle capillaries. J Cell Biol 1968; 37:277-299.
52. Noguchi Y, Shibata Y, Yamamoto T. Endothelial vesicular system in rapid-frozen muscle capillaries revealed by serial sectioning and deep etching. Anat Rec 1987; 217:355-360.
53. Bearer EL, Orci L. Endothelial fenestral diaphragms: A quick-freeze, deep-etch study. J Cell Biol 1985; 100:418-428.
54. Frokjaer-Jensen J. Three-dimensional organization of plasmalemmal vesicles in endothelial cells. An analysis by serial sectioning of frog mesenteric capillaries. J Ultrastruct Res 1980; 73:9-20.
55. Frokjaer-Jensen J. The plasmalemmal vesicular system in striated muscle capillaries and in pericytes. Tissue Cell 1984; 16:31-42.
56. Frokjaer-Jensen J. The endothelial vesicle system in cryofixed frog mesenteric capillaries analysed by ultrathin serial sectioning. J Electron Microsc Tech 1991; 19:291-304.
57. Frokjaer-Jensen J, Wagner RC, Andrews SB et al. Three-dimensional organization of the plasmalemmal vesicular system in directly frozen capillaries of the rete mirabile in the swim bladder of the eel. Cell Tissue Res 1988; 254:17-24.
58. Bundgaard M, Frokjaer-Jensen J, Crone C. Endothelial plasmalemmal vesicles as elements in a system of branching invaginations from the cell surface. Proc Natl Acad Sci USA 1979; 76:6439-6442.
59. Bundgaard M. Vesicular transport in capillary endothelium: Does it occur? Fed Proc 1983; 42:2425-2430.
60. Bundgaard M. The three-dimensional organization of smooth endoplasmic reticulum in capillary endothelia: Its possible role in regulation of free cytosolic calcium. J Struct Biol 1991; 107:76-85.
61. Milici AJ, Watrous NE, Stukenbrok H et al. Transcytosis of albumin in capillary endothelium. J Cell Biol 1987; 105:2603-2612.
62. Ghitescu L, Galis Z, Simionescu M et al. Differentiated uptake and transcytosis of albumin in successive vascular segments. J Submicrosc Cytol Pathol 1988; 20:657-669.
63. Ghinea N, Simionescu N. Anionized and cationized hemeundecapeptides as probes for cell surface charge and permeability studies: Differentiated labeling of endothelial plasmalemmal vesicles. J Cell Biol 1985; 100:606-612.
64. Predescu D, Palade GE. Plasmalemmal vesicles represent the large pore system of continuous microvascular endothelium. Am J Physiol 1993; 265:H725-733.
65. Predescu D, Horvat R, Predescu S et al. Transcytosis in the continuous endothelium of the myocardial microvasculature is inhibited by N-ethylmaleimide. Proc Natl Acad Sci USA 1994; 91:3014-3018.
66. Predescu SA, Predescu DN, Palade GE. Plasmalemmal vesicles function as transcytotic carriers for small proteins in the continuous endothelium. Am J Physiol 1997; 272:H937-949.
67. Schnitzer JE, Allard J, Oh P. NEM inhibits transcytosis, endocytosis, and capillary permeability: Implication of caveolae fusion in endothelia. Am J Physiol 1995; 268:H48-55.
68. Palade GE. Transport in quanta across endothelium in blood capillaries. Anat Rec 1960; 136:254.
69. Palade GE. Blood capillaries of the heart and other organs. Circulation 1961; 24:368-388.

70. Wagner RC, Chen SC. Transcapillary transport of solute by the endothelial vesicular system: Evidence from thin serial section analysis. Microvasc Res 1991; 42:139-150.
71. Ghitescu L, Bendayan M. Transendothelial transport of serum albumin: A quantitative immunocytochemical study. J Cell Biol 1992; 117:745-755.
72. McIntosh DP, Tan XY, Oh P et al. Targeting endothelium and its dynamic caveolae for tissue-specific transcytosis in vivo: A pathway to overcome cell barriers to drug and gene delivery. Proc Natl Acad Sci USA 2002; 99:1996-2001.
73. Predescu D, Predescu S, McQuistan T et al. Transcytosis of alpha1-acidic glycoprotein in the continuous microvascular endothelium. Proc Natl Acad Sci USA 1998; 95:6175-6180.
74. Predescu SA, Predescu DN, Palade GE. Endothelial transcytotic machinery involves supramolecular protein-lipid complexes. Mol Biol Cell 2001; 12:1019-1033.
75. Pascariu M, Bendayan M, Ghitescu L. Correlated endothelial caveolin overexpression and increased transcytosis in experimental diabetes. J Histochem Cytochem 2004; 52:65-76.
76. Carver LA, Schnitzer JE, Anderson RG et al. Role of caveolae and lipid rafts in cancer: Workshop summary and future needs. Cancer Res 2003; 63:6571-6574.
77. McIntosh DP, Schnitzer JE. Caveolae require intact VAMP for targeted transport in vascular endothelium. Am J Physiol 1999; 277:H2222-2232.
78. Feng D, Flaumenhaft R, Bandeira-Melo C et al. Ultrastructural localization of vesicle-associated membrane protein(s) to specialized membrane structures in human pericytes, vascular smooth muscle cells, endothelial cells, neutrophils, and eosinophils. J Histochem Cytochem 2001; 49:293-304.
79. Schnitzer JE, Liu J, Oh P. Endothelial caveolae have the molecular transport machinery for vesicle budding, docking, and fusion including VAMP, NSF, SNAP, annexins, and GTPases. J Biol Chem 1995; 270:14399-14404.
80. Henley JR, Krueger EW, Oswald BJ et al. Dynamin-mediated internalization of caveolae. J Cell Biol 1998; 141:85-99.
81. Oh P, McIntosh DP, Schnitzer JE. Dynamin at the neck of caveolae mediates their budding to form transport vesicles by GTP-driven fission from the plasma membrane of endothelium. J Cell Biol 1998; 141:101-114.
82. Henley JR, Cao H, McNiven MA. Participation of dynamin in the biogenesis of cytoplasmic vesicles. FASEB J 1999; 13(Suppl. 2):S243-247.
83. Yao Q, Chen J, Cao H et al. Caveolin-1 interacts directly with dynamin-2. J Mol Biol 2005; 348:491-501.
84. Predescu SA, Predescu DN, Timblin BK et al. Intersectin regulates fission and internalization of caveolae in endothelial cells. Mol Biol Cell 2003; 14:4997-5010.
85. Schnitzer JE, Oh P, McIntosh DP. Role of GTP hydrolysis in fission of caveolae directly from plasma membranes. Science 1996; 274:239-242.
86. Schnitzer JE, Oh P, McIntosh DP. Role of GTP hydrolysis in fission of caveolae directly from plasma membranes [published erratum appears in Science 1996; 274:1069]. Science 1996; 274:239-242.
87. Parton RG, Joggerst B, Simons K. Regulated internalization of caveolae. J Cell Biol 1994; 127:1199-1215.
88. Smart EJ, Estes K, Anderson RG. Inhibitors that block both the internalization of caveolae and the return of plasmalemmal vesicles. Cold Spring Harb Symp Quant Biol 1995; 60:243-248.
89. Choudhury A, Dominguez M, Puri V et al. Rab proteins mediate Golgi transport of caveola-internalized glycosphingolipids and correct lipid trafficking in Niemann-Pick C cells. J Clin Invest 2002; 109:1541-1550.
90. Mundy DI, Machleidt T, Ying YS et al. Dual control of caveolar membrane traffic by microtubules and the actin cytoskeleton. J Cell Sci 2002; 115:4327-4339.
91. Pelkmans L, Kartenbeck J, Helenius A. Caveolar endocytosis of simian virus 40 reveals a new two-step vesicular-transport pathway to the ER. Nat Cell Biol 2001; 3:473-483.
92. Pelkmans L, Puntener D, Helenius A. Local actin polymerization and dynamin recruitment in SV40-induced internalization of caveolae. Science 2002; 296:535-539.
93. Pelkmans L, Burli T, Zerial M et al. Caveolin-stabilized membrane domains as multifunctional transport and sorting devices in endocytic membrane traffic. Cell 2004; 118:767-780.
94. Sharma P, Sabharanjak S, Mayor S. Endocytosis of lipid rafts: An identity crisis. Semin Cell Dev Biol 2002; 13:205-214.
95. Sharma DK, Brown JC, Choudhury A et al. Selective stimulation of caveolar endocytosis by glycosphingolipids and cholesterol. Mol Biol Cell 2004; 15:3114-3122.
96. Sharma DK, Choudhury A, Singh RD et al. Glycosphingolipids internalized via caveolar-related endocytosis rapidly merge with the clathrin pathway in early endosomes and form microdomains for recycling. J Biol Chem 2003; 278:7564-7572.

97. Kirkham M, Fujita A, Chadda R et al. Ultrastructural identification of uncoated caveolin-independent early endocytic vehicles. J Cell Biol 2005; 168:465-476.
98. Damm EM, Pelkmans L, Kartenbeck J et al. Clathrin- and caveolin-1-independent endocytosis: Entry of simian virus 40 into cells devoid of caveolae. J Cell Biol 2005; 168:477-488.
99. Aoki T, Nomura R, Fujimoto T. Tyrosine phosphorylation of caveolin-1 in the endothelium. Exp Cell Res 1999; 253:629-636.
100. Shajahan AN, Tiruppathi C, Smrcka AV et al. Gbetagamma activation of Src induces caveolae-mediated endocytosis in endothelial cells. J Biol Chem 2004; 279:48055-48062.
101. Shajahan AN, Timblin BK, Sandoval R et al. Role of Src-induced dynamin-2 phosphorylation in caveolae-mediated endocytosis in endothelial cells. J Biol Chem 2004; 279:20392-20400.
102. Vogel SM, Minshall RD, Pilipovic M et al. Albumin uptake and transcytosis in endothelial cells in vivo induced by albumin-binding protein. Am J Physiol Lung Cell Mol Physiol 2001; 281:L1512-1522.
103. Minshall RD, Tiruppathi C, Vogel SM et al. Endothelial cell-surface gp60 activates vesicle formation and trafficking via G(i)-coupled Src kinase signaling pathway. J Cell Biol 2000; 150:1057-1070.
104. John TA, Vogel SM, Minshall RD et al. Evidence for the role of alveolar epithelial gp60 in active transalveolar albumin transport in the rat lung. J Physiol 2001; 533:547-559.
105. Tiruppathi C, Song W, Bergenfeldt M et al. Gp60 activation mediates albumin transcytosis in endothelial cells by tyrosine kinase-dependent pathway. J Biol Chem 1997; 272:25968-25975.
106. Yamaguchi T, Murata Y, Fujiyoshi Y et al. Regulated interaction of endothelin B receptor with caveolin-1. Eur J Biochem 2003; 270:1816-1827.
107. Foster LJ, De Hoog CL, Mann M. Unbiased quantitative proteomics of lipid rafts reveals high specificity for signaling factors. Proc Natl Acad Sci USA 2003; 100:5813-5818.
108. Vogel SM, Easington CR, Minshall RD et al. Evidence of transcellular permeability pathway in microvessels. Microvasc Res 2001; 61:87-101.
109. Wood MR, Wagner RC, Andrews SB et al. Rapidly-frozen, cultured, human endothelial cells: An ultrastructural and morphometric comparison between freshly-frozen and glutaraldehyde prefixed cells. Microcirc Endothel Lymph 1986; 3:323-358.
110. McGuire PG, Twietmeyer TA. Morphology of rapidly frozen aortic endothelial cells. Glutaraldehyde fixation increases the number of caveolae. Circ Res 1983; 53:424-429.
111. Gratton JP, Bernatchez P, Sessa WC. Caveolae and caveolins in the cardiovascular system. Circ Res 2004; 94:1408-1417.
112. Razani B, Combs TP, Wang XB et al. Caveolin-1-deficient mice are lean, resistant to diet-induced obesity, and show hypertriglyceridemia with adipocyte abnormalities. J Biol Chem 2002; 277:8635-8647.
113. Frank PG, Lee H, Park DS et al. Genetic ablation of caveolin-1 confers protection against atherosclerosis. Arterioscler Thromb Vasc Biol 2004; 24:98-105.
114. Frank PG, Lisanti MP. Caveolin-1 and caveolae in atherosclerosis: Differential roles in fatty streak formation and neointimal hyperplasia. Curr Opin Lipidol 2004; 15:523-529.
115. Rizzo V, McIntosh DP, Oh P et al. In situ flow activates endothelial nitric oxide synthase in luminal caveolae of endothelium with rapid caveolin dissociation and calmodulin association. J Biol Chem 1998; 273:34724-34729.
116. Rizzo V, Sung A, Oh P et al. Rapid mechanotransduction in situ at the luminal cell surface of vascular endothelium and its caveolae. J Biol Chem 1998; 273:26323-26329.
117. Arnal JF, Dinh-Xuan AT, Pueyo M et al. Endothelium-derived nitric oxide and vascular physiology and pathology. Cell Mol Life Sci 1999; 55:1078-1087.
118. Hillman N, Cox S, Noble AR et al. Increased numbers of caveolae in retinal endothelium and pericytes in hypertensive diabetic rats. Eye 2001; 15:319-325.
119. Kurozumi T, Imamura T, Tanaka K et al. Effects of hypertension and hypercholesteremia on the permeability of fibrinogen and low density lipoprotein in the coronary artery of rabbits. Immunoelectron-microscopic study. Atherosclerosis 1983; 49:267-276.
120. Kurozumi T, Imamura T, Tanaka K et al. Permeation and deposition of fibrinogen and low-density lipoprotein in the aorta and cerebral artery of rabbits-immuno-electron microscopic study. Br J Exp Pathol 1984; 65:355-364.
121. Sakata N, Ida T, Joshita T et al. Scanning and transmission electron microscopic study on the cerebral arterial endothelium of experimentally hypertensive rats fed an atherogenic diet. Acta Pathol Jpn 1983; 33:1105-1113.
122. Traub O, Berk BC. Laminar shear stress: Mechanisms by which endothelial cells transduce an atheroprotective force. Arterioscler Thromb Vasc Biol 1998; 18:677-685.
123. Wiener J, Loud AV, Giacomelli F et al. Morphometric analysis of hypertension-induced hypertrophy of rat thoracic aorta. Am J Pathol 1977; 88:619-633.

124. Rippe B, Kamiya A, Folkow B. Transcapillary passage of albumin, effects of tissue cooling and of increases in filtration and plasma colloid osmotic pressure. Acta Physiol Scand 1979; 105:171-187.
125. Haraldsson B, Johansson BR. Changes in transcapillary exchange induced by perfusion fixation with glutaraldehyde, followed by measurements of capillary filtration coefficient, diffusion capacity and albumin clearance. Acta Physiol Scand 1985; 124:99-106.
126. Rosengren BI, Carlsson O, Venturoli D et al. Transvascular passage of macromolecules into the peritoneal cavity of normo- and hypothermic rats in vivo: Active or passive transport? J Vasc Res 2004; 41:123-130.
127. Carlsson O, Rosengren BI, Rippe B. Transcytosis inhibitor N-ethylmaleimide increases microvascular permeability in rat muscle. Am J Physiol Heart Circ Physiol 2001; 281:H1728-1733.
128. Rosengren BI, Al Rayyes O, Rippe B. Transendothelial transport of low-density lipoprotein and albumin across the rat peritoneum in vivo: Effects of the transcytosis inhibitors NEM and filipin. J Vasc Res 2002; 39:230-237.
129. Renkin EM, Gustafson-Sgro M, Sibley L. Coupling of albumin flux to volume flow in skin and muscles of anesthetized rats. Am J Physiol 1988; 255:H458-466.
130. Thomsen P, Roepstorff K, Stahlhut M et al. Caveolae are highly immobile plasma membrane microdomains, which are not involved in constitutive endocytic trafficking. Mol Biol Cell 2002; 13:238-250.
131. Le PU, Guay G, Altschuler Y et al. Caveolin-1 is a negative regulator of caveolae-mediated endocytosis to the endoplasmic reticulum. J Biol Chem 2002; 277:3371-3379.
132. Gratton JP, Lin MI, Yu J et al. Selective inhibition of tumor microvascular permeability by cavtratin blocks tumor progression in mice. Cancer Cell 2003; 4:31-39.
133. Bucci M, Gratton JP, Rudic RD et al. In vivo delivery of the caveolin-1 scaffolding domain inhibits nitric oxide synthesis and reduces inflammation. Nat Med 2000; 6:1362-1367.
134. Zhao YY, Liu Y, Stan RV et al. Defects in caveolin-1 cause dilated cardiomyopathy and pulmonary hypertension in knockout mice. Proc Natl Acad Sci USA 2002; 99:11375-11380.
135. Drab M, Verkade P, Elger M et al. Loss of caveolae, vascular dysfunction, and pulmonary defects in caveolin-1 gene-disrupted mice. Science 2001; 293:2449-2452.
136. Razani B, Engelman JA, Wang XB et al. Caveolin-1 null mice are viable but show evidence of hyperproliferative and vascular abnormalities. J Biol Chem 2001; 276:38121-38138.
137. Schubert W, Frank PG, Razani B et al. Caveolae-deficient endothelial cells show defects in the uptake and transport of albumin in vivo. J Biol Chem 2001; 276:48619-48622.
138. Schubert W, Frank PG, Woodman SE et al. Microvascular hyperpermeability in caveolin-1 (-/-) knock-out mice. Treatment with a specific nitric-oxide synthase inhibitor, L-name, restores normal microvascular permeability in Cav-1 null mice. J Biol Chem 2002; 277:40091-40098.
139. Gratton JP, Yu J, Griffith JW et al. Erratum: Cell-permeable peptides improve cellular uptake and therapeutic gene delivery of replication-deficient viruses in cells and in vivo. Nat Med 2003; 9:1221.
140. Gratton JP, Yu J, Griffith JW et al. Cell-permeable peptides improve cellular uptake and therapeutic gene delivery of replication-deficient viruses in cells and in vivo. Nat Med 2003; 9:357-362.
141. Bernatchez PN, Bauer PM, Yu J et al. Dissecting the molecular control of endothelial NO synthase by caveolin-1 using cell-permeable peptides. Proc Natl Acad Sci USA 2005; 102:761-766.
142. Bauer PM, Yu J, Chen Y et al. Endothelial-specific expression of caveolin-1 impairs microvascular permeability and angiogenesis. Proc Natl Acad Sci USA 2005; 102:204-209.
143. Maul GG. Structure and formation of pores in fenestrated capillaries. J Ultrastruct Res 1971; 36:768-782.
144. Apkarian RP. The fine structure of fenestrated adrenocortical capillaries revealed by in-lens field-emission scanning electron microscopy and scanning transmission electron microscopy. Scanning 1997; 19:361-367.
145. Oberling C, Gautier A, Bernhardt W. La structure des capillaires glomerulairevue au microscope electronique. Presse medicale Parisiene 1951; 59:938-940.
146. Pease DC. Electron microscopy of the vascular bed of the kidney cortex. Anat Rec 1955; 121:701-712.
147. Reeves WH, Kanwar YS, Farquhar MG. Assembly of the glomerular filtration surface. Differentiation of anionic sites in glomerular capillaries of newborn rat kidney. J Cell Biol 1980; 85:735-753.
148. Yamada E. The fine structure of the mouse kidney glomerulus. J Biophys Biochem Cytol 1955; 1:551-566.
149. Friederici HH. On the diaphragm across fenestrae of capillary endothelium. J Ultrastruct Res 1969; 27:373-375.
150. Stan RV, Tkachenko E, Niesman IR. PV1 is a key structural component for the formation of the stomatal and fenestral diaphragms. Mol Biol Cell 2004; 15:3615-3630.

151. Stan RV. Multiple PV1 dimers reside in the same stomatal or fenestral diaphragm. Am J Physiol Heart Circ Physiol 2004; 286:H1347-1353.
152. Stan RV, Kubitza M, Palade GE. PV-1 is a component of the fenestral and stomatal diaphragms in fenestrated endothelia. Proc Natl Acad Sci USA 1999; 96:13203-13207.
153. Stan RV, Ghitescu L, Jacobson BS et al. Isolation, cloning, and localization of rat PV-1, a novel endothelial caveolar protein. J Cell Biol 1999; 145:1189-1198.
154. Rostgaard J, Qvortrup K. Electron microscopic demonstrations of filamentous molecular sieve plugs in capillary fenestrae. Microvasc Res 1997; 53:1-13.
155. Granger DN, Granger JP, Brace RA et al. Analysis of the permeability characteristics of cat intestinal capillaries. Circ Res 1979; 44:335-344.
156. Levick JR, Smaje LH. An analysis of the permeability of a fenestra. Microvasc Res 1987; 33:233-256.
157. Simionescu N, Simionescu M, Palade GE. Permeability of intestinal capillaries. Pathway followed by dextrans and glycogens. J Cell Biol 1972; 53:365-392.
158. Milici AJ, Bankston PW. Fetal and neonatal rat intestinal capillaries: Permeability to carbon, ferritin, hemoglobin, and myoglobin. Am J Anat 1982; 165:165-186.
159. Lammert E, Gu G, McLaughlin M et al. Role of VEGF-A in vascularization of pancreatic islets. Curr Biol 2003; 13:1070-1074.
160. Feng D, Nagy JA, Pyne K et al. Pathways of macromolecular extravasation across microvascular endothelium in response to VPF/VEGF and other vasoactive mediators. Microcirculation 1999; 6:23-44.
161. Chen J, Braet F, Brodsky S et al. VEGF-induced mobilization of caveolae and increase in permeability of endothelial cells. Am J Physiol Cell Physiol 2002; 282:C1053-1063.
162. Milici AJ, L'Hernault N, Palade GE. Surface densities of diaphragmed fenestrae and transendothelial channels in different murine capillary beds. Circ Res 1985; 56:709-717.
163. Roberts WG, Palade GE. Endothelial fenestrae and fenestral diaphragms. In: Rissau W, Rubanyi GM, eds. Morphogenesis of Endothelium. Amsterdam: Hardwood Academic Publishers, 2000:23-41.
164. Dvorak AM, Kohn S, Morgan ES et al. The vesiculo-vacuolar organelle (VVO): A distinct endothelial cell structure that provides a transcellular pathway for macromolecular extravasation. J Leukoc Biol 1996; 59:100-115.
165. Kohn S, Nagy JA, Dvorak HF et al. Pathways of macromolecular tracer transport across venules and small veins. Structural basis for the hyperpermeability of tumor blood vessels. Lab Invest 1992; 67:596-607.
166. Vasile E, Dvorak AM, Stan RV et al. Isolation and characterization of caveolae and vesiculo-vacuolar organelles from endothelial cells cultured with VPF/VEGF and from human lung. Mol Biol Cell 2000; 11:121.

CHAPTER 20

Molecular Transfers through Transient Lymphoid Cell-Cell Channels

Mary Poupot, Julie Gertner and Jean-Jacques Fournié*

Abstract

Intercellular communication is inherent to life and has evolved with extremely diversified forms from prokaryotes to the most sophisticated multicellular eukaryotes. Haematopoietic cells from mammalians have adapted a transient and highly versatile means for the direct transfer of molecular information through cell-cell channels. This channel takes place and operates transiently at a highly specialised interface now referred to as the immunological synapse. The oriented transfer of membrane molecules from one cell to another is now referred to as trogocytosis, a cell process discovered recently, but which consequences have been described for decades without clear understanding of its cellular basis. This article will review current state-of-the-art knowledge on trogocytosis, its technological use and physiological significance.

Introduction: Transient Cell-Cell Contacts Take Place at the Immunological Synapse between Lymphoid Cells

Haematopoietic cells are highly mobile, and frequently engage numerous contacts with other cells. To fulfil their immunological functions, these contacts must enable the lymphoid cells to recognize and discriminate normal self tissues and cells from foreign ones, including microbial pathogens, parasites and transformed cells. The site of such meetings is variable, from blood and vasculature to inflamed tissues and lymphoid organs. Three successive phases characterize the cell-cell contact involving lymphoid cells: the attachment/scanning/activation phase, the lymphoid cell response phase and the cell dissociation. However, given the huge number of possible encounters and responses the lymphocyte has to face, cell contacts are usually transient, and adapted to the functions that immune cells will fulfil. In vitro, while a normal T lymphocyte may bind, scan and release an autologous resting T lymphocyte within ten seconds because no activation ensued the scanning of its surface, the same cell may engage a cognate interaction for hours with an antigen-presenting cells, till adequate cytokine secretion takes place. The lymphoid cell surface in tight contact with a scanned cell (we will refer below to this recognized cell entity as to "target cell", regardless of the recognizing lymphoid cells and nature of its response) is also unstable in structure and supramolecular organization.

The immunological synapse (Fig. 1) has been the subject of recent reviews (see for instance refs. 1-4) and will not be discussed further here. We will focus our review on events related locally and temporally to this structure: the intercellular transfer of membrane patches located at the apposed surfaces. These have been described recently,[5,6] and may take place through either large synapses or

*Corresponding Author: Jean-Jacques Fournié—Groupe d'Etude des Antigènes NonConventionnels, département Oncogénèse & Signalisation dans les Cellules Hématopoïétiques, INSERM Unité 563, Centre de Physiopathologie de Toulouse Purpan, Centre de Physiopathologie de Toulouse Purpan, BP3048, Hôpital Purpan, 31024 Toulouse Cedex 03, France. Email: fournie@toulouse.inserm.fr

Cell-Cell Channels, edited by Frantisek Baluska, Dieter Volkmann and Peter W. Barlow.
©2006 Landes Bioscience and Springer Science+Business Media.

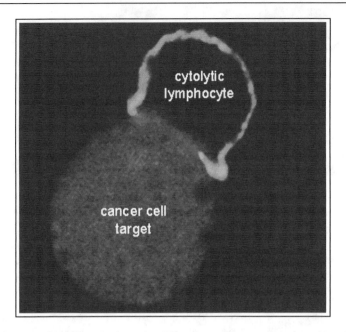

Figure 1. The immunological synapse of a cytolytic T lymphocyte (green membrane) attacking a B-cell lymphoma target(red). Note the concentration of phosphoprotein signals at the center of the synapse (anti phosphoprotein antibody, stained blue) of the killer cell, prior to delivery of its lethal hits. A color version of this figure is available online at www.Eurekah.com.

conversely very narrow structures referred to as nanotubes.[7] Their possible role in cell communication will be discussed below.

First Findings of Intercellular Transfer of Membrane Markers through the SI

Several reports in the 70'-80's had described the presence of unexpected markers on the cell surface of lymphocyte subsets known for not expressing them, such as Ig and allogenic MHC molecules on T cells[8,9] or Mls superantigen on IE cells.[10] Along the same line, but at the time not understood as the same process, was the acquisition of fluorescence derived from labelled cell targets by antigen-specific effector T cells.[11] Although of unclear significance at the time, a major breakthrough in this domain was brought by the demonstration that GFP-MHC molecules derived from antigen presenting cells could be acquired by antigen-specific CD8+ CTL through their immunological synapse.[12] Both TCR and CD28 T-cell surface receptors were found to trigger what by then was thought to be the mere consequence of uptake by internalization of cognate receptor/ligand complexes.[13]

Further, two nearly simultaneous reports described however this process to be of wider occurrence, being also mediated by antigen-specific B lymphocytes[14,15] and not restricted to capture of ligand molecules in T cells, where it encompasses capture of most surface molecules from the antigen cells.[16] Likewise, the ability of other lymphoid cell types to nibble the surface of cells to which they are actively conjugated was soon documented in activated γδT cells[17] and in activated NK cells either in vitro and in vivo.[18-22]

CD4 T lymphocytes were also found able to capture MHC Class II molecules.[23,24] By then, it became clear that most lymphoid cells were fully capable to acquire/nibble/capture/transfer membrane-derived molecules from cells to which they bind (Fig. 2),[25] but despite little understanding of its physiological function, this process was called "trogocytosis" from the greek *trogos* –to nibble-.[26]

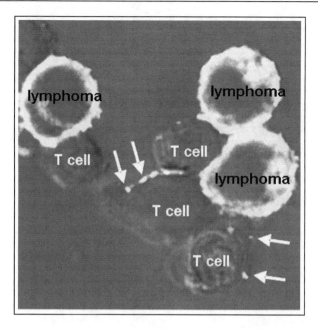

Figure 2. Example of ectopic marker Igμ chain (stained yellow) present on the surface of cytolytic T lymphocyte after it attacked an IgM⁺ B-cell lymphoma target as in Figure 1. This Ig was nibbled from the B cell surface and has a short half-life on the T cell surface (~1-2 hr). A color version of this figure is available online at www.Eurekah.com.

Two Different Ways for Cell-Cell Transfer between Lymphoid Cells

Cell-Cell Channels by Trans-Synaptic Trogocytosis

On the basis of published data, there is global agreement for trogocytosis as an active process which takes place between cells in contact through an immunological synapse, is highly polarized from the target cell to the effector lymphocyte, and only results in transfer of membrane molecules, whatever their chemical structure. Recent work suggests however that under some defined circumstances, the transfer may be bi-directional with effector membrane transfer on the target cell surface,[27] resulting in a formal molecular swapping[28] between conjugated cell partners (Fig. 3). Here however, one might argue that of two cells conjugated through an immunological synapse, each one is the other's target, opening the possibility that bidirectional transfer just reflects each effector cell's activation. Little -if any evidence- supports the possibility that target-derived cytosol molecules are acquired during the immunological synapse, or that any such transfer occurs in absence of cell-cell contact. Clearly, trogocytosis is totally different from capture of soluble secreted exosomes which don't need cell-cell contact to occur and are usually detected once cells which exchange exosomes have been coincubated for 24 hours or more.[29]

How then does membrane material from one cell switches onto the surface of conjugated cells? Much remains to be clarified on this topic, but important clues were provided in the CD8 CTL/ target cell model, where superb electronic images of the conjugates evidenced the presence of bridges localised at the immunological synapse between these cells.[30] There, small membrane continuities between effector and target cells bridge physically both cells through their surfaces, creating small pores. No such observations have ever been made with other lymphoid cell types, but it is now tempting to speculate that similar bridges are formed at most immunological synapses involving almost any lymphoid cell types.

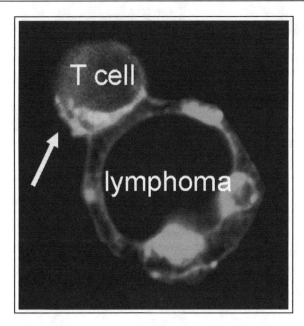

Figure 3. Confocal image of a conjugate T lymphocyte-lymphoma cell undergoing trogocytosis illustrating the large patches of green fluorochrome-labelled membranes swapping from one cell to the other (image taken 5' after cell contact). A color version of this figure is available online at www.Eurekah.com.

Cell-Cell Channels by Post Synaptic Intercellular Nanotubes

Two recent studies have also uncovered another type of cell-cell channels, made of extremely thin tubular structures of 50-200 nm diameter and several tens of μm length, with an F-actin skeleton and a membrane bilayer.[7] Not only the pheochromocytome cell line P12 in the initial study but also human lymphoid cells such as blood NK cells, macrophages, and EBV-transformed B cells make nanotubes[31] (and our unpublished observations, see below). In contrast with the stable immunological synapse of two intimately bound cells however, these highly labile nanotubular structures appear transiently at the dissociation phase of immunological cell conjugates. As one cell moves away, several nanotubes anchored at its previous attachement appear as unfolded from the pseudopode of the amoeboid shape of the motile, crawling lymphocyte, in a "webspider" style (Fig. 4). Thus, short-lived nanotubular networks could be assembled de novo and rapidly erased between post synaptic, moving immune cells. Since such structures actively transport selected endosome-like organelles and membrane proteins or lipids but not cytosolic metabolites,[7] the tubular networks might play some yet unclear role in late cell-cell communication.[32] Unfortunately, adequate experimental methods for analyzing the role and function of nanotubes are scarce, since shear force applied on cells along flow cytometry analysis for instance is inadequate to maintain these labile structures, and their study in vivo does not seem much easier either. Nanotubes are described in greater detail in another chapter of this book.

In addition, the relation of trogocytosis to nanotubular highways remains thus far unknown, although our current understanding of both processes would strongly favour a temporal separation of both entities: trogocytosis and then nanotubular networking. A main characteristic of trogocytosis is its propensity to increase with effector cell activation, and most likely with duration of the immunological synapse. Since this latter reflects the counting process, i.e., the integration of activatory signaling cascades below the synapse, it is thus also the consequence of activating ligand detected on the target cell surface. Therefore, analytic techniques have been developed that monitor trogocytosis as a readout for immunological activation of effector cells.

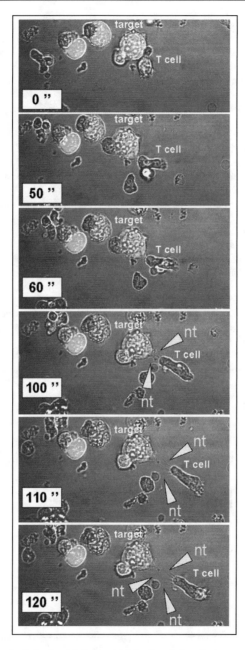

Figure 4. Cell-cell nanotubes (nt) progressively formed at the dissociation of an immunological synapse between a γδT lymphocyte and an anaplastic lymphoma cell target. The intracellular cytolytic granules of γδT lymphocytes are stained red using Lyso-Tracker, while the anaplastic lymphoma cytoplasm is stained green with Calcein for viability and its cell membrane is stained blue using the PTIR271 dye (reagent provided by courtesy of B. Gray, PTIR-Research Labs, Ex, IL). Note also the trogocytosis taking place in the upper left side conjugate, seen as blue membrane accumulated at the attacking γδT cell synapse. A color version of this figure is available online at www.Eurekah.com.

A Mechanistic Model of Trogocytosis

The main features of intercellular transfer through cell-cell channels referred to as trogocytosis involve: its strict localization (and time dependence to immunological synapses), its restriction to membrane material, and the main polarization of transferred molecules from target to the effector cell. In addition, there is of as yet no selectivity in the structural class of membrane molecules that shift from one cell to the other, suggesting the whole lipidic bilayer is involved in this movement. We present below a model for trogocytosis which integrates these features (Fig. 5).

This model involves a progressive acquisition of target cell membrane patches by the effector cell, only once the synapse has been formed. This latter relies upon effector cell activation by the antigen/ activating stimuli scanned at the target surface during the synapse existence. This model has the following implications:

- The structure of the immunological synapse must involve membrane bridges in all kinds of cell conjugates where trogocytosis takes place.
- The whole cell membrane structure may pass in the form of patches and integrates with the same orientation on the effector cell membrane.
- The effector cell actively "sucks" the target cell surface, since its pretreatment with inhibitory drugs abrogate this activity,
- Dissociation may release live cells with surface markers they did not encode for (ectopic markers), and which lifespan may enable to mediate some specific functions (selective binding for example, see below).

This model also puts forward several questions:

- How does the effector cell achieve selectivity for membrane but not cytoplasm molecules in its acquisition process? Alternatively, this could be formulated as: what blocks the molecular diffusion or acquisition from the target cell? Are there molecular filters, physical constraints (size, hydrophobicity, …) induced locally and temporally, and finally is there any selective sorting of the respective partners compartments at the level of these bridges?
- Why is such a membrane acquisition polarized from target to effector? Conceivably although one lymphoid cell (e.g., a T lymphocyte) may recognize activating ligands (e.g., MHC/peptide molecules) on the bound target cell surface, it is likely that in frequent cases, the "target" cell is also another lymphocyte which may also recognize activating ligands on the "effector" cell and thus trigger its own activity en retour. Thus such situations should

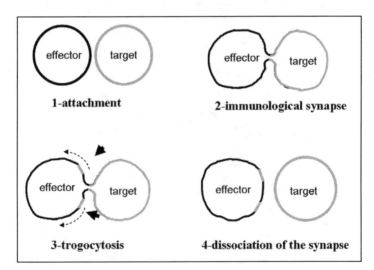

Figure 5. A mechanistic model for trogocytosis.

provoke trogocytosis apparently unpolarized for they just result from the sum of two-reverse direction trogocytosis.

- Is the receptor-mediated activation of one effector cell the driving force that triggered its nibbling activity?
- At dissociation in the model applied to a conjugate of a non cytolytic effector and its target, why and how do both cells achieve separation at the very same place of the bilayer continuity so that they both continue existing without profound membrane chimerism?
- Does some trogocytosis occur nonsynaptically, through other kinds of cell-cell channels? This question may apply to the recently discovered intercellular nanotubes described in this book.

Methods and Techniques for the Monitoring of in Vitro Trogocytosis

Confocal microscopy is a very useful tool to visualize trogocytosis. This technique was applied in the study from both J. Sprent's and F. Batista's groups who provided a remarkable dynamic imaging of T-cell and B-cell mediated antigen-specific trogocytosis, repectively.[12,14] These studies took advantage of GFP-labelled specific antigen which selective capture by specific lymphocytes could be monitored in time and space. Trogocytosis by NK cells was also depicted using this technique.[18]

Quantitative study of results is based on quantification of images with using counts of fluorescence intensity on selected pixels or related image analysis. Although highly informative at the single cell level (see example in Fig. 6), most microscopy-based techniques were however not really adequate for the quantitative studies on large number of cells from whole lymphoid populations.

In this regard, flow cytometry proved an unvaluable tool to rapidly quantify trogocytosis in effector cell subsets. Hudrisier's group took advantage of hydrophobic fluorochromes stably inserted in the cell membranes of antigen-presenting cells for demonstrating that their trogocytosis by antigen-specific CD8 T lymphocytes actually encompassed the capture of integral membrane patches, with transfer of the antigen but also of most membrane proteins and lipids, including the fluorochrome tracer.[16] Using one hour in vitro cocultures of target cells prelabelled with PKH67 and alike fluorochromes which stably insert in membranes[33] with unlabelled effector γδ T or NK cells, we showed that effector cell activation was correlated to their trogocytic acquisition of PKH.[17,20] Experimental settings are quite simple, since cocultures comprise usually two cell types: one as bait and one as hunter, usually mixed in a five-to-one bait-to-hunter cell ratio. The assay is versatile since

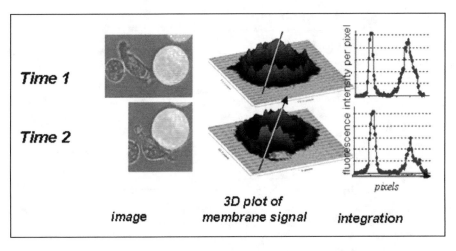

Time 1

Time 2

image   3D plot of membrane signal   integration

Figure 6. Quantification of trogocytosis based on distributions of fluorescence per pixel using successive confocal images of cell conjugates with fluorescent target cell membranes involved in immunological synapses. (J. Gertner et al, unpublished).

highly variable coculture conditions (time, temperature, exogenous stimuli or inhibitor drugs) can be used. Its final readout is the mean of PKH fluorescence intensity (PKH mfi) of the whole effector cell population as measured by straightforward flow cytometry analysis. Statistical significance of the results is ensured by gating the analysis on effector cells only, usually 30,000 cells from which the PKH mfi is calculated. We found most convenient to compare the PKH mfi after 0' (negative control) and 60' of cell coincubation, and with setting the cytometer detection sensitivity such as 1°) the background fluorescence PKH mfi of effector cells prior to cell contact is below the first decade (usually t0' mfi~5), and 2°) stained "bait" cells have their PKH mfi above the fourth decade (usually t0' mfi for targets is > 5,000). While measuring the trogocytosis-induced fluorescence loss from target cells could represent a readout for this test, we found it far less sensitive than the reciprocal measure of its appearance on unlabelled effectors.

Usually, quantification of trogocytosis may be based on arithmetic or geometric mean of green fluorescence intensity from raw data, but also using the fold increase of fluorescence on effector cells by the formula: fold increase = $\log_2$ (PKH mfi at t60'/ PKH mfi at t0'). A typical example for transfer is given above, with increases of one order of magnitude: mfi shifts from 5 to 40. Their graphical representation may conveniently show either raw flow cytometry results (Fig. 7), histograms or heat

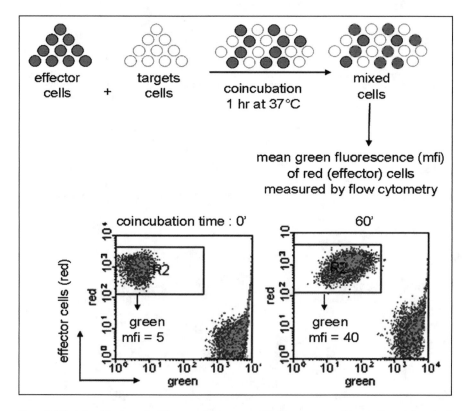

Figure 7. Principle and basic method for measuring trogocytosis by flow cytometry. Target cells must be stained on their membrane by a green (or red) fluorochrome easily monitored with PMT1 or PMT2 on flow cytometers. The analysis is gated on more than 10, 000 effector cells only (R2 box in this example), acquired before (t0') and after (t60') minutes of coincubation. These increased their green fluorescence from 5 to 40 in this example. Note that in the example shown here, the target cells (bright green cells on Y axis did simultaneously trogocytose the red fluorochrome (incorporated prior to coincubation) from the surface of effector cells: bidirectional trogocytosis. A color version of this figure is available online at www.Eurekah.com.

maps of fluorescence intensity or their increase (as fold factors, see § below). Statistical comparisons of sets of data are usually best obtained by Mann-Whitney's log rank sum test, since the distributions of cell fluorescence intensities acquired by bulks of effector and noneffector cells are not always corresponding to normal parametric distributions.

Intercellular Transfer in Activated Lymphoid Cells

In the settings described previously, the usual trogocytosis of a resting T lymphocyte towards autologous cells in absence of any exogenous stimulus is typically almost null, of mfi (t0) = 4.8, mfi (t60) = 4.8 +/- 0.2, corresponding to very little if any significant change in fluorescence. However, these figures are highly divergent as the target cell carries stimulating molecules, like antigen/MHC, or alloantigen, or mitogen for T lymphocytes: typically of mfi (t0) = 4.8, mfi (t60) = 150 +/- 20. Activation of trogocytosis is not solely due to TCR-delivered signals however, as very similar conclusions may be drawn from NK cell stimulation by adequate target cells harbouring activating ligands and lacking inhibitory molecules such as self HLA alleles.[20] In the context of such cell targets, NK sometimes trogocytose extremely high amounts of the target cell surface, typically up to 1-10 % of the membrane material constituting the target cell membrane (Fig. 8). An example is shown in Figure 8, with mfi (t0) = 4.8, mfi (t60) = 1200 when IL2-activated NK cells trogocytose HLA-deficient or allogeneic EBV-transformed B cell line targets.

The first studies in this domain[12,16] have documented this correlate by analysing activated T lymphocytes: stronger amounts of target cell membrane derived fluorescence was acquired by T cells stimulated with increasing peptide antigen concentrations loaded on the APC target, demonstrating the strong dependence of trogocytosis to TCR-driven stimulation signals. Thus effector cell activation was the driving force for the intercellular transfer. Further confirmation of this view came from another study with T cells which show an MHC-unrestricted mode of antigen recognition, the human γδ T lymphocytes. In this model, the T lymphocytes respond to exogenous soluble small nonpeptide antigens referred to as phosphoantigens.[34] Using well established γδ T targets cell lines stained with a membrane fluorochrome and exposed for one hour to the γδ T cells, it was shown that the spontaneous trogocytosis of tumour cells increased by adding either soluble or cell phosphoantigens to the culture medium.[17] Similar conclusions were drawn from NK cell models, where TCR-mediated antigen specific activation is absent. In contrast with the relation of effector-to target in cytolytic assays, where lysis increases with the killer-to-target cell ratio, here the trogocytosis increases on the opposite, with the target-to-effector cell ratio. This is due to the fact that any effector gets more chances to trogocytose and acquire fluorescence with increasing numbers of surrounding fluorescent surfaces, i.e., target cells.

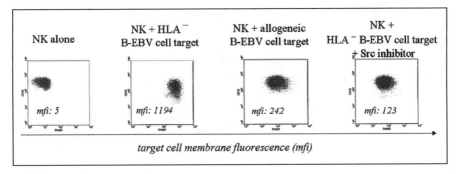

Figure 8. NK cell-mediated trogocytosis of EBV-transformed B-cell targets.

Spontaneous Homotypic Intercellular Transfer in Lymphoid Cancer Cells

As specified above, resting lymphocytes do not mediate trogocytosis in absence of any stimulus. This can be easily demonstrated experimentally by the fully stable intensity of membrane fluorescence (using the PKH67 fluorochrome for instance) of whole blood lymphocytes derived from one donor and coincubated for one hour without exogenous stimulus but with the same cells prelabelled with PKH67. This result is strictly recovered whether any purified cell subset comprising the initial PBMC sample is tested separately. Some cancer cells however do present the unusual ability to strongly trogocytose cells of the same cell line, in absence of exogenous stimulus. This has been examplified with the Daudi Burkitt's lymphoma cell line and other lymphoma and leukaemia cell lines, which spontaneous self trogocytic activity is related to their constitutive autoreactivity. For the Daudi cells, their BCR-driven signals fully control this spontaneous auto-trogocytosis, as indicated by both receptor agonists and selective inhibitors of BCR signal transduction.[35] Although the physiological significance of this finding remains yet undetermined, it points out the possible intercellular transfer of tumoural cell surface markers (oncoproteins) onto surrounding nonmalignant lymphocytes of the same subset and its physiological consequences. Our preliminary results in this topic tend to indicate that such cancer-to-noncancer cell trogocytosis does occur in the context of some hematopoietic malignancies.

Physiological Consequences of Intercellular Transfer by T Lymphocytes

Although different roles for trogocytosis have been proposed, none has thus far been demonstrated formally. The first reports that described intercellular transfers involving T cells, although this was not called trogocytosis by then, showed that thymocytes cells acquired MHC proteins from their surrounding cell environment.[8,36] This was though to reflect a passive process,[37] but later intercellular transfer from antigen-presenting cells[38] was found to be highly active, resistant to cycloheximide,[39] and to further drive to internalization by a surface-receptor-driven endocytic process[13] that involves a highly active cytoskeleton remodelling.[40] In this context, MHCp capture was shown to result in further fratricide killing by neighbouring T cells.[12,16,41] This elimination reaction "en chaine" was thus proposed to account for exhaustion of the immune response. Conceivably also, this homeostatic control could induce tolerance[39] or appearance of antigen-specific Treg cells that acquire alloantigen from antigen-presenting cells and present it to activated syngeneic CD8 T cells, to provide them with death signals.[42] Another alternative was to propose that since T cells compete in vivo for APC,[43] the antigen nibbling by high affinity lymphocytes exhaust it from the APC surface and thus promotes the in vivo selection of mature, high affinity T cell clones.[44] This latter hypothesis is highly speculative however, since trogocytosis does not exhaust the APC surface from its structural components. Most of trogocytosis that was analysed carefully showed that quantitatively, it concerns at best around 1-10% of the cell surface material (most often this is rather of around 1‰ in ag-induced trogocytosis) and not its whole amount, and qualitatively that trogocytosis is not a selective sorting for some few cell surface markers but rather a global transfer of MHC and many other compartments.[16] Accordingly, nonMHC molecules such as costimulatory molecules including CD80, CD86, B7-H1, PD-L2, B7-H3, and B7RP-1 are also acquired through trogocytosis by T lymphocytes and to provide their functional activities to the recipient cells.[45-47] During transendothelial migration, activated T cells also acquire a variety of markers from endothelial cell plasma membranes, possibly to further deliver them to perivascular sites. These acquired endothelial markers include CD31, CD49d, CD54, CD61, and CD62E (along with the lipophilic dye (DiOC-16) used to detect the endothelial cells).[48]

The bad counterpart of nonselective T cell-mediated trogocytosis is also its unfortunate ability to convey viral targeting. Molecules such as OX40L, once acquired by normal reactive CD4 T cells, enable to stimulate latent HIV-1-infected cells to produce viral proteins via OX40 signaling.[49] Likewise, cell contact rapidly induces polarization of the cytoskeleton of HTLV-I infected cell to the cell-cell junction with uninfected T cells, enabling both HTLV-I core protein complexes and its genome to be trogocytosed by the uninfected cell.[50] In line with similar results documented earlier for NK cells[51] (see below), it is highly likely that other lymphotropic viruses such as HIV-1 may similarly subvert normal T cell physiology for their propagation. The detection of HIV-specific T lymphocytes based on the monitoring of their ability to trogocytose an HIV-specific Tat peptide complexed to HLA-GFP has thus been proposed recently.[52]

Physiological Consequences of Intercellular Transfer by NK Cells: Protective Role of Acquired HLA Class I Alleles, Cytokine Receptor and Deleterious Role of Acquired Viral Receptors

In NK cells, which surface receptors may transduce either activatory or inhibitory signaling cascades according to the recognized ligands,[53] the unselective acquisition of extremely high amounts of target cell surface markers may have opposite consequences. Since self-MHC class I alleles are protective against NK cell lysis, their acquisition strongly influenced the NK cell activity in vivo, such as ligand uptake,[22] partial inactivation of cytotoxic activity by transduction of the inhibitory signals from the detected protective MHC molecules of ectopic origin in vivo[21] and in vitro.[20] Interestingly enough, a recent study has demonstrated that target cells involved in immunological synapses with NK cells are also able to acquire their NK receptors (KIR molecules), and that these molecules were associated with intracellular signling in the recipient cells, suggesting that in this setting, the trogocytosis was bidirectional.[27]

Although it remains to be shown that in this case the target cell was not itself an activated lymphocyte recognizing stimulating ligands borne at the NK cell surface, this finding indicated that some synapses may be privileged sites for bidirectional information and molecules exchanges. In this regard, IL15 represents a unique illustration of a possible evolutive role of trogocytosis. IL15 is a well known survival and maturation cytokine for NK cell differentiation.[54] Thus presence of its heterotrimeric receptor IL15R ($\alpha\beta\gamma$ chain complex) at the cell surface of maturing NK cells[55] is of the uttermost importance to their development.[56] However, the mechanism by which IL-15Rα mediates IL-15 functions is unique among cytokines. Although initially, IL-15Rα was thought to be a component of this heterotrimeric receptor complex required for mediating signalling, it barely acts as such actually since it is dispensable for or IL-15 action in vivo and is solely expressed on monocytes/dendritic cells but not by NK cells! This chain serves rather to capture and present IL-15 in trans through immunological synapses to cells expressing the IL-15R$\beta\gamma$ signalling component[57] and involved in a direct cell-cell contact channel (trogocytosis). IL-15 trans-presentation to NK cells by trogocytosis operates in vivo and increase their maturation and lytic activity, thereby augmenting the tumour immune surveillance mechanism.[58] This explains the broad, nonlymphoid expression pattern of IL-15Rα and its dispensable expression by recipient lymphocytes in vivo. Hence trogocytosis-mediated cytokine delivery by multicomponent acquisition represents a new concept for lymphocyte developmental biology and haematology.[59]

The darker face of NK cell-mediated trogocytosis of target cells is also due to the maintenance of functionality for cell surface receptors acquired from the target cells by the attacking NK cell. While NK cells establish immunological synapses with EBV-infected B cells prior to delivering their lytic granules, NK cells acquire their CD21 receptor for EBV infection.[51] This enables further binding of EBV to NK cells surface to which no attachment was otherwise possible, and permitted NK cell infections by EBV in vitro, which could account for clinical findings of neoplastic EBV⁺ NK histotypes in malignant infectious mononucleosis[60] (Fig. 10).

Technological Consequences of Trogocytosis

As depicted above, whatever the effector/target conjugate considered, trogocytosis increases with activation. It is thus straightforward to use a trogocytosis assay involving a defined effector lymphocyte and a weakly trogocytosed "bait" cell line for the screening of new stimuli to the above lymphocyte. In the example shown below, the B-cell mediated trogocytosis of a bait cell line without exogenous stimulus is referred to as baseline: mfi (t0) = 4.8, mfi (t60) = 16, while mfi (t60) = 52 when a B-cell-specific stimulus was added to the coincubation well (Fig. 9). Accordingly, trogocytosis increases with: stimulus concentration, strength, bait cell number etc..., or conversely decreases with inhibitors of effector cells (see Src inhibitor Fig. 8). This property has been used to compare the

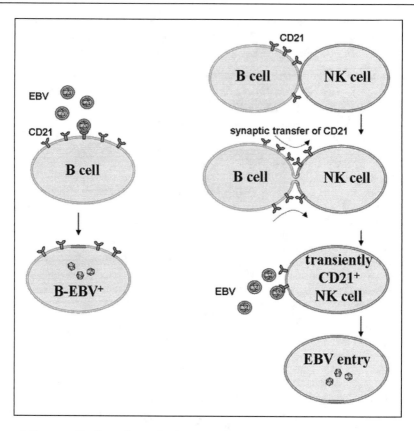

Figure 9. Exogenous B cell-specific stimulus (anti Igμ) selectively increases the trogocytosis mediated by B lymphocytes. Left) total PBMC (lymphocyte gate) were coincubated with PKH67-labelled Burkitt's lymphoma cell line (bait cells). Right) trogocytosis measured by green (PKH67) fluorescence intensity of the gated CD20⁺ B cells in the specified conditions. Note that nonB cells did not increase their trogocytosis in the stimulated assay, as compared to control t60'. A color version of this figure is available online at www.Eurekah.com.

potency of different phosphoantigens which selectively activated human $\gamma\delta$ T cells, and the resulting ranking fully matched the results of other classic functional assays for these lymphocytes.[17]

Trogotypes for Immuno-Monitoring the Lymphoid Cell Reactivity to Cancer

Among the possible conceptual application of assays based on trogocytosis to measure lymphocyte reactivity to soluble or cell antigens presented in the previous paragraph, one was indeed to monitor the lymphoid cell reactivity to cancer cells. The ability of the immune system to "see "tumour cells in a cancer patient is currently a crucial question to oncologists.[61] It is important because if cancer cells are seen but the immune response is qualitatively inadequate, this prompts the clinician for other kinds of therapeutic directions than if the cancer cells are totally ignored by the immune system (in which case better issues may be expected by reenabling its detection). We thus monitored lymphoid cell contact with cancer cells through trogocytosis by a two-step process wherein the tumour cell are labelled with PKH67, prior to 60' (or 0') coincubation with whole un-separated PBMC from the patients, the labelling of PBMC subsets with anti-CD lymphoid cell typing reagents and measuring each subset-specific trogocytosis of the tumour cell. This techniques enables the direct and individual quantification of all cell interactions with the proposed bait cell, and is highly versatile as it is amenable to declinations as various as (i) identification of a reactive blood cell

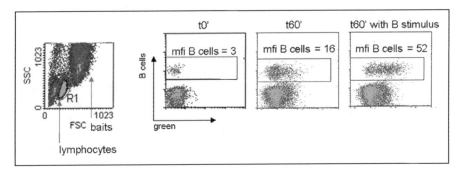

Figure 10. Model for trans-synaptic infections of NK cells by EBV. Left) conventional mode of EBV infection. Right) trogocytosis-promoted infection.

subset, (ii) identification of the possible contact cell, (iii) screening and/or identification of exogenous stimuli enabling the trogocytosis and its activation.[62] By multiplexing this analysis to one human PBMC donor tested against several cancer cells simultaneously, one can produce a global mapping of the whole PBMC trogocytic ability to a large panel representative of several different cancer cells. This map, defined as the trogotype (Fig. 10), constitutes a stable and characteristic profile for any individual's ability to "see" cancer cells. As required above for clinical issues, we are currently analysing the temporal changes associated in the trogotype of defined cancer patients monitored through longitudinal follow-ups of their clinical evolution, possibly meaningful for their treatment.

Transynaptic Acquisition of Functional Markers in Oncology

Novel studies about haematological but also in nonhaematological cells now appear more and more frequently, that describe intercellular transfers mediated through cell contacts, with features corresponding to trogocytosis as described in this article. This concerned the intercellular transfer of chemokine receptor CCR5 from HIV-infected macrophages which may lead to HIV infection of tissues without endogenous CCR5 expression,[63] but also of the intercellular transfer of cell transporters such as the P-glycoprotein (P-gp) by human cancer cells.[64] In this latter case, the transfer was demonstrated in vitro and in vivo, and led to acquisition of the multidrug resistance (MDR) phenotype by human neuroblastoma cells. This example documented trogocytosis through cell-cell contact distinct from the conventional immunological synapse, but clearly evidenced the selective advantage acquired by the recipient cell upon acquisition of the functional molecules. The trogocytosis here enabled recipient cells to resist to chemotherapy, and this selective advantage enabled investigators to evidence the progressive spread of P-gp recipient cells through a whole cancer cell population that does not encode for this marker. Former reports had indeed mentioned the increased resistance

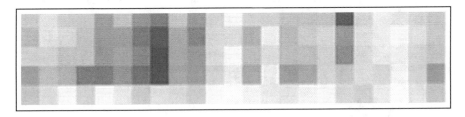

Figure 11. Trogotype of a healthy adult donor to cancer cells. 24 different cancer cell lines (columns) were tested for trogocytosis by 5 distinct lymphoid cell subsets (lines) from this donor. Intensity of the trogocytosis here shown as Δ mfi (t60'minus t0') from raw flow cytometry data is represented as a heatmap of green colour scale. A color version of this figure is available online at www.Eurekah.com.

to toxic molecules in a mixed coculture with resistant and sensitive cells[65,66] or in vivo around the tumour mass.[67,68] The finding of proteins with unexpected patterns of cell expression, notably in tissues known for not expressing the corresponding genes is thus likely to increase in the near future. These findings, in the light of our current understanding of trogocytosis, should also induce for more careful appraisal of the proteome profilings currently undertaken to characterize cancer cells in their local microenvironment. Of great importance will be the careful assignment of their original producing cell, which is also appearing as a witness of past interactions with the tumours.

Concluding Remarks

This article as reviewed current features associated with a recently discovered field of cell biology, namely the frequent acquisition of ectopic and functional cell markers through immunological and possibly other nonimmunological cell synapses. Indeed cell-cell channels have always been considered a central aspect of cell biology. With haematopoietic cells and trogocytosis however, these appear to provide a more and more versatile way of cell communication, with its so far not sufficiently appreciated consequences in cell biology. At the higher scale of whole, mixed cell populations, tissues and organism biology, these cell-cell transfers are very likely to lead to the appearance of complex and transient phenotypes not present in each of the selective cell components of the population taken separately. This new combinatorial dimension of system biology may thus uncover the appearance of complex and transient phenotypes which careful analysis at the single cell level will be hardly evaluated using our current tools of biological science. Novel techniques will need to be designed to bring such complexity into a tractable scientific arena.

Acknowledgements

The author's work is supported by INSERM, MENSR and l'ARC.

References

1. Kupfer A, Kupfer H. Imaging immune cell interactions and functions: SMACs and the Immunological Synapse. Semin Immunol 2003; 15:295-300.
2. Gascoigne NR, Zal T. Molecular interactions at the T cell-antigen-presenting cell interface. Curr Opin Immunol 2004; 16:114-119.
3. Krogsgaard M, Huppa JB, Purbhoo MA et al. Linking molecular and cellular events in T-cell activation and synapse formation. Semin Immunol 2003; 15:307-315.
4. Davis DM, Igakura T, McCann FE et al. The protean immune cell synapse: A supramolecular structure with many functions. Semin Immunol 2003; 15:317-324.
5. Hudrisier D, Bongrand P. Intercellular transfer of antigen-presenting cell determinants onto T cells: Molecular mechanisms and biological significance. FASEB J 2002; 16:477-486.
6. Davis DM. Assembly of the immunological synapse for T cells and NK cells. Trends Immunol 2002; 23:356-363.
7. Rustom A, Saffrich R, Markovic I et al. Nanotubular highways for intercellular organelle transport. Science 2004; 303:1007-1010.
8. Nagy Z, Elliott BE, Nabholz M et al. Specific binding of alloantigens to T cells activated in the mixed lymphocyte reaction. J Exp Med 1976; 143:648-659.
9. Nagy Z, Elliott BE, Nabholz M. Specific binding of K- and I-region products of the H-2 complex to activated thymus-derived (T) cells belonging to different Ly subclasses. J Exp Med 1976; 144:1545-1553.
10. Speiser DE, Schneider R, Hengartner H et al. Clonal deletion of self-reactive T cells in irradiation bone marrow chimeras and neonatally tolerant mice. Evidence for intercellular transfer of Mlsa. J Exp Med 1989; 170:595-600.
11. Sellin D, Wallach DF, Fischer H. Intercellular communication in cell-mediated cytotoxicity. Fluorescein transfer between H-2 d target cells and H-2 b lymphocytes in vitro. Eur J Immunol 1971; 1:453-458.
12. Huang JF, Yang Y, Sepulveda H et al. TCR-Mediated internalization of peptide-MHC complexes acquired by T cells. Science 1999; 286:952-954.
13. Hwang I, Huang JF, Kishimoto H et al. T cells can use either T cell receptor or CD28 receptors to absorb and internalize cell surface molecules derived from antigen-presenting cells. J Exp Med 2000; 191:1137-1148.

14. Batista FD, Iber D, Neuberger MS. B cells acquire antigen from target cells after synapse formation. Nature 2001; 411:489-494.
15. Batista FD, Neuberger MS. B cells extract and present immobilized antigen: Implications for affinity discrimination. EMBO J 2000; 19:513-520.
16. Hudrisier D, Riond J, Mazarguil H et al. Cutting edge: CTLs rapidly capture membrane fragments from target cells in a TCR signaling-dependent manner. J Immunol 2001; 166:3645-3649.
17. Espinosa E, Tabiasco J, Hudrisier D et al. Synaptic transfer by human gamma delta T cells stimulated with soluble or cellular antigens. J Immunol 2002; 168:6336-6343.
18. Carlin LM, Eleme K, McCann FE et al. Intercellular transfer and supramolecular organization of human leukocyte antigen C at inhibitory natural killer cell immune synapses. J Exp Med 2001; 194:1507-1517.
19. McCann FE, Suhling K, Carlin LM et al. Imaging immune surveillance by T cells and NK cells. Immunol Rev 2002; 189:179-192.
20. Tabiasco J, Espinosa E, Hudrisier D et al. Active trans-synaptic capture of membrane fragments by natural killer cells. Eur J Immunol 2002; 32:1502-1508.
21. Sjostrom A, Eriksson M, Cerboni C et al. Acquisition of external major histocompatibility complex class I molecules by natural killer cells expressing inhibitory Ly49 receptors. J Exp Med 2001; 194:1519-1530.
22. Zimmer J, Ioannidis V, Held W. H-2D ligand expression by Ly49A+ natural killer (NK) cells precludes ligand uptake from environmental cells: Implications for NK cell function. J Exp Med 2001; 194:1531-1539.
23. Undale AH, van den Elsen PJ, Celis E. Antigen-independent acquisition of MHC class II molecules by human T lymphocytes. Int Immunol 2004; 16:1523-1533.
24. Wetzel SA, McKeithan TW, Parker DC. Peptide-specific intercellular transfer of MHC class II to CD4+ T cells directly from the immunological synapse upon cellular dissociation. J Immunol 2005; 174:80-89.
25. Poupot M, Pont F, Fournie JJ. Profiling blood lymphocyte interactions with cancer cells uncovers the innate reactivity of human gamma delta T cells to anaplastic large cell lymphoma. J Immunol 2005; 174:1717-1722.
26. Joly E, Hudrisier D. What is trogocytosis and what is its purpose? Nat Immunol 2003; 4:815.
27. Vanherberghen B, Andersson K, Carlin LM et al. Human and murine inhibitory natural killer cell receptors transfer from natural killer cells to target cells. Proc Natl Acad Sci USA 2004; 101(48):16873-16878.
28. Sprent J. Swapping molecules during cell-cell interactions. Sci STKE 2005; 2005(273):pe8.
29. Amigorena S. Exosomes derived from dendritic cells. J Soc Biol 2001; 195:25-27.
30. Stinchcombe JC, Bossi G, Booth S et al. The immunological synapse of CTL contains a secretory domain and membrane bridges. Immunity 2001; 15:751-761.
31. Onfelt B, Nedvetzki S, Yanagi K et al. Cutting edge: Membrane nanotubes connect immune cells. J Immunol 2004; 173:1511-1513.
32. Onfelt B, Davis DM. Can membrane nanotubes facilitate communication between immune cells? Biochem Soc Trans 2004; 32:676-678.
33. Slezak SE, Horan PK. Cell-mediated cytotoxicity. A highly sensitive and informative flow cytometric assay. J Immunol Methods 1989; 117:205-214.
34. Poupot M, Fournie JJ. Nonpeptide antigens activating human Vgamma9/Vdelta2 T lymphocytes. Immunol Lett 2004; 95:129-138.
35. Poupot M, Fournie JJ. Spontaneous membrane transfer through homotypic synapses between lymphoma cells. J Immunol 2003; 171:2517-2523.
36. Sharrow SO, Mathieson BJ, Singer A. Cell surface appearance of unexpected host MHC determinants on thymocytes from radiation bone marrow chimeras. J Immunol 1981; 126:1327-1335.
37. Lorber MI, Loken MR, Stall AM et al. I-A antigens on cloned alloreactive murine T lymphocytes are acquired passively. J Immunol 1982; 128:2798-2803.
38. Merkenschlager M. Tracing interactions of thymocytes with individual stromal cell partners. Eur J Immunol 1996; 26:892-896.
39. Patel DM, Arnold PY, White GA et al. Class II MHC/peptide complexes are released from APC and are acquired by T cell responders during specific antigen recognition. J Immunol 1999; 163:5201-5210.
40. Hwang I, Sprent J. Role of the actin cytoskeleton in T cell absorption and internalization of ligands from APC. J Immunol 2001; 166:5099-5107.
41. Hanon E, Stinchcombe JC, Saito M et al. Fratricide among CD8(+) T lymphocytes naturally infected with human T cell lymphotropic virus type I. Immunity 2000; 13:657-664.

42. Zhang ZX, Yang L, Young KJ et al. Identification of a previously unknown antigen-specific regulatory T cell and its mechanism of suppression. Nat Med 2000; 6:782-789.
43. Kedl RM, Rees WA, Hildeman DA et al. T cells compete for access to antigen-bearing antigen-presenting cells. J Exp Med 2000; 192:1105-1113.
44. Kedl RM, Schaefer BC, Kappler JW et al. T cells down-modulate peptide-MHC complexes on APCs in vivo. Nat Immunol 2002; 3:27-32.
45. Sabzevari H, Kantor J, Jaigirdar A et al. Acquisition of CD80 (B7-1) by T cells. J Immunol 2001; 166:2505-2513.
46. Tatari-Calderone Z, Semnani RT, Nutman TB et al. Acquisition of CD80 by human T cells at early stages of activation: Functional involvement of CD80 acquisition in T cell to T cell interaction. J Immunol 2002; 169:6162-6169.
47. Ferlazzo G, Semino C, Meta M et al. T lymphocytes express B7 family molecules following interaction with dendritic cells and acquire bystander costimulatory properties. Eur J Immunol 2002; 32:3092-3101.
48. Brezinschek RI, Oppenheimer-Marks N, Lipsky PE. Activated T cells acquire endothelial cell surface determinants during transendothelial migration. J Immunol 1999; 162:1677-1684.
49. Baba E, Takahashi Y, Lichtenfeld J et al. Functional CD4 T cells after intercellular molecular transfer of 0X40 ligand. J Immunol 2001; 167:875-883.
50. Igakura T, Stinchcombe JC, Goon PK et al. Spread of HTLV-I between lymphocytes by virus-induced polarization of the cytoskeleton. Science 2003; 299:1713-1716.
51. Tabiasco J, Vercellone A, Meggetto F et al. Acquisition of viral receptor by NK cells through immunological synapse. J Immunol 2003; 170:5993-5998.
52. Tomaru U, Yamano Y, Nagai M et al. Detection of virus-specific T cells and CD8+ T-cell epitopes by acquisition of peptide-HLA-GFP complexes: Analysis of T-cell phenotype and function in chronic viral infections. Nat Med 2003; 9:469-476.
53. Vivier E, Nunes JA, Vely F. Natural killer cell signaling pathways. Science 2004; 306:1517-1519.
54. Becknell B, Caligiuri MA. Interleukin-2, interleukin-15, and their roles in human natural killer cells. Adv Immunol 2005; 86:209-239.
55. Colucci F, Caligiuri MA, Di Santo JP. What does it take to make a natural killer? Nat Rev Immunol 2003; 3:413-425.
56. Schluns KS, Lefrancois L. Cytokine control of memory T-cell development and survival. Nat Rev Immunol 2003; 3:269-279.
57. Dubois S, Mariner J, Waldmann TA et al. IL-15Ralpha recycles and presents IL-15 In trans to neighboring cells. Immunity 2002; 17:537-547.
58. Kobayashi H, Dubois S, Sato N et al. Role of trans-cellular IL-15 presentation in the activation of NK cell-mediated killing, which leads to enhanced tumor immunosurveillance. Blood 2005; 105:721-727.
59. Schluns KS, Stoklasek T, Lefrancois L. The roles of interleukin-15 receptor alpha: Trans-presentation, receptor component, or both? Int J Biochem Cell Biol 2005; 37:1567-1571.
60. Trempat P, Tabiasco J, Andre P et al. Evidence for early infection of nonneoplastic natural killer cells by Epstein-Barr virus. J Virol 2002; 76:11139-11142.
61. Pardoll D. Does the immune system see tumors as foreign or self? Annu Rev Immunol 2003; 21:807-839.
62. Poupot M, Pont F, Fournie JJ. Inventors; INSERM patent. Global Scan for Cell Interactions. INPI Paris 2004.
63. Mack M, Kleinschmidt A, Bruhl H et al. Transfer of the chemokine receptor CCR5 between cells by membrane-derived microparticles: A mechanism for cellular human immunodeficiency virus 1 infection. Nat Med 2000; 6:769-775.
64. Levchenko A, Mehta BM, Niu X et al. Intercellular transfer of P-glycoprotein mediates acquired multidrug resistance in tumor cells. Proc Natl Acad Sci USA 2005; 102:1933-1938.
65. Tofilon PJ, Buckley N, Deen DF. Effect of cell-cell interactions on drug sensitivity and growth of drug-sensitive and -resistant tumor cells in spheroids. Science 1984; 226:862-864.
66. Frankfurt OS, Seckinger D, Sugarbaker EV. Intercellular transfer of drug resistance. Cancer Res 1991; 51:1190-1195.
67. Miller BE, Machemer T, Lehotan M et al. Tumor subpopulation interactions affecting melphalan sensitivity in palpable mouse mammary tumors. Cancer Res 1991; 51:4378-4387.
68. Petrylak DP, Scher HI, Reuter V et al. P-glycoprotein expression in primary and metastatic transitional cell carcinoma of the bladder. Ann Oncol 1994; 5:835-840.

CHAPTER 21

Cell-Cell Transport of Homeoproteins:
With or Without Channels?

Alain Joliot* and Alain Prochiantz

Abstract

Homeoproteins are a class of transcription factors that have the unusual property of intercellular transfer, both in metaphytes and in metazoans. Here we discuss the cellular mechanisms and the function of this transfer.

Introduction

Intercellular communication is critical in all aspects of life, from infection by bacterial pathogens to the spatial organization of multi-cellular organisms. Multiple mechanisms have thus arisen during evolution to achieve this task. Many of them are based on ligand receptor binding followed by the activation of intracellular transduction cascades. However, direct protein intercellular transfer, leading to the nuclear or cytosolic of the transferred protein, accumulation was also described. In metaphytes, the latter process (symplasmic transfer) is achieved through specialized inter-cytoplasmic channels called plasmodesmata. The discovery of thin extensions connecting animal cells, cytonemes,[1] and more recently tunneling nanotubes[2] (TNTs) that allow the passage of larger organelles (endosomes, mitochondria), suggest that this mode of transfer also takes place in metazoans.

In the recent years, we have shown that homeoproteins, a family of transcription factors, can transfer between animal cells. In this review, we shall describe the cellular mechanisms involved and discuss the differences and similarities between plants and animals.

Intercellular Transfer of Homeoproteins in Animals

Cellular Aspects

Homeoprotein intercellular transfer between animal cells has first been described in coculture models[3] where the protein transiently expressed in a donor cell type (donor cells) is detected in the other cell type (recipient cells). The transfer leads to the accumulation of the intact homeoprotein primarily in the cytosol and nucleus of the recipient cells, and not in a vesicular compartment. Although homeoprotein transfer does not require close cellular contacts, such as plasma membrane apposition, it is mainly restricted to cells in the vicinity of the donor cell, demonstrating high capture efficiency. Homeoproteins are retrieved in the 100 000 g supernatant of the conditioned media.[3] They are thus secreted under a soluble form in spite of the fact that are devoid of classical N-terminus secretion signal sequence, suggesting an unconventional mechanism of secretion. Secretion of the protein is tightly correlated with its accumulation in intra-vesicular compartments,[3] with the physical chemical properties of caveolae, enriched in cholesterol and glycosphingolipids.[4] Unexpectedly, the transit of the protein through the nucleus seems required for its subsequent addressing to a secretion vesicular compartment.[5]

*Corresponding Author: Alain Joliot—Homeoprotein Cell Biology, CNRS UMR8542, Ecole Normale Supérieure, 46 rue d'Ulm, 75230 Paris Cedex 05, France. Email: joliot@biologie.ens.fr

Cell-Cell Channels, edited by Frantisek Baluska, Dieter Volkmann and Peter W. Barlow.
©2006 Landes Bioscience and Springer Science+Business Media.

These unique features of homeoprotein intercellular transfer do no formally rule out the involvement of TNTs. Due to the extreme fragility of these structures, their breakdown during the collection of the conditioned media, leading to extra-cellular leakage, cannot be precluded. Indeed, homeoproteins also associate with the external face of vesicles[4] that might be transported through TNTs. However, we consider the TNT hypothesis as unlikely on the following premises. First, an important characteristic of channel-based communication is that transported intracellular proteins never reach the extra-cellular compartment. Second, mutational analysis demonstrates that homeoprotein secretion is correlated with their passage into a luminal compartment.[3] Third, recombinant homeoproteins when directly added into the extra-cellular medium are efficiently internalized by live cells and accumulate in their cytosol and nucleus.[6,7] Finally, internalization-deficient homeoprotein mutants have lost their transfer ability.[3] We thus favor the idea that homeoprotein intercellular transfer is a two steps mechanism, involving vesicular secretion into the medium and internalization through an endocytosis-independent mechanism that still need to be fully characterized (for a review see ref. 8).

Molecular Aspects

Homeoproteins are defined by the structure of their DNA binding domain, the homeodomain. We have demonstrated that the latter is driving protein transfer,[5] suggesting that this property might be shared by several members of this family of transcription factors. However, the presence of a homeodomain can not univocally predict transfer as few bona fide homeodomains do not transfer[5] (and unpublished data). Within the 60 amino acid long sequence of the homeodomain, we have identified two specific motifs that have a critical role in internalization and secretion, respectively.[3,9] The existence of two distinct sequences for secretion and internalization reinforces the idea that the two steps are based on distinct mechanisms. Indeed, secretion, but not internalization,[10] is inhibited at low temperature and the internalization sequence corresponding to the third helix of the homeodomain (also known as Penetratin[11]), is efficiently internalized,[12] but is not secreted in absence of the secretion sequence[9] (and unpublished data).

Phylogenesis of Homeoprotein Intercellular Transfer

Homeoproteins represent an interesting paradigm to approach the diversity of intercellular transfer mechanisms. First, they are present in all eukaryote organisms (unicellular and multi-cellular). Secondly, they are transferred both in metazoans and in metaphytes. In the latter case, it is widely accepted that transfer occurs through plasmodesmata. Recently, we have demonstrated that the transfer mechanisms in metaphytes and metazoans share striking similarities, despite radically different multi-cellular organizations. The homeodomain of Knotted1 (KN1), a maize homeoprotein showing strong intercellular transfer in metaphytes,[13,14] is efficiently transferred between animal cells.[5] In both phyla, the homeodomain is necessary and sufficient for this process.[5,15] Most strikingly, a same mutation within the homeodomain abolishes transfer in plants[13,14] and secretion—therefore transfer—in animal cells in culture.[5] It is also noteworthy that in the same experimental conditions the homeodomain of the plant Bellringer homeoprotein is not transferred, neither between plant cells[15] nor between animal cells.[5] Taken together, these results strongly argue for the existence of similarities between intercellular transfer mechanisms at stake in both embranchments.

Although we can not rule out that intercellular transfer has arisen independently during evolution in different organisms and phyla, the latter results obtained with wild type and mutated homeoproteins strongly suggest the existence of a common ancestral mechanism. Direct transfer of soluble proteins, including transcription factors,[16] is used by some pathogens during the phase of bacteria infection. As for unicellular eukaryotes, the action of homeoproteins is often associated with cell fusion during mating,[17,18] an extreme case of direct intercellular transfer! More generally, it can be suggested that channel-based and non channel-based intercellular transfer are two related mechanisms. Indeed, members of the Hsc70[19] and of the thioredoxin[20] families transfer through plasmodesmata and are secreted by animal cells by unconventional mechanisms.[21-23] In metaphytes, the presence of a cellulose cell wall, which strongly impairs the diffusion of large molecules, could have spatially restricted intercellular communication pathways at the level of plasmodesmata, but these pathways might be heterogeneous. A good evidence of this heterogeneity is the existence of conditions in which the transfer of small molecules (10 kD FITC-Dextran) or proteins (GFP, LEAFY) is inhibited, whereas other proteins, including KN1, are still able to transfer.[13,14,24,25] Thus, a provocative hypothesis based on the similar behaviors of some homeoproteins in plants and animals, is that unconventional secretion and internalization pathways in animal cells and intercellular channels are in part functionally similar and evolutionary equivalent.

The similarities of homeoprotein intercellular transfer mechanisms between plants and animals can be at different steps. First, the protein has to be addressed at the plasma membrane. Since homeoproteins are mainly localized in the nucleus, a conserved intracellular pathway could allow their targeting towards the plasma membrane. We have shown that, in animal cells, homeoproteins must pass through the nucleus to gain access to a vesicular secretion compartment.[5] A mutation within the nuclear localization signal (nls) of KN1 homeodomain strongly decreases nuclear accumulation and secretion of the protein in the animal model, both effects being reverted by the addition of an ectopic nls (SV40 nls). Similarly, the deletion of its nls also abolished the transfer of KN1 homeodomain in plants.[15] In strong contrast to the nuclear accumulation of an other type of transcription factor, SHR, antagonizes the symplasmic transfer of this protein in plants.[26]

Once at the plasma membrane, intercellular transfer proper takes place. In metaphytes, KN1 transfer correlates with an increase in the size exclusion limit of the plamodesmata, illustrated by the cotransfer of FITC-Dextran.[13] In animals, homeoprotein transfer requires secretion and internalization. Although both series of events seem to have little in common, recombinant plant KN1 homeodomain added in the extra-cellular medium is efficiently internalized by animal cells,[5] a process clearly distinct from the intra cellular addressing of the protein to the plasma membrane. It will be interesting to test in plants the transfer of wild type animal homeoproteins and of internalization mutants.

Function of Intercellular Transfer

Direct protein transfer must preserve the maintenance of distinct cellular properties. This implies that the transfer must be precisely regulated, both quantitatively and qualitatively. Transfer selectivity is illustrated by the behavior of the soluble reporter protein GFP which does not pass through TNTs[2] and shows limited diffusion through plasmodesmata, in comparison with actively transported molecules (e.g., KN1, viral movement proteins). In the case of homeoproteins, regions outside of the homeodomain can regulate the intercellular transfer between animal cells. Phosphorylation of a small serine-rich domain by protein kinase CK2 in its N-terminal strongly inhibits the secretion of Engrailed-2 homeoprotein.[27] In the rat embryonic mesencephalon, endogenous Engrailed homeoprotein is highly phosphorylated,[27] suggesting that its transfer is actively regulated in vivo. Due to the low degree of sequence conservation outside the homeodomain, each homeoprotein may have its own mechanism of transfer regulation. In plants, unidirectional KN1 passage through plasmodesmata between epidermis and mesophyll also demonstrates transfer regulation.[28]

What is the function of homeoprotein transfer in animals? An obvious hypothesis is that the protein has retained its transcriptional function, as observed in plants. Indeed, homeoproteins added in the extra-cellular medium accumulate in the nuclear compartment and are able to regulate the expression of specific target genes, similarly to the intracellular synthesized protein.[29-31] However, we have not been able to detect a direct transcriptional activity of homeoproteins following intercellular transfer. A possible explanation is the low amount of transferred protein in our ex vivo culture model[3] (5% of the intracellular pool). In a physiological situation, an amplification loop and the presence of appropriate nuclear coactivators may amplify the initial amount of protein or promote its paracrine activity. In fact, several homeoproteins activate their own expression, directly or indirectly (for a review see ref. 32).

A growing number of studies demonstrate that homeoproteins also act as translational regulators.[33-35] Recently we have demonstrated that local application of an extra-cellular gradient of Engrailed-2, in the nM range, regulates axonal guidance in retinal ganglion cells.[36] This effect follows Engrailed-2 internalization by the growth cones and is translation-dependent, with no transcription involved. This illustrates how translational regulation could be an important mode of action of transferred homeoproteins. In addition, the access to distinct sub-cellular compartments (e.g., the growth cone of neurons) through internalization could reflect a specific function of the transferred homeoprotein compared to resident homeoproteins. It is also possible that homeoproteins acquire new functions following their transfer, either through conformational or post-translational modification during this process.

Although the physiological function of direct transfer between animal cells still remains poorly understood, we speculate that it presents several specificities compared to ligand/receptor mode of signalization. First, it does not obligatorily require the presence of cell surface receptors. Secondly, although the signal may need an amplification step, similarly to ligand/receptor dependent pathways, second messenger cascades would not play a central role. Transcriptional and translational regulations are indeed good candidates for signal amplification, as observed for retinoids, a class of small hydrophobic molecules that directly transfer between cells.[37] Finally, this new mode of cell-cell signalization would broaden the spectrum of cellular targets that can respond to paracrine stimuli.

References

1. Ramirez-Weber FA, Kornberg TB. Cytonemes: Cellular processes that project to the principal signaling center in Drosophila imaginal discs. Cell 1999; 97:599-607.
2. Rustom A, Saffrich R, Markovic I et al. Nanotubular highways for intercellular organelle transport. Science 2004; 303:1007-1010.
3. Joliot A, Maizel A, Rosenberg D et al. Identification of a signal sequence necessary for the unconventional secretion of Engrailed homeoprotein. Curr Biol 1998; 8:856-863.
4. Joliot A, Trembleau A, Raposo G et al. Association of Engrailed homeoproteins with vesicles presenting caveolae-like properties. Development 1997; 124:1865-1875.
5. Tassetto M, Maizel A, Osorio J et al. Plant and animal homeodomains use convergent mechanisms for intercellular transfer. EMBO Rep 2005; 6:1-6.
6. Chatelin L, Volovitch M, Joliot AH et al. Transcription factor Hoxa-5 is taken up by cells in culture and conveyed to their nuclei. Mech Dev 1996; 55:111-117.
7. Joliot A, Pernelle C, Deagostini-Bazin H et al. Antennapedia homeobox peptide regulates neural morphogenesis. Proc Natl Acad Sci USA 1991; 88:1864-1868.
8. Joliot A, Prochiantz A. Transduction peptides: From technology to physiology. Nat Cell Biol 2004; 6:189-196.
9. Maizel A, Bensaude O, Prochiantz A et al. A short region of its homeodomain is necessary for Engrailed nuclear export and secretion. Development 1999; 126:3183-3190.
10. Joliot AH, Triller A, Volovitch M et al. alpha-2,8-Polysialic acid is the neuronal surface receptor of antennapedia homeobox peptide. New Biol 1991; 3:1121-1134.
11. Dupont E, Joliot A, Prochiantz A. Penetratins. Pharmacology and Toxicology Series. CRC Press, 2002:23-51.
12. Derossi D, Joliot AH, Chassaing G et al. The third helix of Antennapedia homeodomain translocates through biological membranes. J Biol Chem 1994; 269:10444-10450.
13. Lucas WJ, Bouché-Pillon S, Jackson DP et al. Selective trafficking of KNOTTED1 homeodomain protein and its mRNA through plasmodesmata. Science 1995; 270:1980-1983.
14. Kim JY, Yuan Z, Cilia M et al. Intercellular trafficking of a KNOTTED1 green fluorescent protein fusion in the leaf and shoot meristem of Arabidopsis. Proc Natl Acad Sci USA 2002; 99:4103-4108.
15. Kim JY, Rim Y, Wang J et al. A novel cell-to-cell trafficking assay indicates that the KNOX homeodomain is necessary and sufficient for intercellular protein and mRNA trafficking. Genes Dev 2005; 19:788-793.
16. Szurek B, Rossier O, Hause G et al. Type III-dependent translocation of the Xanthomonas AvrBs3 protein into the plant cell. Mol Microbiol 2002; 46:13-23.
17. Zhao H, Lu M, Singh R et al. Ectopic expression of a Chlamydomonas mt+-specific homeodomain protein in mt- gametes initiates zygote development without gamete fusion. Genes Dev 2001; 15:2767-2777.
18. Casselton LA, Olesnicky NS. Molecular genetics of mating recognition in basidiomycete fungi. Microbiol Mol Biol Rev 1998; 62:55-70.

19. Aoki K, Kragler F, Xoconostle-Cazares B et al. A subclass of plant heat shock cognate 70 chaperones carries a motif that facilitates trafficking through plasmodesmata. Proc Natl Acad Sci USA 2002; 99:16342-16347.
20. Ishiwatari Y, Fujiwara T, McFarland KC et al. Rice phloem thioredoxin h has the capacity to mediate its own cell-to-cell transport through plasmodesmata. Planta 1998; 205:12-22.
21. Broquet AH, Thomas G, Masliah J et al. Expression of the molecular chaperone Hsp70 in detergent-resistant microdomains correlates with its membrane delivery and release. J Biol Chem 2003; 278:21601-21606.
22. Guzhova I, Kislyakova K, Moskaliova O et al. In vitro studies show that Hsp70 can be released by glia and that exogenous Hsp70 can enhance neuronal stress tolerance. Brain Res 2001; 914:66-73.
23. Rubartelli A, Bajetto A, Allavena G et al. Secretion of thioredoxin by normal and neoplastic cells through a leaderless secretory pathway. J Biol Chem 1992; 267:24161-24164.
24. Kragler F, Monzer J, Xoconostle-Cazares B et al. Peptide antagonists of the plasmodesmal macromolecular trafficking pathway. EMBO J 2000; 19:2856-2868.
25. Lee JY, Yoo BC, Rojas MR et al. Selective trafficking of noncell-autonomous proteins mediated by NtNCAPP1. Science 2003; 299:392-396.
26. Gallagher KL, Paquette AJ, Nakajima K et al. Mechanisms regulating SHORT-ROOT intercellular movement. Curr Biol 2004; 14:1847-1851.
27. Maizel A, Tassetto M, Filhol O et al. Engrailed homeoprotein secretion is a regulated process. Development 2002; 129:3545-3553.
28. Kim JY, Yuan Z, Jackson D. Developmental regulation and significance of KNOX protein trafficking in Arabidopsis. Development 2003; 130:4351-4362.
29. Noguchi H, Kaneto H, Weir GC et al. PDX-1 protein containing its own antennapedia-like protein transduction domain can transduce pancreatic duct and islet cells. Diabetes 2003; 52:1732-1737.
30. Mainguy G, Montesinos ML, Lesaffre B et al. An induction gene trap for identifying a homeoprotein-regulated locus. Nat Biotechnol 2000; 18:746-749.
31. Amsellem S, Pflumio F, Bardinet D et al. Ex vivo expansion of human hematopoietic stem cells by direct delivery of the HOXB4 homeoprotein. Nat Med 2003; 9:1423-1427.
32. Kiecker C, Lumsden A. Compartments and their boundaries in vertebrate brain development. Nat Rev Neurosci 2005; 6:553-564.
33. Nedelec S, Foucher I, Brunet I et al. Emx2 homeodomain transcription factor interacts with eukaryotic translation initiation factor 4E (eIF4E) in the axons of olfactory sensory neurons. Proc Natl Acad Sci USA 2004; 101:10815-10820.
34. Topisirovic I, Culjkovic B, Cohen N et al. The proline-rich homeodomain protein, PRH, is a tissue-specific inhibitor of eIF4E-dependent cyclin D1 mRNA transport and growth. EMBO J 2003; 22:689-703.
35. Niessing D, Blanke S, Jackle H. Bicoid associates with the 5'-cap-bound complex of caudal mRNA and represses translation. Genes Dev 2002; 16:2576-2582.
36. Brunet I, Weinl C, Piper A et al. Engrailed-2 guides retinal axons. Nature 2005; In Press.
37. Aranda A, Pascual A. Nuclear hormone receptors and gene expression. Physiol Rev 2001; 81:1269-1304.

CHAPTER 22

Virological Synapse for Cell-Cell Spread of Viruses

Eduardo Garcia and Vincent Piguet*

Abstract

Cell-to-cell spread of retroviruses via virological synapse (VS) contributes to overall progression of disease. VS are specialized pathogen-induced cellular structures that facilitate cell-to-cell transfer of HIV-1 and HTLV-1. VS provide a mechanistic explanation for cell-associated retroviral replication. While VS share some common features with neurological or immunological synapses, they also exhibit important differences. The role of VS might not be limited to human retroviruses and the emerging role of a plant synapse suggests that VS might well be conserved structures for cell-cell spreading of both animal and plant viruses. Dissection of the VS is just at its beginning, but already offers ample information and fascinating insights into mechanisms of viral replication and cell-to-cell communication.

Neural, Immunological and Virological Synapse

The complex functioning of biological systems requires the capacity of cells to interact in a synchronized manner. The capacity of cells to come in close contact with one another enables rapid exchange of information through directed secretion. In complex systems such as the nervous and immune systems, characteristic rearrangements of plasma membrane proteins appear at the cell-cell junction, called synapse. A synapse is defined as "a stable adhesive junction across which information is relayed by directed secretion".[1]

The concept of the neural synapse (NS) was first introduced over a century ago and was depicted as a stable structure organized and specialized in intercellular signaling between neurons. Plasma membranes of the pre and post-synaptic neurons are contiguous and information is conveyed to the downstream cell via secretion of neurotransmitters. In order to generate a favorable microenvironment, stabilization of synapse by scaffolding proteins, mainly cadherins and other adhesion molecules, is required (reviewed in ref. 1).

In the immune system, interactions between T cells and antigen presenting cells (APC) are essential for an effective adaptive immune response. By analogy with the nervous system, these specialized interactions occur via an immunological synapse (IS). The concept of the IS has been extended to several types of cell-cell interactions within the context of the immune system (signaling via receptor engagement, lytic granules, directed secretion of cytokines) since its first description 20 years ago (reviewed in refs. 2,3). Although the IS shares many similarities with the NS, it also differs in two aspects. First, the panel of receptors and adhesion proteins recruited to the IS diverges from those in the neural synapse: integrins play the central role in stabilizing IS. Second, the establishment of an IS is a dynamic process between moving cells, whereas the neural synapse is long-lived. Therefore, in

*Corresponding Author: Vincent Piguet—Department of Dermatology and Venereology, University Hospital of Geneva, 4-747, 24 Rue Micheli-du-Crest, 1211 Geneva, Switzerland. Email: vincent.piguet@medecine.unige.ch

Cell-Cell Channels, edited by Frantisek Baluska, Dieter Volkmann and Peter W. Barlow. ©2006 Landes Bioscience and Springer Science+Business Media.

order to permit immune responses to take place, ISs need to be assembled and disassembled quickly. An example is CTL-mediated killing, where a single effector cell has been shown to contact sequentially target cells through several stable IS[4] (for reviews see in refs. 1,5-7).

In recent years, the concept of the synapse has been further extended to cell-cell contacts during viral replication. To initiate an infection, viruses need to gain access to the replicative machinery of the host cell. In the cell-free virus model, viruses do so by crossing the plasma membrane of the target cell after binding to surface receptors. Nevertheless, some viruses use direct passage from cell-to-cell to spread within their host achieving, in the process, protection from neutralizing antibodies[8] and complement as well as higher kinetics of replication (reviewed in ref. 9). Recent articles have described virological synapses (VS) for two retroviruses, human T cell leukemia virus type 1 (HTLV-1) and human immunodeficiency virus type 1 (HIV-1)[10-15] (reviewed in ref. 16). VS, like their neural and immunological counterparts, suit the minimal criteria that define a synapse: both pre and post-synaptic cells implied in cell-cell contact remain discrete cells (no plasma membrane fusion), a stable adhesive connection is established between the two cells and directed transmission of information (viral genome) occurs from the infected cell (presynaptic cell) to the uninfected cell (post-synaptic).

Virological Synapse during Retroviral Infection

Although viral cell-to-cell transfer has been identified many years ago,[9,17-20] we gained only recently some insight into the mechanisms of this mode of viral transmission. Cell-free HTLV-1 ineffectively infects T lymphocytes and spreads within and between individuals via cell-to-cell transfer. With the partial unraveling of the mechanisms involved in HTLV-1 dissemination from lymphocyte to lymphocyte via VS,[10,21,22] puzzling questions, such as HTLV-1 cell tropism, regardless of the ubiquitous expression of its surface receptor, have found satisfying explanations.

Other retroviruses, such as HIV-1 and SIV, also use VS to propagate within their respective hosts. Efficient HIV-1 infection requires permissive target cells to be located in close vicinity in order to initiate infection and subsequent spreading throughout different tissues. At least three modes of propagation have been described for HIV-1. Firstly, cell-free transmission of HIV-1 is well characterized. Cell-free HIV-1 binds surface receptors/coreceptors (CD4/CCR5 and CXCR4) of permissive cells before fusing with the plasma membrane of the target cell and following the subsequent steps of the viral replication cycle.[23-25] Secondly, HIV is able to propagate through infection in trans. Cells such as dendritic cells (DC) capture virions through viral binding to cell-surface receptors such as C-type lectins. HIV-1⁺ DCs, not necessarily infected themselves, then present the virus to target cells in trans via a VS or an Infectious Synapse.[26-30] Thirdly, HIV-1-infected cells (also termed effector cells) are able to transmit the virus to uninfected target cells, without the previous requirement of virus budding in the extracellular milieu, illustrating direct cell-to-cell viral transmission through a VS.[12] Until now, three types of VS have been described for HIV-1: the DC-T cell VS, also referred to as "Infectious Synapse",[11,13-15] the T cell-T cell VS[12] and the mononuclear cell-mucosal epithelial VS, implicated in HIV transcytosis through mucosal epithelia.[31-33]

The use of VS for viral transmission is probably not limited to retroviruses and is exploited by other intracellular pathogens in order to disseminate through their host. Early in vitro experiments show a VS-like structure possibly contributing to SARS-coronavirus (SARS-CoV) dissemination from DCs to target cells.[34]

As the concept of infectious or virological synapse is further applied to other organisms, such as plants,[35] VS emerges as a general mechanism of cell-to-cell transmission for many pathogens and parasites.

Virological Synapses during HIV Infection

Dendritic Cell-T Cell Infectious Synapse during HIV Sexual Transmission

In model systems of sexual transmission, myeloid dermal DCs and Langerhans cells (LC) play a central role in the early steps of HIV-1 propagation (reviewed in refs. 36-40). DCs locate to the skin and mucosal tissues in an immature state (iDC) until coming across pathogen-derived antigens. DC activation and differentiation into mature APC[41-43] results from contact with different stimuli such as bacterial products,[44] TNF family ligands,[45,46] double-stranded[47] and single-stranded RNA.[48] Migration of mature DCs (mDC) from the periphery to secondary lymphoid organs is strongly associated with maturation and allows DCs to encounter antigen-specific T cells in order to initiate adequate immune responses.[49-53] Although HIV-1 infects CD4+T cells more effectively, LC and other DC types support low levels of viral replication, both in vivo and in vitro.[54-58] DC are also able to capture HIV-1 in an infectious form and transfer such virions to target CD4+T cells without the need of virus replication within the effector cell (here the DC)[13,59,60] (reviewed in refs. 37,61). Recognitions of adhesion molecules inserted in the viral envelope [62,63] or binding through lectin receptors, such as DC-SIGN, mannose receptor or langerin, allow DCs to bind HIV-1 efficiently.[28-30,64,65] The C-type lectin DC-SIGN (CD209), strongly expressed in iDCs, plays a crucial role in capture and transfer of HIV-1 to T cells in trans.[64,65] DC-SIGN was shown to mediate VS (or rather infectious synapse) formation in vitro between DCs and autologous resting T cells, favoring transfer of a CXCR4-using HIV-1.[14] As a major attachment factor on DCs, DC-SIGN has been shown to bind many viruses such as HIV-1, HIV-2, simian immunodeficiency virus (SIV),[66,67] Dengue virus,[68] Cytomegalovirus (CMV),[69] Ebola virus[70] and SARS-CoV.[34]

Professional APCs play a central role in antigen processing. As the archetypal APC, DCs are rich in degradative compartments.[71] Nevertheless, efficient digestion of HIV-1 occurs in DCs, but a small fraction DC-SIGN-internalized virus remains infectious for extended periods of time[13,64,72,73] and can be transferred in trans to target cells. The characteristic DC lysosomal degradative functions are activated upon DC maturation.[43,74] Several studies suggest that HIV-1-induced maturation is only partial and might fail to induce a full activation of the lysosomal system.[75,76] HIV-loaded DCs retains a population of infectious virus within an intracellular compartment that, until recently, was poorly described. Surprisingly, dissection of non replicating (CXCR4-using) HIV-1 trafficking pathways in monocyte-derived DCs revealed that, virus does not accumulate in lysosomes after capture but in a novel mildly acidic nonconventional compartment distinct from the classical late endosome/multivesicular body (MVB). This novel endosome targeted by HIV after capture by DCs is enriched in specific tetraspanins (CD81 and CD9) but contains only little CD63 (marker of MVB) and virtually no LAMP-1 (marker of lysosomes).[15] This tetraspanin rich compartment targeted by HIV-1 after capture by DCs is also rich in MVB. This is reminiscent of the situation in macrophages, a DC-related cell type, where HIV-1 assembles in late endosomes exploiting the machinery implicated in MVB biogenesis. Viral release from macrophages happens subsequently by exocytosis.[77-81] Although the tetraspanin rich endosome targeted by HIV in DCs[15] resembles the structures where HIV assembles in macrophages,[81] the location and mechanisms of HIV-1 replication and budding within DCs remain to be characterized.[82]

Importantly, both HIV-infected and HIV-pulsed DCs are able to transmit a strong infection to T cells in trans.[59,60,83,84] The recent depiction of a VS formed between uninfected T lymphocytes and DCs pulsed with fluorescently tagged HIV-1 has shed some light on the molecular processes at play.[11] The DC-T cell VS has also been termed "Infectious Synapse". In the DC-T cell situation the dendritic cell is not necessarily replicating virus and is transferring HIV to a target cell in trans, whereas in the T cell-T cells VS both cells (pre and postsynaptic) are productively infected. For the purpose of clarity in this review we will use the term VS also in the case of the DC-T cell Infectious Synapse. In DC-T cells conjugates, virions polarize to the contact surface between the adjacent cells. Simultaneously, HIV-1 receptors (CD4) and coreceptors (CXCR4/CCR5) seem to be at least partially enriched on the T cell side of the junction with the DC[11] (EG and VP, unpublished observations). VS formation is possibly initiated by normal cellular interactions in which T cells "scan" DC in an antigen-independent fashion, searching for the cognate peptide presented by the APC.[85] Upon contact with T cells, internalized HIV-1 relocates rapidly to the VS in which the tetraspanins CD81 and CD9 are also redistributed.[15] Given the apparent role of CD81 as an element of the IS[86,87] (reviewed in see refs. 6,88), HIV-1 subverts a pathway involved in IS formation and T cell activation to spread from DCs to uninfected CD4+T cells.[15] On the T cell side of the synapse, engagement of the CD81 receptor might also play a role in increasing viral gene expression.[89]

The dissection of the DC-T cell VS is still ongoing and many questions remain to be answered. Is VS formation relevant in the context of sexual transmission of HIV-1? Shown to facilitate nonreplicative HIV-1/SIV transfer in DC-T cell conjugates,[11,13-15] DC-T cell VS usage by HIV-1 has to be confirmed with replicative CCR5-using strains. What is the relationship between the DC-T cell immunological synapse and the DC-T cell VS? The molecular basis of DC-T cell VS assembly remains poorly understood. Interference studies using receptor-blocking antibodies, inhibitors of cellular processes involved in cytoskeletal rearrangements and signaling, and RNA interference of surface receptor expression are ongoing in order to address this issue.

HIV-1 T Cell-T Cell Virological Synapse

Upon cell-to-cell contact, HIV-1-infected T cells are able to induce rapid clustering of viral receptors on uninfected T cells.[90-92] The molecular interactions behind this process were recently detailed and led to the description of an HIV-1 induced VS between T cells.[12] Interactions between HIV-1 Env protein on the effector cell with CD4 and CXC chemokine receptor 4 (CXCR4) on the naïve T cell are essential to induce a fast actin-dependent recruitment of viral receptors and lymphocyte-associated antigen 1 (LFA-1) to the VS.[12] F-actin disassembly/reassembly is central to the mobilization of all players within the T cell VS, as demonstrated by inhibitors for both processes.[12] Indeed, stable antigen-independent clusters between CD4+T cells seldom occur when compared with antigen-dependant DC-T cell clusters. Therefore, stabilization of T cell-T cell contacts must be triggered by a specific signal. In the case of HIV-1 VS, Env seems to function as the triggering signal. Blocking antibodies and chemical inhibitors preventing Env binding to CD4 and CXCR4 on the naïve T cell reduce T cell VS formation as well as T cell-T cell conjugates.[12]

Virological Synapse and HIV-1 Transcytosis across Mucosal Epithelia

Mucosal epithelia are the first line of defense of the human body against sexual transmission of HIV-1. The virus needs to circumvent this obstacle in order to gain a foothold within a new individual. In addition to capture by DCs or Dendritic Cells residing in mucosal epithelia, transcytosis of infectious virions across epithelial cells at mucosal sites of exposure may well be a strategy used by HIV-1. Early studies showed convincingly that transcytosis with cell-associated HIV-1 was much more efficient than transcytosis of cell-free virions through epithelial cell layers.[19,32,33] Virological synapses, in which HIV-1-infected blood mononuclear cells establish contacts with mucosal epithelial cells, were recently described, providing a likely explanation for this cell-to-cell vial transmission.[31] In this context, HIV-1 buds locally from the effector cell, followed by endocytosis and transcytosis without fusion from the apical to the serosal pole of epithelial cells.[93] Infection grants HIV-1-loaded cells the ability to interact with epithelial cells by upregulating the expression of surface adhesion molecules[94] and by the presence of the viral envelope proteins gp120 and gp41. Epithelial cells also take part in VS formation and stabilization as well as in proper initiation of HIV-1 transcytosis. The heparan sulfate proteoglycan (HSPG) agrin, present in the scaffolding complexes of neural and immunological synapses,[1,95] serves as an HIV-1 attachment receptor through gp41-binding, reinforcing virion interactions with its previously described endocytic receptor galactosyl ceramide (GalCer).[31] Nevertheless, this is not sufficient to initiate HIV-1 trancytosis and additional signals supplied by the synaptic scaffold are crucial. Stable interactions between epithelial cells and HIV-1-infected PBMCs result partially from epithelial expression of the RGD-dependant Beta-1 integrin. Contacts between RGD-containing molecules, either at the surface of HIV-1-infected PBMCs or released as soluble factors,[96] with Beta-1 integrins potentially initiate the signaling pathways leading to an efficient HIV-1 trancytosis and its subsequent spread throughout the host.[31]

These three examples of HIV-1 VS demonstrate that VS play a central role in HIV cell-to-cell transmission. The benefit of VS for HIV spread is observed so far in vitro, but suggests an important function for VS in vivo.

Virological Synapse for HTLV-1 Replication

HTLV-1 is an oncogenic retrovirus spreading from infected T lymphocytes to uninfected T lymphocytes through VS, with little if any contribution from cell-free virions.[97] Upon cell-to-cell

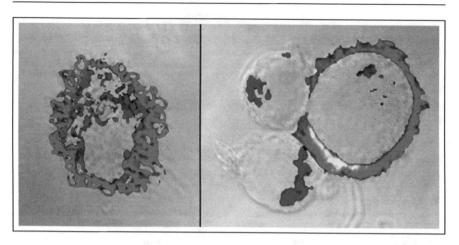

Figure 1. DC-T cell HIV-1 Virological Synapse. Left) Immature Dendritic Cells (DC) were incubated with HIV-1 for 24 hrs at 37°C. HIV-1 accumulates in an intracellular "viral endosome". Right) Lipopolysaccharide-matured Dendritic Cells (DC) were incubated with HIV-1 for 2 hrs at 37°C. Upon encountering Jurkat CD4+ T cells, HIV-1 is redistributed from this intracellular compartment to the zone of contact (infectious synapse) between the DC and the CD4+ T cell (D center and right). Immunological synapse marker MHC-II (HLA-DR) does not appear enriched in the infectious synapse. (Green: Immunostaining of HIV-1 p24^{gag} ; Red: HLA-DR; Blue: Lamp-1)

contact, HTLV-1 Env and Gag proteins polarize in the effector cell (presynaptic cell). On the post-synaptic side, talin polarizes as well at the site of cell-cell interaction and within minutes of synapse formation. Subsequently, HTLV-1 Gag protein transfer through VS is closely followed by HTLV-1 RNA genome transmission to the post-synaptic cell.[10] Interestingly, HTLV-1 T cell VS shares a common feature with the CTL-mediated IS: in both cases, the microtubules organizing center (MTOC) polarizes toward the cell-cell junction within the effector cell.[10,98] Recognition of the cognate peptide and engagement of the TCR are responsible for MTOC movement in the CTL-mediated IS, while in the HTLV-1-induced VS polarization occurs regardless of the potential antigen presented.[10] The molecular basis underlining HTLV-1 T cell VS formation have partially been revealed. Using an antibody-coated bead-cell assay used previously to analyze T cell activation[99-101] followed by interfering experiments, engagement of the intercellular adhesion molecule-1 (ICAM-1) on the effector cell (presynaptic cell) by lymphocyte function-associated antigen-1 (LFA-1) (on the postsynaptic side) was shown to be a crucial signal causing microtubules to polarize to the VS.[21] VS formation is also facilitated by viral encoded proteins such as HTLV-1 transcriptional activator protein (Tax).[22] Tax resides in the nucleus of unconjugated HTLV-1-infected T cells.[102,103] Upon contact with naïve T cells, Tax is found at the site of contact between cells and around the MTOC, in association with the cis-Golgi apparatus.[22] Transient transfection of Jurkat cells with Tax demonstrated a facilitating role for Tax in cell-cell contact-induced MTOC polarization, suggesting that Tax synergizes with ICAM-1 engagement to cause microtubule reorientation during VS formation.[22] Finally, the recent identification of HTLV-1 receptor, glucose transport protein 1 (GLUT-1),[104] will certainly lead to further understanding of the mechanisms involved in HTLV-1 T cell VS formation.

Emerging Role for a Plant Virological Synapse

Passage of intracellular pathogens, such as viruses, bacteria and parasites, between animal cells has been an area of intense scrutiny (reviewed in refs. 9,105,106). Thus it is likely that the concept of virological synapse or rather infectious synapse might be extended beyond animal viruses described above. Recently, the concept of synapse, including the VS has been extended to plants.[35] Plant viruses are known to take advantage of plasmodesmata to gain access to the next cell.

Plasmodesmata are cytoplasmic channels formed and maintained between neighboring plant cells[107,108] that selectively allow passage of macromolecules as well as viral particles. In a physiological context, plant synapses share limited similarities with the mammalian neuronal as well as immunological synapse, allowing plants to deal with pathogen attacks, as well as establishing symbiotic interactions, by polarizing the endocytic and secretory machineries towards the intruding organisms (reviewed in ref. 35). The use of a VS-like structure in plants, implicating genetic transfer from one discrete cell to another has been recently demonstrated in the case of Tobacco Mosaic Virus (TMV), supporting the concept of VS in plants.[109] Unlike HIV-1 DC-T cell VS that originates in tetraspanin rich multivesicular endosomes (MVB),[15] TMV replication originates in the endoplasmic reticulum, before cell-to-cell propagation across plasmodesmata.[109] There are significant differences between the VS of mammalian viruses when compared to VS-like structures in plants. Plasmodesmata are membrane linked pores in plant cell walls that provide continuity between adjacent cells, whereas in the immune system contacts between cells are transient and do not necessitate the formation of a pore. Nevertheless, cell-to-cell propagation of TMV through a plant VS-like structure is very reminiscent of the VS of mammalian retroviruses.

Conclusions

The identification and characterization of the virological synapse provides a satisfying explanation for cell-cell spread of retroviruses within the immune system. VS contribute to stealthy retroviral replication as these viruses hop from cell-to-cell across VS without possibility of neutralization by the immune system. Plant viruses use a plant VS-like structure, indicating that VS are conserved evolutionary structures facilitating replication of animal as well as plant viruses. For each virus and cellular context VS present themselves differently. Only in-depth study of VS in its various forms will provide us with a useful knowledge that may potentially allow us to interrupt cell-cell viral spread.

Acknowledgements

We thank Frantisek Baluska for helpful discussions during the preparation of this manuscript and Allison Piguet for proofreading the manuscript. This work was supported by the Swiss National Science Foundation grant No 3345-67200.01, Leenaards Foundation, NCCR oncology and the Geneva Cancer League to VP. VP is the recipient of a "Professor SNF" position (PP00A—68785).

References

1. Dustin ML, Colman DR. Neural and immunological synaptic relations. Science 2002; 298:785-789.
2. Norcross MA. A synaptic basis for T-lymphocyte activation. Ann Immunol (Paris) 1984; 135D:113-134.
3. Paul WE et al. Regulation of B-lymphocyte activation, proliferation, and differentiation. Ann NY Acad Sci 1987; 505:82-89.
4. Bossi G et al. The secretory synapse: The secrets of a serial killer. Immunol Rev 2002; 189:152-160.
5. Davis DM, Dustin ML. What is the importance of the immunological synapse? Trends Immunol 2004; 25:323-327.
6. Taner SB et al. Control of immune responses by trafficking cell surface proteins, vesicles and lipid rafts to and from the immunological synapse. Traffic 2004; 5:651-661.
7. Huppa JB, Davis MM. T-cell-antigen recognition and the immunological synapse. Nat Rev Immunol 2003; 3:973-983.
8. Ganesh L et al. Infection of specific dendritic cells by CCR5-tropic HIV-1 promotes cell-mediated transmission of virus resistant to broadly neutralizing antibodies. J Virol 2004; 78:11980-11987.
9. Johnson DC, Huber MT. Directed egress of animal viruses promotes cell-to-cell spread. J Virol 2002; 76:1-8.
10. Igakura T et al. Spread of HTLV-I between lymphocytes by virus-induced polarization of the cytoskeleton. Science 2003; 299:1713-1716.
11. McDonald D et al. Recruitment of HIV and its receptors to dendritic cell-T cell junctions. Science 2003; 300:1295-1297.
12. Jolly C, Kashefi K, Hollinshead M et al. HIV-1 cell-to-cell transfer across an Env-induced, actin-dependent synapse. J Exp Med 2004; 199:283-93.

13. Turville SG et al. Immunodeficiency virus uptake, turnover, and 2-phase transfer in human dendritic cells. Blood 2004; 103:2170-2179.
14. Arrighi JF et al. DC-SIGN-mediated infectious synapse formation enhances X4 HIV-1 transmission from dendritic cells to T cells. J Exp Med 2004; 200:1279-1288.
15. Garcia E et al. HIV-1 trafficking to the dendritic cell-T-cell infectious synapse uses a pathway of tetraspanin sorting to the immunological synapse. Traffic 2005; 6:488-501.
16. Piguet V, Sattentau Q. Dangerous liaisons at the virological synapse. J Clin Invest 2004; 114:605-610.
17. Gupta P, Balachandran R, Ho M et al. Cell-to-cell transmission of human immunodeficiency virus type 1 in the presence of azidothymidine and neutralizing antibody. J Virol 1989; 63:2361-2365.
18. Sato H, Orenstein J, Dimitrov D et al. Cell-to-cell spread of HIV-1 occurs within minutes and may not involve the participation of virus particles. Virology 1992; 186:712-724.
19. Phillips DM. The role of cell-to-cell transmission in HIV infection. AIDS 1994; 8:719-731.
20. Carr JM, Hocking H, Li P et al. Rapid and efficient cell-to-cell transmission of human immunodeficiency virus infection from monocyte-derived macrophages to peripheral blood lymphocytes. Virology 1999; 265:319-329.
21. Barnard AL, Igakura T, Tanaka Y et al. Engagement of specific T-cell surface molecules regulates cytoskeletal polarization in HTLV-1-infected lymphocytes. Blood 2005; 106:988-995.
22. Nejmeddine M, Barnard AL, Tanaka Y et al. Human T-lymphotropic virus, type-1, tax protein triggers microtubule reorientation in the virological synapse. J Biol Chem 2005; 280:29653-29660.
23. Pierson TC, Doms RW. HIV-1 entry and its inhibition. Curr Top Microbiol Immunol 2003; 281:1-27.
24. Kilby JM, Eron JJ. Novel therapies based on mechanisms of HIV-1 cell entry. N Engl J Med 2003; 348:2228-2238.
25. Stebbing J, Gazzard B, Douek DC. Where does HIV live? N Engl J Med 2004; 350:1872-1880.
26. Geijtenbeek TB et al. Identification of DC-SIGN, a novel dendritic cell-specific ICAM-3 receptor that supports primary immune responses. Cell 2000; 100:575-585.
27. Bobardt MD et al. Syndecan captures, protects, and transmits HIV to T lymphocytes. Immunity 2003; 18:27-39.
28. Nguyen DG, Hildreth JE. Involvement of macrophage mannose receptor in the binding and transmission of HIV by macrophages. Eur J Immunol 2003; 33:483-493.
29. Turville SG et al. Diversity of receptors binding HIV on dendritic cell subsets. Nat Immunol 2002; 3:975-983.
30. Hu Q et al. Blockade of attachment and fusion receptors inhibits HIV-1 infection of human cervical tissue. J Exp Med 2004; 199:1065-1075.
31. Alfsen A, Yu H, Magerus-Chatinet A et al. HIV-1-infected blood mononuclear cells form an integrin- and agrin-dependent viral synapse to induce efficient HIV-1 transcytosis across epithelial cell monolayer. Mol Biol Cell 2005; 16:In press.
32. Bomsel M. Transcytosis of infectious human immunodeficiency virus across a tight human epithelial cell line barrier. Nat Med 1997; 3:42-47.
33. Bomsel M et al. Intracellular neutralization of HIV transcytosis across tight epithelial barriers by anti-HIV envelope protein dIgA or IgM. Immunity 1998; 9:277-287.
34. Yang ZY et al. pH-dependent entry of severe acute respiratory syndrome coronavirus is mediated by the spike glycoprotein and enhanced by dendritic cell transfer through DC-SIGN. J Virol 2004; 78:5642-5650.
35. Baluska F, Volkmann D, Menzel D. Plant synapses: Actin-based domains for cell-to-cell communication. Trends Plant Sci 2005; 10:106-111.
36. Shattock RJ, Moore JP. Inhibiting sexual transmission of HIV-1 infection. Nat Rev Microbiol 2003; 1:25-34.
37. Pope M, Haase AT. Transmission, acute HIV-1 infection and the quest for strategies to prevent infection. Nat Med 2003; 9:847-852.
38. Steinman RM et al. The interaction of immunodeficiency viruses with dendritic cells. Curr Top Microbiol Immunol 2003; 276:1-30.
39. Turville S, Wilkinson J, Cameron P et al. The role of dendritic cell C-type lectin receptors in HIV pathogenesis. J Leukoc Biol 2003; 74:710-718.
40. Geijtenbeek TB, van Kooyk Y. DC-SIGN: A novel HIV receptor on DCs that mediates HIV-1 transmission. Curr Top Microbiol Immunol 2003; 276:31-54.
41. Chow A, Toomre D, Garrett W et al. Dendritic cell maturation triggers retrograde MHC class II transport from lysosomes to the plasma membrane. Nature 2002; 418:988-994.
42. Boes M et al. T-cell engagement of dendritic cells rapidly rearranges MHC class II transport. Nature 2002; 418:983-988.

43. Trombetta ES, Ebersold M, Garrett W et al. Activation of lysosomal function during dendritic cell maturation. Science 2003; 299:1400-1403.
44. Rescigno M, Granucci F, Citterio S et al. Coordinated events during bacteria-induced DC maturation. Immunol Today 1999; 20:200-203.
45. Caux C, Dezutter-Dambuyant C, Schmitt D et al. GM-CSF and TNF-alpha cooperate in the generation of dendritic Langerhans cells. Nature 1992; 360:258-261.
46. Rescigno M et al. Fas engagement induces the maturation of dendritic cells (DCs), the release of interleukin (IL)-1beta, and the production of interferon gamma in the absence of IL-12 during DC-T cell cognate interaction: A new role for Fas ligand in inflammatory responses. J Exp Med 2000; 192:1661-1668.
47. Cella M, Engering A, Pinet V et al. Inflammatory stimuli induce accumulation of MHC class II complexes on dendritic cells. Nature 1997; 388:782-787.
48. Heil F et al. Species-specific recognition of single-stranded RNA via toll-like receptor 7 and 8. Science 2004; 303:1526-1529.
49. Banchereau J, Steinman RM. Dendritic cells and the control of immunity. Nature 1998; 392:245-252.
50. Banchereau J et al. Immunobiology of dendritic cells. Annu Rev Immunol 2000; 18:767-811.
51. Lanzavecchia A, Sallusto F. Regulation of T cell immunity by dendritic cells. Cell 2001; 106:263-266.
52. Stoll S, Delon J, Brotz TM et al. Dynamic imaging of T cell-dendritic cell interactions in lymph nodes. Science 2002; 296:1873-1876.
53. Mempel TR, Henrickson SE, Von Andrian UH. T-cell priming by dendritic cells in lymph nodes occurs in three distinct phases. Nature 2004; 427:154-159.
54. Tschachler E et al. Epidermal Langerhans cells—a target for HTLV-III/LAV infection. J Invest Dermatol 1987; 88:233-237.
55. Ringler DJ et al. Cellular localization of simian immunodeficiency virus in lymphoid tissues. I. Immunohistochemistry and electron microscopy. Am J Pathol 1989; 134:373-383.
56. Stahl-Hennig C et al. Rapid infection of oral mucosal-associated lymphoid tissue with simian immunodeficiency virus. Science 1999; 285:1261-1265.
57. Hu J, Gardner MB, Miller CJ. Simian immunodeficiency virus rapidly penetrates the cervicovaginal mucosa after intravaginal inoculation and infects intraepithelial dendritic cells. J Virol 2000; 74:6087-6095.
58. Smed-Sorensen A et al. Differential susceptibility to human immunodeficiency virus type 1 infection of myeloid and plasmacytoid dendritic cells. J Virol 2005; 79:8861-8869.
59. Blauvelt A et al. Productive infection of dendritic cells by HIV-1 and their ability to capture virus are mediated through separate pathways. J Clin Invest 1997; 100:2043-2053.
60. Lore K, Smed-Sorensen A, Vasudevan J et al. Myeloid and plasmacytoid dendritic cells transfer HIV-1 preferentially to antigen-specific CD4+ T cells. J Exp Med 2005; 201:2023-2033.
61. Piguet V, Blauvelt A. Essential roles for dendritic cells in the pathogenesis and potential treatment of HIV disease. J Invest Dermatol 2002; 119:365-369
62. Tsunetsugu-Yokota Y et al. Efficient virus transmission from dendritic cells to CD4+ T cells in response to antigen depends on close contact through adhesion molecules. Virology 1997; 239:259-268.
63. Tardif MR, Tremblay MJ. Presence of host ICAM-1 in human immunodeficiency virus type 1 virions increases productive infection of CD4+ T lymphocytes by favoring cytosolic delivery of viral material. J Virol 2003; 77:12299-12309.
64. Geijtenbeek TB et al. DC-SIGN, a dendritic cell-specific HIV-1-binding protein that enhances trans-infection of T cells. Cell 2000; 100:587-597.
65. Arrighi JF et al. Lentivirus-mediated RNA interference of DC-SIGN expression inhibits human immunodeficiency virus transmission from dendritic cells to T cells. J Virol 2004; 78:10848-10855.
66. Bashirova AA et al. A dendritic cell-specific intercellular adhesion molecule 3-grabbing nonintegrin (DC-SIGN)-related protein is highly expressed on human liver sinusoidal endothelial cells and promotes HIV-1 infection. J Exp Med 2001; 193:671-678.
67. Pohlmann S et al. DC-SIGN interactions with human immunodeficiency virus: Virus binding and transfer are dissociable functions. J Virol 2001; 75:10523-10526.
68. Tassaneetrithep B et al. DC-SIGN (CD209) mediates dengue virus infection of human dendritic cells. J Exp Med 2003; 197:823-829.
69. Halary F et al. Human cytomegalovirus binding to DC-SIGN is required for dendritic cell infection and target cell trans-infection. Immunity 2002; 17:653-664.
70. Alvarez CP et al. C-type lectins DC-SIGN and L-SIGN mediate cellular entry by Ebola virus in cis and in trans. J Virol 2002; 76:6841-6844.

71. Mellman I, Steinman RM. Dendritic cells: Specialized and regulated antigen processing machines. Cell 2001; 106:255-258.

72. Kwon DS, Gregorio G, Bitton N et al. DC-SIGN-mediated internalization of HIV is required for trans-enhancement of T cell infection. Immunity 2002; 16:135-144.

73. Moris A et al. DC-SIGN promotes exogenous MHC-I-restricted HIV-1 antigen presentation. Blood 2004; 103:2648-2654.

74. Delamarre L, Pack M, Chang H et al. Differential lysosomal proteolysis in antigen-presenting cells determines antigen fate. Science 2005; 307:1630-1634.

75. Granelli-Piperno A, Golebiowska A, Trumpfheller C et al. HIV-1-infected monocyte-derived dendritic cells do not undergo maturation but can elicit IL-10 production and T cell regulation. Proc Natl Acad Sci USA 2004; 101:7669-7674.

76. Fantuzzi L, Purificato C, Donato K et al. Human immunodeficiency virus type 1 gp120 induces abnormal maturation and functional alterations of dendritic cells: A novel mechanism for AIDS pathogenesis. J Virol 2004; 78:9763-9772.

77. Garrus JE et al. Tsg101 and the vacuolar protein sorting pathway are essential for HIV-1 budding. Cell 2001; 107:55-65.

78. Martin-Serrano J, Zang T, Bieniasz PD. HIV-1 and Ebola virus encode small peptide motifs that recruit Tsg101 to sites of particle assembly to facilitate egress. Nat Med 2001; 7:1313-1319.

79. Strack B, Calistri A, Craig S et al. AIP1/ALIX is a binding partner for HIV-1 p6 and EIAV p9 functioning in virus budding. Cell 2003; 114:689-699.

80. von Schwedler UK et al. The protein network of HIV budding. Cell 2003; 114:701-713.

81. Pelchen-Matthews A, Kramer B, Marsh M. Infectious HIV-1 assembles in late endosomes in primary macrophages. J Cell Biol 2003; 162:443-455.

82. Kramer B, Pelchen-Matthews A, Daneka M et al. HIV interaction with endosomes in macrophages and dendritic cells. Blood Cells Mol Dis 2005; In press.

83. Cameron PU et al. Dendritic cells exposed to human immunodeficiency virus type-1 transmit a vigorous cytopathic infection to CD4+ T cells. Science 1992; 257:383-387.

84. Pope M et al. Conjugates of dendritic cells and memory T lymphocytes from skin facilitate productive infection with HIV-1. Cell 1994; 78:389-398.

85. Revy P, Sospedra M, Barbour B et al. Functional antigen-independent synapses formed between T cells and dendritic cells. Nat Immunol 2001; 2:925-931.

86. Miyazaki T, Muller U, Campbell KS. Normal development but differentially altered proliferative responses of lymphocytes in mice lacking CD81. EMBO J 1997; 16:4217-4225.

87. Mittelbrunn M, Yanez-Mo M, Sancho D et al. Cutting edge: Dynamic redistribution of tetraspanin CD81 at the central zone of the immune synapse in both T lymphocytes and APC. J Immunol 2002; 169:6691-6695.

88. Levy S, Shoham T. The tetraspanin web modulates immune-signalling complexes. Nat Rev Immunol 2005; 5:136-148.

89. Tardif MR, Tremblay MJ. Tetraspanin CD81 provides a costimulatory signal resulting in increased human immunodeficiency virus type 1 gene expression in primary CD4+ T lymphocytes through NF-kappaB, NFAT, and AP-1 transduction pathways. J Virol 2005; 79:4316-4328.

90. Phillips DM, Bourinbaiar AS. Mechanism of HIV spread from lymphocytes to epithelia. Virology 1992; 186:261-273.

91. Sattentau QJ, Moore JP. The role of CD4 in HIV binding and entry. Phil Trans R Soc Lond B Biol Sci 1993; 342:59-66.

92. Fais S et al. Unidirectional budding of HIV-1 at the site of cell-to-cell contact is associated with copolarization of intercellular adhesion molecules and HIV-1 viral matrix protein. AIDS 1995; 9:329-335.

93. Bomsel M, Alfsen A. Entry of viruses through the epithelial barrier: Pathogenic trickery. Nat Rev Mol Cell Biol 2003; 4:57-68.

94. Shattock RJ, Burger D, Dayer JM et al. Enhanced HIV replication in monocytic cells following engagement of adhesion molecules and contact with stimulated T cells. Res Virol 1996; 147:171-179.

95. Bezakova G, Ruegg MA. New insights into the roles of agrin. Nat Rev Mol Cell Biol 2003; 4:295-308.

96. Rusnati M, Presta M. HIV-1 Tat protein and endothelium: From protein/cell interaction to AIDS-associated pathologies. Angiogenesis 2002; 5:141-151.

97. Bangham CR. The immune control and cell-to-cell spread of human T-lymphotropic virus type 1. J Gen Virol 2003; 84:3177-3189.

98. Stinchcombe JC, Bossi G, Booth S et al. The immunological synapse of CTL contains a secretory domain and membrane bridges. Immunity 2001; 15:751-761.

99. Mescher MF. Surface contact requirements for activation of cytotoxic T lymphocytes. J Immunol 1992; 149:2402-2405.
100. Lowin-Kropf B, Shapiro VS, Weiss A. Cytoskeletal polarization of T cells is regulated by an immunoreceptor tyrosine-based activation motif-dependent mechanism. J Cell Biol 1998; 140:861-871.
101. Sedwick CE et al. TCR, LFA-1, and CD28 play unique and complementary roles in signaling T cell cytoskeletal reorganization. J Immunol 1999; 162:1367-1375.
102. Bex F, McDowall A, Burny A et al. The human T-cell leukemia virus type 1 transactivator protein Tax colocalizes in unique nuclear structures with NF-kappaB proteins. J Virol 1997; 71:3484-3497.
103. Semmes OJ, Jeang KT. Localization of human T-cell leukemia virus type 1 tax to subnuclear compartments that overlap with interchromatin speckles. J Virol 1996; 70:6347-6357.
104. Manel N et al. The ubiquitous glucose transporter GLUT-1 is a receptor for HTLV. Cell 2003; 115:449-459.
105. Cossart P, Sansonetti PJ. Bacterial invasion: The paradigms of enteroinvasive pathogens. Science 2004; 304:242-248.
106. Sibley LD. Intracellular parasite invasion strategies. Science 2004; 304:248-253.
107. Zambryski P, Crawford K. Plasmodesmata: Gatekeepers for cell-to-cell transport of developmental signals in plants. Annu Rev Cell Dev Biol 2000; 16:393-421.
108. Oparka KJ. Getting the message across: How do plant cells exchange macromolecular complexes? Trends Plant Sci 2004; 9:33-41.
109. Kawakami S, Watanabe Y, Beachy RN. Tobacco mosaic virus infection spreads cell to cell as intact replication complexes. Proc Natl Acad Sci USA 2004; 101:6291-6296.

CHAPTER 23

Cell-Cell Fusion:
Transient Channels Leading to Plasma Membrane Merger

William A. Mohler*

Abstract

Despite the diversity of intercellular connections that are the subject of this book, most eukaryotic cells retain their distinct character as mononucleated compartments. Their membranes describe morphologically separate cytoplasms, while electrical connectivity and low-flux intercellular exchange of components occurs through small or selective channels between neighboring cell surfaces. However, in many instances throughout eukaryotes, pairs or groups of cells make a developmental decision to completely fuse their plasma membranes, allowing wholesale exchange and mixing of membranous, cytoplasmic and nuclear components. The products of these fusion events are either cell hybrids, in which chromosomes are combined into a single nucleus, or syncytia, wherein distinct nuclei are maintained within a single cytoplasm and plasma membrane (Fig. 1). While limited to very specific instances in the life cycle of any given organism, these precise cell fusions lead to a diverse set of dramatic developmental transitions: from formation of a new zygote, to construction of the musculoskeletal system, to refinement of the optical transparency of the developing eye. In addition, it appears possible to repair damaged cells, such as neurons, through the fusion of severed cellular fragments.[1,2] This chapter will survey the various contexts for developmental cell fusion, examining the scant but growing knowledge of the molecules that initiate membrane permeability and removal of cell boundaries between merging partner cells. The understanding that is beginning to emerge suggests that cell-fusion channels or pores are transient affairs, both as structural antecedents of fully merged cell membranes, and possibly as replaceable molecular machines that were reinvented often through the course of evolution to drive a similar process by a variety of mechanisms.

Introduction

The predominant case for formation of cell hybrids in nature is the fertilization or conjugation of haploid gametes to form diploid zygotes. This of course underlies the initiation of development by sexual reproduction in the vast majority of plant and animal species. It also plays a role in the development of many unicellular eukaryotes. Nonsexual somatic cell hybrids have been described as well, notably during the engraftment of stem cells into rodent hepatic and neural tissues after experimental cellular transfer.[3-10] Chemically induced fusion of cells in vitro typically results in death for the majority of resulting hybrids, although viable cell hybrids are often selected experimentally and expanded as clones.[11-13] It is unclear how prevalent the formation of somatic cell hybrids may be in the course of normal development and homeostasis. However, it has been reported that many cancer cell types are either fusogenic or are products of aberrant cell fusion,[14,15] and the formation of

*Corresponding Author: William A. Mohler—Department of Genetics and Developmental Biology, University of Connecticut Health Center, 263 Farmington Ave., Farmington, Connecticut 06030-3301, U.S.A. Email: wmohler@neuron.uchc.edu

Cell-Cell Channels, edited by Frantisek Baluska, Dieter Volkmann and Peter W. Barlow.
©2006 Landes Bioscience and Springer Science+Business Media.

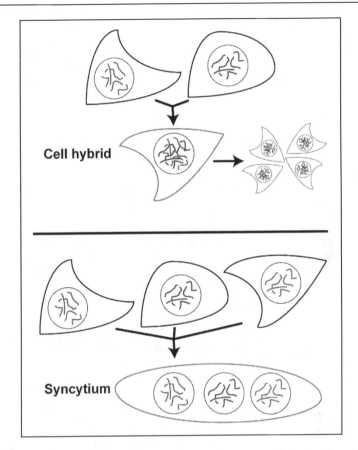

Figure 1. Cell hybrids and syncytia, two possible products of cell-cell fusion. A cell hybrid combines the content of both nuclei into a single nucleus, giving rise to a mitotically viable cell. A syncytium retains the distinction between nuclei within the giant cell formed by fusion.

rare yet viable cell hybrids can therefore have major impacts upon the health and lifespan of individual animals.

In contrast to cell hybrids, syncytia are most often formed during the terminal differentiation of specific cell types. In most cases, multinucleated giant cells are postmitotic and nonmotile, assuming a variety of roles in mature tissues. Reasoned teleologically, the functions performed by syncytial cells must be best accomplished by a single cellular compartment that is too large to be maintained by the gene-expression potential of a single diploid nucleus. Increases in gene copy-number can also be accomplished through endoreduplication of DNA.[16-18] However, multinucleation by cell fusion allows for establishment of pattern and form via the migration and proliferation of the mononucleate precursor cells before they fuse to form the final product. Fusion also allows the giant cell to be expanded or regenerated by later addition of new precursors to an existing syncytium. In addition, the distribution of individual nuclei through the giant cell permits local differences in gene expression to define sub-cellular specializations, such as the post-synaptic neuromuscular junction of skeletal muscle fibers.[19,20]

Below, several cases of developmental cell fusion are surveyed, and we summarize the current knowledge of molecules controlling actual plasma membrane merger and discuss the possibility that the same effect may be brought about by a variety of mechanisms. To focus on the process of membrane fusion, per se, we will avoid the discussion of the diverse regulatory pathways leading to the fusion-competent differentiated state in each of these cell types. In contrast to viral membrane-fusion pores and other forms of stable cell-cell channels covered in this book, very little is known about the biophysical properties of cell-fusion-initiating pores. We therefore also address the hypothetical involvement of initial cell-membrane fusion structures similar to those found in viral infection or specialized cell-cell channels between nonfusing cell types.

Fertilization (Mouse)

Sperm-egg fusion requires several steps prior to the actual fusion of gamete cell membranes, including binding of the sperm to the zona pelucida and initiation of the acrosome reaction in the sperm. Only sperm that have penetrated the zona, by secretion of degradative enzymes, and have extended an acrosomal process are competent to fuse with the egg. Izumo, an immunoglobulin-superfamily (IgSF) membrane glycoprotein, has been found newly presented on the surface of acrosome-reacted mouse sperm (Fig. 2).[21] When Izumo is either blocked by specific antibodies or deleted genetically, mouse sperm become completely incapable of fertilizing wild-type eggs, despite retaining their ability to penetrate the zona, elaborate a normal acrosomal process, and bind to the egg membrane. Furthermore, once experimentally introduced into an activated egg, Izumo-mutant sperm contribute normally to post-fusion zygotic development. Izumo, therefore, stands out among a collection of molecules previously implicated in fertilization,[22-28] as it appears completely and specifically required for the fusion competence of sperm during fertilization.

On the egg plasma membrane, a tetraspanin molecule, CD9, is also absolutely required for fusion of the mouse egg membrane with wild-type sperm.[29-31] Some peptide sequences in CD9 that are known to be required for sperm-egg fusion have also been shown to mediate binding of CD9 to IgSF molecules.[32] Both CD9 and Izumo are encoded in the human genome, as well, suggesting that their role in sperm-egg fusion may be evolutionarily conserved in mammals other than mice. Yet, it is still a matter of conjecture whether the two molecules interact directly in trans to effect the membrane fusion reaction, or whether either molecule plays a direct role in the formation of a permeability pore between the two fusing cells.

Fertilization (Nematode)

Nematode sperm are unflagellated cells that move by amoeboid crawling. Current knowledge, via genetic studies of *Caenorhabditis elegans* fertility mutants, suggests that they also differ from mammalian sperm in their mechanism of sperm-egg fusion. Three different sperm-encoded proteins, SPE-9 (an EGF-repeat-containing membrane protein, TRP-3/SPE-41 (a TRPC-type calcium channel), and SPE-38 (a novel tetraspan membrane protein), are required specifically for sperm interaction and/or fusion with the egg.[33-37] A recessive mutation in any of the three genes yields sperm that activate and migrate normally but fail to fertilize eggs. TRP-3 is interesting as it is perhaps the only ion-channel protein so far implicated specifically in a cell fusion event.[35] SPE-38, although containing four membrane spanning domains, does not encode a homologue of the mammalian CD9, but it is structurally similar to the tetraspan protein PRM1 that regulates yeast cell fusion (see below).[33] It is worth noting that EFF-1, while required for many tissue cell fusions in the worm (see below), is not required for sperm-egg fusion.[38] This indicates two quite distinct mechanisms of membrane fusion in nematode gametes and somatic cells.

Myoblasts (Mouse)

In arthropods and vertebrates, precursors of skeletal muscle tissue fuse together to form multinucleated myotubes (Fig. 3) and ultimately muscle fibers, each of which can contain thousands of nuclei. This is the most prevalent form of cell fusion in most animals, and it produces long tube-shaped cells in which a continuously reiterated lattice of contractile filaments can extend uninterrupted over many centimeters in length. In addition, the neurally stimulated excitation/contraction response of this "spring" is controlled by one motor-neuron synapse at a single neuromuscular junction on the large cell membrane. Both of these properties of myofibers suggest selective advantages that concur with the universal incidence of multinucleated muscle in phyla

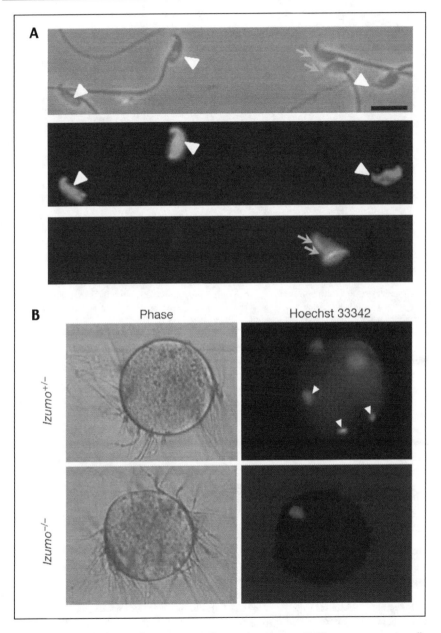

Figure 2. Izumo enables fusion of acrosome-reacted sperm into the egg. A) The acrosome reaction allows presentation of Izumo on the surface of mouse sperm. Unreacted sperm have green fluorescence from acrosin-promoter::green fluorescent protein (arrows, bottom), while reacted sperm lose the green signal. Acrosome-reacted sperm (arrowheads, middle) react with antibodies to Izumo. B) Sperm lacking Izumo fail to fuse with wild-type eggs. Eggs have been loaded with the fluorescent DNA dye Hoechst 33342 to highlight successfully fused sperm. Reprinted with permission from: Inoue et al. Nature 2005; 434(7030):234-238; ©2005 Nature Publishing Group (http://www.nature.com/).[21] A color version of this figure is available online at www.Eurekah.com.

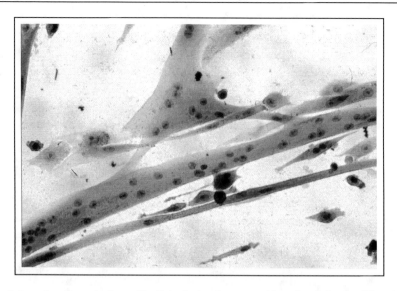

Figure 3. Formation of myotubes by myoblast fusion in vitro. A mouse myoblast culture induced to differentiation is shown stained with hematoxylin and eosin. Many mononucleated cells have fused to form binucleate or multinucleated myotubes, the precursors of mature muscle fibers.

having large muscles and jointed "lever-action" movements. Through the proliferation and fusion of normally quiescent satellite cells, muscle fibers can be repaired or regenerated, even late in life in mammals.[39] Interestingly, a microtubule-binding compound, myoseverin, has been shown to induce the fission of mouse myotubes in vitro, producing mononucleated fragments that could resume proliferation.[40] This suggests that even multinucleated terminally differentiated cells in mammals may have much of the capacity for regeneration exhibited by similar cells in amphibians.

Several IgSF, cadherin, and integrin proteins have been implicated in mouse myoblast fusion. However, each of the molecules tested directly by genetic deletion in mice has been found not to be required for multinucleation of muscle fibers.[41-45] Three members of the ADAM family of proteins, meltrin-α, β and γ were found to be induced in differentiating myoblasts, and antibody or antisense inhibition of meltrin-α were reported to inhibit myoblast fusion in vitro.[46] However, mouse knockouts of each of the meltrin genes (and several other ADAM-coding genes) show no apparent defects in myoblast fusion in vivo.[47,48] Thus, the highly regulated and specific mechanism by which differentiating mammalian myoblasts fuse their membranes is still unknown.

Myoblasts (Fruit Fly)

Mutations in a variety of genes have been reported to block the fusion of myoblasts during formation of embryonic body-wall muscle fibers in *Drosophila melanogaster*. These include known components of the muscle contractile apparatus,[49,50] as well as actin-associated, cytoplasmic, and integral membrane proteins (reviewed in refs. 51,52). Electron microscopy on staged embryos has led to a model for myoblast fusion involving trafficking of electron-dense vesicles, formation of electron-dense membrane plaques, and ultimate membrane fusion at the site of contact between fusing cells (Fig. 4).[53] The hypothesis that these vesicles and plaques contain components required for fusion is supported to some degree by EM images of the fusion-arrested membranes of mutant embryos.[53,54] However, none of the genetically-identified proteins has yet been reported localized to these structures, and the functional sequence and relationship between vesicle traffic, plaque, and pore formation has not been reported in live cells.

Three genes that are required for normal *Drosophila* myoblast fusion encode integral membrane IgSF proteins: Dumfounded/Kin of Irre (Duf/Kirre), Roughest/Irregular chiasm (Rst/IrreC), and

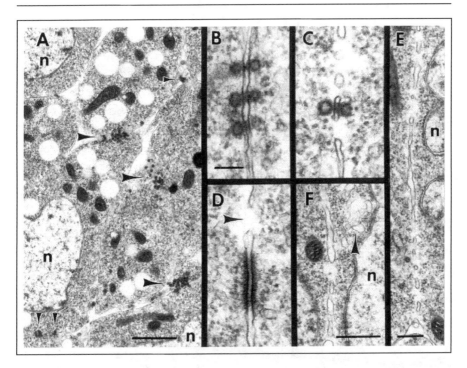

Figure 4. Ultrastructure of intermediate steps in *Drosophila* myoblast fusion. A) Myoblasts in early stage of fusion. Note prefusion complexes at points of cell-cell contact (arrowheads); n indicates myoblast nuclei. B) Three sets of paired vesicles. Note electron-dense material in the extracellular space between pairs of vesicles. C) Paired vesicles oriented across a vesiculating pair of plasma membranes. D) An electron-dense plaque near a region of actively fusing membrane; note fusion pore (arrow). E) Fusion pores in a vesiculating plasma membrane. The cytoplasm within and beneath the pore is free of staining material such as ribosomes. F) Later stage vesiculating plasma membrane. The membrane sacs have increased in width and a group of irregular clear vesicles is present (arrowhead). Bars: (A) 1 μm; (B-D) 100 nm; (E) 250 μm; (F) 500 μm. Figure and legend are reproduced from: J Cell Biol 1997; 136(6):1249-1261, by copyright permission of the Rockefeller University Press.

Sticks and Stones (SNS).[55-57] Duf/Kirre and Rst/IrreC are expressed in founder cells, early-differentiating myocytes that recruit fusion-competent myoblasts (fcm) for two rounds of developmentally regulated fusion to form mature muscle fibers. SNS, in contrast, is expressed only on the fcm cells. Models involving interaction of these receptors to produce chemoattractive and fusion-effecting signals are suggested by the reciprocal expression of these molecules on pairs of cells destined to fuse. However, while SNS can interact heterotypically with either Duf/Kirre or Rst/IrreC to mediate efficient adhesion and aggregation of transfected *Drosophila* S2 cells in culture, these interactions cannot elicit the membrane fusion with which they are associated in vivo.[58] At this point, no molecule or combination of molecules known to be required for *Drosophila* myogenic fusion has been shown to be sufficient for membrane fusion when ectopically expressed. This suggests either that the actual fusion mechanism involves fusogenic molecules not yet identified, or that it combines the action of the known players in a way that has not been reconstituted in heterologous cell types.

Placental Syncytiotrophoblast Cells (Human and Mouse)

During uterine implantation of the early embryo in placental mammals, the trophoblast cells of the conceptus proliferate and invade the maternal endometrium. Subsequently, many of these cells fuse to form multinucleated syncytiotrophoblast cells at the interface of the maturing placenta with maternal tissue.[59] The syncytiotrophoblasts may serve as a more selective barrier or regulator of nutrient, metabolite, gas, waste, and immune interchange between the circulatory systems of mother and fetus than is possible with a typical epithelium of mononucleated cells. Interestingly, in some species, fusion of fetal and maternal cells has been observed by electron microscopy.[60]

Syncytin-1 and -2 are developmentally regulated genes encoding human genomic copies of retroviral envelope glycoproteins, identified through hybridization and expression screening of placental cDNA libraries and via in silico genomic analysis of human genome sequences.[61-64] Syncytin expression is selectively induced during the development of the placenta, and forced overexpression of syncytins in cultured cells can induce fusion of normally mononucleated cells (Fig. 5). Thus syncytins appear individually sufficient to induce receptor-dependent cell membrane fusion, similar to homologous proteins acting during infection by enveloped viruses. However, syncytin-1 and -2 are encoded exclusively in a subset of recently evolved primate genomes.[63,65] This has precluded genetic tests of their necessity in syncytiotrophoblast fusion. Furthermore, if syncytin-1 and -2 do underlie the fusion mechanism in primate placenta, the molecules must have been adapted to the task recently in evolution (apparently within the past 40 million years), and some different fusogen(s) must drive cell fusion in other mammal species with fusing placental cells.

Recently, two murine viral envelope genes have been discovered, syncytin-A and -B, that are evolutionarily distinct from primate syncytins, yet appear to serve the same function in the syncytiotrophoblasts of mice and related rodents.[66] This suggests that placental cell fusion may

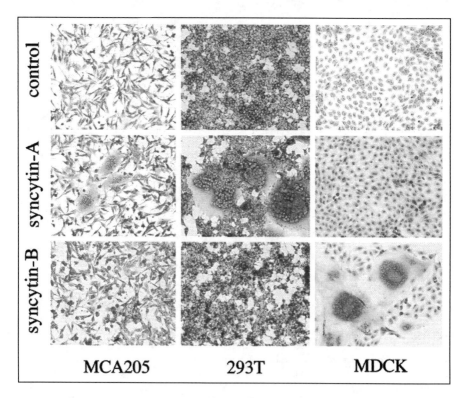

Figure 5. Cell-cell fusion induced by expression of murine syncytins. Overexpression of syncytins A and B induce multinucleation of cultured mononucleated cell lines. Each protein has a different degree of fusogenic activity in murine (MCA205), human (293T), and canine (MDCK) cells, suggesting that each uses a different receptor on the host surface. Figure is reprinted with permission from: Dupressoir et al. Proc Natl Acad Sci USA102(3):725-730, ©2005 National Academy of Science, U.S.A.[66]

involve a mechanism allowing fortuitous substitution of fusogenic proteins, including independent cooption of several different viral genes. In the mouse model system, the opportunity now exists to prove or disprove, through targeted gene knockouts, the necessity of syncytin genes in placental cell fusion. Given that the four known syncytins are not encoded in many species with fusing syncytiotrophoblast cells, the alternative possibility still remains tenable: that nonsyncytin molecules may comprise a more evolutionarily conserved cell fusion mechanism common to all placental mammals.

Lens Fiber Cells (Mouse and Other Vertebrates)

Cells of the crystalline lens also fuse extensively in the development of the amphibian, avian, and mammalian eye.[67-70] Fusion appears to occur between hexagonal fiber cells that are already terminally differentiated, producing progressively more and more cytoplasmic communication between cells with increasing age (Fig. 6). However, fused cells maintain their form as individual hexagonal prisms over much of their length. The central core of the lens, where fusion is most frequent, is also the region where fiber cells are often enucleated. Thus, fusion may enable formation of a clarified syncytium, in which nucleated cells at the periphery of the central zone are able to sustain the viability of enucleated central fibers through free exchange of cytoplasm. Presumably, this overall structure of the lens enhances transmission and refraction that produce an image on the retina. As yet, no molecules involved in this process have been described.

Macrophages/Osteoclasts (Mouse)

Cells of the mammalian monocyte-macrophage lineage form two major types of multinucleated cells: macrophage giant cells in many tissues and osteoclasts in bone.[71] In each case, the physiological roles of these cells involves endocytosis and resorption of relatively large objects, including cell corpses, invading pathogens, foreign bodies, and chunks of mineralized bone. The increased size achieved through cell fusion presumably affords the extra membrane surface area and endosome/lysosome volume to achieve these tasks. In addition, it is conceivable that a larger cell can more safely distance the nuclei and cell body from the noxious degradative cocktails used to dispose of objects within the engulfment apparatus.

MFR and CD44 are cell surface IgSF proteins implicated in the mechanism of macrophage and osteoclast fusion.[72,73] The expression of each is induced transiently at the onset of fusion in

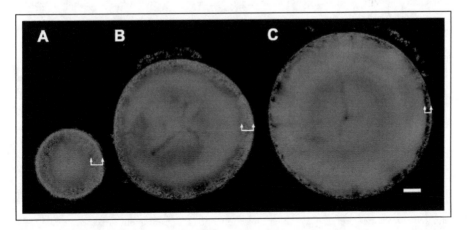

Figure 6. Progressive cell fusion in the growing mouse lens. A syncytial central region expands during postnatal development. Fluorescence within living lenses from one-day-old (A), one-month-old (B) and six-month-old (C) mice expressing a variegated GFP transgene was imaged in the equatorial plane by confocal microscopy. Although the uniformly labeled lens core expands with age, the thickness of the variegated layer at the periphery (arrows) thins slightly over this period. Bar, 250 μm. Figure and legend are reprinted with permission from: Shestopalov and Bassnett. J Cell Sci 116(Pt 20):4191-4199; ©2003 the Company of Biologists Ltd.[70]

macrophage cultures, and antibody blockade of MFR disrupts macrophage fusion in vitro. MFR is known to be a receptor for the constitutively expressed IgSF protein CD47,[74] while CD44 has no known ligand. Despite their correlation with fusogenicity of macrophages, no genetic evidence for the requirement or sufficiency of these molecules in cell fusion has been reported. Recently, a seven-transmembrane receptor, DC-STAMP, was found to be required for macrophage/osteoclast fusion. Knockout mice lacking DC-STAMP fail to form multinucleated osteoclasts or macrophage-derived giant cells, and they display a mild osteopetrotic phenotype.[75] Interestingly, the *DC-STAMP* mutation does not affect the mRNA levels of *MFR, CD44, CD47, E-cadherin,* or *meltrin-α,* all molecules hypothesized to contribute to the fusion mechanism. What the ligand for DC-STAMP is and what the actual components of the macrophage/osteoclast fusion mechanism might be are still unknown. It is interesting to note, however, that mononucleated *DC-STAMP*[-/-] osteoclasts appear to mediate reasonably normal bone development in the mouse, even without achieving the size of normal fused osteoclast giant cells.

Implanted Stem Cells (Mouse and Human)

Recent studies in mice have shown that implanted bone-marrow-derived cells undergo tissue-specific differentiation in host tissues via formation of cell-hybrids between donor and host cells.[3,5-8,10] Such donor-host fusion has even been reported in the neurons of human bone-marrow transplant recipients.[9] In studies of hepatic disease in mice, where a selective advantage is conferred upon donor/host hybrids, fusion-based engraftment has been shown to rescue the viability of an entire organ.[5,76] In a converse experiment, committed neuronal precursor cells cocultured with embryonic stem cells have been shown to produce cell-hybrids that regain pluripotent stem-cell character.[4] The result of these cell-hybridizations in reprogramming gene expression is not unexpected, given previous evidence of the plasticity of nuclei in heterokaryons.[77] But the cell fusions that are revealed between normally mononucleated cell types in the transplant studies is surprising. Although apparently quite rare events, these cell fusions might be hypothesized to proceed via similar mechanisms to the more robust developmentally programmed fusions in other tissues. Given that bone-marrow-derived lineages give rise to fusing macrophages and osteoclast giant cells, it has been reasonably hypothesized that the fusion mechanism responsible is conferred by a macrophage-like activity in the implanted bone-marrow cells. However, without further knowledge of the fusion mechanism of monocyte-derived cells (see above), it will be difficult to test this hypothesis directly.

Epithelia (Nematode)

One third of all somatic nuclei in adult *Caenorhabditis elegans* are found within multinucleated cells.[78-82] Beginning halfway through embryogenesis, a stereotyped sequence of cell fusions between specific partner cells produces 44 adult syncytia ranging in size from 2 nuclei to 139. A number of tissues contain fused cells, including muscle cells in the pharynx (but not body wall), and cells of the somatic gonad and excretory system. The best-studied fusions occur among polarized epithelial cells in the epidermis and specialized organs derived from epidermal precursor cells, including the vulva, the male tail, and the lateral seam epidermis.[78,83,84] Although apparently not required for viability of the animal in culture, these cell fusions are essential for achieving the sleek morphology and reproductive proficiency of the wild type.[85] The invariance of these fusion events has allowed detailed observation of structural intermediates formed during the merger of two cells. Three-dimensional time-lapse imaging of epidermal fusions in the embryo indicates that the opening between fusing cells originates at a point on the apical edge of the lateral-membrane interface of two cells.[86] The opening then widens as a single growing aperture, in part via vesiculation of the conjoined membranes, and displaces intercellular junctions that can remain intact even while the lipid bilayers retreat (Fig. 7). The initial permeabilization of the membrane to cytoplasmic diffusion precedes the visible widening of the opening by 5-10 minutes (W. Mohler, unpublished observations).

The integral membrane protein EFF-1 appears to lie at the heart of the membrane fusion mechanism in epidermal, vulval, and pharyngeal cells.[38,85,87] Loss-of-function mutations in the *eff-1* gene block essentially all epidermal cell fusions, without disrupting the ability of epidermal cells to function in other respects as an intact skin epithelium.[85] Defects in the pharynx of *eff-1*

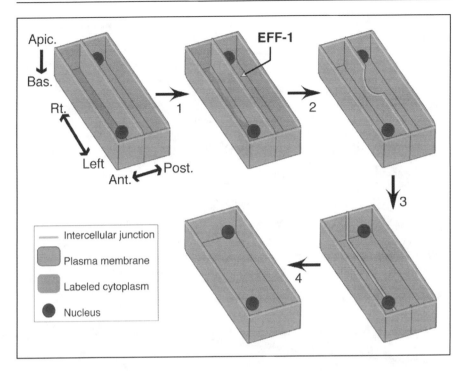

Figure 7. Structural intermediates and action of EFF-1 in fusion of *C. elegans* epidermal cells. Two neighboring cells are first joined by formation of an EFF-1-induced pore allowing cytoplasmic diffusion. The opening in the membrane later widens to completely join the two cytoplasms. Finally, the intercellular junction dissolves producing the fully formed syncytium.

mutants[87] also seem confined to cells' ability to fuse, as the organ forms and functions to feed the worm. In contrast, misexpression of EFF-1 in nonfusing cell types can be very lethal, causing inappropriate fusions that severely disrupt the normal course of development.[85,87] Thus EFF-1 is both absolutely necessary for developmentally programmed fusions and quite sufficient to induce fusion in nonfusion-fated cells. This combination of attributes set it apart, for now, from all other molecules implicated in developmental cell fusion. But EFF-1 does not account for all somatic cell fusions in *C. elegans*. Neither sperm-egg, nor anchor cell-uterus, nor seam-seam cell fusion in the worm appears to require EFF-1, indicating that two or more distinct molecular mechanisms must be at work in producing all of the fused cell in the animal.

EFF-1 is also unique because its sequence does not match any known proteins or protein families in non-nematode organisms.[38,85] Predicted to be a single-span transmembrane protein, EFF-1 has a large N-terminal domain and a small cytoplasmic C-terminus. Fluorescently tagged EFF-1 remains in cytoplasmic pools until a fusion-fated membrane contact forms between epidermal precursor cells. Once fusion-fated partners touch, EFF-1 rapidly accumulates at the point of cell contact. Furthermore, the pattern of this localization among groups of cells suggests that EFF-1 plasma-membrane localization is dependent upon EFF-1 expression in both partner cells; thus EFF-1 may interact homophilically between the surfaces of fusing membranes.[38] In the extracellular domain of EFF-1 lies a short extracellular hydrophobic peptide that is hydropathically similar to fusion peptides and fusion loops in enveloped virus fusion proteins.[88-94] Mutations in the EHP abrogate cell fusion activity, implicating this region as critical to EFF-1 function. However, EHP-mutant EFF-1 is not properly localized to cell-fusion-fated plasma membranes.[38] Thus, it is unclear whether the EHP functions in targeting EFF-1 to the cell surface, in protein interactions that retain the protein at the membrane, or in formation of a membrane fusion pore itself (the function ascribed to fusion peptides in viral fusogens).[88-94]

No other genes have been discovered in *C. elegans* with the fusion-defective viable phenotype characteristic of loss of *eff-1* function. Three possible scenarios explain this finding: (1) EFF-1 is the only required component of the fusion mechanism; (2) other components of the fusion mechanism function redundantly, and yield no phenotype when mutated singly; or (3) other nonredundant fusion components function pleiotropically in additional aspects of development, and therefore give lethal loss-of-function phenotypes. Interestingly, loss of subunits of the vacuolar ATPase (V-ATPase) complex causes ectopic cell fusions in *C. elegans* embryos, among other phenotypic defects.[95] The V-ATPase complex is known to play critical roles in secretory pathways of mammalian and yeast cells.[96] Extracellular barriers to fusion normally appear during the formation of tissues during *C. elegans* development,[38] and V-ATPase function is known to affect extracellular protein activity.[97] It is therefore possible that the promiscuous, delayed cell fusions caused by loss of the V-ATPase occur as a result of defective extracelluar matrix formation.

Primary Mesenchyme (Sea Urchin)

The primary mesenchyme cells of echinoderm embryos produce the pluteus larva's skeleton of calcified spicules. In doing so, they migrate to distinct positions in the blastocoel cavity, extend thin processes to one another, and fuse their plasma membranes to create a syncytial network (Fig. 8).[98,99]

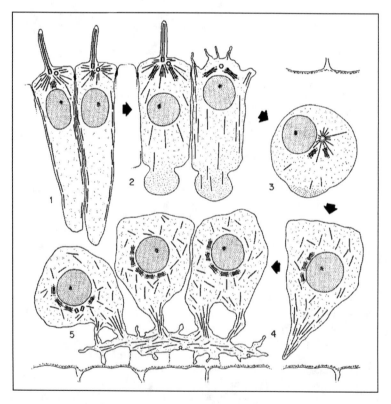

Figure 8. Migration and fusion of primary mesenchyme cells in the sea urchin embryo. Morphologies of the mononucleated precursors and syncytial cables are indicated, including nuclei, Golgi bodies, and microtubules. Figure is reproduced from: Gibbins JR et al. J Cell Biol 1969; 41(1):201-226, by copyright permission of The Rockefeller University Press.[99]

Within the cytoplasm of this syncytium, mineralized calcium is deposited to form the calcified spicules.[99] Cultured primary mesenchyme cells retain their ability to fuse.[100] Yet they are unable to fuse with fusion-competent blastocoelar cells, suggesting that two distinct molecular mechanisms are at work in forming separate syncytial tissues simultaneously in the same same space.[98] Molecular players in either of these mechanisms are still unknown.

Embryonic Blastomeres (Leech)

An exception to the rule of fusion as part of terminal differentiation appears in the early stages of embryogenesis in the leech *Helobdella robusta*.[101] Here three large endodermal precursor cells undergo partial fusion to form a syncytium, which later gives rise to recellularized descendents lining the digestive tract. Experiments employing protein-synthesis inhibitors indicate that each cell in a fusion-pair must express new proteins for cell fusion to occur,[102] but the molecules required for membrane fusion remain unknown.

Haploid Mating (Yeast)

Pheromone-induced conjugation of haploid *Saccharomyces cerevisiae* cells involves cell-wall remodeling in advance of plasma membrane fusion. Mutations in a number of single genes yield mating-defective yeast that undergo normal morphological changes but fail to fuse, as the cell wall continues to separate the cell membranes.[103-105] These genes are apparently involved in degradation of the cell wall, and not in membrane fusion per se. Mutants in the *PRM1* gene, in contrast, are often blocked at a step after cell-wall remodeling (Fig. 9).[106] Still, Prm1 protein appears not to be completely required for fusion, because nearly half of *prm1* null-mutant cell pairs fuse successfully. Recently, it was found that many *prm1*-mutant cells undergo lysis that is specifically associated with the membrane fusion step of mating.[107] Thus, Prm1 appears to play a role in stabilizing the cell-fusion interface in the two cell membranes, permitting safe and efficient membrane merger by a mechanism whose components remain unknown.

Fungal Hyphae

Many fungi propagate as multinucleated hyphae, which can form anastomoses through fusion. In some species, it is hypothesized that fusion-dependent exchange of nuclei can stabilize the genome of such a multinucleated organism, even when the individual nuclei have varying genomic content, a phenonmenon termed heterokaryosis.[108] In the genetically tractable species *Neurospora crassa*, the genes *so/ham-1* and *ham-2* have been implicated in the process of hyphal fusion,[109,110] and the HAM-2 protein, which has orthologs in animals and other fungi, is predicted to contain three membrane-spanning domains.[110] As in many other model systems, however, the nature of the membrane-fusion mechanism is still unclear.

Summary of Molecules Driving Cell Fusion: Obstacles to Their Discovery

To summarize the current understanding of molecules that drive cell membrane fusion in development, we know very little. Why is this? First, the fusion competent contact between cells is transient, and by definition self-destructive. This precludes biochemical preparation of membrane extracts specifically enriched in cell-membrane fusion activity. Second, the biology of each and every cell in an organism, including all nonfusing cell types, involves millions of intracellular membrane fusion events during a cell's lifetime. A fusion competent cell may fuse plasma membranes only once or twice with neighbors during development. Thus purifying and reconstituting a fusion mechanism, even from pure populations of fusing cells, is a "needle in a haystack" task. Third, although the fusion-fated cell must certainly change its membrane-protein content to become fusion competent, it typically changes its gene expression and proteome in many other ways while differentiating into a functioning myofiber, osteoclast, or lens fiber. This makes candidate-gene testing a more difficult task.

Loss-of-function genetics and expression-cloning, combined with functional-genome/proteome profiling have yielded the best current candidates for cell membrane fusogens, as well as several proteins that regulate the activities of still-unknown fusion proteins. However, forward genetic

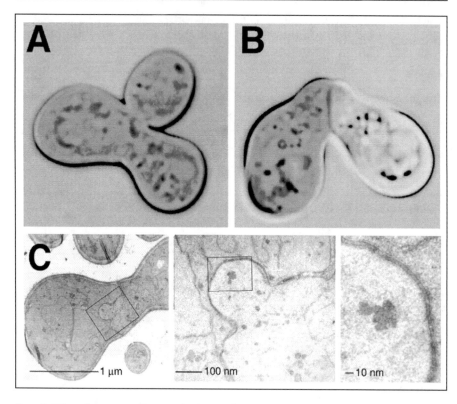

Figure 9. Failure of yeast gamete fusion in the absence of Prm1. Mating yeast haploids in the wild type (A) form zygotes with mixed cytoplasm (uniform fluorescence). Haploids with *prm1* mutations. B) fail to permeabilize their membranes to allow cytoplasmic mixing. C) Intact plasma membranes persists in electron micrographs, despite dissolution of the cell walls. Sequentially magnified regions are indicated. Figures are reproduced from: Heiman MG, Walter P. J Cell Biol 2000; 151(3):719-730, by copyright permission of The Rockefeller University Press.[106]

screening may be frustrated by either redundancy or pleiotropy of components of the fusion mechanism. Furthermore, confirmation of the importance of required proteins in a tightly regulated fusion machine ultimately will require reconstitution of that machine from its component parts. This work has only begun, and we can hope that lessons learned from the first apparent successes will guide the discovery of new cell fusion proteins either by similar screening approaches or by homology/analogy between confirmed fusogens and membrane proteins found in other systems.

Structural Origins of Cell Fusion Channels?

The best-understood examples of exoplasmic membrane fusion are those of enveloped viruses and virus-infected cells. Prevailing models, based on biochemical studies of the known fusogen molecules and electophysiological measurements of the early fusion pores, invoke a "fence" of oligomerized integral membrane proteins that span the intermembrane space and surround a lipid-lined pore with an inner diameter of 7-8 nm.[111-116] This initial pore structure soon widens and eventually spreads to entirely join the two cytoplasms (Fig. 10). Synctyins, as clear orthologs of viral fusion proteins, fit implicitly within such models for pore formation. Lacking understanding of the key molecules in most other developmental fusions, we cannot yet address the structures of the pores in any detail. Nevertheless, time-lapse imaging in of EFF-1-induced fusions in *C. elegans* embryos indicates that (as in viral fusion) a small permeable pore forms to initiate the opening between cells,

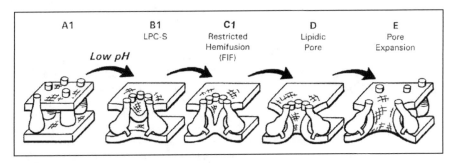

Figure 10. Model of pH-dependent membrane fusion by the influenza virus hemagglutnin protein. Aggregating hemagglutinin proteins deploy fusion peptides to form a fusion pore through the virus and host cell membranes. Figure is reproduced from: Chernomordik LV et al. J Cell Biol 1998; 140(6):1369-1382, by copyright permission of The Rockefeller University Press.[112]

an aperture that can be seen visibly widening only several minutes later.[38,86] Based on diffusion of fluorescently labeled globular proteins from the egg to sperm cytoplasm, it has been concluded that the mouse sperm-egg fusion pore is at least 8nm in diameter.[117] However, like the identities of the pore forming molecules, an upper limit to the actual pore size has yet to be determined.

Another plausible model has arisen with the discovery of intercellular nanotubules (or nanotubes) in a variety of cultured animal cell types.[118-120] These thin membrane bridges appear to form quite frequently between nonfusing cells, allowing exchange of organelles via active transport along cytoskeletal filaments, but preventing appreciable diffusion of cytoplasm between cells. If nanotubules are indeed as prevalent in vivo as the in vitro evidence would indicate, cell fusions could simply be instances in which a nanotubule dilates to fully merge the two connected cells. If this were the case, then the cell fusion mechanism may not involve specialized membrane fusion events at all, and could rely on ubiquitous nanotubule-forming fusogens. The activities specific to syncytium-forming cells could instead be dilation-promoting molecules that force this transition from a diffusion-impermeable tubule to a large fusion aperture. The example of EFF-1 makes it clear that developmental fusion can involve single molecules that are both required and sufficient to form syncytia. Whether they form virus-like pores, expand nanotubular channels, or employ some other mechanism involving lipases or other catalytic enaymes, fusing cell types are likely to express proteins that specifically enable a precise topological change that is unfavored in the constitutive composition of the plasma membrane.

Evolution and Cell Fusion

Has the mechanism of cell fusion been reinvented multiple times during evolution? To date, none of the membrane proteins implicated in the fusion of one cell type have been found to function in another system. This fact may simply reflect a dearth of knowledge, as molecules verifiably essential to the fusion mechanism are unknown in most examples of cell fusion (see above). Then again, the syncytins and EFF-1 are the only two cases where current data strongly support the model of a developmentally regulated fusogenic (sufficient) membrane protein, and each of these molecules appears restricted phyletically to only a few million years of evolution between closely related animal genomes. The syncytins, as clear orthologs of retroviral envelope proteins, present a ready explanation for this novelty, since their coding sequences may have simply arrived as mobile genetic elements that fell under the control of placenta-specific *cis*-regulatory DNA sequences. EFF-1, in contrast, is not obviously related to any known viral proteins. However, its tight sequence conservation within the nematodes and apparent absence in other phyla suggest a sudden origin in the worm genome, possibly also by transfer from a viral or other transposable source. If so, then it is remarkable how the various *cis* regulatory sequences controlling *eff-1* expression have been tuned to drive transcription in the precise and complex spatiotemporal pattern that occurs during development.[85,121]

It is possible that developmental cell fusogens have been adopted in specific cases over only short spans of evolution, and that the fusion mechanisms of different cell types are radically different, even within the same organism. Mismatched fusion machines in distinct cell types—as are indicated in nematodes and echinoderms—may permit the formation of separate syncytial tissues without promiscuous cross-fusion of cell types. Cell fusion, a cellular function that is not required for cellular viability and is quite specialized to discrete developmental contexts, might be more tolerably lost and replaced than other more highly conserved molecular systems: at least in *C. elegans*, fusing not at all is safer than fusing too much.[38,85,87] Moreover, the task of inducing membrane-fusion, without building a stable intercellular connection, could require somewhat less molecular specificity than other cell-cell channels, allowing selection from a variety of protein types that can do the job. Still, we might expect that long-conserved syncytial structures, such as skeletal muscle, that have become integral to the body plans of multiple phyla, may be formed by mechanisms that have remained little-changed during evolution.

References

1. Deriemer SA, Elliott EJ, Macagno ER et al. Morphological evidence that regenerating axons can fuse with severed axon segments. Brain Res 1983; 272(1):157-161.
2. Shi R, Borgens RB, Blight AR. Functional reconnection of severed mammalian spinal cord axons with polyethylene glycol. J Neurotrauma 1999; 16(8):727-738.
3. Alvarez-Dolado M, Pardal R, Garcia-Verdugo JM et al. Fusion of bone-marrow-derived cells with Purkinje neurons, cardiomyocytes and hepatocytes. Nature 2003; 425(6961):968-973.
4. Ying QL, Nichols J, Evans EP et al. Changing potency by spontaneous fusion. Nature 2002; 416(6880):545-548.
5. Wang X, Willenbring H, Akkari Y et al. Cell fusion is the principal source of bone-marrow-derived hepatocytes. Nature 2003; 422(6934):897-901.
6. Terada N, Hamazaki T, Oka M et al. Bone marrow cells adopt the phenotype of other cells by spontaneous cell fusion. Nature 2002; 416(6880):542-545.
7. Wurmser AE, Gage FH. Stem cells: Cell fusion causes confusion. Nature 2002; 416(6880):485-487.
8. Weimann JM, Johansson CB, Trejo A et al. Stable reprogrammed heterokaryons form spontaneously in Purkinje neurons after bone marrow transplant. Nat Cell Biol 2003; 5(11):959-966.
9. Weimann JM, Charlton CA, Brazelton TR et al. Contribution of transplanted bone marrow cells to Purkinje neurons in human adult brains. Proc Natl Acad Sci USA 2003; 100(4):2088-2093.
10. Medvinsky A, Smith A. Stem cells: Fusion brings down barriers. Nature 2003; 422(6934):823-825.
11. Wang HS, Niewczas V, de SNHR et al. Cytogenetic characteristics of 26 polyethylene glycol-induced human-hamster hybrid cell lines. Cytogenet Cell Genet 1979; 24(4):233-244.
12. Chu EH, Powell SS. Selective systems in somatic cell genetics. Adv Hum Genet 1976; 7:189-258.
13. Antczak DF. Monoclonal antibodies: Technology and potential use. J Am Vet Med Assoc 1982; 181(10):1005-1010.
14. Hart IR. Tumor cell hybridization and neoplastic progression. Symp Fundam Cancer Res 1983; 36:133-143.
15. Duelli D, Lazebnik Y. Cell fusion: A hidden enemy? Cancer Cell 2003; 3(5):445-448.
16. Hedgecock EM, White JG. Polyploid tissues in the nematode Caenorhabditis elegans. Dev Biol 1985; 107(1):128-133.
17. Zybina EV, Zybina TG. Polytene chromosomes in mammalian cells. Int Rev Cytol 1996; 165:53-119.
18. Royzman I, Orr-Weaver TL. S phase and differential DNA replication during Drosophila oogenesis. Genes Cells 1998; 3(12):767-776.
19. Moscoso LM, Merlie JP, Sanes JR. N-CAM, 43K-rapsyn, and S-laminin mRNAs are concentrated at synaptic sites in muscle fibers. Mol Cell Neurosci 1995; 6(1):80-89.
20. Schaeffer L, de Kerchove d'Exaerde A, Changeux JP. Targeting transcription to the neuromuscular synapse. Neuron 2001; 31(1):15-22.
21. Inoue N, Ikawa M, Isotani A et al. The immunoglobulin superfamily protein Izumo is required for sperm to fuse with eggs. Nature 2005; 434(7030):234-238.
22. Manandhar G, Toshimori K. Exposure of sperm head equatorin after acrosome reaction and its fate after fertilization in mice. Biol Reprod 2001; 65(5):1425-1436.
23. Nishimura H, Cho C, Branciforte DR et al. Analysis of loss of adhesive function in sperm lacking cyritestin or fertilin beta. Dev Biol 2001; 233(1):204-213.
24. Blobel CP, Wolfsberg TG, Turck CW et al. A potential fusion peptide and an integrin ligand domain in a protein active in sperm-egg fusion. Nature 1992; 356(6366):248-252.

25. Hao Z, Wolkowicz MJ, Shetty J et al. SAMP32, a testis-specific, isoantigenic sperm acrosomal membrane-associated protein. Biol Reprod 2002; 66(3):735-744.

26. Ilayperuma I. Identification of the 48-kDa G11 protein from guinea pig testes as sperad. J Exp Zool 2002; 293(6):617-623.

27. Anderson DJ, Abbott AF, Jack RM. The role of complement component C3b and its receptors in sperm-oocyte interaction. Proc Natl Acad Sci USA 1993; 90(21):10051-10055.

28. Rochwerger L, Cohen DJ, Cuasnicu PS. Mammalian sperm-egg fusion: The rat egg has complementary sites for a sperm protein that mediates gamete fusion. Dev Biol 1992; 153(1):83-90.

29. Kaji K, Oda S, Shikano T et al. The gamete fusion process is defective in eggs of CD9-deficient mice. Nat Genet 2000; 24(3):279-282.

30. Le Naour F, Rubinstein E, Jasmin C et al. Severely reduced female fertility in CD9-deficient mice. Science 2000; 287(5451):319-321.

31. Miyado K, Yamada G, Yamada S et al. Requirement of CD9 on the egg plasma membrane for fertilization. Science 2000; 287(5451):321-324.

32. Ellerman DA, Ha C, Primakoff P et al. Direct binding of the ligand PSG17 to CD9 requires a CD9 site essential for sperm-egg fusion. Mol Biol Cell 2003; 14(12):5098-5103.

33. Chatterjee I, Richmond A, Putiri E et al. The Caenorhabditis elegans spe-38 gene encodes a novel four-pass integral membrane protein required for sperm function at fertilization. Development 2005; 132(12):2795-2808.

34. Putiri E, Zannoni S, Kadandale P et al. Functional domains and temperature - sensitive mutations in SPE-9, an EGF repeat-containing protein required for fertility in Caenorhabditis elegans. Dev Biol 2004; 272(2):448-459.

35. Xu XZ, Sternberg PW. A C. elegans sperm TRP protein required for sperm-egg interactions during fertilization. Cell 2003; 114(3):285-297.

36. Singson A, Mercer KB, L'Hernault SW. The C. elegans spe-9 gene encodes a sperm transmembrane protein that contains EGF-like repeats and is required for fertilization. Cell 1998; 93(1):71-79.

37. Zannoni S, L'Hernault SW, Singson AW. Dynamic localization of SPE-9 in sperm: A protein required for sperm-oocyte interactions in Caenorhabditis elegans. BMC Dev Biol 2003; 3(1):10.

38. del Campo JJ, Opoku-Serebuoh E, Isaacson AB et al. Fusogenic activity of EFF-1 is regulated via dynamic localization in fusing somatic cells of C. elegans. Curr Biol 2005; 15(5):413-423.

39. Collins CA, Olsen I, Zammit PS et al. Stem cell function, self-renewal, and behavioral heterogeneity of cells from the adult muscle satellite cell niche. Cell 2005; 122(2):289-301.

40. Rosania GR, Chang YT, Perez O et al. Myoseverin, a microtubule-binding molecule with novel cellular effects. Nat Biotechnol 2000; 18(3):304-308.

41. Taverna D, Disatnik MH, Rayburn H et al. Dystrophic muscle in mice chimeric for expression of alpha5 integrin. J Cell Biol 1998; 143(3):849-859.

42. Yang JT, Rando TA, Mohler WA et al. Genetic analysis of alpha 4 integrin functions in the development of mouse skeletal muscle. J Cell Biol 1996; 135(3):829-835.

43. Charlton CA, Mohler WA, Radice GL et al. Fusion competence of myoblasts rendered genetically null for N-cadherin in culture. J Cell Biol 1997; 138(2):331-336.

44. Charlton CA, Mohler WA, Blau HM. Neural cell adhesion molecule (NCAM) and myoblast fusion. Dev Biol 2000; 221(1):112-119.

45. Hollnagel A, Grund C, Franke WW et al. The cell adhesion molecule M-cadherin is not essential for muscle development and regeneration. Mol Cell Biol 2002; 22(13):4760-4770.

46. Yagami-Hiromasa T, Sato T, Kurisaki T et al. A metalloprotease-disintegrin participating in myoblast fusion. Nature 1995; 377(6550):652-656.

47. Kurohara K, Komatsu K, Kurisaki T et al. Essential roles of Meltrin beta (ADAM19) in heart development. Dev Biol 2004; 267(1):14-28.

48. Kurisaki T, Masuda A, Sudo K et al. Phenotypic analysis of Meltrin alpha (ADAM12)-deficient mice: Involvement of Meltrin alpha in adipogenesis and myogenesis. Mol Cell Biol 2003; 23(1):55-61.

49. Zhang Y, Featherstone D, Davis W et al. Drosophila D-titin is required for myoblast fusion and skeletal muscle striation. J Cell Sci 2000; 113(Pt 17):3103-3115.

50. Menon SD, Chia W. Drosophila rolling pebbles: A multidomain protein required for myoblast fusion that recruits D-Titin in response to the myoblast attractant Dumbfounded. Dev Cell 2001; 1(5):691-703.

51. Chen EH, Olson EN. Towards a molecular pathway for myoblast fusion in Drosophila. Trends Cell Biol 2004; 14(8):452-460.

52. Chen EH, Olson EN. Unveiling the mechanisms of cell-cell fusion. Science 2005; 308(5720):369-373.

53. Doberstein SK, Fetter RD, Mehta AY et al. Genetic analysis of myoblast fusion: Blown fuse is required for progression beyond the prefusion complex. J Cell Biol 1997; 136(6):1249-1261.

54. Schroter RH, Lier S, Holz A et al. Kette and blown fuse interact genetically during the second fusion step of myogenesis in Drosophila. Development 2004; 131(18):4501-4509.
55. Ruiz-Gomez M, Coutts N, Price A et al. Drosophila dumbfounded: A myoblast attractant essential for fusion. Cell 2000; 102(2):189-198.
56. Bour BA, Chakravarti M, West JM et al. Drosophila SNS, a member of the immunoglobulin superfamily that is essential for myoblast fusion. Genes Dev 2000; 14(12):1498-1511.
57. Strunkelnberg M, Bonengel B, Moda LM et al. Rst and its paralogue kirre act redundantly during embryonic muscle development in Drosophila. Development 2001; 128(21):4229-4239.
58. Galletta BJ, Chakravarti M, Banerjee R et al. SNS: Adhesive properties, localization requirements and ectodomain dependence in S2 cells and embryonic myoblasts. Mech Dev 2004; 121(12):1455-1468.
59. Robertson WB. Pathology of the pregnant uterus. In: Fox H, ed. Obstetrical and Gynaecological Pathology. 3rd ed. London: Churchill Livingstone 1987:1149-1176.
60. Wooding FBP. Role of binucleate cells in fetomaternal cell fusion at implantation in the sheep. American Journal of Anatomy 1984; 170:233-250.
61. Blond JL, Beseme F, Duret L et al. Molecular characterization and placental expression of HERV-W, a new human endogenous retrovirus family. J Virol 1999; 73(2):1175-1185.
62. Blond JL, Lavillette D, Cheynet V et al. An envelope glycoprotein of the human endogenous retrovirus HERV-W is expressed in the human placenta and fuses cells expressing the type D mammalian retrovirus receptor. J Virol 2000; 74(7):3321-3329.
63. Blaise S, de Parseval N, Benit L et al. Genomewide screening for fusogenic human endogenous retrovirus envelopes identifies syncytin 2, a gene conserved on primate evolution. Proc Natl Acad Sci USA 2003; 100(22):13013-13018.
64. Mi S, Lee X, Li X-p et al. Syncytin is a captive retroviral envelope protein involved in human placental morphogenesis. Nature 2000; 403:785-789.
65. Mallet F, Bouton O, Prudhomme S et al. The endogenous retroviral locus ERVWE1 is a bona fide gene involved in hominoid placental physiology. Proc Natl Acad Sci USA 2004; 101(6):1731-1736.
66. Dupressoir A, Marceau G, Vernochet C et al. Syncytin-A and syncytin-B, two fusogenic placenta-specific murine envelope genes of retroviral origin conserved in Muridae. Proc Natl Acad Sci USA 2005; 102(3):725-730.
67. Kuszak JR, Macsai MS, Bloom KJ et al. Cell-to-cell fusion of lens fiber cells in situ: Correlative light, scanning electron microscopic, and freeze-fracture studies. J Ultrastruct Res 1985; 93(3):144-160.
68. Kuszak JR, Ennesser CA, Bertram BA et al. The contribution of cell-to-cell fusion to the ordered structure of the crystalline lens. Lens Eye Toxic Res 1989; 6(4):639-673.
69. Shestopalov VI, Bassnett S. Expression of autofluorescent proteins reveals a novel protein permeable pathway between cells in the lens core. J Cell Sci 2000; 113(Pt 11):1913-1921.
70. Shestopalov VI, Bassnett S. Development of a macromolecular diffusion pathway in the lens. J Cell Sci 2003; 116(Pt 20):4191-4199.
71. Sutton JS, Weiss L. Transformation of monocytes in tissue culture into macrophages, epithelioid cells, and multinucleated giant cells. An electron microscope study. J Cell Biol 1966; 28(2):303-332.
72. Saginario C, Sterling H, Beckers C et al. MFR, a putative receptor mediating the fusion of macrophages. Mol Cell Biol 1998; 18(11):6213-6223.
73. Sterling H, Saginario C, Vignery A. CD44 occupancy prevents macrophage multinucleation. J Cell Biol 1998; 143(3):837-847.
74. Han X, Sterling H, Chen Y et al. CD47, a ligand for the macrophage fusion receptor, participates in macrophage multinucleation. J Biol Chem 2000; 275(48):37984-37992.
75. Yagi M, Miyamoto T, Sawatani Y et al. DC-STAMP is essential for cell-cell fusion in osteoclasts and foreign body giant cells. J Exp Med 2005; 202(3):345-351.
76. Lagasse E, Connors H, Al-Dhalimy M et al. Purified hematopoietic stem cells can differentiate into hepatocytes in vivo. Nat Med 2000; 6(11):1229-1234.
77. Blau HM, Pavlath GK, Hardeman EC et al. Plasticity of the differentiated state. Science 1985; 230(4727):758-766.
78. Podbilewicz B, White JG. Cell fusions in the developing epithelial of C. elegans. Dev Biol 1994; 161(2):408-424.
79. Sulston JE, Schierenberg E, White JG et al. The embryonic cell lineage of the nematode Caenorhabditis elegans. Dev Biol 1983; 100(1):64-119.
80. Sulston JE, Horvitz HR. Post-embryonic cell lineages of the nematode, Caenorhabditis elegans. Dev Biol 1977; 56(1):110-156.
81. Shemer G, Podbilewicz B. Fusomorphogenesis: Cell fusion in organ formation. Dev Dyn 2000; 218(1):30-51.

82. Yochem J, Gu T, Han M. A new marker for mosaic analysis in Caenorhabditis elegans indicates a fusion between hyp6 and hyp7, two major components of the hypodermis. Genetics 1998; 149(3):1323-1334.
83. Nguyen CQ, Hall DH, Yang Y et al. Morphogenesis of the Caenorhabditis elegans male tail tip. Dev Biol 1999; 207(1):86-106.
84. Sharma-Kishore R, White JG, Southgate E et al. Formation of the vulva in Caenorhabditis elegans: A paradigm for organogenesis. Development 1999; 126(4):691-699.
85. Mohler WA, Shemer G, del Campo JJ et al. The type I membrane protein EFF-1 is essential for developmental cell fusion. Dev Cell 2002; 2(3):355-362.
86. Mohler WA, Simske JS, Williams-Masson EM et al. Dynamics and ultrastructure of developmental cell fusions in the Caenorhabditis elegans hypodermis. Curr Biol 1998; 8(19):1087-1090.
87. Shemer G, Suissa M, Kolotuev I et al. EFF-1 is sufficient to initiate and execute tissue-specific cell fusion in C. elegans. Curr Biol 2004; 14(17):1587-1591.
88. Gamblin SJ, Haire LF, Russell RJ et al. The structure and receptor binding properties of the 1918 influenza hemagglutinin. Science 2004; 303(5665):1838-1842.
89. Gibbons DL, Vaney MC, Roussel A et al. Conformational change and protein-protein interactions of the fusion protein of Semliki Forest virus. Nature 2004; 427(6972):320-325.
90. Stevens J, Corper AL, Basler CF et al. Structure of the uncleaved human H1 hemagglutinin from the extinct 1918 influenza virus. Science 2004; 303(5665):1866-1870.
91. Modis Y, Ogata S, Clements D et al. Structure of the dengue virus envelope protein after membrane fusion. Nature 2004; 427(6972):313-319.
92. Bullough PA, Hughson FM, Skehel JJ et al. Structure of influenza haemagglutinin at the pH of membrane fusion. Nature 1994; 371(6492):37-43.
93. Nieva JL, Agirre A. Are fusion peptides a good model to study viral cell fusion? Biochim Biophys Acta 2003; 1614(1):104-115.
94. Harter C, James P, Bächi T et al. Hydrophobic binding of the ectodomain of influenza hemagglutinin to membranes occurs through the "fusion peptide". J Biol Chem 1989; 264:6459-6464.
95. Kontani K, Moskowitz IP, Rothman JH. Repression of cell-cell fusion by components of the C. elegans vacuolar ATPase complex. Dev Cell 2005; 8(5):787-794.
96. Nishi T, Forgac M. The vacuolar (H+)-ATPases—nature's most versatile proton pumps. Nat Rev Mol Cell Biol 2002; 3(2):94-103.
97. Maquoi E, Peyrollier K, Noel A et al. Regulation of membrane-type 1 matrix metalloproteinase activity by vacuolar H+-ATPases. Biochem J 2003; 373(Pt 1):19-24.
98. Hodor PG, Ettensohn CA. The dynamics and regulation of mesenchymal cell fusion in the sea urchin embryo. Dev Biol 1998; 199(1):111-124.
99. Gibbins JR, Tilney LG, Porter KR. Microtubules in the formation and development of the primary mesenchyme in Arbacia punctulata. I. The distribution of microtubules. J Cell Biol 1969; 41(1):201-226.
100. Karp GC, Solursh M. In vitro fusion and separation of sea urchin primary mesenchyme cells. Exp Cell Res 1985; 158(2):554-557.
101. Liu NL, Isaksen DE, Smith CM et al. Movements and stepwise fusion of endodermal precursor cells in leech. Dev Genes Evol 1998; 208(3):117-127.
102. Isaksen DE, Liu NJ, Weisblat DA. Inductive regulation of cell fusion in leech. Development 1999; 126(15):3381-3390.
103. White JM, Rose MD. Yeast mating: Getting close to membrane merger. Curr Biol 2001; 11(1):R16-20.
104. McCaffrey G, Clay FJ, Kelsay K et al. Identification and regulation of a gene required for cell fusion during mating of the yeast Saccharomyces cerevisiae. Mol Cell Biol 1987; 7(8):2680-2690.
105. Trueheart J, Boeke JD, Fink GR. Two genes required for cell fusion during yeast conjugation: Evidence for a pheromone-induced surface protein. Mol Cell Biol 1987; 7(7):2316-2328.
106. Heiman MG, Walter P. Prm1p, a pheromone-regulated multispanning membrane protein, facilitates plasma membrane fusion during yeast mating. J Cell Biol 2000; 151(3):719-730.
107. Jin H, Carlile C, Nolan S et al. Prm1 prevents contact-dependent lysis of yeast mating pairs. Eukaryot Cell 2004; 3(6):1664-1673.
108. Bever JD, Wang M. Arbuscular mycorrhizal fungi: Hyphal fusion and multigenomic structure. Nature 2005; 433(7022):E3-4, (discussion E4).
109. Fleissner A, Sarkar S, Jacobson DJ et al. The so locus is required for vegetative cell fusion and postfertilization events in Neurospora crassa. Eukaryot Cell 2005; 4(5):920-930.
110. Xiang Q, Rasmussen C, Glass NL. The ham-2 locus, encoding a putative transmembrane protein, is required for hyphal fusion in Neurospora crassa. Genetics 2002; 160(1):169-180.

111. Leikina E, Chernomordik LV. Reversible merger of membranes at the early stage of influenza hemagglutinin-mediated fusion. Mol Biol Cell 2000; 11:2359-2371.
112. Chernomordik LV, Frolov VA, Leikina E et al. The pathway of membrane fusion catalyzed by influenza hemagglutinin: Restriction of lipids, hemifusion, and lipidic fusion pore formation. J Cell Biol 1998; 140:1369-1382.
113. Zimmerberg J, Blumenthal R, Sarkar DP et al. Restricted movement of lipid and aqueous dyes through pores formed by influenza hemagglutinin during cell fusion. J Cell Biol 1994; 127(6 Pt 2):1885-1894.
114. Plonsky I, Zimmerberg J. The initial fusion pore induced by baculovirus GP64 is large and forms quickly. J Cell Biol 1996; 135(6 Pt 2):1831-1839.
115. Bonnafous P, Stegmann T. Membrane perturbation and fusion pore formation in influenza hemagglutinin-mediated membrane fusion. A new model for fusion. J Biol Chem 2000; 275(9):6160-6166.
116. Zimmerberg J. Hole-istic medicine. Science 1999; 284(5419):1475, 1477.
117. Jones KT, Soeller C, Cannell MB. The passage of Ca2+ and fluorescent markers between the sperm and egg after fusion in the mouse. Development 1998; 125(23):4627-4635.
118. Rustom A, Saffrich R, Markovic I et al. Nanotubular highways for intercellular organelle transport. Science 2004; 303(5660):1007-1010.
119. Onfelt B, Davis DM. Can membrane nanotubes facilitate communication between immune cells? Biochem Soc Trans 2004; 32(Pt 5):676-678.
120. Onfelt B, Nedvetzki S, Yanagi K et al. Cutting edge: Membrane nanotubes connect immune cells. J Immunol 2004; 173(3):1511-1513.
121. Opoku-Serebuoh E. Transcriptional regulation of the eff-1 gene [Ph.D.]. Farmington, CT: Genetics and Developmental Biology, University of Connecticut Health Center; 2005.

Index